Metals and Genetics

Metals and Genetics

Edited by

Bibudhendra Sarkar

The Hospital for Sick Children and
The University of Toronto
Toronto, Ontario, Canada

Kluwer Academic / Plenum Publishers
New York, Boston, Dordrecht, London, Moscow

Library of Congress Cataloging-in-Publication Data

Metals and genetics / edited by Bibudhendra Sarkar.
    p.   cm.
    "Proceedings of the Second International Symposium on Metals and
Genetics, held May 26-29, 1998, in Toronto, Ontario, Canada"--T.p.
verso.
    Includes bibliographical references and index.
    ISBN 0-306-46101-3
    1. Metals--Toxicology--Congresses. 2. Metals--Carcinogenicity-
-Congresses. 3. Metalloproteins--Physiological effect--Congresses.
4. Mutagens--Congresses. I. Sarkar, Bibudhendra.
II. International Symposium on "Metals and Genetics" (2nd : 1998 :
Toronto, Ont.)
    [DNLM: 1. Metals--metabolism congresses. 2. Metals--toxicity
congresses. 3. Metal Metabolism, Inborn Errors--chemically induced
congresses. 4. Mutagenesis congresses. 5. Gene Expression
Regulation congresses.   QU 130 M5876 1999]
RA1231.M52M49 1999
615.9'253--dc21
DNLM/DLC
for Library of Congress   98-49080
                                        CIP

RA
1231
.M52
M49
1999

Proceedings of the Second International Symposium on Metals and Genetics,
held May 26 – 29, 1998, in Toronto, Ontario, Canada

ISBN 0-306-46101-3

© 1999 Kluwer Academic / Plenum Publishers, New York
233 Spring Street, New York, N.Y. 10013

10 9 8 7 6 5 4 3 2 1

A C.I.P. record for this book is available from the Library of Congress.

Printed in the United States of America

PREFACE

During the past few years, major scientific discoveries have greatly contributed to our understanding of the relationship between metals and genetics. The fields which have contributed to this area range from Clinical Medicine and Genetics to Biochemistry and Chemistry. The aim of this book is to bring together investigators from these diverse fields to reflect on the broad implications of direct and indirect interactions of metals and genetic components.

The volume begins with a tribute to the late Karen Wetterhahn, an outstanding scientist in the field, who will be sadly missed by her friends and colleagues because of her untimely death. The book has 28 chapters contributed by scientists who are internationally known for their expertise and outstanding research. The subject matters are divided into five major sections. The first section discusses genetic response to environmental exposure to metals. Potentially devastating health crises have been reported in recent years from several parts of the world, which stem from environmental exposure to metals. In this section, authors report their findings on the effects and influence of metals in gene expression and their consequences to human health. The section on metal carcinogenesis and metal caused DNA damage, presents the latest advances in our knowledge of the molecular mechanisms of metal-induced mutagenesis and carcinogenesis. This topic is at the very heart of our understanding of how cancer may be caused by various metals. Recent studies have revealed that metals may play a role in several neurodegenerative diseases. A section is devoted to metals and neurodegenerative diseases. The identification of several disease genes related to metals is a major breakthrough in recent years. The section on genetics and biochemistry of metal-related diseases presents authoritative accounts of current information on both inherited and acquired metal-related diseases. They have discussed the genetics and pathophysiology of these diseases and reported the cloning, expression, purification and characterization of the gene products. Exciting results have been reported recently describing regulation of gene expression by metals. Chapters in the section on metals and gene regulation report recent advances dealing with the mechanism of metal-mediated regulation and control of gene expression.

It is difficult to compile all the relevant topics to cover all areas of a rapidly advancing field of Metals and Genetics. Only selected areas are chosen to highlight the recent important advances. This book will be an invaluable resource material to researchers in the field of Inorganic Biochemistry, Environmental Chemistry, Genetics, Molecular Biology, Physiology, Pharmacology, Toxicology and Medicine.

Bibudhendra Sarkar

ACKNOWLEDGMENTS

This volume contains the proceedings of the Second International Symposium on Metals and Genetics, held May 26–29, 1998 under the auspices of The Hospital for Sick Children, Toronto and The Faculty of Medicine, University of Toronto, Toronto, Canada.

We are indebted to Nira Jayanetti, who assisted in the planning and coordinating the Symposium. Thanks are due to Bonnie McKee for consultations, Rose Templeton for Symposium assistance and Loretta LeBlanc for preparing the manuscripts for editorial processes.

We gratefully acknowledge the following for financial support: The Hospital for Sick Children Foundation, The Research Institute of The Hospital for Sick Children, The Faculty of Medicine, University of Toronto, Department of Biochemistry, University of Toronto, The Connaught Committee, University of Toronto, The Samuel Lunenfeld Charitable Foundation, Ontario Hydro, National Cancer Institute of Canada, National Institute of Environmental Health Sciences (USA), Nickel Producers Environmental Research Association, Burroughs Wellcome Fund (USA), DiaMed Lab Supplies and Novartis Pharma Canada Inc.

Bibudhendra Sarkar

CONTENTS

Metal Carcinogenesis and Metal Caused DNA Damage

Metals and Neurodegenerative Diseases

Genetics and Biochemistry of Metal-Related Diseases

Metals and Gene Regulation

EULOGY FOR KAREN WETTERHAHN

As an Introduction to the Session "Genetic Response to Environmental Exposure to Metals" for the Second International Symposium on Metals and Genetics

William A. Suk

Director
Office of Program Development
National Institute of Environmental Health Sciences

I want to thank my colleagues on the International Organizing Committee for this session in tribute to Karen Wetterhahn. I only wish it wasn't necessary! I knew Karen for about 10 years. I met her through Max Costa. That's the way it was with Karen. You always knew someone through her or met someone who knew her. In that way we were all touched by her in life and, now, unfortunately in death. I am a better person for having known her. She was the most unassuming, nice person I knew. Scientific and academic credentials aside, she was truly a warm and sincere friend and colleague.

Karen was exposed to dimethyl mercury, an insidious neurotoxin, in August of 1996. She died on June 8, 1997. She leaves a husband, a son, 15 year old Leon, Jr., and a 13-year-old daughter, Charlotte.

I last saw her in November 1996 in Kuala Lumpur at the Pacific Basin Conference on Hazardous Waste Research. I had lunch with Karen and her son, Leon, Jr. We had skied together in Italy some years before; had dinner in Tucson in between. We talked science; we talked family; we talked about all kinds of things. She was fun and interesting.

Karen established a major paradigm in metal toxicology: the uptake-reduction model. She was the first to appreciate the importance of the oxidation state in the metabolism and toxicology of chromium. Karen was one of the best in the field of metal toxicology; but she was unsurpassed when dealing with chromium.

Karen's career began as an Assistant Professor at Dartmouth in 1976. She later served as Dean of Graduate Studies and as Associate Dean of Faculty in the Sciences. She was also recognized as the Albert Bradley 3rd Century Professor.

It is my understanding that she was most proud of two accomplishments: (1) the founding and development of the Dartmouth Women in Science Program (WISP) in 1987;

1

and (2) the development and receipt of the Dartmouth Superfund Program Grant in 1995 which focused on heavy metals.

WISP was developed in response to the disproportionately high drop-out rate of women in the sciences during their college and post-graduate careers. This program developed a number of strategies to address this issue, most prominent of which is a mentoring program in which first-year undergraduate women are given individual research opportunities in laboratories and receive mentorship by faculty and students to begin their college science careers. This program has now been emulated by dozens of universities around the U.S. I only wish that my eight-year-old daughter, Annie, would have had the opportunity to have met Karen.

The Dartmouth Superfund Program grant is the single largest grant ever received by that university and represents an interdisciplinary approach to examining the impact of heavy metals on the environment and human health. This approach was initiated by Karen and I at a lunch in San Diego during an American Chemical Society meeting.

Karen Wetterhahn

She called the grant "a microcosm of the life-science issue". "I like it as a model for how we can bring together people from different departments and schools. It's intellectually coherent, but faculty and students from different areas are coming together. So many of the large complex problems we face must be solved by the interdisciplinary approach." She got it right! So that this insight will not be lost, NIEHS through its Superfund Basic Research Program is going to establish a Fellowship funding for young women scientists working on some aspect of metals research.

There is an irony here: the dangers of heavy metals claimed her interest, then claimed her life.

Her scientific contributions will be well documented, but she will be most remembered for her smile, her gee-whiz approach to things new and different, and her gentleness.

METALLOREGULATION OF SOFT METAL RESISTANCE PUMPS

Chun Xu and Barry P. Rosen[*]

Department of Biochemistry and Molecular Biology
Wayne State University School of Medicine
Detroit, Michigan 48201

1. INTRODUCTION

Many metals are required for cellular growth. Others, such as cadmium, lead, arsenic and antimony, are toxic. Some, such as zinc and copper, are both micronutrients but toxic in excess. For these processes cells must contain both uptake and efflux systems. Perhaps due to geochemical sources of metals, bacterial resistances to metals probably arose early in evolution. Before the modern use of antibiotics, bacterial metal resistance determinants encoding extrusion systems were more widespread than antibiotic resistances (Datta and Hughes, 1983).

Expression of resistances is frequently regulated at the transcriptional or allosteric level. In this chapter the regulation of the pumps that confer resistance to ions of soft metals in bacteria will be described. In this context the expression *soft metal* will be used as an operational term for metal ions that coordinate with sulfur and nitrogen ligands in proteins and include light transition metals such as Cu(I) and Zn(II), heavy transition metals such as Ag(I), Hg(II) and Cd(II) and the metalloids As(III) and Sb(III). In contrast, hard metals are those of groups I and II such as Na^+, K^+, Mg^{2+} and Ca^{2+} form ionic bonds with such as amino acid carboxylates. While the expression of most soft metal pumps is regulated, there are families of metalloregulatory proteins that control different types of ion pumps, and families of pumps that are controlled by different types of regulatory proteins. For example, ArsR, the first identified member of the ArsR family of metal-regulated repressors, controls expression of the *ars* operon, which encodes a novel soft metal pump for extrusion of As(III) or Sb(III), whileCadC, another ArsR family member, controls expresssion of the *cad* operon that encodes a Cd(II)-translocating P-type ATPase. Yet a third ArsR family member, SmtB,

---

[*]To whom requests for reprints should be addressed. Phone: (313) 577-1512; Fax: (313) 577-2765; email: brosen@med.wayne.edu

Metals and Genetics, edited by Sarkar.
Kluwer Academic / Plenum Publishers, New York, 1999.

controls expression of a bacterial metallothionine—a non-pump metal resistance. The expression of genes for three different soft metal P-type ATPases, CadA for Cd(II), CopA for Cu(I) and ZntA for Zn(II), are each controlled by a different type of regulatory element. Each of these cases will be described below in more detail.

2. METALLOREGULATION BY ARSENIC AND ANTIMONY

2.1. ArsA

Resistance to arsenite and antimonite in bacteria has been found to be mediated by a novel soft metal pump. The arsenical pump of the *E. coli* plasmid R773 is encoded by the *ars* (arsenical resistance) operon (Kaur and Rosen, 1992). The operon consists of two regulatory genes (*arsR and arsD*) and three structural genes, (*arsA, arsB, and arsC*). The pump consists of two types of subunits, the products of the *arsA* and *arsB* genes (Figure 1). The ArsA subunit is a 63 kDa anion-stimulated ATPase (Hsu and Rosen, 1989) that remains anchored to the inner membrane by interaction with ArsB (Dey et al., 1994), which is an intrinsic membrane protein that most likely forms the anion conducting pathway (Dey et al., 1994; Kuroda et al., 1997). Expression of a third structural gene, *arsC*, expands the resistance spectrum to allow for arsenate resistance. The 16 kDa ArsC enzyme is an arsenate reductase that reduces arsenate (As(V)) to arsenite (III) (Gladysheva et al., 1994; Oden et al., 1994).

From its sequence ArsA most likely arose through gene duplication and fusion of a gene half the size of the existing *arsA* gene (Chen et al., 1986). Both the N-terminal (A1) half and the C-terminal (A2) half contain a consensus nucleotide binding site that have been shown by site directed mutagenesis to be required for resistance, ATPase activity and arsenite transport (Karkaria et al., 1990; Kaur and Rosen, 1992). Recent results demonstrate that the two nucleotide binding sites must interact to promote catalysis (Li et al., 1996; Li and Rosen, 1998).

Figure 1. The Ars As(III)/Sb(III)-translocating ATPase. Top: The *ars* operon of plasmid R773 has five genes, *arsRDABC*, driven from a single promoter (p$_{ars}$). Middle: The phylogenetic relationship of the amino acid sequences of ArsA (left) and ArsB (right) homologues of prokaryotes, archaea and eukaryotes are shown. All dendograms were made using DNasis (Hitachi). ArsA homologues: Distantly relatives of ArsA include proteins involved in nitrogen fixation (Jones et al., 1993) and cell division (de Boer et al., 1991). Homologues proposed to be involved in resistance to As(III)/Sb(III) include the proteins encoded by plasmids R773 and R46, which are the result of gene duplication and fusion, producing A1 and A2 halves, each of which has a consensus nucleotide binding sequence. The protein from the archaea *H. salinarum* has the same duplicated domain structure. All of the other ArsA homologues have single domain structures with a single nucleotide binding site and are approximately half the size of the R773 enzyme. Of these single domain proteins, only the human homologue has been implicated in As(III) resistance (Kurdi-Haidar et al., 1996). ArsB homologues: There are two branches to this family. The lower branch are all bacterial proteins closely related to R773 ArsB. The other branch includes more distantly related membrane proteins from bacteria (*Bacillus subtilis*), eukaryotes (*Saccharomyces cerevisiae*) and archae (*Archaeoglobus fulgidus*). The yeast ACR3 (Bobrowicz et al., 1997; Wysocki et al., 1997) and bacterial skin proteins (Takemaru et al., 1995) have been demonstrated to be required for metalloid resistance. Bottom: The products of the *arsA* and *arsB* form a pump that extrudes arsenite or antimonite from the cells, affording metalloid resistance. In this model the pump is postulated to have two ArsA subunits and one ArsB subunit, consistent with the known dimerization of ArsA. The pump would have four consensus nucleotide binding sites, two A1 and two A2 sites.

ArsA is allosterically activated by binding of Sb(III) or As(III) to a triad of sulfurs from Cys117, Cys172 and Cys422 (Bhattacharjee et al., 1995). Although distant from each other in the primary sequence of ArsA, from the results of crosslinking experiments with the bifunctional reagent dibromobimane, the three cysteine residues have been shown to be located within 3 to 6 Å of each other (Bhattacharjee and Rosen, 1996). More recently X-ray absorption spectroscopy of ArsA saturated with As(III) has shown that there are only As-S interactions that are all at 2.25 Å (R. Scott, H. Bhattacharjee and B.P. Rosen, unpublished). This suggests a model of the allosteric site in ArsA in which As(III) or

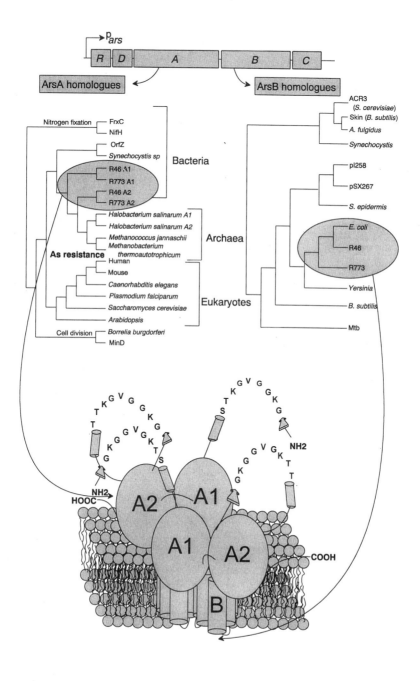

Sb(III) interact with these three cysteines in a trigonal pyramidal geometry, forming a novel soft metal-thiol cage (Figure 2).

Many homologues of both ArsA and ArsB have been identified in prokaryotes, eukaryotes and archaea (Figure 1). Although the functions of most are unknown, the human ArsA homologue has been shown to be involved in soft metal resistance (Kurdi-Haidar et al., 1996). A more distant ArsA homologue is product of the *nifH* gene, the iron protein subunit of the nitrogenase complex, for which a crystal structure has been reported (Georgiadis et al., 1992). In the absence of soft metal the A1 and A2 domains are independent of each other, held together by a flexible linker peptide. The two nucleotide binding sites do not interact, and the enzyme has low catalytic activity. Filling of the allosteric site by formation of the As-S structure brings together the two halves of the protein. This forms an interface between the two nucleotide sites and an acceleration of catalysis.

2.2. ArsR

The ArsR repressor encoded by the gram-negative plasmid R773 belongs to a novel family of small metalloregulatory proteins (Wu and Rosen, 1991; Wu and Rosen, 1993; Shi et al., 1994) that includes subgroups of proteins that respond to As(III)/Sb(III) [ArsR], Cd(II)/Zn(II) [CadC] (Yoon et al., 1991) or Zn(II) [SmtB] (Morby et al., 1993) (Figure 3). ArsR is a *trans*-acting regulatory protein that binds as a dimer to an operator and represses *ars* operon transcription (Wu and Rosen, 1993). Inducibility results from the release of the ArsR from the operator upon binding of As(III) or Sb(III) (Wu and Rosen, 1991). Gel mobility shift assays showed that the formation of DNA-ArsR protein complexes (Wu and Rosen, 1993).

As a metal-repressive repressor protein that negatively regulates the transcription of a detoxification and/or metal efflux operon, each member of the ArsR family should have at least three domains, a metal binding domain, a DNA-binding domain, and a dimerization domain. ArsR has 117 amino acid residues, in which a putative helix-turn-helix DNA-binding motif from approximately residues 33 to 54 was suggested (Shi et al., 1994). Mutations in this putative DNA binding domain produced ArsRs that no longer repressed *in vivo* nor bound to DNA *in vitro*. Alteration of residues Cys32 and Cys34, located just at the beginning of the putative DNA binding domain, resulted ArsRs that were still capable of repression but had lost inducibility (Shi et al., 1994). This result indicated that Cys32 and Cys34 are involved in metal binding. As(III) and Sb(III) are normally three coordinate, suggesting that a third residue might be involved in metal binding in ArsR.

To identify additional residues, mutations in this region were introducing by site-directed mutagenesis, but no other residues required for induction were found (Shi et al., 1996). However, Cys37 was identified as the third ligand to the metalloid by other methods, even though a C37S ArsR bound normally to the *ars* operator, repressed transcription and was inducible by metalloids. First, by X-ray absorption spectroscopy three As-S interactions at 2.25 Å. This indicated that the third ligand was a cysteine residue. Second, the interaction of ArsRs with As(III) was examined using a phenylarsine oxide affinity resin. Phenylarsine oxide is the most effective inducer of the *ars* operon. Since As is bound to the phenyl group, phenylarsine oxide can form only two bonds to ArsR, so binding to an affinity column with phenylarsine oxide should require only two cysteine residues. ArsRs containing any two of the three cysteine residues of Cys32, Cys34 or Cys37 bound to the resin, while alteration of any two of the three resulted in loss of binding. These results suggest that the thiolates of Cys32, Cys34 and Cys37 form a three coordinate AsS_3 site (Figure 2). From small molecule crystallography of As(III) trithiols, each sulfur should be located 2.25 Å from the As atom and 3.2 Å from each other. In an SbS_3 structure the distances should be 2.45 Å and 3.5 Å,

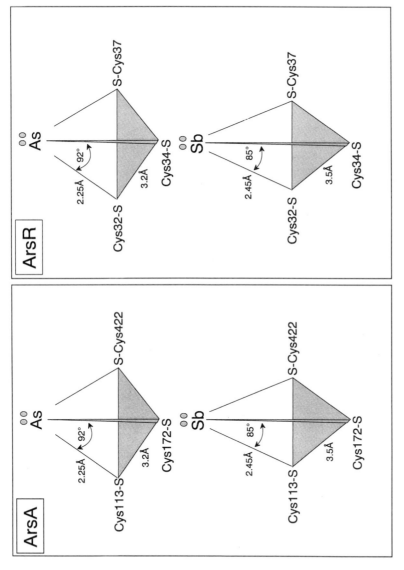

Figure 2. Geometry of the As(III)/Sb(III) binding domains in ArsA and ArsR. In ArsA (left) the allosteric site contains Cys113, Cys172 and Cys422. These residues are located distant from each other in the primary sequence of the protein. Thus formation of the soft metal site requires folding of the protein to bring those residues in spatial proximity of each other. In contrast, in ArsR (right), the three cysteine residues Cys32, Cys34 and Cys37 of the inducer binding site are co-linear in the primary sequence. The trigonal pyramidal structures contain three-coordinately liganded sulfur thiolates, with As(III) or Sb(III) at the apex. The bond angles and distances are postulated from crystallographic analysis of small molecules containing As–S or Sb–S bonds.

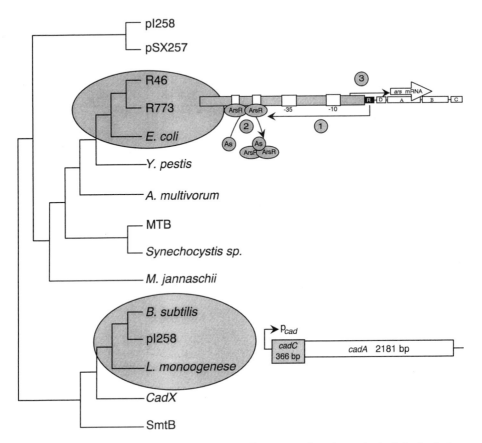

Figure 3. The ArsR family of metalloregulatory proteins. There are two branches to the ArsR family. In the upper branch are repressors either shown or postulated to be As(III)/Sb(III) responsive, illustrated by ArsR, the regulator of the *ars* operon of *E. coli* plasmid R773 (right). The lower branch include Cd(II)/Zn(II)/Pb(II) responsive repressors, illustrated by CadC, the regulator of the *cad* operon of *S. aureus* plasmid pI258 (right).

respectively. While Cys37 may be a ligand, is not required for induction; binding to Cys32 and Cys34 are sufficient to produce the conformational change that results in release of the repressor from the DNA and induction.

Recently the crystal structure of SmtB, an ArsR homologue, has been determined (Cook et al., 1998). SmtB is a Zn(II) regulated repressor (Morby et al., 1993), but the structure was determined in the absence of Zn(II). When ArsR was modeled on SmtB, the soft metal binding site could be seen to be within the first helix of the helix-turn-helix structure that most likely forms the DNA binding domain (Figure 4). Cys32 is located just before the helix, while Cys34 and Cys37 are within the helix. The sulfur atoms of Cys32 and Cys34 are close enough to each other to coordinate with As(III) or Sb(III). However, the sulfur of Cys37 cannot be placed within less than 5 Å of the other two sulfurs without severe distortion of the α helix. We propose that metal binding melts the first helix of the DNA binding domain, so that the repressor dissociates from the operator site, resulting in induction.

The 117-residue ArsR is a functional homodimer of 26 kDa (Wu and Rosen, 1991). To investigate the nature of the dimerization domain, a series of 5' and 3' deletions of *arsR* were constructed (Xu and Rosen, 1997). These allowed analysis of dimer formation using

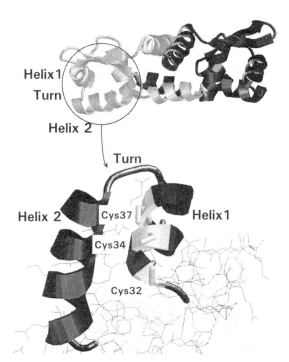

Figure 4. DNA and inducer binding domains of the ArsR aporepressor. The structure of ArsR was modeled on the crystal structure of the SmtB repressor (Cook et al., 1998). The DNA binding domain is composed of a helix (residues 33–40) - turn (residues 41–43) – helix (residues 44–55). The inducer binding domain contains Cys32, Cys34 and Cys37. In the metal-bound repressor the three sulfur atoms must be within 3.2–3.5 Å of each other. This is not possible without distortion of the first helix, suggesting that induction involves disruption of that helix.

a combination of the yeast two-hybrid system, *in vivo* regulation of reporter gene expression in *E. coli* and biochemical analysis of the purified truncated ArsRs *in vitro*. ArsRs with C-terminal truncations from residue 90 were still able to form homodimers capable of metalloregulation. Larger C-terminal truncations were unable to form dimers and did not repress. From the N-terminal side deletion of codons for the first 8 amino acids of ArsR resulted in repressors that retained the ability to dimerize and repress. In contrast, removal of the first 11 residues of ArsR abolished both dimerization and repression. These results are consistent with dimerization being essential for repression. Further, it suggests that the first 10 residues and last 28 residues are not required for ArsR function. A core sequence of about 80 residues has all of the information necessary for dimerization, repression and metal recognition (Xu and Rosen, 1997).

The model of ArsR based on the SmtB structure is consistent with the experimental data (Figure 5). The dimerization domain can be seen to consist of an interface of four α helices, two from each subunit. Two N-terminal helices interact with each other, and two C-terminal helices interact with each other. The N-terminal helix is composed of residues 1–13, but the experimental data indicate that only 9–13 are required for dimerization, and the model clearly shows that the regions of the two N-terminal helices overlap only in those residues. The C-terminal helix is composed of residues 75–92, of which only 75–88 may be required for dimerization. Similarly, the regions of the two C-terminal helices overlap only in those residues.

2.3. ArsD

In the *ars* operon of plasmid R773, the second gene, *arsD*, encodes a second soft metal regulated *trans*-acting repressor (Wu and Rosen, 1993). ArsR and ArsD share no se-

N-terminal

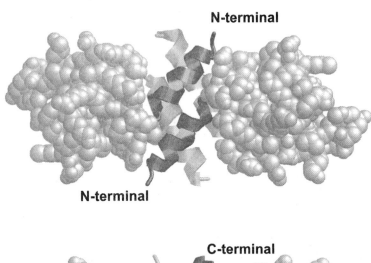

N-terminal

C-terminal

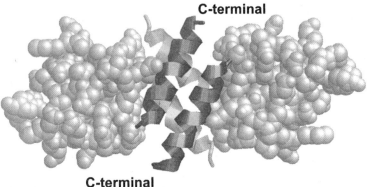

C-terminal

Figure 5. Dimerization domain of the ArsR aporepressor. The dimerization domain is composed of four α-helices, two contributed by residues 1–13 of each monomer (top) and two by residues 75–92 of each monomer (bottom). The two N-terminal helices interact with each other in an antiparallel manner, as do the two C-terminal helices.

quence similarity, although both are approximately the same size and are both functional homodimers (Chen and Rosen, 1997). The results of DNase I footprinting analysis revealed that ArsD binds to the same site on the *ars* promoter element as the ArsR but with two orders of magnitude lower affinity. It was also found that binding of arsenite, phenylarsine oxide or antimonite each produced dissociation of ArsD from the *ars* DNA promoter. This is the same soft metal inducer specificity as ArsR, but *in vivo* repression by ArsR can be fully relieved with 10 µM sodium arsenite, while ArsD repression requires approximately 100 µM sodium arsenite for induction. Thus ArsR has higher affinity for both DNA and inducer than ArsD. From these differences it is reasonable to conclude that the two repressors act at the same site but at different times. They form a regulatory circuit that keeps expression of the *ars* operon within a prescribed range (Figure 6). ArsR repressor controls the basal level of operon expression, while ArsD controls the upper level of *ars* expression. In the absence of an inducer, constitutively expressed ArsR binds to the *ars* operator with high affinity, repressing transcription. A low concentration of soft metal inducer produces dissociation of ArsR, resulting in transcription of the operon. However, high level expression of *arsB* is itself deleterious to the cell (Wu et al., 1992). To prevent

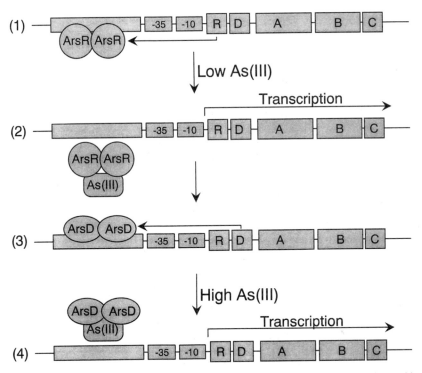

Figure 6. Arsenic homeostasis in *E. coli*. A regulatory circuit provides homeostasis in the sensing a wide range of environmental arsenicals and antimonials. (1) In the absence of inducer transcription is repressed by a basal level of ArsR synthesis. (2) In the presence of low to moderate concentrations of inducer the ArsR-inducer complex dissociates from the DNA, resulting in transcription of the *ars* operon. (3) When the concentration of ArsD increases sufficiently to allow this low affinity DNA binding protein to bind to the operator site, transcription is again repressed. (4) In the presence of high concentrations of inducer the ArsD-inducer complex dissociates from the DNA, resulting in a further increase in *ars* transcription.

this the amount of ArsD increases in proportion to the amount of ArsB; as the intracellular concentration of ArsD exceeds its K_D for *ars* DNA, ArsD would fill the *ars* operator site vacated by ArsR, repressing further transcription of the operon and synthesis of ArsB. Since the affinity of ArsD for inducer is relatively low, the intracellular concentration of soft metalloid would be unable to prevent its binding. However, if the extracellular concentration of metalloid increased further, it would saturate ArsD, resulting in additional transcription. Thus action of the two repressors forms a homeostatic regulatory circuit that maintains balance between detoxification of the metalloid and expression of the pump genes.

3. METALLOREGULATION BY ZINC AND CADMIUM

3.1. SmtB

The ArsR homologue SmtB, a *trans*-acting repressor encoded by a divergently transcribed gene *smtB* of a metallothionein (MT) locus, *smt*, from the cyanobacteria Synechococcus PCC7942 and PCC 6301 is required for Zn(II)-repressive expression of the metallothionein gene *smtA* (Morby et al., 1993). Thus, although this Zn(II) resistance mechanism does not involve soft metal transport, its regulatory aspects are similar to the Ars pump. Recombinant SmtB has been expressed in *Escherichia coli* and purified (Erbe et al., 1995). Based on gel mobility shift assays, methylation interference and molecular mass calculations, it was proposed that recombinant SmtB binds to the *smt* DNA promoter in multimeric fashion. At low concentrations of SmtB a single complex formed, and with higher concentrations of recombinant SmtB multiple complexes were detected in gel mobility shift assays. It was also shown that SmtB is capable of directly interacting with Zn(II), with inhibition of complex formation in the presence of various metals including zinc, copper, cadmium, cobalt, nickel and chromium. Although SmtB is a member of the ArsR family of metalloregulatory proteins, it does not contain the cysteines that form the metal binding site in ArsR. Within the helix-turn-helix domain it has only a single cysteine residue corresponding to Cys32 of ArsR. On the other hand, from the results of site-directed mutagenesis studies, the vicinal histidine pair His105 and His106 in the C-terminal region was suggested to be part of the metal binding site (Turner et al., 1996). More recenlty purified SmtB was shown to bind two molecules of Zn(II) per monomer, suggesting that there may be multiple metal binding sites (Kar et al., 1997). The average affinity for Zn(II) was 3.4×10^{-6} M, and binding of the Zn(II) to the repressor produced a conformational change that made the dimer more compact.

As discussed above, SmtB has been crystallized as the zinc- and DNA-free dimer (Cook et al., 1998). A Hg(II) derivative was used to solve the structure of the aporepressor. The structure contained four Hg(II), two per monomer that have been hypothesized to be at the Zn(II) sites. Two of the sites are formed by residues contributed solely by each monomer, and the other two sites are formed by residues contributed by both monomers at their shared interface. From consideration of the structure and the results of metal binding studies, a model similar to that discussed above for ArsR has been proposed for SmtB regulation of *smt* operon. Upon binding of soft metal ions to SmtB, there is stronger self-association of the monomers, making the dimeric repressor more compact. The metal-bound conformation would no longer bind to the operator site on the DNA, leading to transcription of *smtA* (Kar et al., 1997). However, if the metal-bound repressor assumes a considerably more compact conformation than the aporepressor, then the question arises whether the heavy atom derivative is isomorphous with the aporepressor or the induced form. To confirm the location of the inducer binding sites, it will be necessary to determine the structure of the Zn(II) bound protein.

3.2. CadC

The Cd(II)/Zn(II) resistance (*cad*) operon from the staphylococcal plasmid pI258 has two genes, *cadA* and *cadC* (Nucifora et al., 1989). CadA is a Cd(II) efflux pump that is a member of the superfamily of cation-translocating P-type ATPases (Lutsenko and Kaplan, 1995) (Figure 7). P-type ATPases are found in all organisms, with all biologically important cations transported by these types of pumps. These ATPases have a consensus ATP binding domain and a conserved aspartate residue that becomes phosphorylated during the catalytic cycle. There are two branches in the superfamily. Members of one branch transport hard metal cations, the best known of which are the sodium pump, which exchanges intracellular Na^+ for extracellular K^+ and is the major electrogenic pump of mammalian cells, the H^+/K^+-ATPase, which exchanges intracellular H^+ for extracellular K^+ and is responsible for gastric acidification, and the calcium extrusion pumps of plasma membrane and sarcoplasmic reticulum. The other subfamily of the P-type ATPases, termed

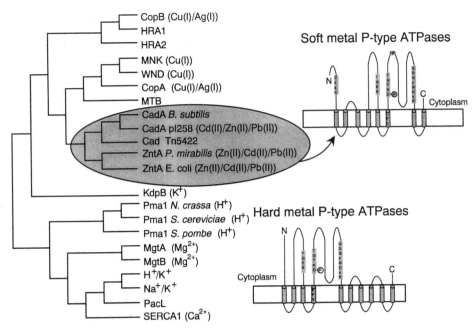

Figure 7. Superfamily of P-type ATPases: Phylogenetic relationship of the amino acid sequences of hard and soft metal translocating P-type ATPases. The superfamily has two branches: soft metals (top) and hard metal (bottom) ATPases. Proteins (known functions are identified, otherwise reading frames are identified solely from DNA sequence analysis). **Soft metal ATPases**: CopB, *Enterococcus hirae* Cu(I)/Ag(I) pump; HRA1 and HRA2, *E. coli* reading frames; MNK, human Menkes protein, WND, human Wilson protein, CopA, *Enterococcus hirae* Cu(I)/Ag(I) uptake pump, MTB, *Mycobacterium tuberculosis* reading frame, CadA, Cd(II) pumps from *Bacillus subitilis, Staphylococcus aureus* plasmid pI258 and *Staphylococcus aureus* transposon Tn5422; ZntA, Zn(II) pumps from *Proteus mirabilis* and *Escherichia coli*, CtaA, *Synechococcus* 7942 reading frame; Mtb2 **Hard metal ATPases**: KdpB, *Escherichia coli* K⁺ uptake pump; Pma1, fungal plasma membrane H⁺ pumps; MgtA and MgtB, *Salmonella typhimurium* Mg²⁺ uptake pumps; H⁺/K⁺, mammalian gastric H⁺ pump; N⁺/K⁺, mammalian Na⁺ pump; PacL, *Synechococcus* reading frame; SERCA1, mammalian sarcoplasmic reticulum Ca²⁺ pump.

CPx-type ATPases (Solioz and Vulpe, 1996) or soft metal ATPases (Rensing et al., 1997), include CadA, a Cd(II) pump, and ZntA, a Zn(II) pump. Although ZntA is physiologically a Zn(II) pump, it also pumps Cd(II) and provides resistance to it (Rensing et al., 1997). Both CadA and ZntA produce Pb(II) resistance and are almost certainly Pb(II) pumps as well (C. Rensing and B.P. Rosen, unpublished).

The *cadC* gene product is the transcriptional regulator of the *cad* operon (Yoon and Silver, 1991). The operator site for CadC binding to the *cad* promoter has been mapped (Endo and Silver, 1995). There are two retarded species in gel mobility shift assays that are dissociated in the presence of high concentrations of Cd(II), Bi(III) or Pb(II) but not Zn(II). CadC is a 122 residue protein, and is a member of the ArsR family (Figure 3). CadC has the potential metal binding motif $ELC_{58}VC_{60}DL$, which is identical to the ArsR soft metalloid binding site. However, CadC is not induced by As(III) or Sb(III) (Endo and Silver, 1995). Neither CadC nor ArsR have a vicinal histidine pair corresponding to the putative Zn(II) binding site of SmtB. However, CadC does have an N-terminal sequence absent in ArsR. This sequence contains two cysteine residues, Cys7 and Cys11, that, together with Cys58 and Cys60, may form a four-coordinate structure with Cd(II).

4. METALLOREGULATION BY COPPER AND SILVER—CopY AND CopZ

The *cop* operon that confers copper resistance in gram-positive bacteria *Enterococcus hirae* consists of the four genes *copY*, *copZ*, *copA*, and *copB* (Odermatt et al., 1992). CopA and CopB are two homologous P-type ATPases; CopA serves in copper uptake, while CopB catalyzes extrusion (Odermatt et al., 1993). Since the products of a single operon catalyze both uptake and extrusion, copper homeostasis would be predicted to require a complicated set of regulatory mechanisms. Indeed, expression of these two copper-translocating ATPases is controlled by two *trans*-acting metalloregulatory proteins, CopY and CopZ, that can be induced by copper, cadmium or silver ions (Odermatt and Solioz, 1995). CopY has 145 amino acids and contains the soft metal binding motif $CXCX_4CXC$. Disruption of *copY* led to constitutive overexpression of CopB. CopY in *trans* repressed this overexpression, suggesting that CopY acts as a metal-fist repressor (Odermatt and Solioz, 1995). *In vitro* purified CopY acts as a copper-inducible repressor (Strausak and Solioz, 1997). DNase I footprinting demonstrated that purified CopY bound to two discrete sites in the promoter/operator region of the *cop* operon, and the CopY-DNA interaction could be disrupted by copper. CopY is a homodimer that interacts with the operator in two steps in a copper-sensitive fashion.

Control of the *cop* operon of *E. hirae* also requires CopZ (Odermatt and Solioz, 1995). CopZ has only 69 amino acids and contains the conserved soft metal binding motif GMXCXXC. Disruption of *copY* resulted in overexpression of both CopA and CopB, indicating that CopY is a repressor. In contrast, in *copZ*-disrupted strains, transcription of the *cop* operon was repressed, and *copZ* function could be complemented in *trans*, suggesting that CopZ is a *trans*-acting activator. Although there is no evidence of an involvement of CopZ in derepression of the *cop* operon *in vitro*, a model for the observed biphasic regulation was proposed in which derepression of the *cop* operon could be achieved in two ways: by copper binding to CopY, which releases the repressor from the operator under conditions of excessive copper, or through activation of transcription by CopZ under copper-limiting conditions (Strausak and Solioz, 1997). Further studies will be required to elucidate the details of the function of CopZ in the regulation of *cop* operon.

5. METALLOREGULATION BY MERCURY: MerR AND MerD

Regulation of expression of a number of mercury resistance (*mer*) genetic determinants has been studied in detail. All *mer* operons have a primary regulatory gene *merR*. Some also have a coregulatory gene *merD*. MerR regulates mercury resistance both positively and negatively (O'Halloran et al., 1989). In the absence of Hg(II), MerR represses transcription by binding to the *mer* operator region. Hg(II) binding to the MerR-DNA complex activates transcription by inducing changes in the conformation of the DNA that makes the promoter a better template for RNA polymerase (Frantz and O'Halloran, 1990; Ansari et al., 1995). Genetic and biochemical studies of MerR have defined the functional domains of MerR. The DNA-binding domain is associated with a putative helix-turn-helix motif in the highly conserved amino terminal region of the protein. The key elements of the mercury-binding domain were elucidated by site-directed mutagenesis and heterodimer formation of the *Bacillus* sp. RC607 MerR protein (Helmann et al., 1990). Three cysteine residues, Cys79, Cys114 and Cys123 are involved the binding of one Hg(II) ion per

MerR dimer. Cys79 from one subunit and Cys114 and Cys123 from the other form a three coordinate species with Hg(II) (Helmann et al., 1990). By ^{199}Hg NMR, Hg(II) was shown to be centrally bound to the three cysteine thiolates in a trigonal planar geometry (Utschig et al., 1995).

The *merD* gene is the most promoter distal in *mer* operons of gram negative bacteria. Like the ArsR/ArsD regulatory protein pair of the R773 *ars* operon, both MerR and MerD bind to a common operator site on DNA, but MerD has considerably lower affinity than MerR (Mukhopadhyay et al., 1991). From the similarity of the amino terminus of MerD with that of MerR, MerD has been suggested to be a coregulator that fine tunes expression of the *mer* operon (Brown et al., 1986; Mukhopadhyay et al., 1991).

REFERENCES

Ansari, A.Z., J.E. Bradner and T.V. O'Halloran (1995). "DNA-bend modulation in a repressor-to-activator switching mechanism." Nature 374, 371–375.

Bhattacharjee, H., J. Li, M.Y. Ksenzenko and B.P. Rosen (1995). "Role of cysteinyl residues in metalloactivation of the oxyanion- translocating ArsA ATPase." J Biol Chem 270, 11245–11250.

Bhattacharjee, H. and B.P. Rosen (1996). "Spatial proximity of Cys113, Cys172, and Cys422 in the metalloactivation domain of the ArsA ATPase." J Biol Chem 271, 24465–24470.

Bobrowicz, P., R. Wysocki, G. Owsianik, A. Goffeau and S. Ulaszewski (1997). "Isolation of three contiguous genes, ACR1, ACR2 and ACR3, involved in resistance to arsenic compounds in the yeast *Saccharomyces cerevisiae*." Yeast 13, 819–828.

Brown, N.L., T.K. Misra, J.N. Winnie, A. Schmidt, M. Seiff and S. Silver (1986). "The nucleotide sequence of the mercuric resistance operons of plasmid R100 and transposon Tn501: further evidence for *mer* genes which enhance the activity of the mercuric ion detoxification system." Mol Gen Genet 202, 143–151.

Chen, C.M., T.K. Misra, S. Silver and B.P. Rosen (1986). "Nucleotide sequence of the structural genes for an anion pump. The plasmid-encoded arsenical resistance operon." J Biol Chem 261, 15030–15038.

Chen, Y. and B.P. Rosen (1997). "Metalloregulatory properties of the ArsD repressor." J Biol Chem 272, 14257–14262.

Cook, W.J., S.R. Kar, K.B. Taylor and L.M. Hall (1998). "Crystal structure of the cyanobacterial metallothionein repressor SmtB: a model for metalloregulatory proteins." J Mol Biol 275, 337–346.

Datta, N. and V.M. Hughes (1983). "Plasmids of the same Inc groups in Enterobacteria before and after the medical use of antibiotics." Nature 306, 616–617.

de Boer, P.A., R.E. Crossley, A.R. Hand and L.I. Rothfield (1991). "The MinD protein is a membrane ATPase required for the correct placement of the *Escherichia coli* division site." Embo J 10, 4371–4380.

Dey, S., D. Dou and B.P. Rosen (1994). "ATP-dependent arsenite transport in everted membrane vesicles of *Escherichia coli*." J Biol Chem 269, 25442–25446.

Dey, S., D. Dou, L.S. Tisa and B.P. Rosen (1994). "Interaction of the catalytic and the membrane subunits of an oxyanion- translocating ATPase." Arch Biochem Biophys 311, 418–424.

Endo, G. and S. Silver (1995). "CadC, the transcriptional regulatory protein of the cadmium resistance system of *Staphylococcus aureus* plasmid pI258." J Bacteriol 177, 4437–4441.

Erbe, J.L., K.B. Taylor and L.M. Hall (1995). "Metalloregulation of the cyanobacterial *smt* locus: identification of SmtB binding sites and direct interaction with metals." Nucleic Acids Res 23, 2472–2478.

Frantz, B. and T.V. O'Halloran (1990). "DNA distortion accompanies transcriptional activation by the metal- responsive gene-regulatory protein MerR." Biochemistry 29, 4747–4751.

Georgiadis, M.M., H. Komiya, C. P., D. Woo, J.J. Kornuc and D.C. Rees (1992). "Crystallographic structure of the nitrogenase iron protein from *Azotobacter vinelandii*." Science 257, 1653–1659.

Gladysheva, T.B., K.L. Oden and B.P. Rosen (1994). "Properties of the arsenate reductase of plasmid R773." Biochemistry 33, 7288–7293.

Helmann, J.D., B.T. Ballard and C.T. Walsh (1990). "The MerR metalloregulatory protein binds mercuric ion as a tricoordinate, metal-bridged dimer." Science 247, 946–948.

Hsu, C.M. and B.P. Rosen (1989). "Characterization of the catalytic subunit of an anion pump." J Biol Chem 264, 17349–17354.

Jones, R., P. Woodley, A. Birkmann-Zinoni and R.L. Robson (1993). "The *nifH* gene encoding the Fe protein component of the molybdenum nitrogenase from *Azotobacter chroococcum*." Gene 123, 145–146.

Kar, S.R., A.C. Adams, J. Lebowitz, K.B. Taylor and L.M. Hall (1997). "The cyanobacterial repressor SmtB is predominantly a dimer and binds two Zn^{2+} ions per subunit." Biochemistry 36, 15343–15348.

Karkaria, C.E., C.M. Chen and B.P. Rosen (1990). "Mutagenesis of a nucleotide-binding site of an anion-translocating ATPase." J Biol Chem 265, 7832–7836.

Kaur, P. and B.P. Rosen (1992). "Mutagenesis of the C-terminal nucleotide-binding site of an anion- translocating ATPase." J Biol Chem 267, 19272–19277.

Kaur, P. and B.P. Rosen (1992). "Plasmid-encoded resistance to arsenic and antimony." Plasmid 27, 29–40.

Kurdi-Haidar, B., S. Aebi, D. Heath, R.E. Enns, P. Naredi, D.K. Hom and S.B. Howell (1996). "Isolation of the ATP-binding human homolog of the arsA component of the bacterial arsenite transporter." Genomics 36, 486–491.

Kuroda, M., S. Dey, O.I. Sanders and B.P. Rosen (1997). "Alternate energy coupling of ArsB, the membrane subunit of the Ars anion-translocating ATPase." J Biol Chem 272, 326–331.

Li, J., S. Liu and B.P. Rosen (1996). "Interaction of ATP binding sites in the ArsA ATPase, the catalytic subunit of the Ars pump." J Biol Chem 271, 25247–25252.

Li, J. and B.P. Rosen (1998). "Steric limitations in the interaction of the ATP binding domains of the ArsA AT-Pase." J Biol Chem 273, 6796–6800 .

Lutsenko, S. and J.H. Kaplan (1995). "Organization of P-type ATPases: significance of structural diversity." Biochemistry 34, 15607–15613.

Morby, A.P., J.S. Turner, J.W. Huckle and N.J. Robinson (1993). "SmtB is a metal-dependent repressor of the cyanobacterial metallothionein gene smtA: identification of a Zn inhibited DNA-protein complex." Nucleic Acids Res 21, 921–925.

Mukhopadhyay, D., H.R. Yu, G. Nucifora and T.K. Misra (1991). "Purification and functional characterization of MerD. A coregulator of the mercury resistance operon in gram-negative bacteria." J Biol Chem 266, 18538–18542.

Nucifora, G., L. Chu, T.K. Misra and S. Silver (1989). "Cadmium resistance from Staphylococcus aureus plasmid pI258 cadA gene results from a cadmium-efflux ATPase." Proc Natl Acad Sci U S A 86, 3544–3548.

Oden, K.L., T.B. Gladysheva and B.P. Rosen (1994). "Arsenate reduction mediated by the plasmid-encoded ArsC protein is coupled to glutathione." Mol Microbiol 12, 301–306.

Odermatt, A. and M. Solioz (1995). "Two trans-acting metalloregulatory proteins controlling expression of the copper-ATPases of Enterococcus hirae." J Biol Chem 270, 4349–4354.

Odermatt, A., H. Suter, R. Krapf and M. Solioz (1992). "An ATPase operon involved in copper resistance by Enterococcus hirae." Ann N Y Acad Sci 671, 484–486.

Odermatt, A., H. Suter, R. Krapf and M. Solioz (1993). "Primary structure of two P-type ATPases involved in copper homeostasis in Enterococcus hirae." J Biol Chem 268, 12775–12779.

O'Halloran, T.V., B. Frantz, M.K. Shin, D.M. Ralston and J.G. Wright (1989). "The MerR heavy metal receptor mediates positive activation in a topologically novel transcription complex." Cell 56, 119–129.

Rensing, C., B. Mitra and B.P. Rosen (1997). "The zntA gene of Escherichia coli encodes a Zn(II)-translocating P-type ATPase." Proc Natl Acad Sci U S A 94, 14326–14331.

Shi, W., J. Dong, R.A. Scott, M.Y. Ksenzenko and B.P. Rosen (1996). "The role of arsenic-thiol interactions in metalloregulation of the ars operon." J Biol Chem 271, 9291–9297.

Shi, W., J. Wu and B.P. Rosen (1994). "Identification of a putative metal binding site in a new family of metalloregulatory proteins." J Biol Chem 269, 19826–19829.

Solioz, M. and C. Vulpe (1996). "CPx-type ATPases: a class of P-type ATPases that pump heavy metals." Trends Biochem Sci 21, 237–241.

Strausak, D. and M. Solioz (1997). "CopY is a copper-inducible repressor of the Enterococcus hirae copper AT-Pases." J Biol Chem 272, 8932–8936.

Takemaru, K., M. Mizuno, T. Sato, M. Takeuchi and Y. Kobayashi (1995). "Complete nucleotide sequence of a skin element excised by DNA rearrangement during sporulation in Bacillus subtilis." Microbiology 141, 323–327.

Turner, J.S., P.D. Glands, A.C. Samson and N.J. Robinson (1996). "Zn^{2+}-sensing by the cyanobacterial metallothionein repressor SmtB: different motifs mediate metal-induced protein-DNA dissociation." Nucleic Acids Res 24, 3714–3721.

Utschig, L.M., J.W. Bryson and T.V. O'Halloran (1995). "Mercury-199 NMR of the metal receptor site in MerR and its protein-DNA complex." Science 268, 380–385.

Wu, J. and B.P. Rosen (1991). "The ArsR protein is a trans-acting regulatory protein." Mol Microbiol 5, 1331–1336.

Wu, J. and B.P. Rosen (1993). "The arsD gene encodes a second trans-acting regulatory protein of the plasmid-encoded arsenical resistance operon." Mol Microbiol 8, 615–623.

Wu, J. and B.P. Rosen (1993). "Metalloregulated expression of the ars operon." J Biol Chem 268, 52–58.

Wu, J., L.S. Tisa and B.P. Rosen (1992). "Membrane topology of the ArsB protein, the membrane subunit of an anion-translocating ATPase." J Biol Chem 267, 12570–12576.

Wysocki, R., P. Bobrowicz and S. Ulaszewski (1997). "The *Saccharomyces cerevisiae* ACR3 gene encodes a putative membrane protein involved in arsenite transport." J Biol Chem 272, 30061–30066.

Xu, C. and B.P. Rosen (1997). "Dimerization is essential for DNA binding and repression by the ArsR metalloregulatory protein of *Escherichia coli*." J Biol Chem 272, 15734–15738.

Yoon, K.P., T.K. Misra and S. Silver (1991). "Regulation of the *cadA* cadmium resistance determinant of *Staphylococcus aureus* plasmid pI258." J Bacteriol 173, 7643–7649.

Yoon, K.P. and S. Silver (1991). "A second gene in the *Staphylococcus aureus cadA* cadmium resistance determinant of plasmid pI258." J Bacteriol 173, 7636–7642.

CLASTOGENIC EFFECTS AND INFLUENCE OF INORGANIC ARSENIC ON DNA-REPAIR IN MAMMALIAN SYSTEMS

R. Nilsson,[1] A. T. Natarajan,[2] A. Hartwig,[3] F. Dulout,[4] M. E. De la Rosa,[5] and M. Vahter[6]

[1]Department of Genetic and Cellular Toxicology
Wallenberg Laboratory, Stockholm University
S-106 91 Stockholm, Sweden
[2]Department of Radiation Genetics and Mutagenesis
University of Leiden
The Netherlands
[3]Department of Biology and Chemistry
University of Bremen
Federal Republic of Germany
[4]CIGEBA, Centro de Investigaciones en Genetica Basica y Aplicada
Universidad Nacional de la Plata, Argentina
[5]Depto. de Quimica Inorganica y Nuclear
Divisio de Quimica, Facultad de Quimica
Universidad Nacional Autonoma de México (UNAM)
México, D.F.
[6]Institute of Environmental Medicine
Karolinska Institute
Stockholm, Sweden

SUMMARY

The mode of carcinogenic action as well as metabolic disposition are both important decision elements when choosing the appropriate approach for conducting high-to-low dose human cancer risk assessment with respect to inorganic arsenic (As). Conflicting opinions exist whether or not a dose threshold should be applied for this human carcinogen, and/or if metabolic overloading with respect to methylation to monomethylarsonic acid (MMA) and dimethylarsinic acid (DMA) occurs at elevated exposures that might potentiate the carcinogenic action of As. Unfortunately, progress has been hampered by the

Metals and Genetics, edited by Sarkar.
Kluwer Academic / Plenum Publishers, New York, 1999.

fact that neoplasia cannot be induced consistently by As in experimental animals, and that no appropriate animal model has yet been found that adequately mimics its metabolism in most humans. The latter also includes the Chimpanzee, in as much as we have demonstrated that this primate does not metabolize As.

As inhibits DNA-repair, and our investigations on the effect of As(III) on UV-induced nucleotide excision repair in repair proficient human cells, and partly repair deficient xeroderma pigmentosum cultivated human fibroblasts indicate incision to be the most sensitive step. Inhibition occurs at concentrations that are relevant with respect to cancer induction in humans. At higher concentrations ligation is also inhibited. The global genome repair as well as the transcription-coupled repair pathway are both affected, and exposure to As may, thus, induce a partially repair deficient condition similar to that seen in xeroderma, where additional promotive action such as keratotic changes will determine the site of tumor appearance.

Although As does not induce point mutations per se, it is pro-mutagenic in many *in vitro* systems, including human cell lines. It is a clastogen and an aneugen in cultured mammalian cells, as well as in peripheral lymphocytes from exposed humans from northwestern Argentina: In an isolated native Indian settlement in the Andean region, as well as in a mestizo population in the lower Chaco region, Salta Province, where both populations are exposed to As in drinking water at about the same level (200µg/L), there was a highly significant increase in the frequency of micronuclei (MN) as well as of aneuploidy in lymphocytes from exposed individuals, but no notable effects were found on sister chromatid exchanges, specific chromosomal translocations, or on cell cycle progression. As supported by fluorescence *in situ* hybridization (FISH), at least a proportion of MN appears to originate from whole chromosome loss. The incidence of MN was similar to that found previously by us for women from a heavily contaminated area around the Srednogorie copper smelter, Bulgaria, and these findings are also supported by similar findings with respect to exfoliated bladder cells in an exposed cohort in Nevada. The induction of MN appears to constitute the most sensitive toxicological endpoint for exposure to As so far been described. These findings support the hypothesis, that As exerts its carcinogenic effects by an indirect mechanism, that may exhibit a dose threshold.

The exposed native population in the Andes had a different metabolic disposition than previously found for populations in Europe, U.S., and Japan, in as much as these individuals mainly excrete arsenic as DMA and inorganic As, with very little MMA. Further, whereas none of the typical signs of systemic As intoxication could be seen among the Indians studied in the Puna region—who have been exposed for a very long time in history—hyperkeratoses and skin cancer were present among the investigated mestizos exposed to about the same concentration of As in drinking water from the Chaco area. The observed ethnic differences most probably have a genetic background.

1. BACKGROUND

1.1. Arsenic Induced Cancers

Inorganic arsenic (**As**) is an established human carcinogen that upon chronic inhalation mainly induces small cell or squamous cell cancer of the lung (Newman et al., 1976; Pershagen et al., 1987; IPCS/WHO, 1981). Upon chronic oral intake, As may cause severe toxic effects including cancer of skin, bladder, and possibly liver as well as adverse effects on the cardiovascular system (IPCS/WHO, 1981; U.S.EPA, 1988; Bates et al., 1992). It has, on the other hand, so far not been possible to consistently induce neoplasia by As in experimental animals. However, it was recently reported, that treatment with As of v-Ha-ras transgenic mice—an animal model for skin cancer induction—caused an increased incidence of papillomas (Luster et al., 1995).

Since "der Reichensteiner Krankheit" was described from the county of Glatz (Klodzko), Silesia, at the end of last century, As has been known to cause pigmentation changes, palmo-plantar keratosis and skin cancers upon ingestion of As containing drinking water as well as from treatment with pharmaceuticals like "Fowler's solution". During the early part of this century, observations of chronic arsenism were made in the Cordoba province of Argentina as well as from the Antofagasta area in Chile (Early literature reviewed by Neubauer, 1947). Later, arsenism due to contaminated drinking water was described in the Lagunera area of the Mexican state Coahuila (Cebrian et al., 1983), and an endemic arsenic poisoning in a population in Taiwan was discovered (Tseng et al., 1968; Tseng, 1977). The As induced neoplasias of the skin mostly include Bowen's disease and basal cell carcinoma, but occasionally squamous cell carcinoma may develop. In addition to the dermal effects, a vascular occlusive disorder ("Blackfoot disease") was described in Taiwan, that in severe cases results in gangrene of the lower extremities.

1.2. Exposure to Arsenic

Being far from a forensic curiosity of the past, at the present time millions of humans all over the world are chronically exposed to As at sufficiently high levels to cause severe toxic effects, including cancer. Occupational exposure to arsenic is almost invariably associated with the smelting of copper sulphide ores, but since arsenopyrite ($FeAsS_2$) is also commonly present in other sulphide ores, high exposures to arsenic may occur in the mining and production of gold, tin, tungsten, lead and zinc. In terms of number of individuals involved, the most extensive problems have been caused in developing countries by the drilling of wells for drinking water into geological strata rich in arsenic without adequate prior geochemical analysis, e.g. as was the case in Hungary (Börzsönyi et al., 1992), in the Xingjiang Uyghur Autonomous Region, China (Lianfang and Jianzhong, 1994), as well as in West Bengal, India. In West Bengal, six districts encompassing an area of approximately 34,000 km^2 and harboring a population of about 30 million are affected. An estimated 200,000 people already have arsenic-induced skin lesions, including skin cancer. Evidently arsenic derives from ground water percolating sediments rich in arsenopyrite. The wells reach a depth anywhere between 20 to 150 meters, and some wells have levels as high as 3,200 µg/L (Mondal et al., 1996). Physicians have reported, that the problem is at least as severe in adjoining parts of Bangladesh (Tsushima, S., 1997).

Emissions and waste disposal of As associated with current or past mining, smelting, and refining of copper, gold, lead, nickel, tin as well as of zinc, constitutes a health problem which is not well recognized in many areas all over the world. Acid leaching of tailings from sulphide ore mining results in contamination of water and soil. As demonstrated in Cornwall and Devon of Great Britain, where non-ferrous metal mining and processing already began in Roman times, abandoned mining areas for As-containing sulphide ores, as well as locations that once harbored copper smelters, are often heavily polluted by As (Mitchell and Barr, 1995). Emissions from a copper smelter in Bulgaria was found to increase the incidence of chromosomal aberrations in residents living in the surrounding area (Nilsson et al., 1993). Unfortunately, assessment of exposure to arsenic with respect to the general population to arsenic is sketchy, or inadequate in most countries.

1.3. Pharmacokinetics

In mammals, most of the ingested inorganic arsenic is rapidly eliminated as methylated metabolites via the kidneys. However, high levels of arsenic are maintained for longer periods of time in bone, skin, hair and nails of exposed humans. Following ingestion, or inhalation of inorganic arsenic, the major forms of arsenic excreted in human urine are the methylated products dimethylarsinic acid (DMA) (60–80%), and monomethylarsonic acid (MMA) (10–20%), the remaining fraction (10–30%) being in the form of inorganic arsenic. Consumers of marine fish, shellfish, and crustaceans, in addition, excrete significant quantities of arsenobetaine and related compounds (Norin and Vahter, 1981). While cadmium, mercury and lead are excreted in milk to a significant degree, Concha et al. (1998a) found that the total average concentration of arsenic in human breast milk (<2 µg/L) remained low in spite of a relatively high exposure via drinking water. The rat, mouse hamster, and rabbit differ from most humans in that very low levels of MMA is excreted (reviewed by Vahter, and Marafante, 1988), while the guinea pig (Healy et al., 1997), the chimpanzee (Vahter et al., 1995a), and some New World monkeys (Vahter et al.,1982; Zakharayan et al., 1996) are unable to methylate As.

Uptake of arsenite by cultured human fibroblasts is fairly slow reaching a maximum after 18 h with a 15-fold accumulation with respect to the concentration in the medium (Hartwig et al., 1997). The *in vivo* methylation of inorganic arsenic occurs mainly in the liver, but Healy et al. (1998) have recently demonstrated a high methylating activity is also present in testis, kidney and lung of the mouse. Before methylation can take place, pentavalent arsenic must be reduced to trivalent arsenic, a reaction that requires glutathione. Methylation, catalyzed by methyltransferases, proceeds in two steps where S-adenosyl methionine acts as methyl donor. Monomethylarsonic acid (MMA), formed in the first step is subsequently methylated to dimethylarsinic acid (DMA) (Buchet and Lauwerys, 1985; Vahter and Marafante, 1988; Stylbo et al., 1995). The methyltransferases have been partially purified from the liver of rabbit and Rhesus monkey (Zakharayan et al., 1995, 1996). In rabbits, choline, methionine, or protein deficient diets will result in decreased methylation and such defieciences will, presumably, result in an increase of the body burden of inorganic arsenic (Vahter and Marafante, 1987).

Under low-level exposures to arsenic, there seems to exist a balance between the amount entering the body and the amount being excreted. At high doses (>1mg/kg b.w.; Vahter, 1983) methylation efficiency in rodents decreases with increasing dose levels, and it has been proposed, that this may have toxicological importance in terms of a reduction also of the capacity of humans to handle arsenic at elevated levels of intake (Valentine et al., 1979; Buchet et al., 1981; U.S.EPA, 1988). However, studying a population of mixed ethnic origin in northern Chile exposed to up to 600 µg/L in drinking water, Hopenhayn-Rich and coworkers (1996a) failed to find any evidence of saturation of the methylating capacity.

1.4. Genotoxic Activity

Results from bacterial mutation tests—like the Ames' test—have largely been negative. In some studies clastogenic effects as well as SCEs have been observed *in vitro*. While arsenic compounds have not been shown to cause point mutations, it has a co-mutagenic effect in bacteria and in cultivated mammalian cells with different types of chemical mutagens as well as with UV. A co-mutagenic effect has also been demonstrated in mice *in vivo* (Sram, 1976). Further, arsenite induces malignant transformation in rodent cells (Rossman, 1994).

It has been reported, that arsenic induces sister chromatid exchanges (SCEs) in humans (Burgdorf et. al., 1977; Wen et al., 1981; Lerda, 1994). However, Nordenson et al. (1978, 1979) could not confirm these observations, but described an increased incidence of chromosome aberrations in arsenic exposed individuals. Although no differences in the incidence of chromosome deletions and chromatid deletions could be found, the frequency of dicentrics, rings, and translocations in lymphocytes was reported to be higher in 13 exposed individuals from the Lagunera, Mexico, in comparison with 11 persons from a control area. However, this increase was statistically not significant. These investigators also found an inhibition of cell-cycle kinetics (Ostrosky-Wegman et al., 1991).

1.5. Quantitative Cancer Risk Assessment

Metabolic disposition as well as mode of carcinogenic action are both important decision elements when choosing the appropriate approach for conducting high-to-low dose human cancer risk assessment for inorganic arsenic. With respect to ubiquitous exposures lower than those causing overt clinical signs of intoxication—but involving much larger segments of the general population—conflicting opinions exist with respect to how quantitative risk assessment of As should be best performed, especially for cancer induction. U.S.EPA used linear extrapolation of epidemiological data from the highly exposed population in Taiwan (Tseng et al., 1968) to obtain a carcinogen potency factor (slope factor) of 1.8 (mg/kg and day)$^{-1}$ for life time risk (U.S.EPA, 1988). Based on this concept, WHO (1993) has lowered the provisional drinking water quality guideline for As from 50 μg/L to 10μg/L. If implemented, this new standard will have serious economic implications in terms of investments in water purification, and/or for providing alternative water supplies. However, before such investments are made, common sense would dictate that the scientific soundness of the underlying principles must first be thoroughly assessed. It should be noted, that the intake of As for parts of the Japanese population exceeds the WHO recommendations, apparently without negative consequences.

Genetically based defects in the human DNA-repair systems are in general associated with a propensity to develop cancer at various sites, as e.g. in xeroderma pigmentosum, Fanconi's anemia, Bloom's syndrome, and ataxia telangiectasia (Hanawalt, 1996). If the mechanism of cancer induction by arsenic is linked to interference with DNA repair—as was suggested by some researchers (Li and Rossman, 1989; Nilsson et al., 1993, Hartwig et al., 1997)—as for cancer promoters it would be possible to define a "safe" exposure threshold level with respect to the carcinogenic effects of arsenic. Some additional indirect support for this notion is available: although results from cancer promotion studies in rodents have been equivocal (Kroes et al. 1974; Shirachi et al., 1983), there is, on the other hand, epidemiological evidence that As acts as a cancer promoter in humans. Thus, a strong synergistic effect from the simultaneous exposure to arsenic and smoking has been found in smelter workers (Pershagen et al. 1981). Further, As was reported to increase the incidence of lung cancer in tin miners simultaneously exposed to significant levels of radon as well as to arsenic (Taylor et al., 1989). In comparison with normal cells, the mutational spectrum will be different in cells with deficient DNA-repair, and treatment with arsenite has been found to alter the mutational spectrum induced by UV in Chinese hamster ovary cells *in vitro* (Yang et al., 1992). For this reason, an assessment of the mode of action of arsenic on the genetic material is of considerable importance.

1.6. This Project

The results presented below represent a multidisciplinary effort from several research groups in Europe and Latin America to gain further insight into the mechanism of action of this enigmatic carcinogen with the aim of obtaining a more adequate data base for quantitative cancer risk assessment. The research on the metabolism as well as mode of carcinogenic action of As has been greatly hampered by the lack of a suitable animal model. In search of a species that to a greater extent resembles humans in this respect, the metabolic disposition of As was studied in the chimpanzee. However, this human primate did not fulfill our expectations, forcing us to rely heavily on the study of human material. To test the hypothesis that As induces genotoxic events by an indirect mode of action, investigations were conducted with respect to the effects on DNA repair, somatic recombination as well as on clastogenic events in human cells *in vitro* as well as *in vivo*. To explain an ethnically related differential sensitivity with respect to induction of skin cancer, the metabolism of As was investigated in exposed populations in northwestern Argentina.

2. RESULTS

2.1. The Metabolic Disposition of Arsenic in the Chimpanzee

As mentioned above, no adequate animal model is available for the study of cancer induction by As. Further, whereas a rodent like the rabbit efficiently converts inorganic As to DMA, Vahter and coworkers (Vahter et al., 1982) demonstrated that the primitive New World marmoset monkey (*Callithrix jacchus*) totally lacked the ability to methylate As. Caucasians and Japanese exhibit a methylating capacity that lies somewhere between these extremes.

To investigate if the chimpanzee constitutes an appropriate model for the study of As-metabolism in man, [73]As-arsenate was given in drinking water to two male chimpanzees at White Sands Research Center, Alamogordo, NM. Urine, feces, and blood were collected, and [73]As determined in whole blood, plasma, and RBC by gamma counting. Plasma samples were analyzed for As metabolites by ion-exchange chromatography, as well as by thin-layer chromatography on cellulose plates. Similarly to what was previously found for the Marmoset (Vahter et al., 1982), no formation of methylated products from inorganic As were detected in the chimpanzee. The biphasic kinetics of As-elimination from plasma (Figure 1) was found to be very similar in both species of monkey with an initial rapid elimination phase followed by a slower rate of elimination with a $t_{1/2}$ of about 170 h (Vahter et al., 1995a). Zakharyan et al. (1996) have demonstrated, that whereas the Marmoset and Tamarin monkeys lack As-methyltransferases, the liver of the Rhesus monkey contained ample amounts of these enzymes.

2.2. Genotoxic Events Induced by Arsenic in Model Systems

2.2.1. Effects of Arsenic on DNA Repair. As is reduced to arsenite in the mammalian organism, and arsenite (but not arsenate) is highly selective in reacting preferentially with vicinal dithiol groups in enzymic proteins, which is an important feature of some DNA binding proteins, transcription factors, and DNA repair enzymes. As was mentioned initially, hereditary defects with respect to DNA repair is generally associated with an increased propensity to develop cancer. However, the Cockayne's syndrome (CS), is

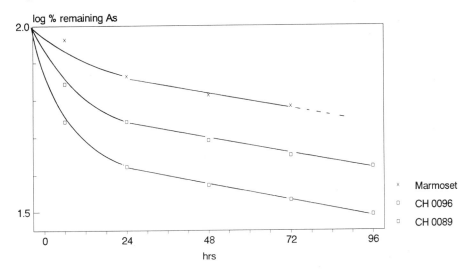

Figure 1. Whole-body retention of [73]As (based on excretion data) in two chimpanzees administered a single i.v. dose of [73]As-arsenate (Vahter et al., 1995a). For comparison, pooled data from four Marmoset monkeys exposed to [74]As-arsenate are included (Vahter and Marafante, 1985).

characterized by a specific defect in transcription-coupled repair that does not predispose the patients to UV-induced skin cancers that e.g. are typical in xeroderma pigmentosum complementation group C (XP-C). While the CS cells are competent with respect to global genome repair, XP-C has transcription coupled repair, but lacks global genome repair covering large domains of the genome (Venema et al., 1990). It has been reported, that As inhibits DNA ligases (Li and Rossman, 1989; Li Chen et al., 1992). However, other steps than ligation may also be involved (Okui and Fujiwara, 1986; Snyder et al., 1989), and we have conducted further investigations to define which part of the repair process is most readily affected (Hartwig et al., 1997).

The action of As was studied (Hartwig et al., 1997) in two repair proficient (VH16, VH25), and in one partially repair deficient human fibroblast cell line (XP21ROC; XPC). To investigate the effects of As on the incision and ligation in the course of DNA repair events, VH16 fibroblasts were grown as monolayers, incubated with $NaAsO_2$ and irradiated with UVC. Strand breaks were quantified by alkaline unwinding (Ahnström and Erixon, 1981), where single- and double stranded DNA are separated on hydroxyapatite columns. To increase sensitivity, repair kinetics were measured in presence of hydroxyurea which slows down repair, and causes repair patches stay open for a longer period of time. In presence of arsenite there was a marked reduction in the frequency of DNA strand breaks. However, once incisions had been made, the ligation of repair patches was delayed as well.

Recognition and incision by endonucleases are rate-limiting in nucleotide excision repair. However, the incision events can be trapped by DNA polymerase inhibitors, such as aphidicolin or cytosine arabinoside (Erixon, 1986). Thus, in the presence of hydroxyurea and aphidicoline, the level of accumulated strand breaks in VH16 human fibroblasts induced by UV was decreased to approximately 70% of the control value already at 2.5 μM arsenite, and to 65% at 10 μM (Figure 2). To investigate the dose-response for effects on ligation, UV-induced DNA strand breaks were accumulated in the presence of aphidicolin and hydroxyurea, after which all four nucleosides were added to the medium to allow repair patches to be ligated. Under these conditions, there was virtually a com-

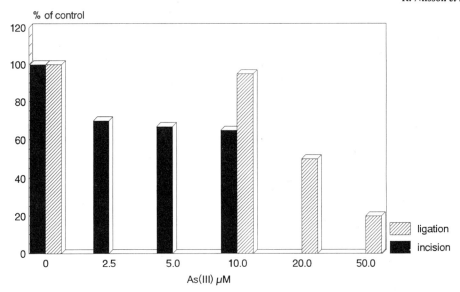

Figure 2. Dose-dependent inhibition by arsenite of the incision (_) as well as ligation (cross-hatched) frequency of single strand breaks induced by UV in human fibroblasts. VH16 cells were pre-incubated with 15 μM aphidocolin and 10 mM hydroxyurea for 1 h, UV irradiated with 0.2 J/m^2 and allowed to repair for 1 h (incision), or 2 h (ligation) in the presence of the inhibitors. To study the effect of As, cells were preincubated with arsenite for 18 h (to ensure uptake) before the addition of inhibitors, and all subsequent steps were carried out in the presence of As. To monitor ligation, after repair for 2 h in presence of the inhibitors, 100 μM of all four nucleosides were added for 15 min to allow ligation of the repair patches (Adapted from Hartwig et al., 1997).

plete and rapid resealing of breaks in the absence of arsenite. As demonstrated in Figure 2, ligation was not markedly affected at 10μM, but was decreased to 50% at 20 μM, and to about 20% at 50 μM As. Thus, under these conditions, the ligation step of DNA nucleotide excision repair is inhibited only at cytotoxic concentrations of As. Depression of incision by As reached a plateau at 65–70% of the control value. The failure to induce further inhibition above 5 μM As(III) seems to reflect differences in the efficiency of the repair of the cyclobutane dimers as compared to pyrimidine-(6–4)-pyrimidone photoproducts.

The study of UV-induced repair replication was carried out by using fibroblasts labeled with ^{14}C-thymidine, and subsequently incubated with arsenite for 18 h. Prior to UVC irradiation, fluorodeoxyuridine and bromodeoxyuridine were added in the presence of arsenite. DNA, isolated from irradiated cells, was fractionated by CsCl gradient centrifugation, and each fraction analyzed for ^{14}C- and ^{3}H-activities. At the position of unreplicated parental DNA, pooled fractions with the highest ^{3}H-content were mixed with alkali and fractionation repeated on a CsCl gradient. The DNA fractions were then analyzed for their ^{14}C and ^{3}H contents. The amount of repair replication was given as the ratio of ^{3}H repair label to ^{14}C prelabel activity. In the presence of 10 μM repair synthesis was reduced by 20–30% in repair competent VH16 fibroblasts and to 48–50% in VH25 cells. In XPC cells which are not able to perform nucleotide excision repair in the overall genome after UV irradiation, but which have the capacity to repair the transcribed strand of active genes, a 50–60% reduction in repair replication was observed.

2.2.2. Induction of Chromosomal Aberrations in Cultured Mammalian Cells.
Inhibitors of DNA repair processes have been found to potentiate the yield of chromosomal aberrations induced by various clastogens (Natarajan et al., 1982; Preston, 1982).

Table 1. Effects of arsenite on X-ray induced chromosomal aberrations in purified human peripheral blood lymphocytes

Treatment	No cells scored	Dicentrics	Rings	Deletions
As	200	0	0	7
1 Gy	200	22	6	12
1 Gy + As	200	45	10	19
2 Gy	200	49	12	25
2 Gy + As	200	85	18	38

Preincubation in 5 μM Na-arsenite for 2 h in serum free medium prior to irradiation, followed by 30 min postincubation and subsequent transfer to complete medium and fixation at 72 h

Thus, one important justification for conducting cytogenetic analysis in As treated cells was that an enhanced frequency of chromosomal aberrations would be expected if As inhibits DNA-repair.

The modulating effect of sodium arsenite induced by UV and X-rays was investigated in cultured human lymphocytes and fibroblasts (Jha et al., 1992). For determination of micronuclei (MN), the cells were cultivated and cytochalasin B was added to inhibit cytokinesis, thus creating binucleated cells (Ramalho et al., 1988a). In fibroblasts the frequency of chromosomal aberrations was evaluated by scoring the incidence of MN in binucleated cells. Sodium arsenite was found to increase the frequencies of UV-induced MN and SCEs in a synergistic way, and was also found to increase the frequency of X-ray induced dicentrics and acentric fragments in isolated human lymphocytes treated *in vitro* (Table 1). These findings were subsequently corroborated in human lymphocytes where arsenite was found to increase the frequencies of SCEs and chromosomal aberrations induced by the alkylating agent diepoxybutane (Wiencke and Yager, 1992). The aneugenic effect of As on Chinese hamster ovary cells *in vitro* is demonstrated in Table 2, where a dose dependent increase in the incidence of lagging chromosomes is evident at As-levels ranging from 2.5 to 10 μM. An elevated incidence of aneuploidy was also accompanied by an increase in frequencies of chromosome breaks and exchanges.

2.2.3. Effects on Somatic Recombination. Recombination between strands in homologous as well as non homologous (e.g., illegitimate) chromosomes constitutes an alternative mechanism for cancer induction. Thus, this type of event has been shown to cause loss of heterozygosity in genes for different genetic diseases, including cancer (Sengstag, 1994), e.g. in retinoblastoma, and therapy with topoisomerase inhibitors may give rise to secondary acute myelogenous leukemia (reviewed by Anderson and Berger, 1994) as a result of site-specific recombination. Using a series of mutant clones exhibiting intragenic duplications of the *hprt* gene of Chinese hamster V79 cells (Zhang and Jenssen, 1994, Daré et al., 1996), it was found that As per se does not seem to induce homologous somatic recombination in mammalian cells (Jenssen et al., unpublished results). The same was found to be true also in the *Drosophila* wing spot test assay for somatic recombination. However, in contrast to the co-mutagenic effect found in bacterial or mammalian cell systems, somatic recombinations in this *in vivo* system induced by alkylating agents as well as X-rays were suppressed in the presence of As (De la Rosa Duque et al., 1994). Cytosine arabinoside was found to mimic As in this respect (De la Rosa Duque and Cruces Martinez, 1996).

Table 2. Anaphase-telophase alterations in CHO cells treated with sodium arsenite (1,000 cells scored per dose level).

Treatment (μM As)	Lagging chromosomes	Lagging fragments	Chromatin bridges	Progression rate index
Controls	5	1	1	36
2.5	27	9	13	30
5.0	29	5	13	30
10.0	40	17	14	22

Cell cycle progression analysis was carried out by counting metaphases in the first (M_1), second (M_2), and third or subsequent (M_3) division, and the Progression Rate Index (PRI) was given by the expression PRI = ($1M_1 + 2M_2 + 3M_3$)/number of cells scored.

2.3. Human Studies

2.3.1. Studied Populations. In an exploratory study, lymphocytes from 5 women living in close vicinity to the copper smelter at *Srednogorie*, Bulgaria, and 4 female controls in an unexposed region (Bunovo) were sampled for cytogenetic analysis. Exposure of the general population here generally occurs by inhalation, but also by ingestion of locally produced food products, like vegetables. Whereas drinking water, that is piped from the mountains, has a relatively low degree of contamination, various locally produced food products shows an approximately 10-fold increase in arsenic content (Petrov et al., 1990). A source of uncertainty in the interpretation of the cytogenetic data data from this population (Nilsson et al., 1993), is that these individuals—in addition to arsenic—are highly exposed to other heavy metals, in particular to selenium.

In order to avoid mixed exposures, subsequent efforts were directed towards monitoring populations in an area that is devoid of industrial pollution as well as exposure to agricultural pesticides, and where As intake occurred solely by drinking water. Our studies were mainly conducted at four sites in the Salta province of northwestern Argentina, including *San Antonio de los Cobres* and *Salta Forestal* in the western part of the Salta Province, as well as *Taco Pozo* in the province of Chaco, with Rosario de Lerma situated close to Salta City as a control area. The average As levels in drinking water ranged from 0.7 μg/L (control area) to 205 μg/L. Except for the analysis of drinking water at Salta Forestal, all chemical analyses were conducted at the Karolinska Institute, Stockholm (Vahter et al., 1995b).

The main sampling effort was conducted in San Antonio de los Cobres, a village located 3,800 m above the sea level in the arid Puna region of the middle Andes. In this area the volcanic bedrock has a high content of arsenic associated with pyrite minerals containing iron and copper, and all surface waters in and around the village are more or less highly contaminated by arsenic. One thermal spring in the vicinity of San Antonio (Pompeya) contained 5,870 μg/L of As, and the level in the river flowing through the settlement was determined at 779 μg/L. The village is served by one single central municipal water supply derived from an aquifer close to the river bed. Sampling at 4 different sites gave an average of 205 μg/l (range 192–219 μg/L). Soup was found to contain 336 μg/L on the average, and the total daily intake of As was estimated to be in the range of 270–370 μg per day. Only low levels of lead and cadmium was detected in the water. Table 3 gives the median values for total As in drinking water, blood, and urine for women and children in the different investigated communities in northwestern Argentina of native Indian as well as of mestizo origin.

Table 3. Total arsenic concentrations (μg/L) in drinking water, blood, and urine in women and children exposed to inorganic As in drinking water in Salta province, northwestern Argentina

Community	Water	Blood	Urine
Andean communities			
San Antonio de los Cobres (women)	205	7.6	258
San Antonio de los Cobres (children)	205	9.1	323
Santa Rosa de los Pastos Grandes (women)	31	1.5	50
Olacapto (women)	14	1.3	23
Tolar Grande (women)	2.5	1.3	18
Chaco Region			
Taco Pozo (women)	214	11	347
Taco Pozo (children)	214	9.1	440
Control Area			
Rosario de Lerma (children)	0.7	0.8	1.3

The inhabitants are mainly descendants from Quechua and Aymara speaking Indians migrating into this region during the Inca civilization with only minor mixing with Caucasian elements. The economy of the region is based on borax mining, breeding of llamas, goats and sheep, and some rudimentary agriculture, and there is no source of industrial pollution. 15 Women and 10 children of native Indian origin were selected from San Antonio, as well as 13 women and 11 children from *Salta Forestal*, closer to Salta City, with moderately elevated arsenic levels in drinking water (around 100 μg/L). In addition, 15 women from three small settlements around San Antonio with lower As levels in drinking water (Santa Rosa de los Pastos Grandes, Olacpato, Tolar Grande), and a cohort consisting of 12 mestizo women from *Taco Pozo* Salta in the Chaco area were investigated with respect to metabolic disposition. As a reference, 10 women and 12 children of similar ethnic characteristics were selected from a control area, *Rosario de Lerma,* near Salta City. For investigation of As load and metabolism, samples of the local drinking water, food, urine and blood were collected, and a detailed assessment was also made with respect to water and food intakes, smoking habits, use of coca leaves and alcohol. The concentration of As in drinking water in Taco Pozo was found to be almost exactly the same as in San Antonio, i.e. around 200 μg/L. Whereas no such symptoms were seen at the latter locality, keratotic changes as well as malignancies typical of chronic arsenism were seen among the population in Taco Pozo.

2.3.2. Cytogenetic Analysis in Exposed Humans. Chromosomal aberrations reflect a genotoxic effect on the whole genome in contrast to point mutations, which only cover a small target (e.g. 45 kb for the HPRT gene), and they constitute reliable end points for the detection of damage of the genetic material. The recent introduction of the fluorescent *in situ* hybridization (FISH) technique using chromosome-specific DNA probes has greatly increased the resolution of detecting and evaluating structural chromosomal aberrations, especially of translocations in human and rodent cells. In addition, the FISH technique makes it possible to study the nature of induced micronuclei in diverse cell types. Frequencies of structural chromosomal aberrations in human lymphocytes have been used as fairly accurate biological dosimeters for exposure to ionizing radiations (Ramalho et al., 1985a; Ramalho et al., 1988; Natarajan et al., 1996).

Table 4. Micronuclei (MN) and sister chromatid exchanges (SCEs) in exposed donors from San Antonio de los Cobres, Salta Forestal, and Rosario de Lerma (controls); MN and As content (ppm) in drinking water (Argentina) as well as in hair and toe nails from Bulgarian women in Srednogorie and unexposed controls

Group (As level in water)	Micronuclei per 1000 cells	SCEs per cell	
San Antonio de los Cobres (205 µg/L)			
Exposed women	41±4.9	5.7±1.3	
Exposed children	35±4.6	4.4±1.1	
Salta Forestal (≈ 100 µg/L)			
Exposed women	10±1.3	5.2±0.12	
Exposed children	19±3.5	4.0±0.22	
Rosario de Lerma (0.7 µg/L)			
Control women	8.5±3.4	5.5±1.3	
Control children	5.6±1.6	4.6±1.2	
		As in hair (µg/g)	As in toe nails (µg/kg)
Srednogorie, Bulgaria			
Exposed women	45±9.5	3.18±1.01	6.7±1.8
Control women	15±2.1	0.07±0.02	0.35±0.07

Micronuclei (MN) were determined as described under *Sec. 2.2.2.*, and sister chromatid exchanges (SCEs) were scored by standard techniques. For fluorescence *in situ* hybridization (FISH), lymphocytes were cultivated, and Bluescribe DNA libraries specific for chromosomes 1,3,18 and X (representing about 1/5 of the human genome) were used. For assessing aneuploidy, centromeric probes were employed (Dulout et al., 1996).

Table 4 demonstrates the incidence of MN and SCEs in peripheral lymphocytes exposed women from the Salta Province of Argentina, as well as from a small cohort of women residing in the vicinity of the Srednogorie copper smelter, Bulgaria. As seen from these results, there was a highly significant increase in the frequency of MN, but not of SCEs in comparison with controls. Cell cycle progression analysis also did not reveal any significant differences (data not shown). Preliminary results from the As-exposed population in West Bengal, and where exposures are much higher, revealed an increase in the incidence of MN which is about double that found for the most exposed individuals in Argentina or Bulgaria. Individuals exhibiting dermatological effects exhibited the highest levels of MN. Unfortunately, no detailed analytical data were supplied, but a range of 700–1,200 µg/L has been reported for the wells used. (Natarajan, unpublished results).

The results from the FISH analysis are presented in Table 5. In the exposed population from San Antonio, a significantly higher frequency of aneuploidy was found (0.2%) in comparison with the control population (0%). In 3 out of 12 subjects from San Antonio under study, around 0.4% translocations were obtained in 1539 cells scored. The mean value of translocations for the exposed group was 0.1%, whereas in the control group composed of 17 individuals (Rosario de Lerma), no translocations were observed. Also

Table 5. Exposed subjects from San Antonio de los Cobres and Rosario de Lerma (controls). Analysis of structural and numerical chromosome aberrations using DNA libraries specific for chromosomes 1, 3, 18, X, and FISH technique

Subjects studied	Number of cells scored	Translocations; (%)(whole genome)	Aneuploidy; (%) (chromosomes involved)
Control group (n=17)	6287	0	0
Exposed group (n=12)			
3	1214	0	0.25 (X)
2	815	0	0.25 (1)
1	410	0	0.25 (3)
2	770	0.4	0.25 (18)
1	830	0	0.24 (1 and 3)
2	1436	0	0.14 (3 and 18)
1	769	0.4	0

when compared to historical control data, (about 5 per 1000 cells for translocations), this incidence in the exposed group cannot be considered as statistically significant.

2.3.3. Ethnic Differences with Respect to Pharmacokinetic Parameters and Toxicological Response. Our study of the native Indian population in the Puna region of the Andes, as well as of a mestizo population in the lower Chaco region, Salta Province, northwestern Argentina (Vahter et al., 1995b; Concha et al., 1998b), both of which are highly exposed to As in drinking water at about the same level (200µg/L), demonstrated a different metabolic disposition than previously found for populations in Europe, U.S., and Japan. Thus, native Indians from San Antonio excrete arsenic mainly as DMA (median value 74%) and inorganic As (median value 25%), with very little MMA (median value 2.2%). Individuals from the surrounding area with the same ethnic characteristics, but exposed to low concentrations of As exhibited the same metabolic disposition. However, whereas none of the typical signs of systemic As intoxication could be seen among the Indians studied in e.g. San Antonio—who have been exposed for a very long time in history—hyperkeratoses and skin cancer were present among the investigated mestizos from Taco Pozo in the Chaco region who were exposed to As in drinking water at almost exactly the same concentration (about 200 µg/L). In Figure 3 the approximate relative distribution of arsenic containing compounds excreted in urine is shown for the chimpanzee, Caucausians/Asians, Indians from San Antonio, as well as for the rabbit.

3. DISCUSSION

The reported investigations provide additional support of the notion, that As exerts its genotoxic effect by an indirect mode of action. Thus, we could confirm previous observations that As inhibits DNA repair, and demonstrate incision to constitute the most sensi-

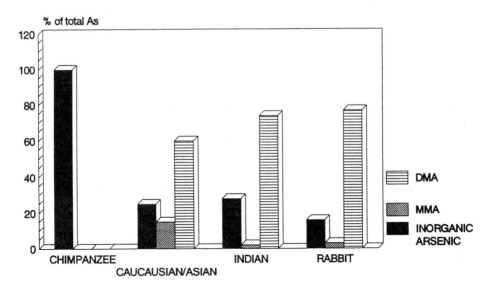

Figure 3. Relative distribution (% of total) of inorganic arsenic, monomethylarsonic acid (MMA), and dimethyl-methylarsinic acid (DMA) in urine from exposed chimpanzees (Vahter et al., 1995a , 1988). Caucasians and Asians (Vahter and Marafante, 1988), Indians from San Antonio de los Cobres (Vahter et al., 1995b), as well as from rabbits (Vahter and Marafante, 1988).

tive step of the nucleotide excision repair pathway. Substantial inhibition was observed at 188 µg/L (2.5 µM), a level that is clearly relevant with respect to e.g. the bladder epithelium in highly exposed humans. Based on a ratio between the level of As in urine vs. drinking water of 1.3 (Vahter et al. 1995b), and an As concentration of 600 µg/L for the highly exposed part of the Taiwanese population for which an increased incidence of bladder cancer has been reported, the concentration of inorganic As (25% of total) in urine would be of the same magnitude as was found to inhibit excision in our *in vitro* studies.

In skin, accumulation of As reaches several mg/kg (see Table 4), but it is not known to which concentration of As the target cells for neoplastic transformation are exposed. The inhibition by incision was only partial, reaching a maximum of 30–40%. The failure to induce further inhibition above 5 µM As(III) seems to reflect differences in the efficiency of the repair of the cyclobutane pyrimidine dimers as compared to pyrimidine-(6–4)-pyrimidone photoproducts. In Fig. 2 strand breaks were allowed to accumulate for only 1 h. During this period of time human cells remove some 50% of the (6–4) photoproducts, but only a small fraction of the cyclobutane pyrimidine dimers (Wood, 1989).

Transcription coupled (specific for transcriptionally active genes), as well as global genome repair pathways are affected, and exposure to As may, thus, induce a partially repair deficient condition similar to that seen in xeroderma, but where additional promotive action such as keratotic changes will determine the site of tumor appearance. Nucleotide excision repair (NER) is the most general repair pathway responsible for the removal of bulky DNA damage induced by UV and many environmental mutagens like polyaromatic hydrocarbons. However, the major repair pathway for endogenously induced DNA damage is the base excision repair pathway (BER) where the metabolism of molecular oxygen constitutes one major source of endogenously induced DNA damage. BER—that is also involved in the elimination of small DNA-adducts, e.g. those from ethylating agents—exhibits several differences in comparison with NER with respect to the damage recognition/incision step, and its sensitivity to As is not yet known. The inhibition by As on somatic recombination found in *Drosophila* seem to indicate, that both nucleotide excision repair and somatic recombination involve similar As-sensitive factors.

Exposure to As drastically increases the incidence of micronuclei in human cells *in vitro* as well as *in vivo* in lymphocytes derived from ethnically diverse populations in Bulgaria (Nilsson et al., 1993), Argentina (Dulout et al., 1996), and India. These observations are in agreement with the finding of an increased frequency of micronuclei in exfoliated bladder cells in a group of 18 individuals in Nevada, who had been exposed to high levels of As in drinking water (Warner et al., 1994). An increase in the level of these aberrations apparently constitutes the most sensitive toxicological endpoint of chronic As-exposure of humans so far described, and can be detected in absence of any other clinical signs of As-poisoning. Although this does not exclude other modes of action, our findings of the induction of chromosomal aberrations in humans exposed to As in drinking water is consistent with the effects on DNA repair. When lesions (breaks) are induced by X-rays in cultured human cells in G_0 in the presence of As, ligation is inhibited giving an increased number of unjoined breaks, leading to high frequencies of dicentrics when the block is removed. Disturbances of somatic recombination, or on the p53 tumor suppressor gene function and failure of cell cycle checkpoints leading to genomic instability constitutes other possible mechanisms.

At levels of 2.5–10 μM of arsenite we could readily reproduce in cultivated Chinese hamster cells the aneugenic effect found in exposed humans in *vitro* (Table 2). However, the claim by Vega et al. (1995), that such effects can be induced in human lymphocytes down to an arsenite concentration of 10^{-10} μM lacks credibility in view of the fact, that the plasma As concentration was about 0.01 μM in our controls (Tables 3–5). Nor could we confirm previously reported increased incidences of SCEs (Burgdorf et. al., 1977; Wen et al., 1981; Lerda, 1994), or depression of cell cycle progression (Ostrosky-Wegman et al., 1991) in peripheral lymphocytes from exposed individuals. As has to be present during DNA replication to have an influence on ligation, leading to SCEs. *In vivo*, the circulating lymphocytes are in a non-cycling stage, explaining why no effects on the frequencies of SCEs could be detected. The aneugenic effect may be related to reported findings, that As causes loss of thick cables of actin filaments at low doses in cultivated cells, and severe loss of microtubules and inhibition of cytoskeletal protein synthesis at higher exposure levels (Li and Chou, 1992). Such an effect on the spindle also represents a non-stochastic mechanism expected to exhibit a dose-threshold. However, the aneugenic effects of As may be unrelated to the induction of human cancer.

The pro-mutagenic action of As may have some other important implications. The potentiating action of As on genotoxic carcinogens such as tobacco smoke and radon has already been mentioned. Based on the extraordinary high exposures to indoor radon in Sweden, the Swedish Cancer Committee (1984) noted, that an increase of lung cancer incidence in the Swedish population of about 50% would be expected in comparison with the socio-economically comparable Danish population for which exposure to radon is negligible. However, no significant differences in cancer rates could, in fact, be observed, and the upper limit of the true excess risk in Sweden due to radon would seem to lower by a factor ten. The risk for lung cancer associated with exposure to radon daughters have been mainly based on studies of uranium miners who are exposed to significant levels of As as well as to other potential promoters present in this environment (Hunter, 1971). Consequently, due to such synergism it is likely that risk factors for lung cancer derived from investigations of uranium miners are not appropriate when deriving risk estimates for radon exposure in homes, because it involves a considerable overestimation of risk.

There are indications that mammals have an ability to adapt to the *acute* toxicity of inorganic arsenic. Among the "arsenic eaters" in Styria (Steiermark), Austria, the daily dose of arsenic were reported to have reached a level of 300–400 mg, which is 3–4 times the lethal dose under normal conditions (Lewin, 1929). The mechanism of this adaptation is not well understood, but it has been suggested that it reflects an induction of glutathione or As-binding proteins in the gastrointestinal tract, or is due to induction of the enzymes capable of methylating inorganic arsenic. However, the arsenite methyltransferase activities in the tissues of mice was not induced by administering As in drinking water at 25 or 2,500 µg/L for 31 or 91 days (Healy et al., 1998). Concha et al. (1998b) interpreted their findings from the studied Argentine populations in support of such an induction: Thus, in children from San Antonio, the fraction of inorganic As decreased, while the fraction of DMA increased with increasing total exposure to As.

So far, there has been no evidence for adaptation to the toxic effects from chronic exposure to arsenic in the populations so far studied, either in Taiwan or India. However, the high level exposure of these populations commenced at a more recent date. Thus, the deep wells in Taiwan were constructed during the period 1900–1910 (U.S.EPA, 1988), and in West Bengal the As containing tube wells were drilled at the beginning of the 1960s. The descendants of the Inca Indians from San Antonio studied by us have, on the other hand, been exposed to As for a very long period of time in history. They seem to possess an increased methylating capacity, as well as a certain resistance also with respect to the chronic effects of this element, probably as a consequence of selective mechanisms. To which extent this apparent resistance can be coupled to the more efficient methylation of As found in this population is not quite clear. The metabolic disposition of woman of mixed ethnic origin from Taco Pozo (Concha et al., 1998b) was similar to those from San Antonio, but a few individuals were found that had a excretion pattern approaching that found for Caucasians. Unfortunately, the number of donors was small, and we had no opportunity to sample the urine from sensitive individuals from Taco Pozo who exhibited As induced skin lesions. Although the metabolic pattern, in general, resembled those for Caucasians, Hopenhayn-Rich and coworkers (1996a, 1996b) subsequently found large interindividual variations for a mixed Indian population (mainly Atacameños) in northern Chile. In this context not only the methylating capacity may be important, but possibly also a genetic polymorphism with respect to the energy-dependent arsenite-efflux pump discovered in Chinese hamster cells by Rossman's group (Wang et al., 1996).

ACKNOWLEDGMENTS

The reported investigations were supported by grants from the National Swedish Chemicals Inspectorate and by the Commission of the European Communities, Directorate General XII, Contracts CII-CT93-0305, EV5V-CT94-0479.

REFERENCES

Ahnström, G., and Erixon, K. (1981), Measurement of strand breaks by alkaline denaturation and hydoxyapatite chromatography. *In*: Friedberg, E.C., and Hanawalt, P.C., (Eds.) "DNA Repair—A Laboratory Manual of Research Procedures.", Vol. 1, Marcel Dekker Inc., New York, pp. 403–418.

Anderson, R.D. and Berger, N.A. (1994) Mutagenicity and carcinogenicity of topoisomerase-interactive agents. Mutat. Research, 309, 109–142.

Bates, M.N., Smith, A.H., and Hopenhayn-Rich, C. (1992). Arsenic ingestion and internal cancers: A review. Am J. Epidemiol. 135, 462–476.

Börzsönyi, M., Bereczky, A., Rudnai, P., Csanady, M., and Horvath, A. (1992). Epidemiological studies on human subjects exposed to arsenic in drinking water in Southeast Hungary. Arch. Toxicol. 66, 77–78.

Buchet, J.P., Lauwerys, R., and Roels, H. (1981): Urinary excretion of inorganic arsenic and its metabolites after repeated ingestion of sodium metaarsenite by volunteers. Int. Arch. Occup. Environ. Health, 48, 111–118.

Buchet, J.P., and Lauwerys, R. (1985) Study of inorganic arsenic methylation by rat liver in vitro: Relevance for the interpretation of observations in man. Arch. Toxicol. 57, 125–129.

Burgdorf, W., Kurvink, K. och Cervenka, J. (1977). Elevated sister chromatid exchange rate in lymphocytes of subjects treated with arsenic. Human Genet. 36, 69–72.

Cebrian, M.E., A. Albores, A., M. Aguilar, M., and Blakely, E. (1983). Chronic arsenic poisoning in the north of Mexico. Human Toxicol. 2, 121–133.

Concha, G., Vogler, G., Nermell, B., and Vahter, M. (1998a). Low-level arsenic excretion in breast milk of native Andean woman exposed to high levels of arsenic in the drinking water. Int. Arch. Occup. Environm. Health 71, 42–46.

Concha, G., Nermell, B., and Vahter, M. (1998b). Metabolism of inorganic arsenic in children with high chronic arsenic exposure in nothern Argentina. Env. Heath Persp. in press.

Daré, E., Zhang, L-H. and Jenssen, D. (1996) Characterization of insertion mutants exhibiting exon duplications in the hprt gene of Chinese hamster V79 cells. Som. Cell Mol. Genetics 22:3, 201–210.

De la Rosa, M.E., Magnusson, J., Ramel, C., and Nilsson, R. (1994). Modulating influence of inorganic arsenic on the recombinogenic and mutagenic action of ionizing radiation and alkylating agents in Drosophila melanogaster. Mutat. Res. 318, 65–71.

De la Rosa, M.E., and Cruces Martinez, M.P. (1996). On the mode of action of inorganic arsenic on somatic re-combination induced by mutagenic agents in Drosophila; evidence for a common denominator in DNA re-pair and somatic recombination. Int. Soc. for the Study of Xenobiotics Meeting, San Diego, October 19–25, 1996, Poster session.

Dulout, F., Grillo, C., Seoane, A., Maderna, C., Nilsson, R., Vahter, M., Darroudi, F., and Natarajan, A.T. (1996). Chromosomal aberrations in peripheral lymphocytes from native Andean women and children from North-western Argentina exposed to arsenic in drinking water. Mutat. Res. 370, 151–158.

Erixon, K. (1986). Differential regulation of base and nucleotide excision repair in mammalian cells. In: Simic, M.G., Grossman, L., and Upton, C., (Eds.) "Mechanism of DNA Damage and Repair." Plenum, N.Y. pp. 159–170.

Hanawalt, P.C. (1996). Role of transcription-coupled DNA repair in susceptibility to environmental carcinogene-sis. Env. Health Persp. 104, Suppl. 3, 547–551.

Hartwig, A., Gröblinghoff, U.D., Beyersmann, D., Natarajan, A.T. Filon, R., Mullenders, L. (1997). Interaction of arsenic(III) with nucleotide excision repair in UV-irradiated human fibroblasts. Carcinogenesis 18, 399–405.

Healy, S.M., Zakharayan, R., and Aposhian, H.V. (1997). Enzymatic methylation of arsenic compounds. IV. In vi-tro and in vivo deficiency of the methylation or arsenite and monomethylarsonic acid in the guinea pig. Mutat. Res.—Reviews in Mutation Research 386, 229–239.

Healy, S.M., Casarez, E.A., Ayala-Fierro, F., and Aposhian, H.V. (1998). Enzymatic methylation of arsenic com-pounds. V. Arsenite methyltransferase activity in tissues of mice. Toxicol. Appl. Pharmacol. 148, 65–70.

Hopenhayn-Rich, C., Biggs, M.L., Smith, A.H., Kalman, D.A., and Moore, L.E. (1996a). Methylation study of a population environmentally exposed to arsenic in drinking water. Env. Health Persp. 104, 620–628.

Hopenhayn-Rich, C., Biggs, M.L., Kalman, D.A., Moore, L.E., and Smith, A.H. (1996b). Arsenic methylation pat-terns before and after changing from high to lower concentrations of arsenic in drinking water. Env. Health Persp. 104, 1200–1207.

Hunter, D. (1971). The Diseases of Occupation. The English University Press Ltd., London, p. 815.

IPCS/WHO (1981), International Programme on Chemical Safety; Environmental Health Criteria No. 18, Arsenic, WHO, Geneva.

Jha, A.N., Noditi, M. , Nilsson, R., and Natarajan, A.T. (1992). Genotoxic effects of sodium arsenite on human cells. Mutat. Res. 284, 215–221.

Kroes, R., van Logten, M.J., Berkvens, J.M., de Vries, T., and van Esch, G.J. (1974), Study on the carcinogenicity of lead arsenate and sodium arsenate and on the possible synergistic effect of diethylnitrosamine. Food Cosm. Toxicol. 12, 671–679.

Lerda, D. (1994). Sister-chromatid exchange (SCE) among individuals chronically exposed to arsenic in drinking water. Mutat. Res. 312, 111–120.

Lewin, L. Gifte und Vergiftungen. Lehrbuch der Toxikologie. Verlag von Georg Stilke, Berlin 1929, pp. 192–195.

Li, W., and Chou, I.-N. (1992). Effects of sodium arsenite on the cytoskeleton and cellular glutathione levels in cultured cells. Toxicol. Appl. Pharmacol. 114, 132–139.

Li, J.-H. and Rossman, T.G. (1989) Inhibition of DNA ligase activity by arsenite: A possible mechanism of its comutagenesis. Mol. Toxicol. 2, 1–9.

Lianfang, W., and Jianzhong, H. (1994). Chronic arsenism from drinking water in some areas of Xinjiang, China. *in:* Nriagu, J.O. (Ed.). "Arsenic in the Environment. Part II. Human Health and Ecosystem Effects". John Wiley, New York. pp. 159–172.

Lee-Chen, S.F., Yu, C.T., and Jan, K.Y. (1992). Effect of arsenite on the repair of UV-irradiated Chinese hamster ovary cells. Mutagenesis 7, 51–55.

Luster, M.I., Wilmer, J.L., Germolec, D.R., et al. (1995). Role of keratinocyte-derived cytokines in chemical toxicity. Toxicol. Letters. 82–83, 471–476.

Mitchell, P., and Barr, D. (1995). The nature and significance of public exposure to arsenic: a review of its relevance to South West England. Env. Geochem. and Health 17, 57–82.

Mandal, B.K., Chowdhury, T.R., Samanta, G., et al. (1996). Arsenic in ground water in seven districts of West Bengal, India—The biggest arsenic calamity in the world. Current Science 70, 976–986.

Natarajan, A.T., Csukas, I., Degrassi, F., van Zeeland, A.A., Palitti, F., Tanzarella, C., de Saliva, R., and Fiore, M. (1982). Influence of inhibition of repair enzymes on the induction of chromosomal aberrations by physical and chemical agents. In: Natarajan, A.T. et al., (Eds.) "Progress in Mutation Research." Vol. 4, Elsevier, Amsterdam, pp 47–59.

Natarajan, A.T., Boei, J.W.A., Darroudi, F., Van Diemen, P.C.M., Dulout, F., Hande, M;.P., and Ramalho, A.T. (1996). Current cytogenetic methods for detecting exposure and effects of mutagens and carcinogens. Env. Health Persp. 104, Suppl. 3, 445–448.

Neubauer, O. (1947). Arsenical cancer: a review. Brit. J. Cancer 1, 192–251.

Newman, J.A., Archer, V.E., Saccomanno, G., Kuschner, M., Auerbach, O., Grondahl, R.D., and Wilson, J.C. (1976). Histological types of bronchogenic carcinoma among members of copper mining and smelting communities. Ann. N.Y. Acad. Sci. 271, 260–268.

Nilsson, R., Jha, A.N., Zaprianov, Z., and Natarajan, A.T. (1993). Chromosomal aberrations in humans exposed to arsenic in the Srednogorie area, Bulgaria. Fresenius Env. Bull. 2, 59–64.

Nordenson, I., Beckman, G., Beckman, L., and Nordström, S. (1978). Occupational and environmental risks in and around a smelter in northern Sweden. II. Chromosomal aberrations in workers exposed to arsenic. Hereditas 88, 47–50.

Nordenson, I., Salmonsson, S., Brun, E., and Beckman, G. (1979). Chromosome aberrations in psoriatic patients treated with arsenic. Hum. Genet. 48, 1–8.

Norin, H., and Vahter, M. (1981). A rapid method for the selective analysis of total urinary metabolites of inorganic arsenic. Scand. J. Work Environm. Health, 7, 38–44.

Okui, T., and Fujiwara, Y. (1986). Inhibition of human excision DNA repair by inorganic arsenic and the co-mutagenic effect in V79 Chinese hamster cells. Mutat. Res. 172, 69–76.

Ostrosky-Wegman, P., Gonsebatt, M.E., Montero, R., Vega, L., Barba, H., Espinosa, J., Palao, A., Cortinas, C., Garcia-Vargas, G., del Razo, L.M., and Cebrian, M. (1991). Lymphocyte proliferation kinetics and genotoxic findings in a pilot study on individuals chronically exposed to arsenic in Mexico. Mutat. Res. 250, 477–482.

Pershagen, G., Wall, S., Taube, A. and Linnman, L. (1981). On the interaction between occupational arsenic exposure and smoking and its relationship to lung cancer. Scand. J. Work Env. Health 7, 302–309.

Pershagen, G., Bergman, F., Klominek, J., Damber, L., and Wall, S. (1987). Histological types of lung cancer among smelter workers exposed to arsenic. Brit. J. Ind. Med. 44, 454–458.

Petrov, I., Zaprianov, Z., Chuldjian, H. et al. (1990). Determination of the content of some toxic elements in environmental and biological samples from selected regions of the country (Bulgaria): A report to the National Committee for Environmental Protection, Sofia [Unpublished]. Cited in: Zelikoff, J.T., Bertin, J. E., Burbacher, T.M., Hunter, E. S., Miller, R.K., Silbergeld, E.K., Tabacova, S., and Rogers, J.M. (1995) Health risks associated with prenatal metal exposure. Symposium overview. Fundam. Appl. Toxicol. 25, 161–170.

Preston, R.J. (1982). The use of inhibitors of DNA repair in the study of the mechanism of induction of chromosome aberrations. Cytogenet. Cell Genet. 33, 20–26.

Ramalho, A., Sunjevaric, I., and Natarajan, A.T. (1988a). Use of frequencies of micronuclei as quantitative indicators of X-ray induced chromosomal aberrations in human peripheral blood lymphocytes: Comparison of two methods. Mutat. Res. 207,141–146.

Ramalho, A.T., Nascimento, A.C.H., and Natarajan, A.T. (1988b). Dose assessments by cytogenetic analysis in the Goiania (Brazil) radiation accident. Rad. Protection Dosimetry 25, 97–100.

Rossman, T.G. (1994). Metal mutagenesis. *In:* Goyer, R.A., and Cherian, M.G. (Eds.). "Toxicology of Metals—Biochemical Aspects." Springer-Verlag, New York, pp. 374–405.

Sengstag, C. (1994). The role of mitotic recombination in carcinogenesis. Critical Revs. Toxicol. 24, 323–353.

Shirachi, D.Y., Johansen, M.G., McGowan, J.P., and Tu, S.-H. (1983). Tumorigenic effect of sodium arsenite in rat kidney. Proc. West. Pharmacol. Soc. 26, 413–415.

Snyder, R.D., Davis, G.F., and Lachmann, P.J. (1989). Inhibition by metals of X-ray and ultraviolet-induced DNA repair in human cells. Biol. Trace Element Res. 21, 389–398.

Sram, R.J. (1976). Relationship between acute and chronic exposures in mutagenicity studies in mice. Mutat. Res. 41, 25–42.

Styblo, M., Yamauchi, H., and Thomas, D.J. (1995). Comparative in vitro methylation of trivalent and pentavalent arsenicals. Toxicol. Appl. Pharmacol. 135, 172–178.

Swedish Cancer Committee (1984). "Cancer—Causes and Prevention, Report to the Ministry of Social Affairs from the Cancer Committee." SOU 1984:67, Stockholm, p. 238. (English translation: Taylor and Francis, London, 1992)

Taylor, P.R., Qiao, Y.-L., Schatzkin, A., Yao, S.-X., Lubin, J., Mao, B.-L., Rao, J.-Y., McAdams, M., Xuan,X.-Z., and Li, J.-Y. (1989). Relation of arsenic exposure to lung cancer among tin miners in Yunnan province, China. Brit. J. Ind. Med. 46, 881–886.

Tseng, W.P., Chu, H.M., How, S.W., Fong, J.M., Lin, C.S., and Yeh, S. (1968). Prevalence of skin cancer in an endemic area of chronic arsenicism in Taiwan. J. Natl. Cancer. Inst. 40, 453–463.

Tseng, W.P. (1977). Effects and dose-response relationships of skin cancer and Blackfoot disease with arsenic. Environm. Health Persp. 19, 109–119.

Tsushima, S. (1997). Arsenic contamination in ground water in Bangladesh: an overview. Asia Arsenic Network, Newsletter No.2, March 30, pp. 2–6.

U.S.EPA (1988). U.S. Environmental Protection Agency. "Special Report on Ingested Inorganic Arsenic: Skin Cancer and Nutritional Essentiality." EPA/625/3–87/013. Washington, D.C.

Vahter, M. (1983). Metabolism of inorganic arsenic in relation to chemical form and animal species. Dissertation. Karolinska Institute, Stockholm.

Vahter, M., and Marafante, E. (1985). Reduction and binding of arsenate in marmoset monkeys Arch. Toxicol. 57, 119–124.

Vahter, M., and Marafante, E. (1987). Effects of low dietary intakes of methionine, choline or proteins on the biotransformation of arsenite in the rabbit. Toxicol. Lett. 37, 41–46.

Vahter, M., and Marafante, E. (1988). In vivo methylation and detoxification of arsenic. In: Craig, P.J and Glockling, F., (Eds.) "The Biological Alkylation of Heavy Elements." Royal Soc. Chem. Special Publications No. 66, Oxford, U.K. pp. 105–119.

Vahter, M., Couch, R., Nermell, B., and Nilsson, R. (1995a). Lack of methylation of inorganic arsenic in the chimpanzee. Toxicol. Appl. Pharmacol. 133, 262–268.

Vahter, M. Concha, G., Nermell, B., Nilsson, R., Dulout, F., and Natarajan, A.T. (1995b). A unique metabolism of arsenic in native Andean women. European J. Pharmacol. 293, 455–462.

Valentine, J.L., Kang, H.K., and Spivey, G. (1979): Arsenic levels in human blood, urine, and hair in response to exposure via drinking water. Environm. Res. 20, 24–32.

Vega, L., Gonsebatt, M.E., and Ostrosky-Wegman, P. (1995). Aneugenic effect of sodium arsenite on human lymphocytes in vitro: an individual susceptibility effect detected. Mutat. Res. 334, 365–373.

Venema, J., Mullenders, L.H.F., Natarajan, A.T., van Zeeland, A.A., and Mayne, L. (1990). The genetic defect in Cockayne syndrome is associated with a defect in repair of UV-induced DNA damage in trascriptionally active DNA. Proc. Natl. Acad. Sci. U.S.A. 87, 4707–4711.

Wang, Z., Dey, S., Rosen, B.P., and Rossman, T.G. (1996). Efflux-mediated resistance to arsenicals in arsenic-resistant and -hypersensitive Chinese hamster cells. Toxicol. Appl. Pharmacol. 137, 112–119.

Warner, M.L., Moore, L.E., Smith, M.T., Kalman, D.A., Fanning, E., and Smith, A.H. (1994). Increased micronuclei in exfoliated bladder cells of individuals who chronically ingest arsenic-contaminated water in Nevada. Cancer Epidemiol. Biomarkers & Prevention, 3, 583–590.

Wen, W.-N., Lieu, T.-L., Chang, H.-J., Wuu, S.W., Yau, M.- L., and Jan, K.Y. (1981). Baseline and sodium arsenite- induced sister chromatid exchanges in cultured lymphocytes from patients with Blackfoot disease and healthy persons. Hum. Genet. 59, 201–203.

WHO (1993). "Guidelines for Drinking-Water Quality." 2nd ed. Vol 1, Recommendations. WHO, Geneva, p. 41.

Wiencke, J.K., and Yager, J.W. (1992). Specificity of arsenite in potentiating cytogenetic damage induced by the DNA crosslinking agent diepoxybutane. Environm. Mol. Mutagenesis 19, 195–200.

Wood, R.D. (1989). Repair of pyrimidine dimer ultra violet light photoproducts by human cell extracts. Biochemistry 28, 8287–8292.

Yang, J.-L., Chen, M.-F., Wu, C.-W., and Lee, T-C. (1992). Posttreatment with sodium arsenite alters the mutational spectrum induced by ultraviolet light irradiation in Chinese hamster ovary cells. Env. Mol. Mut. 20, 156–164.

Zakharyan, R., Wu, Y., Bogdan, G.M., and Aposhian, H.V. (1995). Enzymatic methylation of arsenic compounds. I. Assay, partial purification and properties of arsenite methyltransferase and monomethylarsonic acid methyltransferase of rabbit liver. Chem. Res. Toxicol. 8, 1029–1038.

Zakharyan, R., Wildfang, E., and Aposhian, H.V. (1996). Enzymatic methylation of arsenic compounds: III. The marmoset and tamarin, but not the rhesus, monkeys are deficient in methyltransferases that methylate inorganic arsenic. Toxicol. Appl. Pharmacol. 140, 77–84.

Zhang, L.-H. and Jenssen, D. (1994) Studies on intrachromosomal recombination in V79/SP5 Chinese hamster cells upon exposure to different agents related to carcinogenicity. Carcinogenesis 15, 2303–2310.

3

GROUNDWATER ARSENIC CONTAMINATION AND SUFFERINGS OF PEOPLE IN WEST BENGAL, INDIA AND BANGLADESH

Status Report Up to March, 1998

Badal K. Mandal,[1] Bhaja K. Biswas,[1] Ratan K. Dhar,[1] Tarit Roy Chowdhury,[1] Gautam Samanta,[1] Gautam K. Basu,[1] Chitta R. Chanda,[1] Kshitish C. Saha,[1] Dipankar Chakraborti,[1*] Saiful Kabir,[2] and Sibtosh Roy[2]

[1]School of Environmental Studies
Jadavpur University
Calcutta - 700 032, India
[2]Dhaka Community Hospital Trust
1089 Malibagh Chowdhurypara
Dhaka - 1219, Bangladesh

1. INTRODUCTION

Groundwater is becoming the major source of drinking water around the world, especially in developing countries, to avoid microbial and chemical contamination from surface water. Another reason of wide use of groundwater is that because of its easy access and economic viability. Not only is groundwater being used for drinking, but for farmers in many developing countries like India, Bangladesh groundwater is their main source of irrigation. These countries have achieved a green revolution with the help of underground water. Earlier, India and Bangladesh could get only one crop a year, and that too was rain dependent. But now usually 3/4 crops in a year are common and the source of water for irrigation lies underground. Irrigation in West Bengal and Bangladesh using groundwater was first started around the sixties. In both these countries, there is no groundwater withdrawal regulation. As a result, groundwater exploitation goes on unchecked. In Bangladesh and West Bengal more than 95% of the Rural Water Supply Schemes (RWSS) depend on underground water. Dhaka (population abut 11 million) is the only city in the world where more than 97% of the domestic water requirement comes from underground water sources.

---

* Corresponding author: Tel: 91 33 4735233; Fax: 91 33 4734266; Email: dcsoesju@giascl01.vsnl.net.in

Metals and Genetics, edited by Sarkar.
Kluwer Academic / Plenum Publishers, New York, 1999.

The incidents of arsenic contamination in groundwater and sufferings of people from chronic arsenic toxicity resulting from the drinking of the contaminated water around the world are documented now. The clinical manifestations are many but the most commonly observed symptoms which help identify people suffering from chronic arsenic poisoning are arsenical skin lesions. Complications like liver enlargement (hepatomegaly), spleen enlargement (splenomegaly) and fluid in abdomen (ascites) are seen in several cases. Gangrene in limbs, squamous cell carcinoma, basal cell carcinoma, bowen's disease, carcinoma affecting lung, uterus, bladder, genitourinary tract or other sites are often seen in advanced cases. It has also been reported that palmoplanter keratosis may become secondarily infected and ultimately may develop into skin cancer (Astolfi et al., 1981). Bergolio (Mote Quemodo of Cordoba Province, North Argentina) analysed post-mortem over the period 1950–1959 and compared the data for the contaminated province. From the 2335 post-mortem histories of the highly arsenic affected region, 556 patients had died of cancer (23.8%), as against 15% of the entire province of Cordoba (Bergoglio, 1964). Biagini of Cordoba followed 116 patients with clear sign of chronic arsenic disease over a number of years (Biagini, 1972). After 15 years of follow-up study, 78 were found to have died; 24 of them from cancer (i.e., 30.5% of total deaths). It is observed since long that patients who received chronic treatment with arsenical medications had a higher incidence of both basal cell and squamous cell carcinoma of the skin. These arsenical skin cancers commonly occurred in the presence of dermatological manifestations of arsenicism and some had internal neoplasms that were regarded to be of arsenical origin (Naqvi et al., 1994; Evans et al., 1998).

Major incidents of drinking arsenic-contaminated water were noted in Monte Quemodo of Cordoba Province, North Argentina (Astolfi et al., 1981) where 10,000 people were at risk. During 1963–1983 in the Region Lagunera, North Mexico, groundwater contamination and sufferings of people were observed (Cebrain et al., 1983). But figures relating to how many people were actually drinking arsenic-contaminated water and developed skin lesions are not available. However, the population in the region was 2.0 million. During 1968, the Taiwan incident came to limelight and poisoning was observed during the period 1961–1985. The total population of 103,154 in the area was at risk (Tseng et al., 1968; Yeh, 1962). In Antofagasta, Chile, the total population involved was 1,30,000 (Borgono and Greiber, 1971). Several other incidents of groundwater contamination involving smaller population groups have been reported in Minnesata-USA (Feinglass, 1973), Ontario-Canada (Wyllie, 1973), Nova Scotia-Canada (Grantham and Jones, 1977), Nakajo-Japan (Terada et al., 1960), Millad County-Utah (Southwick et al., 1983), Lane County-Oregon (Morton et al., 1976), Lassen County-California (Goldsmith et al., 1972) and Fairbanks-Alaska (Harrington et al., 1978).

During the last decade, more and more cases of arsenic contamination of groundwater and sufferings of people have been reported from South-East Asian Countries. Xinjiang Uygur A. R. China and Inner Mongolia, A. R. China incidents have now surfaced. Although Xinjiang Uygur A. R. China incident of arsenicism was known in 1980, a detailed study report was published only in 1994 (Wang and Huang, 1994). During the period 1980 to 1984, a total of 31,141 inhabitants were examined for arsenic poisoning in 77 villages and towns situated in the alluvial plain of Xinjiang and 523 cases of human poisoning from arsenic were found among those subjects. Inner Mongolia A. R. China incidents were recognised in 1988. According to the various research reports compiled by the end of 1995, arsenic contamination in Inner Mongolia had spread to 655 villages of 11 counties and 1774 patients were confirmed. The survey conducted since 1987 have found that over 300,000 people were drinking arsenic contaminated well-water (Xiao, 1997; Gao, 1997).

The cause of contamination is considered geological. In the background, however, there is the fact that underground water has been vigorously pumped for agriculture and people have to dig wells deeper and deeper to obtain drinking water.

Until 1996, the arsenic calamity in West Bengal, India, was considered to be the biggest calamity in the world (Das et al., 1994; Chatterjee et al., 1995; Das et al., 1995; Das et al., 1996; Mandal et al., 1996; Mandal et al., 1997; Roy Chowdhury et al., 1997; Guha Mazumder et al., 1997; Mandal et al., 1998). The area and population of the arsenic affected districts of West Bengal are about 38,000 sq. km and 38 million respectively. This does not mean all the people are drinking contaminated water and all will suffer. But there is no doubt they are at risk. So far in our preliminary survey, 911 villages in 63 blocks have been identified where people are drinking arsenic contaminated water above 0.05 mg/l. In our earlier report (Mandal et al., 1997) we had made an approximate calculation by extrapolating the data based on 3 years study of Deganga Block in district to say that more than 2 million people are drinking arsenic contaminated water above 0.05 mg/l and more than 200,000 people in these 8 districts may have arsenical skin lesions. In 6 districts we have examined and documented so far 26654 people and 3908 have been identified with arsenical skin lesions.

The groundwater arsenic problem of Bangladesh came to light recently (Dhar et al., 1997; Biswas et al., 1998; Pearce, 1998; Proceedings, 1998). Out of the total 64 districts of Bangladesh (total population 120 million and total area 148,393 sq. km) in 41 districts arsenic has been found in groundwater above 0.05 mg/l, the WHO maximum permissible limit. The total area and population of these 41 districts are 89186 sq. km and 76.9 million respectively. People suffering from arsenical skin lesions have been identified from 22 districts out of 23 districts where we have made a preliminary dermatological investigation. The area and population of these 22 districts are 33077 sq. km and 26.6 million respectively. In these 23 districts, we had examined at random 6464 people and 2187 people have been identified with arsenical skin lesions. Thus it appears that only in West Bengal, India, and Bangladesh together more than 100 million people are at risk from groundwater arsenic contamination.

On West Bengal's arsenic problem, we are working for the last 9 years and on Bangladesh for about last 3 years. It appears from our survey of the last 3 years that Bangladesh's arsenic calamity is more severe compared to that of West Bengal. In a recent report (Pearce, 1998) published after the groundwater arsenic conference in Dhaka from 8–12 February, 1998, it has been stated that World Health Organisation (WHO) predicts that within a few years one in 10 adult death across much of southern Bangladesh could be from cancer triggered by arsenic. Further, it was stated in that report (Pearce, 1998) that World Bank's local chief said "tens of millions of people are at risk.

The present paper will narrate the current status report of the arsenic calamity in West Bengal, India, and Bangladesh. A comparative study will be made to show the magnitude of the calamity in Bangladesh compared to West Bengal.

2. MATERIALS AND METHODS

Flow injection hydride generation atomic absorption spectrometry (FI-HG-AAS) was used for analysis of most of the water, digested hair, nail,skin scale and urine samples. A detailed description of the instrumentation and FI-HG-AAS procedure is described in our earlier publications (Chatterjee et al., 1995; Das et al., 1995; Samanta and Chakraborti, 1997). Total arsenic in water and bore-hole sediments were also analysed by spectrometric method developed by us using Ag-DDTC in $CHCl_3$ with hexamethylene tetramine (Chakraborti et al., 1982). Sediments were also analysed using X-ray fluorescence (XRF), electron probe micro analysis (EPMA), laser microprobe mass analyser (LAMMA) and X-ray powder diffactometry (XRD). The XRF and EPMA instrumentations have been described previously (Chatterjee et al., 1993; Chakraborti et al., 1992). A LAMMA-500 (Ley bold-Heracus colonge, Germany) was used to measure sediments to get positive and negative mass spectra. A detailed description of LAMMA has been given elsewhere (Das et al., 1996; Heines et al., 1979). The sediment samples have also been studied using a Siemens D-500 powder diffractometer (Das et al., 1996) with a Cu-target and excitation conditions of 45 KV, 39 mA, after the sediment samples have been separated using a permanent magnet to distinguish morphological and magnetic properties.

2.1. Reagents and Glassware

All reagents are of Analar quality. A solution of 1.5% $NaBH_4$ (E.Merck, Germany) was prepared in 0.5% NaOH (E.Merck, India) and 5.0 M solution of HCl (E.Merck, India) was used for flow injection analysis. Details of the reagents and glassware are given elsewhere (Chatterjee et al., 1995; Das et al., 1995).

2.2. Sample Collection and Digestion

2.2.1. Water. Water samples were collected from tubewells (bore-wells). The mode of collection and details of bore-wells were described earlier (Chatterjee et al., 1993; Chatterjee et al., 1995).

2.2.2. Hair, Nails, Skin-Scale and Urine. Hair, nail and urine were collected from people who are living in arsenic affected villages. Skin-scales were collected from the arsenic victims having hyperkeratosis and keratosis. The procedure for collection, cleaning and the mode of digestion have been previously described (Chaterjee et al., 1995; Das et al., 1995; Mandal et al., 1996).

2.2.3. Bore-Hole Sediments. Bore-hole sediments were collected from different areas of the affected districts. The mode of collection and method of digestion are reported in our earlier publication (Das et al., 1996).

2.3. Procedure for Determination of Arsenic in Water, Hair, Nail, Skin-Scale, and Arsenic Metabolites in Urine

A detailed procedure for the determination of arsenic in water, hair, nails, skin-scale and sediments has been described in our earlier publications (Chatterjee et al., 1995; Das et al., 1995; Roy Chowdhury et al., 1997). In this study we have determined total arsenic in water both by FI-HG-AAS and spectrophotometrically. Spectrophotometry using our procedure (Chakraborti et al., 1982) is reliable for arsenic from 0.04 mg/l and above whereas by FI-HG-AAS even we can determine arsenic up to 0.003 mg/l with 95% confidence. Dilution of the sample is necessary for an arsenic concentration above 0.05 mg/l in FI-HG-ASS. Arsenic in hair, nails and skin-scales was measured by FI-HG-AAS and the procedure has been described earlier (Das et al., 1995). For urine sample, only the inorganic arsenic and its metabolites were measured by FI-HG-AAS without any chemical treatment. The detail procedure has been described elsewhere (Das et al., 1995; Mandal et al., 1996; Roy Chowdhury et al., 1997).

3. RESULTS AND DISCUSSION

Figure 1(a) shows the arsenic affected 8 districts of West Bengal and 63 Blocks/Thanas where we have found arsenic in groundwater to be more than the WHO maximum permissible limit of arsenic in drinking water (0.05 mg/l). Our sporadic survey of other four districts (Medinipur, Birbhum, Bankura, West Dinajpur) surrounding these 8 arsenic affected districts of West Bengal show the groundwater arsenic is below WHO rec-

BLOCKS / P.S.	BLOCKS / P.S.
(1) BARUIPUR	(33) KALIGANJ
(2) SONARPUR	(34) KARIMPUR I
(3) JOYNAGAR I	(35) KARIMPUR II
(4) MOGRAHAT II	(36) HANSKHALI
(5) BHANGOR I	(37) RANAGHAT II
(6) BHANGOR II	(38) PURBASTHALI I
(7) BUDGE-BUDGE II	(39) PURBASTHALI II
(8) BISHNUPUR I	(40) KALIACHAK I
(9) BISHNUPUR II	(41) KALIACHAK II
(10) CANNING I	(42) KALIACHAK III
(11) BARASAT I	(43) ENGLISHBAZAR
(12) DEGANGA	(44) MANIKCHAK
(13) BASIRHAT II	(45) BELDANGA I
(14) BADURIA	(46) SUTI II
(15) SWARUPNAGAR	(47) RANINAGAR I
(16) HABRA I	(48) BAHARAMPUR
(17) HABRA II	(49) DOMKAL
(18) GAIGHATA	(50) JALANGI
(19) BARASAT II	(51) HARIHARPARA
(20) BASIRHAT I	(51) NAWDA
(21) HASNABAD	(53) RAGHUNATHGANJ II
(22) BANGAON	(54) BHAGAWANGOLA II
(23) BAGDAH	(55) SUTI I
(24) BARRACKPUR I	(56) RANINAGAR II
(25) BALLY-JAGACHA	(57) FARAKKA
(26) BALAGARH	(58) MURSHIDABAD
(27) CHAKDAHA	(Jiaganj-Lalbag)
(28) SHANTIPUR	(59) RAJARHAT
(29) NABADWIP	(60) HAORA
(30) TEHATTA I	(61) HINGALGANJ
(31) TEHATTA II	(62) AMDANGA
(32) HARINGHATA	(63) BARRACKPUR-II

Figure 1a

Figure 1a. Map showing 63 arsenic affected blocks in 8 districts of West Bengal having arsenic in groundwater above 0.05 mg/l.

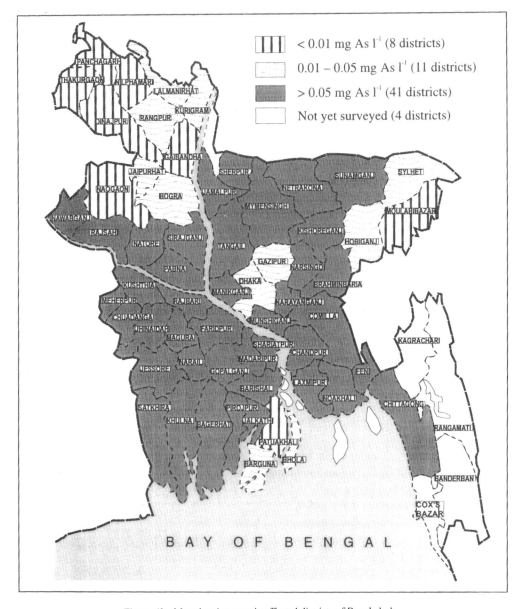

Figure 1b. Map showing arsenic affected districts of Bangladesh.

ommended value. It appears so far that these 8 districts of West Bengal are the possible arsenic contaminated districts. Altogether, in our preliminary survey in 911 villages/wards, we have so far found arsenic in groundwater above 0.05 mg/l to maximum 3.7 mg/l.

So far during our preliminary survey in West Bengal we have surveyed 50 blocks/thanas for arsenic patients and in 44 blocks/thanas we have identified people having arsenical skin lesions. The total number of villages where we have identified arsenic patients in West Bengal are 145. These are a very preliminary surveys and we need to survey in details to know the true picture.

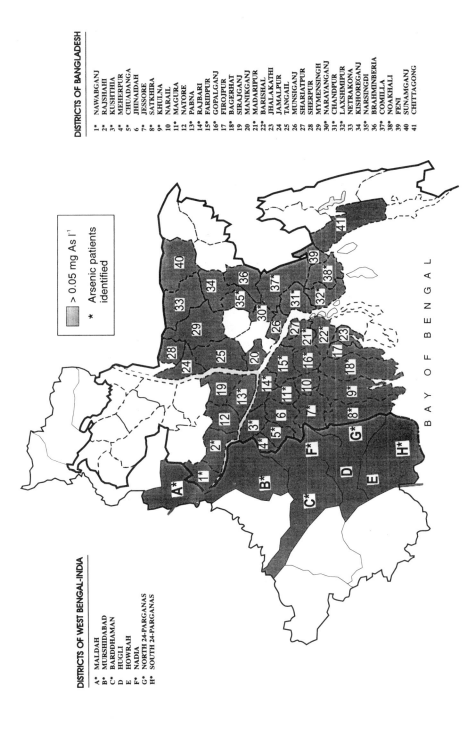

DISTRICTS OF WEST BENGAL-INDIA

A* MALDAH
B* MURSHIDABAD
C* BARDDHAMAN
D   HUGLI
E   HOWRAH
F* NADIA
G* NORTH 24-PARGANAS
H* SOUTH 24-PARGANAS

DISTRICTS OF BANGLADESH

1* NAWABGANJ
2* RAJSHAHI
3* KUSHTHIA
4* MEHERPUR
5* CHUADANGA
6 JHINAIDAH
7* JESSORE
8* SATKHIRA
9* KHULNA
10 NARAIL
11* MAGURA
12 NATORE
13* PABNA
14* RAJBARI
15* FARIDPUR
16* GOPALGANJ
17 FIROJPUR
18* BAGERHAT
19 SIRAJGANJ
20 MANIKGANJ
21* MADARIPUR
22* BARISHAL
23 JHALAKATHI
24 JAMALPUR
25 TANGAIL
26 MUNSIGANJ
27 SHARIATPUR
28 SHERPUR
29 MYMENSINGH
30* NARAYANGANJ
31* CHANDPUR
32* LAXSHMIPUR
33 NETRAKONA
34 KISHOREGANJ
35* NARSINGDI
36 BRAHMINBERIA
37* COMILLA
38* NOAKHALI
39 FENI
40 SUNAMGANJ
41 CHITTAGONG

Legend: > 0.05 mg As l⁻¹

* Arsenic patients identified

BAY OF BENGAL

Figure 1c. Map showing 8 arsenic affected districts of West Bengal, India and Bangladesh and the districts of West Bengal and Bangladesh where arsenic patients were identified.

Figure 1(b) shows that in 41 districts out of total 64 districts of Bangladesh we have found arsenic in groundwater to be more than 0.05 mg/l. There are 296 blocks/thanas in 41 districts and out of them we have surveyed 146 blocks/thanas, and in 109 blocks/thanas we have found arsenic from 0.05 mg/l to maximum 2.1 mg/l. Figure 1(b) also shows that in 52 districts arsenic is above the WHO recommended value i.e. 0.01 mg/l (WHO, 1993). Out of all 64 districts of Bangladesh, we have so far found in our preliminary survey 8 districts where groundwater is safe to drink as per the WHO recommended value. We have not yet been able to survey 4 districts. In Bangladesh, out of 41 districts where we have found arsenic in groundwater above 0.05 mg/l, in 23 districts so far surveyed, we have identified arsenic patients in 22 . There are 135 blocks/thanas in these 23 districts and out of them so far we have surveyed 45 blocks/thanas and found arsenic patients in 44 blocks/thanas. Figure 1(c) shows a combined map of the arsenic affected districts of Bangladesh and West Bengal where groundwater contains arsenic more than 0.05 mg/l. Figure 1(c) clearly indicates that affected area and population number of districts of Bangladesh are much higher compared to West Bengal where groundwater contains arsenic above 0.05 mg/l.

Table 1 shows physical parameters of West Bengal and Bangladesh and their arsenic affected areas. Table 2 shows the physical parameters of districts of West Bengal, India and Bangladesh where we have identified arsenic patients. So far from 8 arsenic affected districts of West Bengal we have collected 42225 water samples and measured the total dissolved arsenic. We have also measured a few thousand samples for both arsenite and arsenate (Chatterjee et al., 1995) and the average ratio of arsenite and arsenate is 1:1. Table 3 shows the results of total arsenic analysis of 42225 samples collected from 8 arsenic affected districts of West Bengal. Table 3 also shows our preliminary survey report of tubewells from 60 districts of Bangladesh. So far we have analysed from these 60 districts 9024 water samples. Table 3 clearly indicates that the presence of arsenic of higher concentration is more prevalent in Bangladesh compared to West Bengal. Figure 2 reports a

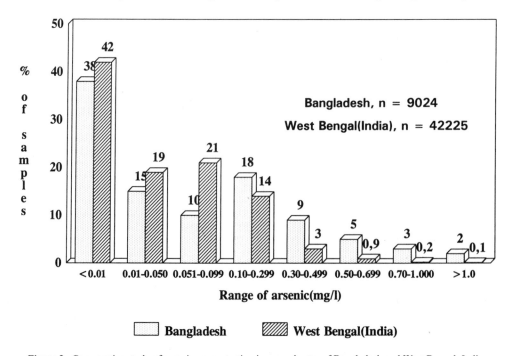

Figure 2. Comparative study of arsenic concentration in groundwater of Bangladesh and West Bengal, India.

Table 1. Physical parameters of Bangladesh and West Bengal, India and their arsenic affected areas

Country	Total area (sq. km)	Total population (million)	Total no. of districts	Total no. of districts surveyed	Total no. of districts where arsenic found above 0.05 mg/l	Total area of the districts where arsenic found above 0.05 mg/l (sq. km)	Total population of the districts where arsenic found above 0.05 mg/l (million)	Total no. of Blocks/Thanas in these districts	Total no. of Blocks/Thanas we have surveyed so far	Total no. of Blocks/Thanas where arsenic found above 0.05 mg/l	Total population of Blocks/Thanas where arsenic found above 0.05 mg/l (million)
Bangladesh	148393	120	64	60	41	89186	76.9	296	146	109	27
West Bengal, India	89192	68	18	8	8	38678	38.4	176	75	63	12

Table 2. Physical parameters of districts of Bangladesh and West Bengal-India where arsenic patients were identified

Country	No of districts surveyed for arsenic patients	No. of districts where arsenic patient identified	Total area of these affected districts (sq. km)	Approx. population of these districts (million)	Total no. of Blocks/Thanas in these districts	No. of Blocks/Thanas we have surveyed for arsenic patients	No. of Blocks/Thanas where we have found arsenic patients	Approx. area of the affected Blocks/Thanas where we have found arsenic patients (sq. km)	Approx. population of the affected Thanas/Blocks where we have found arsenic patients (million)	No. of villages we have surveyed	No. of villages where we have found arsenic patients	Total no. of people we have examined at random	Total no. of patients we have documented
Bangladesh	23	22	33077	26.6	135	45	44	13352	10.8	97	94	6464	2187 (33.8%)
West Bengal, India	6	6	34061	30.3	144	50	44	7652	7.0	475	145	26654	3908 (14.7%)

Table 3. Distribution of arsenic in tubewells water in eight districts of West Bengal India and 60 districts of Bangladesh up to March, 1998

Country	No. of districts	Total no. of tubewell water samples analysed	Distribution of no. of tubewell water samples in different concentration ranges (mg/l) of arsenic							
			<0.01	0.01–0.05	0.051–0.099	0.10–0.299	0.30–0.499	0.50–0.699	0.70–1.0	>1.0
West Bengal, India	8	42225	17636	8234	8656	5953	1267	371	72	36
Bangladesh	41	9024	3447	1381	884	1642	786	460	244	180

Table 4. Parametric presentation of concentration of arsenic in hair, nails and urine collected from the people of arsenic contaminated area of West Bengal, India, and Bangladesh

Parameters	Arsenic content in hair* (mg/kg)		Arsenic content in nail** (mg/kg)		Arsenic content in urine*** (µg/l)	
	West Bengal	Bangladesh	West Bengal	Bangladesh	West Bengal	Bangladesh
No. of valid observations	7135	2342	7381	2340	9995	940
Mean	1.48	4.05	4.56	9.25	272	495
Minimum	0.18	0.28	0.38	0.26	10	24
Maximum	20.34	28.06	44.89	79.49	3221	3086
Standard Deviation	1.55	4.04	3.98	8.73	403	493
% of sample having arsenic above normal/toxic level	47	89	58	98	73	96

*Normal level of arsenic in hair 0.08–0.25 mg/kg with 1 mg/kg being the indication of toxicity. (Arnold et al., 1990).
**Normal level of arsenic in nail 0.43–1.08 mg/kg (Ioanid et al., 1961).
***Normal excretion of arsenic in urine ranges from 5–40 µg/day (1.5 l) (Farmer and Johnson, 1990).

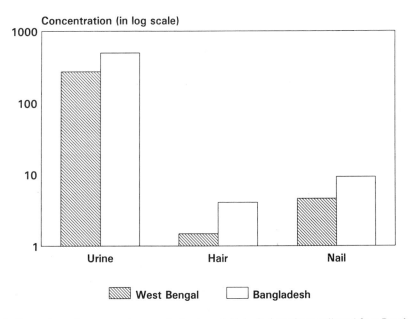

Figure 3. Comparison of mean arsenic concentration in each biological specimen collected from Bangladesh and West Bengal India.

comparative study of percentage of samples against arsenic concentration in Bangladesh and West Bengal.

Analysis of urine is a direct indication of arsenic contamination (Cornelis et al., 1996; Hsu et al., 1997). Total arsenic determination constitutes the unabsorbed form of non-toxic arsenic, as well coming from sea-food, e.g. arsenocholine, arsenobetaine. In our mode of determination of arsenic in urine by FI-HG-AAS, we analyse only inorganic arsenic and its metabolites. Arsenicholine, arsenobetaine do not respond to our mode of determination. The detailed method has been reported earlier (Chatterjee et al., 1995). **Table 4** shows the arsenic content in urine, hair and nail of some people in arsenic affected areas of West Bengal, India and Bangladesh. The mean concentrations of arsenic in urine, hair, and nail of the people of Bangladesh are about 2–3 times higher than the same for West Bengal. The **Figure 3** shows the comparison of mean arsenic concentrations of each specimen. The **Table 4** also indicates that for each specimen in Bangladesh higher percentage of samples contain arsenic above normal level/toxic level than West Bengal. **Table 5** shows the arsenic content in urine, hair and nail of the control population from non-affected area where arsenic in groundwater is less than 0.01 mg/l. During our field survey we frequently observed that in a family despite drinking the same arsenic contaminated water, all members of the family may not be affected. Usually some percentage of the population do not show the skin lesions. This is usually observed for all range of arsenic in drinking water. However, our hair, nail analysis always show elevated level of arsenic. Thus, from the medical point of view, these populations may be sub-clinically affected.

3.1. Study Report on One of the Six Arsenic Affected Districts of West Bengal Where We Have Identified Arsenic Patients

We have been working on arsenic groundwater contamination in 8 districts of West Bengal for about 9 years now. Even after working with 30 people we feel we are still at

Table 5. Parametric presentation of arsenic in urine, hair and nails of control population of West Bengal, India, and Bangladesh

Parameters	Urine (µg/l)		Hair (mg/kg)		Nail (mg/kg)	
	West Bengal	Bangladesh	West Bengal	Bangladesh	West Bengal	Bangladesh
Sample no.	75	62	75	62	75	62
Mean	16	31	0.34	0.41	0.75	0.83
Minimum	10	6	0.22	0.12	0.54	0.09
Maximum	41	94	0.50	0.85	1.07	1.58
Standard deviation	10	20	0.10	0.18	0.11	0.68

the tip of the iceberg. We realised during our survey that it will be wise to study at least one district in details and then we can extrapolate our results to other districts to get an idea of the condition of other districts. We have chosen North-24-Paraganas for three reasons.

1. Our preliminary survey indicates that this region is not as highly affected as Murshidabad, Malda and not as less affected as Bardhaman, South 24-Parganas. It lies between the two extremes.
2. It is easy for communication.
3. We have two groups of field workers in North-24-Paraganas.

As we mentioned earlier, more we are surveying, progressively more arsenic contaminated villages and the suffering of people from arsenicosis are being revealed. We started our preliminary work during 1990 and for the last 4 years we are working on this district in details. **Figure 4(a)-4(e)** show the increase in arsenic affected areas, hence the finding of arsenic patients in the affected blocks. **Table 6** shows the present arsenic status of tubewell water from 22 blocks of North-24-Paraganas which we have surveyed for the last 9 years. **Table 7 and Figure4(e)** show the blocks where we have identified arsenic patients. **Table 7** shows the arsenic in hair, nail, urine and skin scales of some villagers from arsenic affected villages. **Table 7** also shows that 87%, 56% and 80% of the population of North 24-Parganas have arsenic in urine, hair and nail respectively above normal value. An extrapolation of the data provides valuable information of the magnitude of the arsenic calamity in the 6 affected districts where we have found arsenic patients. Although at present 8 districts in West Bengal have been found to contain arsenic above 0.05 mg/l in the groundwater, the 2 districts of Hooghly and Howrah [**Figure 1(a)**] have no information of arsenic patients. In these districts we have analysed only a few hundred samples and have not searched for arsenic patients in the villages.

As for Bangladesh, so far, a detailed study on any one district has yet to be conducted. However, we have undertaken a detailed study on one village of Bangladesh (Biswas et al., 1998). The study report of Samta village indicates the severity of the arsenic problem in Bangladesh (**Table 8**).

3.2. A Preliminary Survey Report of People Having Arsenical Skin Lesions in 6 Districts of West Bengal and 22 Districts of Bangladesh

Almost for the last 9 years we are surveying the arsenic affected villages of 6 districts of West Bengal for arsenic patients. Although at present arsenic in groundwater

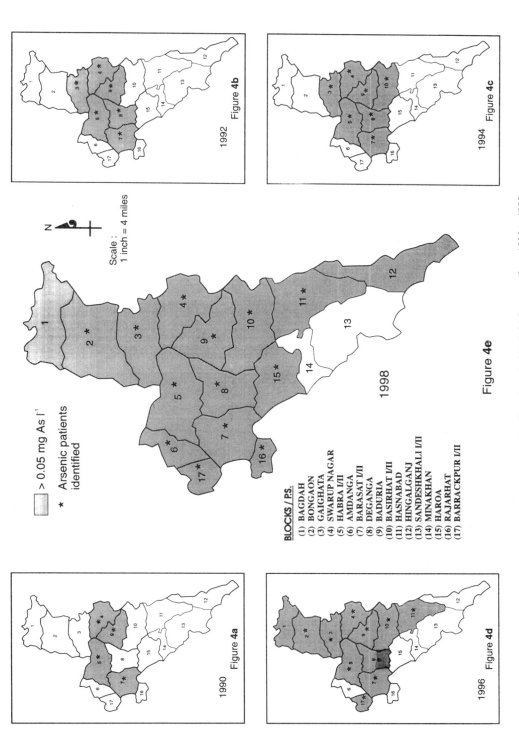

Figure 4. Increase of arsenic affected area in North 24-Parganas from 1990 to 1998.

Table 6. Distribution of arsenic in tubewell water of different blocks in North 24-Parganas district in West Bengal up to March, 1998

Name of blocks	Total no. of samples analysed	No. of samples above 0.01 mg/l	No. of samples above 0.05 mg/l	Distribution of no. of samples in different concentration ranges of arsenic (mg/l)								Whether arsenical patient(s) present
				<0.01	0.01-0.050	0.051-0.099	0.10-0.299	0.30-0.499	0.50-0.699	0.7-1.0	>1.0	
Barasat I	867	482	236	385	246	87	118	20	8	2	1	Yes
Barasat II	441	301	171	140	130	83	78	8	2	–	–	Yes
Habra I	3333	2413	1493	920	1059	795	405	107	46		1	Yes
Habra II	2006	906	577	1100	329	319	238	14	3	2	1	Yes
Basirhat I	330	211	152	119	59	71	57	11	9	3	1	Yes
Basirhat II	2367	1029	571	1338	458	353	180	38				Yes
Gaighata	2766	1669	1163	1097	506	495	585	64	14	5		Yes
Baduria	685	431	345	254	86	106	113	100	18	6	2	Yes
Deganga	8785	5082	3610	3703	1472	1713	1577	203	94	19	4	Yes
Swarupnagar	504	432	276	72	156	102	121	45	4	4		Yes
Hasnabad	555	254	200	301	104	54	95	1				Yes
Bongaon	213	104	49	109	55	22	27					Yes
Bagdah	398	170	71	228	99	46	20	3	2			Yes
Barrackpur I	411	80	32	331	48	17	14	1				Yes
Barrackpur II	785	33	17	752	16	3	12	1			1	Yes
Haroa	1104	451	141	653	310	107	28	5	1			Yes
Rajarhat	811	463	159	348	304	91	68					
Hingalganj	126	24	4	102	20	1	3					
Swandeshkhali I	312			313								
Swandeshkhali II	333			333								
Amdanga	772	190	72	582	118	12	47	13				Yes
Minakhan	315	135		180	135							
North 24-Parganas	28219	14860	9339	13359	5710	4477	3786	634	201	41	11	Yes

Table 7. Concentration of arsenic in urine, hair, nail and skin-scales of exposed
population of North 24-Parganas districts in West Bengal

Parameters	Urine (μg/l)	Hair (mg/kg)	Nail (mg/kg)	Skin Scale (mg/kg)
No of observations	8397	6286	6413	62
Mean	328	2.03	4.25	7.94
Maximum	3108	20.34	44.89	15.51
Minimum	10	0.10	0.25	0.52
Median	187	1.26	2.56	4.11
Standard Deviation	489	4.41	5.32	11.07
% of population above normal/toxic level	87%	56%	80%	

above 0.05 mg/l has been found in 8 districts we could make only a preliminary survey for arsenic patients in villages of 6 districts of West Bengal.

Table 9 shows our study report for the last 9 years. Although we have been working for the last 9 years, we go to the field only once or twice a month. So, compared to the number of villages in the affected blocks, we have covered only a very small segment. Our survey experience reveals that as we survey, an increasing number of villages are identified where people are drinking contaminated water and suffering from arsenicosis. Our detailed study of North-24-Paraganas district also indicates this. Although Table 9 shows that out of 26654 people whom we examined at random in the villages, 3908 (14.7%) have arsenical skin lesions. But this does not mean that in all villages where people are drinking contaminated water, that they will have a similar percentage of affected people.

To understand the actual situation of how many people have arsenical skin lesions in villages, we have studied one Gram Panchayet (GP) Kolsur of Deganga block in details for about a year. Kolsur GP has 4 villages with a population of 16879. Table 10 shows our detailed study report on Kolsur G.P. where we had examined 11,000 available people for arsenical dermatosis. The results show that total 580 people (male, female and children) have arsenical skin lesions (5.3%). Table 10 also indicates that 72.6% tubewell water have arsenic in Kolsur G.P. above 0.05 mg/l. Our study reveals that hundred percent of the people drink tubewell water. So in average 72.6% of people are drinking water above 0.05 mg/l. In average above 90% of the biological species examination show arsenic above normal levels (Table 10). Thus how many of the Kolsur people are subclinically affected is unknown. Children are more susceptible to arsenic toxicity (Abernathy et al., 1989) and normally do not show arsenical skin lesions before 11 years. So how many children are really in danger is not yet known.

Table 11 shows our 2 years survey report in villages of Bangladesh where we have identified arsenic patients. From the Table 11 it appears, in comparison to West Bengal (Table 9) arsenic calamity is severe in Bangladesh. About 33% of the population in arsenic affected villages have arsenical skin lesions. Again this is not the overall picture of affected villages of Bangladesh. The observed high percentage stems from examining the villages where arsenic patients were available. So far, no study has been carried out in Bangladesh like the one we have undertaken in North 24-Paraganas district and Kolsur G.P. in West Bengal. But for Bangladesh we have surveyed in details a seriously arsenic affected Samta village, Sharsa Thana of Jessore district. This is considered one of the severely arsenic affected districts of Bangladesh. Table 8 shows the over all study report of Samta village. It is true from the Table 8 that Samta is highly affected but that does not

Table 8. Physical parameter of Samta village and its water and biological sample status

Total area of Samta village	Total population	Total no. of tubewell	Total no. of tubewell sample analysed	Total no. of tubewell contain arsenic above 0.05 mg/l and range (0.05–1.0 mg/l)	Total no. of hair sample analysed	No. of hair samples contain arsenic above toxic level	Total no. of nail sample analysed	No. of nail sample contain arsenic above normal level	Total no. of urine sample analysed	No. of urine sample contain arsenic above normal level	Total people surveyed for arsenic patient	No. of patients identified
3.2 Sq. km	4841	279	265	242 (91%)	293	280 (95.6%)	236	227 (96%)	301	297 (99%)	600	330 (55%)

Table 9. Nine-year survey for arsenic patients in affected villages of 6 districts of West Bengal

Total no. of districts where arsenic above 0.05 mg/l	Total no. of districts surveyed for arsenic patients	Total no. of blocks surveyed for arsenic patients	Total no. of blocks where arsenic patients found	Total no. of villages surveyed for arsenic patients	Total no. of villages where arsenic concentration above 0.05 mg/l	Total no. of villages where arsenic patients found	Total no. of people surveyed for arsenic patients in 145 villages	Total no. of patients identified
8	6	50	44	911	475	145	26654	3908 (14.7%)

Table 10. Physical parameter of Kolsur village and its water and biological sample status

Total area of Kolsur village	Total population of Kolsur village	Total no. of tubewell in Kolsur village	Total no. of tubewell sample analysed in Kolsur village	Total no. of tubewell contain arsenic above 0.05 mg/l and range (0.05–1.0 mg/l)	Total no. of hair sample analysed	No. of hair samples contained arsenic above toxic level	Total no. of nail sample analysed	No. of nail sample contained arsenic above normal level	Total no. of urine sample analysed	No. of urine sample contained arsenic above normal level	Total people surveyed for arsenic patient	No. of patients identified
14.16 Sq. km	16879	1550	1184	859 (72.6%)	1083	942 (87%)	1103	1047 (95%)	1433	1404 (98%)	11000	580 (5.27%)

Table 11. Two-year survey for arsenic patients in affected villages of 23 districts of Bangladesh

Total no. of districts where arsenic found above 0.05 mg/l	Total no. of districts surveyed for arsenic patients	Total no. of districts where arsenic patients found	Total no. of Thanas surveyed for arsenic patients	Total no. of Thanas where arsenic patients found	Total no. of villages surveyed for arsenic patients	Total no. of villages where arsenic patients found	Total no. of people surveyed for arsenic patients	Total no. of arsenic patients identified	% of arsenic patients identified	Total male patients identified and percentage	Total female patients identified and percentage	Total child patients identified and percentage	Total no. of villages of 23 districts where we have information of arsenic patients	Name of districts where information of arsenic patients other than 22
41	23	22	45	44	97	94	6464	2187	33.8	1099 50.3%	882 40.3%	206 9.4%	20	Jamalpur Jhinaidah Brahminbaria Narail

Table 12. Dermatological features of 4 patients, West Bengal-India and Bangladesh

Country and District	Blocks/Thanas and Village	Age and sex	Melanosis					Keratosis				Others
			Palm Diffuse	Trunk Spotted	Trunk Diffuse	Leuco	Whole body	Palm Spotted	Palm Diffuse	Sole Spotted	Sole Diffuse	
West Bengal, North 24-Parganas	Swarupnagar, Purbapalta	M/54	–	++	–	+	+	+++	+	+++	++	Finger amputated due to cancer (S.C.C.)
West Bengal, North 24-Parganas	Basirhat, Merudandi	M/43	+	++	–	+	+	++	+	++	+	Leg amputated due to Gangrene
Bangladesh, Fakirpara	Faridpur town, Tapakhola	F/45	+	+++	+	+	++	+++	+	+++	+++	
Bangladesh, Nawabganj	Nawabganj town, Rajarampur	M/35	+	++	+	+	–	++	+	++	++	Cancer (S.C.C.)

+ = mild, ++ = moderate, +++ = severe, S.C.C. = Squamous Cell Carcinoma again mean that all arsenic villages of Bangladesh are like Samta. When compared with one highly affected village Fakirpara of Deganga block in West Bengal we found that 23% of total population of Fakirpara have arsenical skin lesions whereas in Samta it is 55%. Thus it appears Bangladesh villages may be more affected. **Photographs 1–4** show four arsenic patients, 2 each from West Bengal, India and Bangladesh. **Table 12** shows the dermatological features of these four patients.

3.3. Socio-Economic Study of Kolsur G.P.

During our survey in the arsenic affected areas both in West Bengal and Bangladesh we have found the villages are more affected where people are more backward. They are suffering from melanosis. We have evidence of many villages whose inhabitants are drinking almost the same concentration of arsenic in drinking water, but the arsenical skin lesions and sufferings of people are much less. These findings can be attributed to their getting better food. A survey made by SOES for socio-economic status of 11,000 out of 16879 people of Kolsur G.P. in Deganga Block with the help of a questionnaire and the following information have been documented.

1. About 3300 families, with an average of 5 members in each family, live in Kolsur G.P.
2. Out of 16879, about 51.13% males, 48.87% females and 22.84% children below 11 years live in huts (80%) made of mud/galvanised tin/bamboo and 20% in brick houses. 90% houses are one roomed and 10% of the houses are 2 roomed and above. 60% of the population use leaves of the trees, dried branch of trees,

Photograph 1. Village: Purbapalta, Block: Swarupnagar, Dist: North 24-Parganas, West Bengal, India. Middle finger amputated due to cancer (squamous cell carcinoma).

Photograph 2. Village: Merudandi, Block: Bashirhat, District: North 24-Parganas, West Bengal, India; Leg amputated due to gangrene.

dried cow dung and straw as the fuel for cooking and the other 40% use kerosine and coal as fuel.

3. 75% of people have their own land for cultivation and the average area owned by them is about 2000 sq. m.

4. Average monthly income of a family is one thousand four hundred rupees (US $ 36).

5. a) 80% families have one earning member
   b) 10% families have two earning members
   c) 2% families have three earning members

6. About 5.57% people take care of their family easily and 94.28% people do so with stress.

7. Literacy rate of the inhabitants is 91% (literacy means they can sign their name).

8. 75% of the residents of this block are farmers depending on agriculture. The type of job villagers are doing: working as daily labourers on others land (29%): business (18%); van driver (0.84%), tailor (1.98%), mechanic (0.65%), shoe-

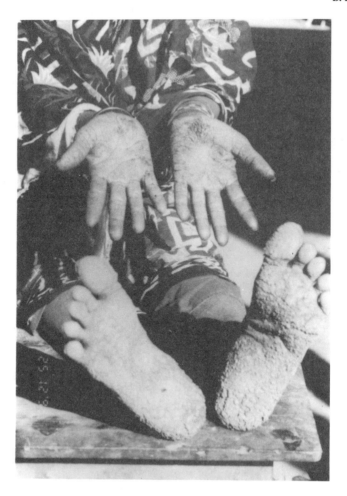

Photograph 3. Village: Tapakhola, Thana: Faridpur Sadar, District: Faridpur, Bangladesh; spotted and diffuse keratosis on palm and sole.

maker, tea-shop owner, goldsmith, hawker, fisherman, milk & newspaper suppliers and others. About 18% males and 35% females do not have any job.

9. Daily labourers work on others' fields for 15 days per month during the rainy season, 19 days per month during winter season and 19 days per month during summer season.

10. For 61 % of the population, the two main meals consist of rice with edible herbs and some vegetables. The cooked food is much too spicy. 39% of the inhabitants take local milk five days per week (200 ml/day), fish average 10 days/month, eggs average 6 days/month and meat average 2 days/month. Handmade bread they take some time as snacks.

11. Almost all residents purchase fish, egg and meat, 94% people purchase vegetables, 56% people purchase rice and cereals from market. Price per kg of fish is about $ 1, meat $2.5 per kg and egg $ 1 per dozen.

12. Practically all residents use tubewell water for drinking, 94% for cooking and 43% for bathing purposes. About 39% of the residents use pond for bathing purposes.

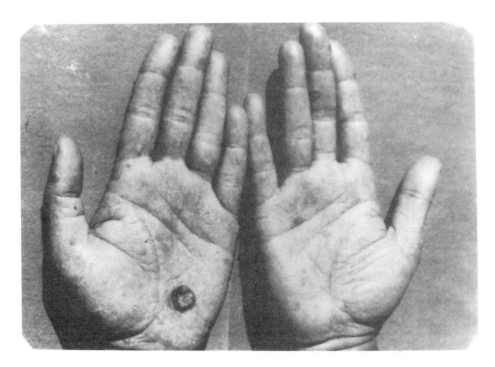

Photograph 4. Village: Rajarampur, Thana: Nawabganj Sadar, District: Nawabganj, Bangladesh; Spotted keratosis and squamous cell carcinoma.

13. Tubewell water intake per day. 54% of the people (age above 20 yr) take 4 liters, 42% people (age below 20 yr) take 3 liters, and 20% (age below 10 yr) take 2 liters.

14. Average arsenic concentration in tubewell water 0.17 mg/l.

15. Generally when the villagers are sick, 41% of the people take recourse to homeopathy, 57% people opt for allopathy, 0.87% people go to empiric medicine and 28% people go for both homeopathy & allopathy.

16. About 47% of the people go to physicians for mild disease and 53% people go for severe diseases.

17. About 6% of the people can afford money to visit a physician and 94% people can do so with difficulty.

18. 90% of male adults take 15 Bidi (country cigarette) per day. An appreciable number of women smoke but we have no data as they are shy to express because of social taboo.

19. Percentage of people having arsenical skin lesion is 1.39% below 11 years of age and 6.42% of the people above that age, out of the 11000 people we have surveyed. Overall, 5.8% people have arsenical skin lesions.

20. In many interior villages, inhabitants are neither aware of their arsenic problem nor the serious nature of the problem. We had asked 1000 villagers in interior villages of Deganga and Kolsur during our sample collection (we first started our work during November,1994) a set of questions. The questions and their answers are shown below:

• Q. Do they know about the arsenic problem in their area and hence in West Bengal?

- Ans. 78% do not know, 10% heard that in tubewell water some poison is present, 12% know about the arsenic in tubewell water but do not know actually what it is.
- Q. What do they think about the skin lesions?
- Ans. 60% say they are just a skin disease and will be cured with ointment: 30% have no idea, 10% realise it is a serious problem.
- Q. Why it happens ?
- Ans. 40% 'Wrath of God' or 'Curse of God'; 60% do not know.

Some narrated their beliefs:

  a) One night he went outside for urinating and the devil urinated on his face and since then he has the arsenical skin manifestation.
  b) When the tubewell was sunk it hit the head of Snake of God, as a result God vomited poison into water.

21. Although the arsenic problem came to notice in 1978 and was officially recognised in 1983 the villagers in the remote villages are still not aware of their arsenical problem. While surveying in Kolsur G.P. (April 1997) we found that in Fakirpara village (total population 764), out of 46 tubewells 41 have arsenic in the range 0.05–0.89 mg/l and 175 people have arsenical skin lesions (22.9%). It is surprising that till 18th April, 1997 the villagers were not aware that they were suffering from arsenic toxicity.

4. SOURCE OF ARSENIC

We have found arsenic in groundwater in West Bengal, India in the area of sediment of Younger Deltaic Deposition (YDD), the same deposition extended eastward covering most of the districts of Bangladesh. Thus we may expect the source of arsenic in affected areas of West Bengal and that of Bangladesh are of similar nature and origin. It has been reported that the source of arsenic is geological (Das et al., 1996; Saha, 1996; Proceedings, 1998).

So far in these 8 districts of West Bengal we have made 112 bore-holes and collected 2235 sediments from each 3.3 meters interval. Out of these 2235 number of samples, 85 show arsenic between 10 and 196 mg/kg. Some selected particles collected from 85 samples show much higher concentration of arsenic (as high as about 3000 mg/kg). Some of the selected particles were examined by EPMA, LAMMA, XRF, X-ray diffraction. All these techniques identified presence of arsenic rich 'pyrite' (Das et al., 1996). The reason why arsenic is coming out with groundwater is not completely clear however there is a possibility that due to heavy groundwater withdrawal the underground aquifer is aerated and oxygen is responsible for degradation of arsenic rich source. We have discussed possible mechanism of oxidation of pyrite and arsenic leaching from the source in our earlier publication (Das et al., 1996). We also feel the role of microbes need to be studied.

5. CONCLUSION

From the overall study of the arsenic affected areas of West Bengal and Bangladesh it appears that Bangladesh arsenic calamity is more severe compared to that of West Bengal. However, we feel an in-depth and more elaborate study is necessary to determine the actual magnitude of the calamity in both these countries.

REFERENCES

Abernathy, C.O., Marcus, W., Chen, C., Gibb, H., White, P. (1989). Report on arsenic workgroup meeting. Office of Drinking Water, Office of Research and Development, USEPA, Memorandum to Cook P., Preuss, P., Office of Regulatory Support and Scientific Management, USEPA.

Arnold, H.L., Odam, R.B., James, W.D. (1990). Disease of the skin clinical dermatology, W.B. Saunders Company, Philadelphia, USA, 8th edn, 121–122.

Astolfi, E., Maccagno, A., Fernandez, J. C. G., Vaccara, R., Stimola, R. (1981). Relation between arsenic in drinking water and skin cancer. Biological Trace Element Research 3, 133–143.

Bergogio, R.M. (1964). Mortalidad por cancer en zones de aguas arsenicales de la provincia de cordoba, Republica Argentina. Prensa Medica Argentina 51, 994–998.

Biagini, R.E. (1972). Hideroarsenicismo cronico y muerte por cancers malignos. La Semana Medica 141, 812–816.

Biswas, B. K., Dhar, R. K., Samanta, G., Mandal, B. K., Chakraborti, D., Faruk, I., Islam, K. S., Chowdhury, M. M., Chowdhury, M. M., Islam, A., Roy, S. (1998), Detailed study report of Samta, one of the arsenic-affected villages of Jessore District, Bangladesh, Current Science 74(2), 134–145.

Borgono, J. M., Greiber, R. (1971). Epidemiological study of arsenicism in the city of Antofagasta. In: Hemphill, D.D. (Ed) Trace Substances in Environmental Health-V. A symposium, Columbia, University of Missouri Press, pp.13–24.

Buchet, J. P., Lauwerys, R. (1994). Inorganic arsenic metabolism in humans. In: Chappell, W. R., Abernathy, C. O., Cothern, C. R, (Eds). Arsenic Exposure and Health, Science and Technology Letters, Northwood, pp 181–189.

Cebrain, M. E., Albores, A., Aguilar, M., Blakely, E. (1983). Chronic arsenic poisoning in the north of Mexico. Human Toxicology 2, 121–133.

Chakraborti, D., Valentova, M. and Sucha, L. (1982). Decomposition of materials containing traces of arsenic and its spectrophotometric determination. Prague Institute of Chemical Technology of Czechoslovakia Analytical Chemistry, H-17, 31–41.

Chakraborti, D., Das, D., Chatterjee, A., Jin, Z., Jiang, S.G. (1992). Direct determination of some heavy metals in urban air particulates by electrothermal atomic absorption spectrometry using Zeeman background correction after simple acid decomposition. Part IV: Application to Calcutta air particulates. Environmental Technology 13, 95–100.

Chatterjee, A., Das, D., Chakraborti, D. (1993). A study of ground water contamination by arsenic in the residential area of Behala, Calcutta due to industrial pollution. Environmental Pollution 80(1), 57–65.

Chatterjee, A., Das, D., Mandal, B. K., Chowdhury, T. R., Samanta, G., Chakraborti, D. (1995). Arsenic in groundwater in six districts of West Bengal, India: the biggest arsenic calamity in the world. Part I: Arsenic species in drinking water and urine of the affected people. The Analyst 120, 643–650.

Cornelis, R., Heintow, B., Herber, R. F. M., Christensen, J. M., Poulsen, O. M., Sabbioni, E., Templeton, D. M., Thomassen, Y., Vather, M., Vesterberg, O. (1996). Sample collection guideline for trace elements in blood and urine. Journal of Trace Elements in Medicine and Biology 10, 103–127.

Das, D., Chatterjee, A., Samanta, G., Mandal, B., Roy Chowdhury, T., Samanta, G., Chowdhury, P. P., Chanda, C., Basu, G., Lodh, D., Nandi, S., Chakraborti, T., Mandal, S., Bhattacharya, S. M., Chakraborti, D. (1994). Arsenic contamination in groundwater in six districts of West Bengal, India: the biggest arsenic calamity in the world. The Analyst 119, 168N–175N.

Das, D., Chatterjee, A., Mandal, B. K., Samanta, G., Chakraborti, D., Chanda, B. (1995). Arsenic in groundwater in six districts of West Bengal, India: the biggest arsenic calamity in the world. Part II: Arsenic concentration in drinking water, hair, nail, urine, skin-scale and liver tissue (Biopsy) of the affected people. The Analyst 120, 917–924.

Das, D., Samanta, G., Mandal, B. K., Roy Chowdhury, T., Chanda, C. R., Chowdhury, P. P., Basu, B. K., Chakraborti, D. (1996). Arsenic in groundwater in six districts of West Bengal, India. Environmental Geochemistry and Health 18, 5–15.

Dhar, R. K., Biswas, B. K., Samanta, G., Mandal, B. K., Chakraborti, D., Roy, S., Jafar, A., Islam, A., Ara, G., Kabir, S., Khan, A. W., Ahmed, S.A., Hadi, A. (1997). Groundwater arsenic calamity in Bangladesh. Current Science 73(1), 48–59.

Evans, S., Cusick, J., Sasieni, P. (1998). Medicinal arsenic, keratosis and bladder cancer. International conference on arsenic pollution of groundwater in Bangladesh: causes, effects and remedies, 8–12, February, Dhaka, Bangladesh, pp 39.

Ferman, J.G., Johnson, L.R. (1990). Assessment of occupational exposure to inorganic arsenic based on urinary concentrations and speciation of arsenic. British Journal of Industrial Medicine 47, 342–348.

Feinglass, E. J. (1973). Arsenic intoxication from well water in the United States. The New England Journal of Medicine 288(16), 828–830.

Gao, C. (1997). The geological setting of arsenic contamination in Inner Mongolia, China. The Association for the Geological collaboration in Japan (Chigaku Dantai Kenkyukai) II, 28–33.

Goldsmith, J. R., Deane, M., Thom, J., Gentry, G. (1972). Evaluation of health implications of elevated arsenic in well water. Water Research 6, 1133–1136.

Grantham, D. A., Jones, J. F. (1977). Arsenic contamination of water wells in Nova Scotia. Journal of American Water Works Association 69, 653–657.

Guha Mazumder, D. N., Das Gupta, J., Santra, A., Pal, A., Ghosh, A., Sarkar, S., Chattopadhaya, N., Chakraborti, D. (1997). Non-cancer effects of chronic arsenicosis with special reference to liver damage. In: Abernathy, C.O., Calderon, R. L., Chappell, W. R. (Eds), "Arsenic: Exposure and Health Effects", Chapman and Hall, New York, pp 112–123.

Harrington, J. M., Middaugh, J. P., Morse, D. L., Housworth, J. (1978). A survey of a population exposed to high concentrations of arsenic in well water in Faribanks, Alaska. American Journal of Epidemiology 108(5), 377–385.

Heines, H. J., Meier, S., Vogt, H., Wechsung, R. (1979). Laser induced mass spectrometry of organic and inorganic compound with LAMMA. 8th Triannaul International Mass Spectrometry conference, Osla. Norway.

Hsu, K. H., Froines, J. R., Chen, C. J. (1997). Studies of arsenic ingestion from drinking water in Northern Taiwan: chemical speciation and urinary metabolites. In: Abernathy, C. O., Calderon, R. L., Chappell, W. R. (Eds), "Arsenic: Exposure and Health Effects", Chapman and Hall, New York, pp 190–209.

Ioanid, N.B., Bors, G., Popa, I., (1961). Beitrage zur kenntnis des normalen Arsengehaltes von Nageln und des Gehaltes in den Fallen von Arsenpolyneuritis. Dtsch. & Gesamte Gerichtt. Med. 52, 90–94

Mandal, B. K., Chowdhury, T. R., Samanta, G., Basu G K, Chowdhury, P. P., Chanda, C. R., Lodh, D., Karan, N. K., Dhar, R. K., Tamili, D. K., Das, D., Saha, K. C., Chakraborti, D. (1996). Arsenic in groundwater in seven districts of West Bengal, India - the biggest arsenic calamity in the world. Current Science 70(2), 976–986.

Mandal, B. K., Chowdhury, T. R., Samanta, G., Basu, G. K., Chowdhury, P. P., Chanda, C. R., Lodh, D., Karan, N. K., Dhar, R. K., Tamili, D. K., Das, D., Saha, K. C., Chakraborti, D. (1997). Chronic arsenic toxicity in West Bengal. Current Science 72(2), 114–117.

Mandal, B. K., Roy Chowdhury, T., Samanta, G., Mukherjee, D. P., Chanda, C. R., Saha, K. C., Chakraborti, D. (1998). Impact of safe water for drinking and cooking on five arsenic-affected families for 2 years in West Bengal, India. The Science of the Total Environment (in press).

Morton, W. E., Starr, G., Pohl, D., Stoner, J., Wagner, S., Weswig, P. (1976). Skin cancer and water arsenic in Lane County Oregon. Cancer 37, 2523–2532.

Naqvi, S.M., Vaishnavi, C., Singh, H. (1994). Toxicity and metabolism of arsenic vertebrates. In: Nriagu, J.O. (Ed) "Arsenic in the Environment, Part II: Human Health and Ecosystem Effects". John Willey and Sons, New York, pp 55–91.

Pearce, F. (1998). Arsenic in the water. The Guardian, London, 19th February, pp 1–3.

Proceedings (1998). International conference on "Arsenic pollution of groundwater in Bangladesh: causes, effects and remedies" 8–12 February, 1998, Dhaka, Bangladesh.

Roy Chowdhury, T. K., Mandal, B. K., Samanta, G., Basu, G. K., Chowdhury, P. P., Chanda, C. R., Karan, N. K., Dhar, R. K., Lodh, D., Das, D., Saha, K. C., Chakraborti, D. (1997). Arsenic in groundwater in seven districts of West Bengal, India: the biggest arsenic calamity in the world. The status report up to August, 1995. In: Abernathy, C.O., Calderon, R.L., Chappell, W.R. (Eds), "Arsenic: Exposure and Health Effects", Chapman and Hall, New York, pp 91–111.

Saha, A.K. (1996). Report on Geology and Geochemistry of arsenic occurrences in groundwater of six districts of West Bengal, submitted to Rajiv Gandhi National Drinking Water Mission, Ministry of Rural Areas and Employment, Government of India, New Delhi, India, Sanctioned No. 11046/24/95 TM.II dt. 31.05.1995.

Samanta, G., Chakraborti, D. (1997). Flow-injection atomic absorption spectrometry for the standardization of arsenic, lead and mercury in environmental and biological standard reference materials. Fresenius Journal of Analytical Chemistry 357, 827–832.

Southwick, J. W., Western, A. E., Beck, M. M., Whitley, T., Isaacs, R., Petajan, J. O., Hansen, C.D. (1983). An epidemiological study of arsenic in drinking water in Millard County Utah. In: Lederer, W. H., Fensterheim, R. L. (Eds) "Arsenic: Industrial, Biomedical, Environmental perspectives", Van Nostrand Reinhold Company, New York, pp 210–225.

Terade, H., Kalsuta, K., Sasagawa, T., Saito, H., Shirata, H., Fukuchi, K., Sekiya, T., Yokoyama, Y., Hirokawa, S., Watanabe, Y., Hasegawa, K., Oshina, T., Sekiguchi, T. (1960). Clinical observation of chronic toxicity by arsenic. Nihon Rinsho 118, 2394–2403.

Tseng, W.P., Chu, H.M., How, S.M., Fong, J.M., Lin, C.S., Yeh, S. (1968). Prevalence of skin cancer in an endemic area of chronic arsenicism in Taiwan. J. Natl Cancer Inst. 40, 453–463.

Wang, L., Huang, J., (1994). Chronic arsenism from drinking water in some areas of Xinjiang, China, In: Nriagu, J.O. (Ed) "Arsenic in the Environment, Part II: Human Health and Ecosystem Effects". John Wiley & Sons, New York, pp 159–172.

WHO, (1993) Guideline for drinking water quality, recommendations (2nd ed), World Health Organisation, Geneva, Vol. 1, p-41.

Wyllie, J. (1973), An investigation of the source of arsenic in a well water. Canadian Public Health Journal 28, 128–135.

Xiao, J. G. (1997), Report from Inner Mongolia, China. Asia Arsenic Network - Newsletter, Japan 2, pp7–9.

Yeh, S. (1962). Relative incidence of skin cancer in Chinese in Taiwan: with special reference to arsenical cancer. The conference on Biology of Cutaneous cancer, Philadelphia, Penns, April 6–110, pp 81–102.

4

THE NATURE AND EXTENT OF ARSENIC-AFFECTED DRINKING WATER IN BANGLADESH

Seth H. Frisbie,[1*] Donald M. Maynard,[1] and Bilqis A. Hoque[2]

[1]The Johnson Company, Inc.
100 State St.
Montpelier, Vermont 05602
[2]Environmental Health Programme
International Centre for Diarrhoeal Disease Research, Bangladesh
GPO Box 128
Dhaka-1000, Bangladesh

1. SUMMARY

The people of Bangladesh used to rely on surface water for drinking, which was often infected with cholera and other life-threatening diseases. To reduce the incidences of these diseases, millions of tubewells were installed in Bangladesh over the last 27 years. This recent transition from surface water to groundwater has significantly reduced deaths from water-borne pathogens; however, recent evidence suggests disease and death from arsenic and potentially other metals in groundwater are impacting large areas of Bangladesh.

In this preliminary assessment the areal and vertical distribution of arsenic and other inorganic chemicals in the groundwater was mapped throughout Bangladesh. The study suggests that a major source of this arsenic may be one or more phosphate minerals containing arsenate as an impurity. Evidence for other potentially toxic heavy metals in groundwater was also discovered. Several appropriate treatment technologies were evaluated.

[*] To whom correspondence should be addressed: (erikamitt@yahoo.com)

Metals and Genetics, edited by Sarkar.
Kluwer Academic / Plenum Publishers, New York, 1999.

2. INTRODUCTION

2.1. Geographic, Demographic, and Economic Overview of Bangladesh

The Peoples' Republic of Bangladesh is a relatively small, intensely populated, and poor country. Bangladesh is located at one of the largest river deltas in the world. The Ganges, Brahmaputra, and Meghna rivers flow through Bangladesh to the Bay of Bengal. Very little of the country is more than 12 m (40 feet) above sea level, and in a normal monsoon season one-third of its cultivated land is flooded. Bangladesh has 120,000,000 people living on 144,000 square kilometers; this is equivalent to having one-half the population of the United States living in an area the size of Wisconsin. The infant mortality rate is 115 per 1,000 live births. There is one doctor per 5,200 people; by comparison the United Kingdom has one doctor per 650 people. The adult literacy rate is 43% for men and 22% for women. The average annual income is equivalent to US $220 per capita. The life expectancy is 55 years (Monan, 1995).

Bangladesh is an agricultural country with the vast majority of people involved in food production. Rice is grown during the rainy season, and is primarily used for domestic consumption. In irrigated areas, a second rice crop is possible, followed by wheat and vegetables in the short, dry winter from November to February. Bangladesh is the world's leading producer of jute, a strong natural fiber used in the carpet and sacking industries. The principal exports of Bangladesh from largest to smallest are garments, jute and its products, shellfish, tea, and leather (Monan, 1995).

2.2. Project Overview

Much of the surface water of Bangladesh is microbially unsafe to drink. Since independence in 1971, approximately 2,500,000 tubewells have been installed to supply microbially safe drinking water to the people of Bangladesh. Unfortunately, vast areas of this 120,000,000 person country contains groundwater with arsenic concentrations above the World Health Organization (WHO) drinking water standard of 0.01 mg/L. Chronic arsenic poisoning attributed to groundwater ingestion was first diagnosed in 1993.

The Government of Bangladesh with funding from the United States Agency for International Development and technical support from The Johnson Company and the International Centre for Diarrhoeal Disease Research, Bangladesh (ICDDR,B) implemented the following study to determine the nature, extent, and treatment of arsenic-affected groundwater in Bangladesh. The purposes of this eight week study in Bangladesh were to:

- provide technical training,
- evaluate analytical chemistry capability,
- determine the extent of arsenic-affected groundwater,
- hypothesize the source of arsenic in groundwater, and
- identify potential options for water treatment.

See USAID (1997) for an extensive discussion of this entire project.

3. MATERIALS AND METHODS

3.1. Groundwater Sampling and Analyses

Groundwater samples were collected from approximately 120 villages throughout Bangladesh. Typically 4 to 6 tubewells per village were sampled. These villages were as

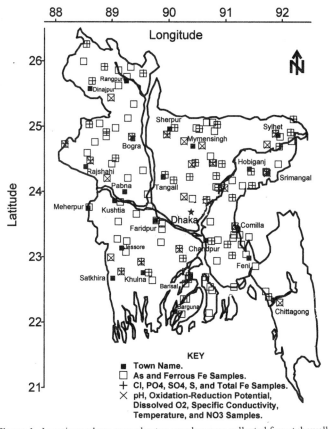

Figure 1. Locations where groundwater samples were collected from tubewells.

evenly distributed throughout the country as possible, given the limited access due to seasonal flooding (see Figure 1). The latitude and longitude of all sample locations were determined using the Global Positioning System.

All groundwater samples were collected (from July 22 to August 14, 1997), preserved, stored, and analyzed using procedures described in *Standard Methods for the Examination of Water and Wastewater* (APHA, AWWA, and WEF, 1995). All analyses were performed at the ICDDR,B Laboratory, unless otherwise stated. Approximately 600 samples were analyzed for total arsenic by the silver diethyldithiocarbamate method and ferrous iron by 1,10-phenanthroline (see Figure 1). Approximately 100 samples were analyzed for chloride by mercuric thiocyanate, phosphate by amino acid, sulfate by barium turbidity, sulfide by methylene blue, and total iron by 1,10-phenanthroline. Approximately 75 samples were analyzed immediately after collection for pH by glass electrode, oxidation-reduction potential by electrode, dissolved oxygen by membrane electrode, specific conductivity by electrode, temperature by thermocouple, and nitrate by cadmium reduction.

3.2. Soil Leaching Study

A total of 31 surface soil samples (0 to 1.2 m or 0 to 4 feet below ground surface, bgs) were collected from random locations throughout Bangladesh from July 22 to August 6, 1997. These samples were stored in coolers packed with ice from the moment of collection until the samples were processed at the ICDDR,B Laboratory between 3 and 5 days later. Each soil sample was homogenized and analyzed for moisture content by evaporation to a constant mass at 105°C. Following desiccation a mass of field-moist soil equivalent to 100 grams of oven-dried soil was delivered to a clean sample jar, distilled water was added to the jar to make the final mass of water equal to 200 grams, and the contents of the jar were mixed for 5 minutes. For example, if a 12 gram sample of field-moist soil weighed 10 grams after drying, then 120 grams of field-moist soil (this initial condition is equivalent to 100 grams of oven-dried soil and 20 grams of water) was delivered to a clean sample jar with 180 grams of distilled water (this final condition is equivalent to 100 grams of oven-dried soil and 200 grams of water; however, the composition of the soil being leached was never altered by oven-drying). After 6 days a 50 mL aliquot of water was removed, filtered through a standard glass fiber filter used to remove total suspended solids (APHA, AWWA, and WEF, 1995), and submitted for total arsenic analysis.

3.3. Bench-Scale Treatability Study for Arsenic Removal from Groundwater

Tubewell water was collected and used immediately for this treatability study without the addition of sample preservatives. The tubewell water was fortified with arsenic to yield 2.0 mg of As(III)/L. One liter aliquots of fortified tubewell water were delivered to 1 L borosilicate glass beakers. Chlorinated lime (a locally available oxidant, $a\text{Ca(OCl)}_2 \cdot b\text{CaCl}_2 \cdot c\text{Ca(OH)}_2 \cdot d\text{H}_2\text{O}$) was added at rates of 4, 6, 10, 15, and 20 mg/L. The solutions were mixed for 1 minute. Ferric chloride hexahydrate (an effective coagulant, $\text{FeCl}_3 \cdot 6\text{H}_2\text{O}$) was added at rates of 100, 150, 200, and 250 mg/L. Each solution was mixed for 1 minute, allowed to settle for 24 hours, then the water column above the precipitate was analyzed for total arsenic.

4. RESULTS AND DISCUSSION

4.1. Evaluation of Analytical Chemistry Capability and Evidence That Other Potentially Toxic Metals Are Widely Distributed in Groundwater

Accurate laboratory results are imperative to understanding and solving the problem of arsenic-affected drinking water in Bangladesh; therefore, a major goal of this project was to determine accuracy and precision of the analytical chemistry results generated by various laboratories studying this problem.

The ICDDR,B Laboratory performed exceptionally well during this study (see Table 1). The recoveries from the blind analysis of all independently prepared standards were within the 100 ± 25% range considered acceptable for routine analytical laboratories (USEPA, 1994). An analytical interference for the determination of ferrous iron in undiluted groundwater by the 1,10-phenanthroline method was identified from suppressed matrix spike recovery (see Table 1). This interference was defeated by diluting all ferrous iron samples 10 times with distilled water before color development. Similarly, all samples submitted for total iron analysis by the 1,10-phenanthroline method were diluted 10 times with distilled water before acidic digestion to defeat this interference; despite this precaution, 24 of 89 samples (27%) failed to develop proper color. Some of these samples

Table 1. The percent recoveries of independent standards and sample matrix spikes analyzed by the ICDDR,B laboratory

Analyte	Independent standard recovery	Sample matrix spike recovery
Arsenic (As)	83%	89 ± 11%
Ferrous iron (Fe^{2+})	93 ± 10%	34 ± 23%*
		96 ± 13% **
Total iron (Fe)	95%	120 ± 12%
Sulfate (SO_4^{2-})	106%	106 ± 20%
Chloride (Cl^-)	114%	90 ± 15%
Phosphate (PO_4^{3-})	88%	84 ± 2%

Results are reported as means ± standard deviations
*without dilution; **with 1 to 10 dilution

generated a whitish colloidal precipitate upon the addition of 1,10-phenanthroline. This analytical interference coupled with the results shown on Figure 8 suggests that one or more of the following potentially toxic metals are widely distributed in groundwater throughout Bangladesh: chromium, zinc, cobalt, nickel, bismuth, cadmium, mercury, and/or silver (APHA, AWWA, and WEF, 1995).

The recovery of independent standards and sample matrix spikes were used to assess other laboratories in Bangladesh (see Table 2). The arsenic results from Laboratory 1 are systematically high by approximately a factor of 4; therefore, a simple calibration error was likely generating results which incorrectly suggest that the arsenic problem is 4 times worse than reality. Laboratories 2 and 3 only recovered approximately 50% of the arsenic from independent standards, such results underestimate the significance to the arsenic problem by a factor of 2.

4.2. The Nature and Extent of Arsenic, Chloride, Phosphate, Sulfate, Sulfide, and Total Iron in Groundwater

This preliminary evaluation of the nature and extent of arsenic and other inorganic chemicals in groundwater is based on the limited number of samples that could be col-

Table 2. The performance of some other laboratories that are evaluating arsenic in Bangladesh

Sample description	Independent laboratory	Result (mg/L)
Standard solution = 1 mg As/L	Laboratory 1	4.891
Distilled water = 0 mg As/L	"	0.002
Sample A (0.25 mg As/L)	"	1.101
Sample A (blind duplicate)	"	1.035
Sample A (blind triplicate)	"	0.266
Sample A plus 6.3 mg As/L	"	24.126
Sample B (0.31 mg As/L)	"	1.109
Standard solution = 1 mg As/L	Laboratory 2	0.533
Sample B (0.31 mg As/L)	"	0.397
Sample B plus 3.3 mg As/L	"	0.884
Standard solution = 1 mg As/L	Laboratory 3	< 0.5
Sample B (0.31 mg As/L)	"	0.30
Sample B plus 7.1 mg As/L	"	> 1.0

lected during our three-week field program. Samples were collected from only 600 of 2,500,000 tubewells in 120 of 68,000 villages; therefore, more extensive studies are required to support or disprove the hypotheses presented in this article.

The concentrations of various inorganic chemicals throughout Bangladesh were mapped to identify areas impacted by arsenic and other toxins, to determine the potential source of arsenic, and to identify possible water treatment requirements. Contour maps of chemical concentration show the aerial extent of arsenic, chloride, phosphate, sulfate, and total iron in groundwater (see Figures 2, 3, 5, 6, 7, 8, and 10). These contour maps were drawn using the average chemical concentration in tubewell water from each village, unless otherwise stated (recall that 4 to 6 tubewells per village were sampled). The vertical distribution of arsenic was evaluated by separating samples collected from "shallow" and "deep" tubewells. Shallow tubewells were arbitrarily assigned a depth less than 30.5 m (100 feet) bgs, and deep tubewells were arbitrarily assigned a depth greater than 30.5 m (100 feet) bgs. This separated the samples into two groups with each group containing approximately 300 samples.

The contour maps of average arsenic concentration in water from shallow and deep tubewells are shown in Figures 2 and 3, respectively. The detection limit for the arsenic method was 0.028 mg/L (defined as the concentration of the most dilute standard used for

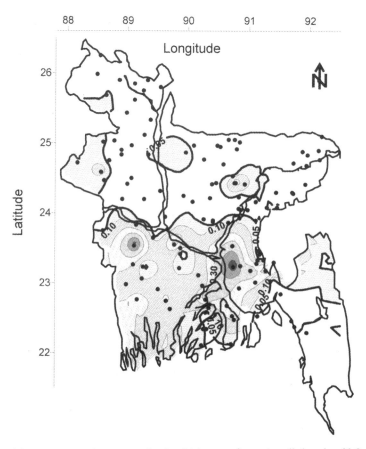

Figure 2. Map of the average arsenic concentration (mg/L) in water from tubewells less than 30.5 m or 100 feet bgs (• = village location).

calibration); therefore, quantitation to the WHO drinking water standard of 0.01 mg/L was not attempted and these contour maps are based on the Bangladesh drinking water standard of 0.05 mg/L. Approximately 50% of the aerial extent of Bangladesh contains groundwater from shallow tubewells with an average arsenic concentration greater than the national drinking water standard (see Figure 2). Approximately 32% of the aerial extent of Bangladesh contains groundwater from deep tubewells with an average arsenic concentration greater than the national drinking water standard (see Figure 3). This wide-spread aerial and vertical distribution suggests the dominant source of the arsenic appears to be one or more geologically deposited minerals. The superimposition of Figures 2 and 3 suggests that drilling deeper tubewells may access water with less arsenic in some locations and drilling shallower tubewells may access water with less arsenic in other locations.

The comparison of 19 pairs of closely spaced (< 100 m or < 328 feet apart) "very deep" (> 91.4 m bgs or > 300 feet bgs) and shallow (< 30.5 m or < 100 feet bgs) tubewells shown in Figure 4 suggests the source of arsenic is hundreds of feet thick in many areas of Bangladesh. This wide-spread vertical distribution supports the hypothesis that the dominant source of arsenic appears to be one or more geologically deposited minerals. The results shown in Figure 4 suggest that in areas with arsenic concentrations greater than the

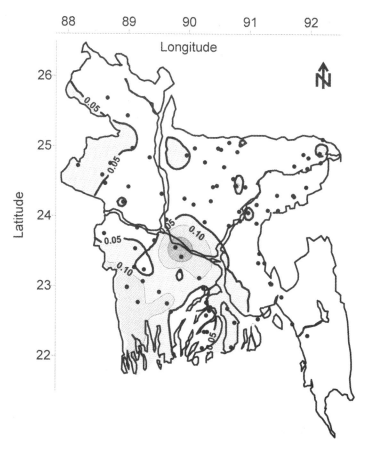

Figure 3. Map of the average arsenic concentration (mg/L) in water from tubewells greater than 30.5 m (100 feet) bgs.

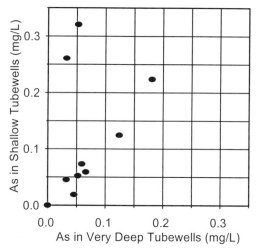

Figure 4. The vertical distribution of arsenic in groundwater.

Bangladesh drinking water standard, drilling deeper tubewells accessed safe water in 2 of 13 cases. The results shown in Figure 4 also suggest that in areas with arsenic concentrations greater than the Bangladesh drinking water standard, drilling shallower tubewells accessed safe water in 3 of 13 cases.

A correlation coefficient matrix for a variety of tubewell water parameters is shown in Table 3 for informational purposes. Interpretation of this data should be performed with caution as it represents the entire country and is probably scale dependant. That is, these correlations may not apply on the district, thana, or village scale.

The contour map of average chloride concentration in water from shallow tubewells is shown in Figure 5. The spacial relationship between chloride and arsenic (see Figures 5 and 2) in connate water (water associated with deposition, Ravenscroft, 1997) suggests the arsenic-leaching mineral or minerals were likely deposited in an estuarine environment. An alternative hypothesis is that the release of arsenic from the solid phase is facilitated by chloride or another component of historical seawater.

Phosphate is commonly used as a fertilizer; therefore, the possibility of this agro-chemical impacting groundwater must be considered before comparing the spacial relationship between phosphate and geologically deposited arsenic. Phosphate fertilizer forms difficultly soluble iron, aluminum, calcium, and magnesium compounds that do not readily leach from most soils (Brady, 1984); however, the phosphate concentration ranged from 1 to 20 mg/L in all 11 groundwater samples collected at greater than 122 m (400

Table 3. Correlation coefficient matrix for a variety of tubewell water parameters

	Arsenic	Sulfate	Sulfide	Chloride	Phosphate	Depth	Total iron
Arsenic	1.00	-0.078	0.059	0.24	0.27	-0.19	0.44
Sulfate		1.00	0.41	0.051	-0.060	-0.14	0.16
Sulfide			1.00	0.062	0.19	-0.097	0.23
Chloride				1.00	0.10	0.028	0.38
Phosphate					1.00	0.16	0.073
Depth						1.00	-0.19
Total iron							1.00

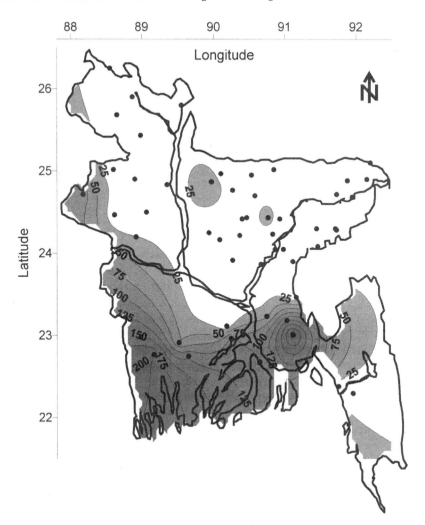

Figure 5. Map of the average chloride concentration (mg/L) in water from tubewells less than 30.5 m (100 feet) bgs.

feet) bgs. Like phosphate, nitrate is associated with agricultural activity. Unlike phosphate, nitrate readily leaches from soils; however, nitrate was detected above 1 mg of NO_3^--N/L in only 5 of 90 groundwater samples. All 5 of the samples with detectable concentrations of nitrate were from wells not greater than 122 m (400 feet) bgs. Four of these 5 samples were from a small area of northwestern Bangladesh (Dinajpur, Phulbari, and Gobandaganj) which had relatively low concentrations of phosphate in groundwater. These findings suggest that the dominant source of phosphate in the groundwater of Bangladesh is geological, not an agricultural leachate.

The contour map of average phosphate concentration in water from shallow tubewells is shown in Figure 6. The spacial relationship between phosphate and arsenic (see Figures 6 and 2) suggests the arsenic-leaching mineral or minerals might contain arsenate isomorphically substituted for phosphate; that is, the major source of arsenic in the groundwater of Bangladesh may be a phosphate or phosphates which have arsenate as an

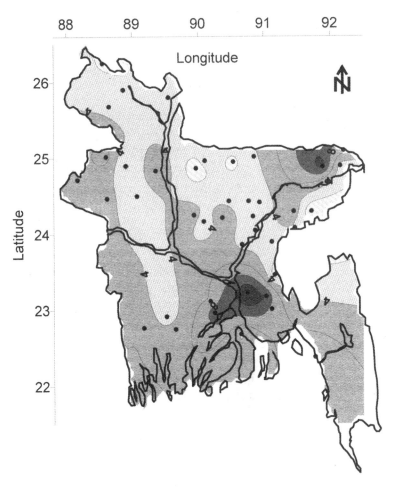

Figure 6. Map of the average phosphate concentration (mg/L) in water from tubewells less than 30.5 m (100 feet) bgs.

impurity. This hypothesis should be confirmed or disproved by a detailed mineralogical evaluation of a statistically significant number of soil samples.

An alternative hypothesis that arsenic is being released from arsenopyrite (FeAsS) or an iron pyrite (FeS_2) has been proposed for the situation in nearby West Bengal, India (Mallick, and Rajagopal, 1995). This hypothesis suggests that arsenic is initially associated with a difficultly soluble pyrite mineral that is underwater in a reducing environment, and the arsenic is released when the pyrite is aerated by lowering the water table during groundwater pumping. This hypothesis is supported by the detailed mineralogical evaluation of one soil sample from West Bengal which suggested that arsenic is associated with iron pyrite (Das, Basu, Chowdhury, and Chakraborty, 1995). If this hypothesis is true for the general situation in Bangladesh, then the concentrations of inorganic sulfur and arsenic should be correlated; however, the concentrations of sulfate and arsenic, and sulfide and arsenic in groundwater from Bangladesh are not correlated (see Table 3).

The contour map of average sulfate concentration in water from shallow tubewells is shown in Figure 7. The spacial relationship between sulfate and arsenic (see Figures 7 and

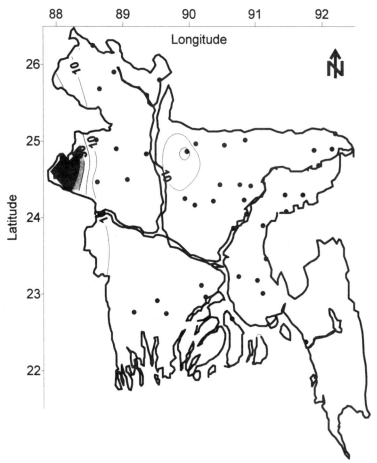

Figure 7. Map of the average sulfate concentration (mg/L) in water from tubewells less than 30.5 m (100 feet) bgs.

2), and that between sulfide and arsenic (not shown) suggests the principal arsenic-leaching mineral or minerals in Bangladesh are not associated with inorganic sulfur in contrast to the presumed situation in West Bengal. If the arsenic-leaching mineral or minerals are not associated with inorganic sulfur, then the hypothesis that groundwater depression from irrigated agriculture facilitates the release of arsenic by aerating arsenopyrite or another pyrite might not represent the general situation in Bangladesh. Nevertheless, the occurrence of sulfate and sulfide in groundwater at the West Bengal / Bangladesh border (see Figure 7) does suggest that pyrites may be an important source of arsenic in West Bengal and limited areas of Bangladesh.

Even if the principal source of dissolved arsenic in West Bengal is pyrites, the general situation in Bangladesh may differ because these two regions have very different geologies. Both West Bengal and Bangladesh receive sediment from the Ganges river basin, but Bangladesh also receives sediment from the Brahmaputra and Meghna Rivers. Each river has different sources of sediment, and most likely different proportions of arsenic, sulfur, and phosphate-bearing minerals. Moreover, Bangladesh is further than West Bengal from the presumed source of pyrites in the Ganges river basin. For these reasons, the percentage of arsenic-containing pyrites in Bangladesh aquifers is likely to differ from, and possibly be much less than that in West Bengal. Furthermore, the environments in which the aquifer sediments were deposited are different in West Bengal and Bangladesh; therefore, the oxidation-reduction conditions, sulfur concentrations, and other important aquifer chemistry parameters are likely different as well. The depositional environments of Bangladesh aquifers commonly include deltaic, estuarine, and riverine sediments, while most of the West Bengal aquifers contain only riverine sediments. Because there are so many substantial differences between the geologies of West Bengal and Bangladesh, it should not be assumed without further study that the source of arsenic in groundwater is the same in both regions.

The composition of all minerals leaching arsenic into the groundwater of Bangladesh should be determined from a detailed evaluation of a statistically significant number of soil samples. If arsenic is released from arsenopyrite or a pyrite, then alternative groundwater management practices should be evaluated and the treatment of drinking water by aquifer oxygenation may prove to be inappropriate. Aquifer oxygenation is the injection of compressed air or oxygen around a well screen to precipitate arsenic and other metals before they are pumped above ground; however, if pyrites are present, then this

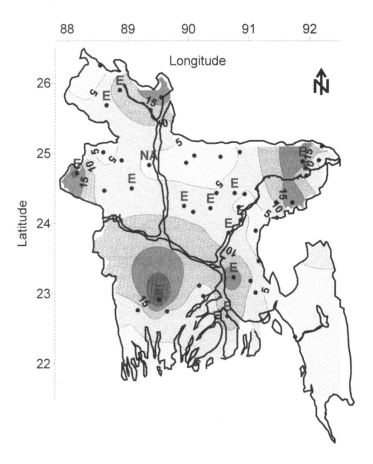

Figure 8. Map of the average total iron concentration (mg/L) in water from tubewells less than 30.5 m (100 feet) bgs.

process may facilitate the release of arsenic to groundwater. Knowledge of the composition of minerals leaching arsenic into groundwater is relatively unimportant if one is only concerned with treating water after it has been pumped above ground.

The contour map of average total iron concentration in water from shallow tubewells is shown in Figure 8. The spacial relationship between total iron and arsenic (see Figures 8 and 2), and the general excess of total iron relative to arsenic suggests that ambient iron might be used to coagulate (coprecipitate) arsenic in a low-input water treatment system (Faust and Ally, 1983). The distribution of analytical interference (see Table 1) for the determination of iron (located on the map with the letter "E") suggests that one or more of the following potentially toxic metals are widely distributed in groundwater throughout Bangladesh: chromium, zinc, cobalt, nickel, bismuth, cadmium, mercury, and/or silver (APHA, AWWA, and WEF, 1995).

Cobalt, nickel, and silver ores are often codeposited with arsenic (Mason and Berry, 1968), and each of these 3 metals are a potential interference to the 1,10-phenanthroline method; therefore, any limited resources available to investigate the occurrence of metals other than arsenic in this water should first include this subgroup of 3 metals (cobalt, nickel, and silver). Subsequent investigation should include the entire group of 8 metals (chromium, zinc, cobalt, nickel, bismuth, cadmium, mercury, and silver). Additional analytes beyond this group of 8 metals should be added as resources allow.

4.3. The Leaching of Arsenic from Surface Soils and Its Potential Impact on Food Crops

The results of the Soil Leaching Study are shown in Figure 9. These results suggest that arsenic readily leaches from many Bangladesh surface soils. Arsenic can be up taken by crops (Brady, 1984); therefore, this human exposure pathway should be evaluated. Rice is often grown in flooded soils analogous to the conditions of this experiment, is the major staple of Bangladeshi diet, and is grown primarily for domestic consumption (Monan,

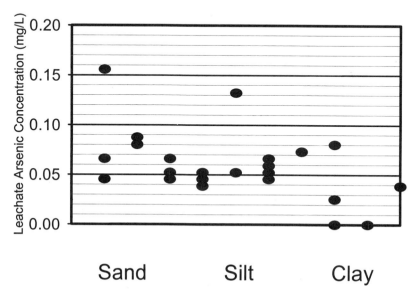

Figure 9. The concentration of arsenic (mg/L) leached into 200 mL of distilled water from 100 grams of surface soil after 6 days.

1995); therefore, the ingestion of arsenic from domestic rice should be specifically evalu-ated as a potential human exposure pathway. Unfortunately, the sample preparation equip-ment required to determine the concentration of metals in biological samples is not currently available in Bangladesh.

4.4. The Potential of Groundwater Monitoring to Reduce the Need for Arsenic Treatment

The contour map of minimum arsenic concentration in water from all (both shallow and deep) tubewells is shown in Figure 10. This contour map was drawn using the lowest ("cleanest") arsenic concentration from the 4 to 6 tubewells sampled in each village. The shaded regions of this map represents areas where no tubewell had an arsenic concentra-tion less than the Bangladesh drinking water standard. Approximately 15% of the aerial extent of Bangladesh contains groundwater with a minimum arsenic concentration greater than the national drinking water standard and will require treatment. This result also sug-gests that 85% of the aerial extent of Bangladesh has access to groundwater that does not require treatment for arsenic removal prior to drinking. An intensive groundwater monitor-ing program identifying suitable drinking water wells within each village would signifi-

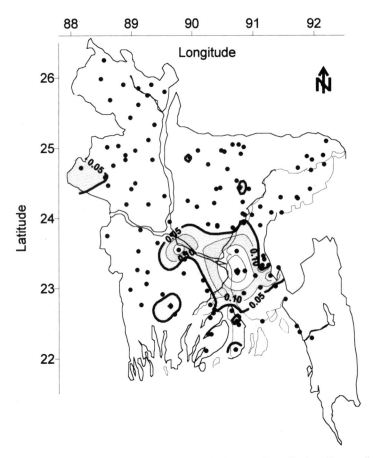

Figure 10. Map of the minimum arsenic concentration (mg/L) in water from all tubewells regardless of depth.

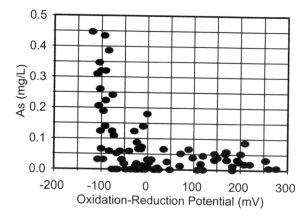

Figure 11. Graph of arsenic concentration (mg/L) versus oxidation-reduction potential in water from all tubewells regardless of depth.

cantly reduce the need for arsenic treatment for the majority of this 120,000,000 person country.

4.5. Implications of the Expected Nature of Arsenic in Groundwater on Treatment

The graphs of arsenic concentration versus oxidation-reduction potential and arsenic concentration versus pH are shown in Figures 11 and 12, respectively. The increase in arsenic concentration at relatively low oxidation-reduction potentials and moderate pH values suggests that soluble As(III), rather than difficultly soluble and potentially colloidal As(V), was the dominate form in the most highly affected tubewells (Ferguson and Gavis, 1972). If the arsenic is in the soluble As(III) oxidation state, then oxidation to difficultly soluble As(V) followed by coagulation, filtration, or sorption is required for effective treatment (Faust and Aly, 1983).

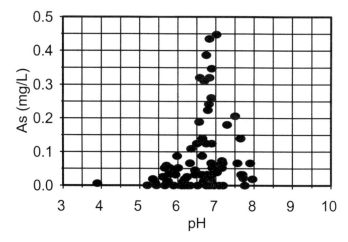

Figure 12. Graph of arsenic concentration (mg/L) versus pH in water from all tubewells regardless of depth.

4.6. The Results of the Bench-Scale Treatability Study for Arsenic Removal from Groundwater

The effect of chlorinated lime (a locally available oxidant) and ferric chloride hexahydrate (an effective coagulant) on the removal of arsenic from groundwater was evaluated at the bench-scale (see Table 4). The tubewell water used in this experiment was fortified to 2.0 mg of As(III)/L, approximately 3 times the total arsenic concentration of the most severely impacted tubewell found during this project. The results shown in Table 4 suggest that oxidation followed by coagulation can reduce relatively large arsenic concentrations in tubewell water to below the WHO drinking standard. The addition of water treatment chemicals can yield pH values outside of the 5.5 to 8.5 drinking water range recommended by WHO (1984); therefore, pH adjustment may be required after coagulation. Limestone ($CaCO_3$) offers inexpensive and effective pH control after coagulation, often without the need of expensive dosing equipment.

4.7. The Conceptual Design of a Pilot-Scale Treatment System for Removing Arsenic and Other Potentially Toxic Metals from Groundwater

The ideal water treatment system for the economic and demographic situation in Bangladesh will effectively remove arsenic and other toxic metals, be inexpensive to build and operate, and be simple to use. Such a system might use atmospheric oxygen as the oxidant and ambient iron as the coagulant; therefore, the long-term expense of purchasing water treatment chemicals would be avoided. Atmospheric oxygen delivered by waterfall or bubble aeration has been routinely used to oxidize As(III) to As(V) in water treatment systems, and should be evaluated in future studies (Faust and Aly, 1983). Ambient iron without the addition of another coagulant might adequately separate precipated arsenic and other toxic metals from water in a large settling tank or an inclined-plate clarifier. Excellent guidance for the construction of low-input water treatment systems for the developing world is provided by the International Reference Centre for Community Water Supply and Sanitation (1988) and Heber (1985).

The apparent reduction in arsenic concentration from 0.16 to <0.002 mg/L shown in Table 5 supports the hypothesis that aeration followed by settling without the addition of

Table 4. The effect of oxidant [chlorinated lime, $aCa(OCl)_2 \cdot bCaCl_2 \cdot cCa(OH)_2 \cdot dH_2O$] and coagulant [$FeCl_3 \cdot 6H_2O$] on the removal of 2.0 mg of As/L

Oxidant (mg/L)	Coagulant (mg/L)	Final arsenic concentration (mg/L)
4	250	0.075
6	250	0.037
10	250	0.006[a]
15	250	0.003[a]
20	250	< 0.002[a]
20	100	0.06
20	150	0.03
20	200	0.02[a]
20	250	< 0.002[a]

Table 5. The influent and effluent water quality of a 1.8 m (6 feet) tall by 4.6 m (15 feet) by 6.1 m (20 feet) water storage tank at the Kishorganj Rural Electrification Board facility that provides drinking water to approximately 300 people

Parameter	Influent	Effluent
Arsenic (mg/L)	0.16	< 0.002[a]
Oxidation-reduction potential (millivolts)	−38	19
pH	7.06	6.47
Conductivity (microsemans)	514	344
Temperature (°C)	27.2	28.2
Total iron (mg/L)	NA[b]	1.4
Sulfate (mg/L)	NA	< 1
Sulfide (mg/L)	NA	< 0.02
Chloride (mg/L)	NA	16
Phosphate (mg/L)	NA	1.3

[a]All arsenic concentrations less than the most dilute standard used for calibration (0.028 mg/L) are estimated.
[b]NA = not analyzed.

coagulant can remove arsenic in tubewell water to below the WHO drinking standard. The apparent increase oxidation-reduction potential suggests that the water was aerated when pumped into the storage tank. The apparent decrease in pH, conductivity, and arsenic concentration suggests that ambient iron was oxidized, hydrolyzed water, and precipitated as a ferric hydroxide coagulant of arsenic. The water pump supplying this large tank does not operate at night due to the diversion of electricity to the capital; therefore, coagulated arsenic has many hours of relatively turbulent-free water to settle each evening.

5. CONCLUSIONS AND RECOMMENDATIONS

An appropriate quality assurance and quality control program should be implemented to determine the accuracy and precision of all analytical results associated with the problem of metal-affected drinking water in Bangladesh. This program should include the routine and blind analysis of independent standards, sample matrix spikes, field duplicates, laboratory duplicates, and blanks. The ICDDR,B Laboratory generated an excellent quality of results during this study; this is especially impressive because the metal analyses were completed without the benefit of an atomic absorption spectrophotometer (see Table 1). The other three laboratories evaluated in this study would presumably benefit from a more rigorous quality assurance and quality control program (see Table 2).

Drinking water should be routinely tested for arsenic and other toxic compounds. Approximately 50% of the aerial extent of Bangladesh contains groundwater from shallow tubewells with an average arsenic concentration greater than the national drinking water standard (see Figure 2). Approximately 32% of the aerial extent of Bangladesh contains groundwater from deep tubewells with an average arsenic concentration greater than the national drinking water standard (see Figure 3). The major source of this arsenic in the Groundwater of Bangladesh is potentially a phosphate mineral or minerals deposited in an estuarine environment. This source appears to be hundreds of feet thick in some areas. The analytical interference for the determination of iron by the 1,10-phenanthroline method (see Table 1) coupled with the results shown on Figure 8 suggests that one or more of the following potentially toxic metals are also widely distributed in groundwater throughout Bangladesh: chromium, zinc, cobalt, nickel, bismuth, cadmium, mercury, and/or silver (APHA, AWWA, and WEF, 1995).

Food crops should also be tested for arsenic and other toxic compounds. Arsenic readily leaches from many Bangladesh surface soils and can be taken up by crops (Brady, 1984). Arsenic exposure from the ingestion of rice and other domestically produced food crops should be evaluated as a potential human exposure pathway.

Field testing kits for the rapid and inexpensive measurement of toxins in drinking water should be immediately developed. Approximately 15% of the aerial extent of Bangladesh contains groundwater with a minimum arsenic concentration greater than the national drinking water standard; therefore, it is hypothesized this is the smallest area of Bangladesh that will require groundwater treatment for arsenic removal prior to drinking (see Figure 10). An intensive groundwater monitoring program identifying suitable drinking water wells within each village would significantly reduce the need for arsenic treatment in this 120,000,000 person country.

An appropriate treatment systems for toxin removal should be developed for areas without access to safe drinking water. Ideally these systems should be effective, inexpensive, and easily operated by an illiterate person. These systems will likely require an oxidant to convert soluble As(III) to difficultly soluble As(V), and a process to allow the settling, filtration, or sorption of arsenic from solution.

ACKNOWLEDGMENTS

This study was performed for the Rural Electrification Board of Bangladesh. It was funded by the United States Agency of International Development (contract number US AID RE III 388-0070). The contract was implemented by National Rural Electric Cooperative Association (NRECA) International, Ltd. of Arlington, Virginia with personnel from The Johnson Company, Inc. of Montpelier, Vermont, USA and the Environmental Health Program of the International Center for Diarrhoeal Disease Research, Bangladesh. The project team offers our sincere thanks to Mr. James Ford of NRECA for providing invaluable leadership and management during the implementation of this ambitious contract. The principal investigators offer our deepest thanks to the dozens of talented and motivated Bengali scientists, engineers, and support staff who made this work a reality—the future of Bangladesh and West Bengal is bright indeed.

REFERENCES

American Public Health Association (APHA), American Water Works Association (AWWA), and Water Environment Federation (WEF). (1995). Standard Methods for the Examination of Water and Wastewater, 19th ed. (Washington, DC: APHA).

Brady, N.C. (1984). The Nature and Properties of Soils, 9th ed. (New York: Macmillian Publishing Company).

Das, D., Basu, G., Chowdhury, T.R., and Chakraborty, D. (1995). Bore-hole soil-sediment analysis of some arsenic affected areas. International Conference on Arsenic in Ground Water: Cause, Effect, and Remedy. School of Environmental Studies, Jadavpur University, Calcutta, India - 700032.

Faust, S.D., and Aly, O.M. (1983). Chemistry of Water Treatment (Boston: Butterworths).

Ferguson, J.F., and Gavis, J. (1972). A review of the arsenic cycle in natural waters. Water Res. *June*, 1259–1274.

Heber, G. (1985). Simple Methods for the Treatment of Drinking Water (Braunschweig/Wiesbaden: Friedr. Vieweg & Sohn).

International Reference Centre for Community Water Supply and Sanitation (1988). Small Community Water Supplies (New York: John Wiley & Sons).

Mallick, S., and Rajagopal, N.R. The mischief of oxygen on groundwater. International Conference on Arsenic in Ground Water: Cause, Effect, and Remedy. School of Environmental Studies, Jadavpur University, Calcutta, India - 700032.

Mason, B., and Berry, L.G. (1968). Elements of Mineralogy (San Francisco: W.H. Freeman and Company).

Monan, J. (1995). Bangladesh the Strength to Succeed (Oxford, UK: Oxfam).

Ravenscroft, P. (1997). An Overview of the Hydrogeology of Bangladesh (Unpublished, British Geological Society).

USAID (1997). Report of the Impact of the Bangladesh Rural Electrification Program on Groundwater Quality (Prepared by the Bangladesh Rural Electrification Board. Funded by US Agency for International Development, Contract Number: US AID RE III 388–0070) (Washington, DC: USEPA).

USEPA (1994). USEPA Contract Laboratory Program National Functional Guidelines for Inorganic Data Review (EPA 540/R-94/013) (Washington, DC: USEPA).

WHO (1984). Guidelines for Drinking Water Quality Vol. 1 Recommendations (Geneva: WHO).

5

CHROMIUM BIOMONITORING

Achievements, Problems, and Future Challenges

Andreas Kortenkamp

The School of Pharmacy
University of London
Centre for Toxicology
29-39 Brunswick Square
London WC1N 1AX, United Kingdom

1. INTRODUCTION

Since David Newman in 1890 published his case report on a chrome pigment worker who suffered from carcinoma of the upper respiratory tract, a multitude of epidemiological studies have been carried out linking inhalation of chromium(VI) compounds with cancers of the airways and lungs (IARC 1990). Today, occupational settings with exposure to chromium(VI) include the primary production of chromates, chromium plating, chromium(VI) pigment manufacture and stainless steel welding. With many of the hazardous workplaces identified and working conditions improved at least in western countries, the major concern is to establish whether exposures close to current occupational exposure limits pose cancer risks. The question is of considerable relevance since in many countries occupational exposure limits have not been defined on a sound toxicological basis. In this context, the biological monitoring of exposed workers is important, not only as a means of verifying compliance with existing health regulations but also as a tool which might allow further refinements of risk estimations.

More than 100 years of research into chromium(VI) cancer risks have yielded important insights into the ways in which the metal compound can cause cancer. Especially the last two decades have seen an explosion of our knowledge of the molecular mechanisms underlying the disease process (for a review see Kortenkamp et al. 1997). Analytical techniques that allow the monitoring of internal exposure to chromium(VI) in body fluids have been developed and critically evaluated. Today, exposure to chromium(VI) at levels well below 50 $\mu g/m^3$ (the occupational exposure limit in a number of countries) can be reliably verified by determining chromium concentrations in urine or plasma. In contrast, experiences with cytogenetic surveillance techniques have been mixed. Studies un-

Metals and Genetics, edited by Sarkar.
Kluwer Academic / Plenum Publishers, New York, 1999.

87

dertaken in the early 1980s with relatively heavily exposed workers have revealed chromosome damage, but the outcome of many of the more recent studies among stainless steel welders was inconclusive or negative. Similarly, the few published reports on DNA damage in the lymphocytes of chromium(VI) exposed workers have produced ambiguous results.

In the present paper we will briefly assess the usefulness of the currently available techniques for chromium(VI) biomonitoring. We will argue that the cytogenetic effects observed in the blood lymphocytes of chromium(VI) exposed workers bear no relationship to the processes which give rise to cancers of the respiratory tract. This undermines the rationale behind cytogenetic surveillance studies in chromium biomonitoring and raises the question as to alternative approaches in chromium biomonitoring. The present contribution will end with a proposal intended to monitor events closer to the target organ, the respiratory tract.

2. A CONCEPTUAL FRAMEWORK FOR CARCINOGEN BIOMONITORING

In carcinogen biomonitoring, attempts are made to trace important stages of the processes involved in neoplasia and to measure parameters that reflect early events in the disease process (Wogan, 1992) (Figure 1).

Measurements of the level of carcinogens or their metabolites in body fluids can give an indication as to how much of the agent or its metabolites has reached the inside of the body and has been distributed to tissues. Once inside cells, some carcinogenic agents are able to interact directly with DNA while others yield DNA-interacting intermediates only after metabolic conversions. On the basis of determinations of DNA damage in certain tissues it may be possible to estimate the dose of an agent which has reached its biological target. Such approaches monitor the biologically effective dose of a carcinogen. The DNA damage caused by the biologically effective dose of a genotoxic carcinogen can be modulated substantially by DNA repair processes. Some damage will eventually be-

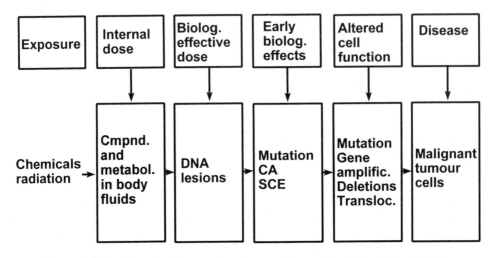

Figure 1. Biological markers in an exposure–disease continuum. Adapted from Wogan (1992).

come fixed as gene mutations, which can be viewed as early biological effects in the process of carcinogenesis. Screening for mutations in marker genes or monitoring other genetic alterations such as chromosome aberrations or sister-chromatid exchanges (SCE's) may give indications as to early biological effects in neoplasia. The later stages of carcinogenesis, namely the processes which give rise to the proliferation and clonal expansion of dormant tumour cells are at present difficult to monitor.

3. CHROMIUM(VI) BIOMONITORING: THE ANALYSIS OF INTERNALISED DOSES

Considerable advances in chromium trace analysis have made it possible to establish the relationships between airborne chromium levels at workplaces and the concentration of the metal in urine, plasma and blood of exposed workers. Urinary chromium is a sensitive and reliable marker of internal exposure to chromium, suitable for the biological monitoring of exposures well below the occupational exposure limits in most industrialised countries. Today, the analytical limit of detection is 0.2 µg/l urine (or 0.13 mg Cr/mg creatinine), approximately a tenth of the upper background level in occupationally unexposed people (2 µg/l urine) (Paustenbach et al. 1997).

On the basis of determinations of chromium in erythrocytes it is possible to distinguish whether exposure to chromium(VI) or chromium(III) has occurred. This is because the chromate anion is able to cross cell membranes readily via anion carrier systems. In contrast, chromium(III) complexes cannot reach the inside of erythrocytes (Kortenkamp et al. 1987). It is therefore not detected in red blood cells if exposure to chromium(III) has occurred. Elevated levels of chromium in plasma or urine however are indicative of exposure to both chromium(III) and chromium(VI).

The work by Angerer and his colleagues was instrumental in establishing that the diagnostic specificity which measurements of chromium in erythrocytes offer with respect to ascertaining chromium(VI) exposure can only be achieved at the expense of diagnostic sensitivity. In their study (Angerer et al. 1987), elevated concentrations of chromium in erythrocytes were only seen at relatively high levels of airborne chromium(VI). In contrast, urinary and plasma chromium concentrations above background levels were observed at considerably lower exposure levels.

It is for these reasons that the determination of chromium in urine is by far most widely used method for the monitoring of internal chromium exposure. It offers the additional advantage of being a more convenient, non-invasive sampling method than the collection of blood.

Variations in renal function have an influence on urinary chromium levels. Such variations are easily corrected for by relating urinary chromium concentrations to the amount of creatinine present in the urine (Araki and Aono, 1989). Other factors which have to be taken into consideration during the design of monitoring programmes include the effect of smoking on urinary chromium. Kalliomäki et al. (1981) and Strindsklev et al. (1993) have found that urinary chromium levels in stainless steel welders who smoke can be two to three times higher than those of their non-smoking colleagues.

When monitoring weakly exposed people it even becomes important to correct for differences in drinking habits and exercise. Bukowski and his colleagues (1991) were able to demonstrate that drinking beer, lack of exercise and diabetes all lead to elevated levels of chromium in urine. Figure 2 compares the levels of urinary chromium measured in a number of occupationally exposed persons with those in non-exposed.

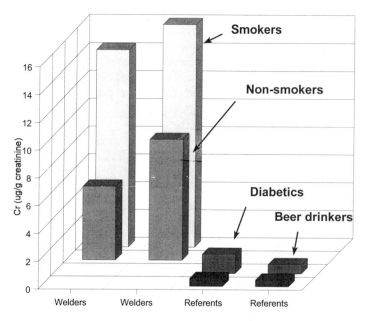

Figure 2. Urinary chromium in occupationally exposed human subjects and in controls. Data from Bukowski *et al.* (1991), Kalliomäki *et al.* (1981) and Strindsklev *et al.* (1993).

4. CHROMIUM(VI) EFFECT MONITORING: CYTOGENETIC SURVEILLANCE STUDIES

The first studies looking for chromosome damage in the blood lymphocytes of chromium(VI) exposed workers appeared in the 1970s. Bigaliev *et al.* (1977) have analysed the number of chromosome aberrations in blood lymphocytes of 132 chromium(VI) production workers and found significant increases (3.6–9% aberrant metaphases) relative to 37 unexposed controls (1.88% ± 0.74%). In a surveillance study among chromium platers Stella *et al.* (1982) observed increased SCE frequencies with levels of 8.08 ± 2.67 per cell versus 6.31 ± 1.56 in controls. The focus of these early studies was on cytogenetics and consequently no attempts were made to relate chromosome damage to internal chromium exposure or even concentrations of the metal compound in the workplace air. The majority of studies published in the 1980s and 1990s provide detailed information about measures of external and internal exposure to chromium, thus enabling to us to evaluate the relative sensitivity of markers of early biological effects and of internal exposure in a variety of occupational settings.

The results of such a meta-analysis (Kortenkamp 1997) are surprising. Although exposure to chromium was readily detected as elevated urinary chromium levels in studies among chromium platers and ferrochromium workers (Sarto *et al.* 1982, Nagaya *et al.* 1986, Choi *et al.* 1987, Sbrana *et al.* 1990, Nagaya *et al.* 1991) there was a striking lack of difference in the number of chromosome aberrations or SCE's between exposed and controls. Overall, the values of SCE's observed in these studies (5–9 SCE's per cell) were well within the range of SCE's reported for control populations (8.12 ± 1.82 SCE's per cell, Ashby and Richardson 1985). Age and smoking status had a strong influence on the level of SCE's.

The results of cytogenetic studies carried out among stainless steel welders using the manual metal arc method (which unlike the other two major welding methods generates welding fumes particularly high in chromium(VI) particulates) show similar trends. As with the studies among platers and ferrochromium workers, urinary chromium proved to be a reliable marker of internal chromium exposure even at levels of airborne chromium(VI) below 10 $\mu g/m^3$. In contrast, the frequency of SCE's appeared to be surprisingly insensitive to exposure to chromium(VI). All but one of the studies (Koshi *et al.* 1984) failed to detect elevated numbers of SCE's in the exposed groups (Littorin *et al.* 1983, Husgafvel *et al.* 1982, Popp *et al.* 1991, Knudsen *et al.* 1992, Jelmert *et al.* 1994). In the studies published after 1990 there is even a tendency for SCE's to be slightly lower in manual metal arc welders than in unexposed controls.

A different picture becomes apparent when chromosome aberrations are considered. Koshi *et al.* (1984), Knudsen *et al.* (1992) and Jelmert *et al.* (1994) were able to observe a statistically significant increase of rare chromosome aberrations such as dicentric chromosomes, translocations, minutes and rings. Importantly, these effects were only seen among the non-smoking manual metal arc welders. Knudsen *et al.* (1992) have emphasised that such aberrations are severe genetic effects which require two independent damaging events to occur.

Overall, the results of these investigations clearly force the conclusion that there is evidence for genotoxic effects among stainless steel welders using the manual metal arc method. It is however important to interpret these data in the context of other cytogenetic surveillance studies. The frequencies of aberrations detected among welders are similar to those observed in unexposed controls of many studies, which is about 1.42 ± 0.96% of aberrant cells (excluding gaps) (Ashby and Richardson 1985). In view of the strong genotoxicity of chromium(VI) compounds (IARC 1990) it is surprising that the effect markers in cytogenetic surveillance studies responded so weakly. This could be due to low levels of exposure with genotoxic effects too small to be detectable among the study populations which were not larger than up to 60 exposed individuals. Alternatively, are blood lymphocytes inappropriate as a surrogate tissue for the monitoring of genetic effects of chromium(VI) in humans? Before we discuss this point further, let us review the outcome of studies undertaken to monitor DNA damage in the blood lymphocytes of chromium(VI) exposed workers.

5. BIOLOGICALLY EFFECTIVE DOSES OF CHROMIUM: DNA DAMAGE IN LYMPHOCYTES

Although the few studies carried out so far to screen DNA damage in the blood lymphocytes of chromium(VI) exposed workers have produced negative results they are instructive because they provide clues as to why chromosome damage might be relatively low in chromium(VI) exposed people.

Gao *et al.* (1994) have assessed oxidative DNA damage in chromium(VI) production workers by monitoring the number of single strand breaks and modified DNA bases (8-hydroxydeoxyguanosine) in lymphocyte DNA. There was no evidence for increased levels of oxidative DNA damage, although the workers' exposure to chromium could be verified using markers of internal exposure such as urinary chromium. Zhitkovich *et al.* (1996) have employed a potassium-SDS precipitation assay in order to analyse DNA-protein cross-links induced by chromium(VI). They analysed samples of lymphocyte DNA from Bulgarian chromium platers but failed to observe significant differences in the levels of

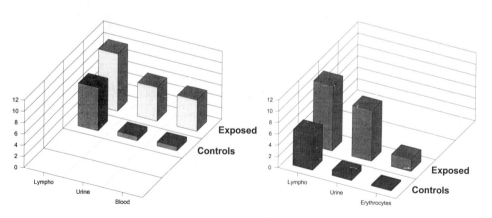

Figure 3. Chromium levels in lymphocytes (lympho, µg/10[11] cells), urine (µg/g creatinine) and whole blood (blood, µg/l) or erythrocytes (µg/100 ml). Data from Gao *et al.* (1994) and Zhitkovich *et al.* (1996).

cross-links between exposed and controls not engaged in plating. Again, elevated concentrations of chromium in the red blood cells and urine of the platers confirmed that exposure to chromium(VI) had occurred.

Gao *et al.* (1994) and Zhitkovitch *et al.* (1996) have carried out determinations of the chromium levels in blood lymphocytes simultaneously with the measurement of DNA damage. The striking observation in both cases was that the differences in lymphocyte chromium levels between exposed and controls were not very pronounced, when at the same time all the other measures of internal exposure were markedly elevated in relation to controls (Figure 3).

The absence of any pronounced differences in the lymphocyte chromium levels of exposed and controls, whatever the reason, could well be the factor which explains the relatively small numbers of SCE's and chromosome aberrations in exposed platers and welders. Thus, lymphocytes may not be an appropriate surrogate tissue for the effect monitoring of chromium exposed individuals.

6. TOXICOKINETICS OF CHROMIUM(VI) FOLLOWING EXPOSURE BY INHALATION

The question whether it makes sense to use blood lymphocytes for chromium(VI) effect monitoring can be addressed in another way by considering where in the body lymphocytes may have the chance to come into contact with chromium(VI). Therefore, how is chromium(VI), once inhaled, distributed in the body?

Postmortem analyses of lung tissue obtained from chromium workers who died of lung cancer have shown that chromium(VI) containing particles stay in the lung for very long periods of time (Hyodo *et al.* 1980, Tsuneta *et al.* 1980). Even years after cessation of exposure most of the chromium can still be found in the respiratory tract. Only small amounts reach liver and kidneys via the bloodstream. The data published by Kishi *et al.* (1987) can be used to illustrate this point: The amount of chromium residing in the lungs of chromium(VI) workers can be estimated as 30–70 mg, whereas the amounts found in

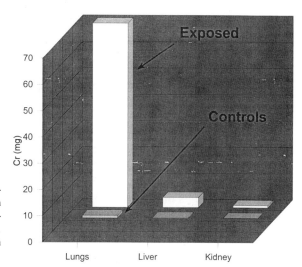

Figure 4. Chromium load in different organs of deceased chromium(VI) production workers and in non-exposed controls. Calculated using the data from Kishi *et al.* (1987) and assuming average organ weights.

the liver and kidney were only 3.8 mg and 0.8 mg, respectively. In comparison, the lungs, livers and kidneys of decedents with no occupational chromium exposure contained 0.08–1.2 mg, or 0.3 mg and 0.07 mg, respectively (Figure 4).

Given these toxicokinetics, we can envisage two ways for lymphocytes to come into contact with chromium(VI) following exposure by inhalation: (1) Lymphocytes, while travelling via the blood, take up chromium(VI) from the plasma that has leached from the lungs; (2) lymphocytes homing to the supporting tissues of the lungs are exposed to chromium before migrating back to the blood stream.

7. LYMPHOCYTE TRAFFIC

Because lymphocytes continuously enter and leave lymphoid and non-lymphoid tissues via the blood, only about 2% of the total lymphocyte pool in the human body reside in the blood at any given time, with an estimated residence time of about 30 minutes (Westermann and Pabst 1990). During one day approximately 500×10^9 lymphocytes travel through the blood, a number which is equivalent to the total lymphocyte population in the human body. However, not every single lymphocyte enters and leaves the blood during one day. There are subsets of the lymphocyte population which reside in certain tissues for long times without migrating through the blood. It is therefore unlikely that the lymphocytes in the blood are representative of the other lymphocytes distributed in the body.

Lymphocytes migrate through most of the organs of the body and have the ability to return to the blood. One established pathway of recirculation from organs to the blood is via the lymph nodes and the thoracic duct, accounting for approximately 5–10% of the lymphocyte population which returns to the blood each day. The spleen is by far the most important organ in lymphocyte recirculation, with 50% of all the recirculated lymphocytes being released by the spleen. In comparison, the daily recirculation of lymphocytes from the lungs back to the blood is negligible (Westermann and Pabst 1990).

Thus, only a very small fraction of the lymphocytes residing in the blood will have had the chance to take up chromium(VI) while travelling through the lungs. The short blood transit time of lymphocytes is likely to be another complicating factor in the uptake of chromium(VI) by lymphocytes and might help explain the small differences in the lymphocyte chromium levels of exposed and unexposed observed by Gao *et al.* (1994) and Zhitkovich *et al.* (1996).

The vast of majority of lymphocytes will have come into contact with chromium(VI) during their migration through the blood. They may take up chromium(VI) that has leached from the respiratory tract into the blood.

These features are important when considering whether chromium-induced genotoxic effects occurring in lymphocytes are in any way predictive of the processes leading to mutations and eventually cancer in the respiratory tract. Given the toxicokinetics of chromium(VI) upon inhalation, it is likely that cells of the respiratory tract are exposed to much higher amounts of chromium(VI) than lymphocytes. It would appear that the likelihood that cancer-initiating events occur in lung cells is considerably higher. Therefore, the absence of any genotoxic effects in blood lymphocytes of exposed human subjects cannot be taken as an indication of absence of lung cancer risks.

8. CHROMIUM(VI) BIOMONITORING: STRENGTHS AND WEAKNESSES

There is no doubt that urinary chromium is a useful marker of exposure to chromium in oxidation states (VI) and (III). At exposure levels likely to be encountered in occupational settings today it provides a sensitive measure of the internalised dose of chromium.

In contrast, the lack of success of cytogenetic surveillance studies using blood lymphocytes as surrogate tissue is surprising, particularly in view of the strong genotoxicity of chromium(VI). We suggest that the reason for the weak responses observed in these studies lies in the disproportionately small differences in lymphocyte chromium levels between exposed and controls. It would be of great importance to establish why the levels of chromium in lymphocytes are relatively high in unexposed controls.

The toxicokinetics of inhaled chromium(VI) and the dynamics of lymphocyte traffic probably mean that lymphocytes are exposed to lower concentrations of chromium(VI) than cells in the respiratory tract. Taken together, these considerations would suggest that blood lymphocytes are not well suited for the monitoring of the biologically effective dose, and of early biological effects arising from exposure to chromium(VI) at levels below 50 $\mu g/m^3$, the current exposure limit in many industrialised countries. At higher exposure levels however, effect monitoring using lymphocytes may well be useful.

This rather sobering conclusion provokes the question as to whether there are alternative approaches to chromium biomonitoring which are more meaningful in relation to lung carcinogenesis.

9. THE NEED TO MONITOR CHROMIUM DEPOSITED IN LUNG TISSUE IN A NON-INVASIVE WAY

The reason why lymphocytes are so popular not only in chromium effect monitoring lies of course in the fact that they are so easily obtained. The opposite is true for tissues of the respiratory tract. The screening of chromium deposited in the lungs of exposed workers can as yet only be carried out during postmortem analyses. There is thus a great need to develop and evaluate non-invasive methods which would allow it to determine chromium levels in the lungs of exposed individuals while they are alive.

We have explored the usefulness of Magnetic Resonance Imaging (MRI) in visualising chromium in lung tissue. MRI involves the absorption of electromagnetic radiation by naturally occurring nuclei such as protons within tissues. The energies normally absorbed by these nuclei are very low. MRI is therefore a gentle, non-invasive probe allowing the monitoring of live organisms. The technique relies on the interactions between a strong magnetic field and a pulse of radio frequency radiation to detect compounds which possess nuclei with magnetic moments (such as hydrogen). Such nuclei align within the externally applied strong magnetic field. Upon application of a radio frequency pulse of appropriate energy, this alignment is disturbed and a magnetic resonance signal is generated. The pulse is called an "excitation" or "90°" pulse since the nuclei are reoriented at 90° to their original position. As the pulse decays, the nuclei "relax" back to their original orientation. The frequency of the resonance signal thus emitted is uniquely determined by the nature of the nucleus under investigation, its chemical environment, and the surrounding magnetic field. If one or more magnetic field gradients are applied across the sample, the signal frequency will also be determined by the spatial positions of the nuclei. This makes it possible to generate a two dimensional image of the sample.

Chromium in oxidation state (III) lends itself particularly well to investigation by MRI. Chromium(III) complexes are paramagnetic and will therefore perturb the proton signals produced during an MRI scan by increasing their relaxation times. Regions of tissue containing chromium(III) can thus be visualised as areas of intense brightness in a MRI scan. The method is sensitive only to chromium(III); chromium(VI) is MRI silent and cannot be visualised. Thus, by using MRI it should be possible to monitor the processes which begin with the slow solubilisation of chromium(VI) containing particles in lung tissue and end with their reduction to chromium(III).

The reductive conversion of chromium(VI) to chromium(III) inside cells is known to lead to the formation of species that interact with DNA. Crucially, chromium(VI) itself is unreactive towards DNA (see the review by Kortenkamp et al. 1997). Hypervalent intermediate oxidation states of chromium, most importantly chromium(V) and chromium(IV), are believed to be able to activate molecular oxygen, thus promoting the formation of highly reactive species which can cause oxidative DNA damage such as single strand breaks and abasic sites (but, intriguingly, not hydroxylated DNA bases) (Kortenkamp et al. 1996). Cross-link type DNA lesions also arise during the reduction of chromium(VI) and there is evidence to suggest that chromium(III) species may participate in the causation of such damage (Zhitkovich et al. 1997).

The reduction of chromium(VI), when occurring outside cells, is likely to prevent the entry of chromium as chromate anions into cells, because the chromium(III) complexes thus produced are unable to cross cell membranes (Kortenkamp et al. 1987). In tissues of the respiratory tract, the extracellular reduction of chromium(VI) is probably to be mediated by glutathione and ascorbate anions. It is conceivable that the slow solubilisation of chromium(VI) particules can lead to the localised exhaustion of extracellular glutathione and ascorbate stores, ultimately facilitating the entry of chromium(VI) into cells of the respiratory tract. Although MRI does not allow to differentiate between the extracellular and intracellular reduction of chromium(VI), this does not *a priori* invalidate its potential usefulness in biomonitoring because it gives a measure of the total amount of solubilised and reduced, i.e. "bioactivated" chromium(VI).

Figure 5 shows MRI scans of a freshly killed male Wistar rat which received a single dose of 100 μg sodium chromate after termination. The agent was delivered to the

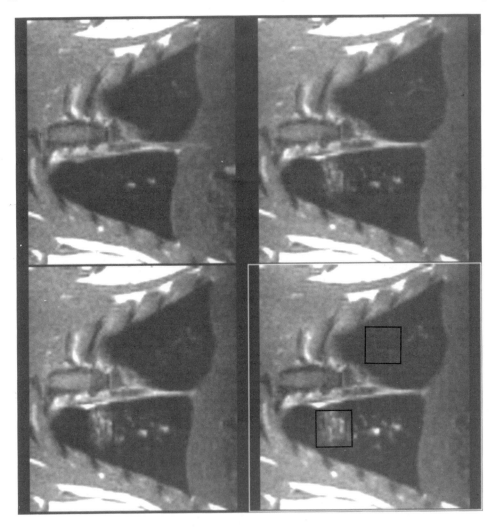

Figure 5. Magnetic Resonance Imaging scans of a freshly killed Wistar rat (top left). After termination, 100 μg sodium chromate were administered to the right lung and the animal scanned after 10 min (top right), 20 min (bottom left) and 30 min (bottom right).

right lung of the animal by means of a catheter, and the animal scanned at regular time intervals (0–30 min). The left lung was left untreated to provide an internal control. Already 10 min after dosing, areas of increased brightness became discernable in the right lung. Their intensity increased further 20 and 30 min after delivery of chromium(VI). These images demonstrate that it is possible to monitor the reduction of chromium(VI) to chromium(III) by using MRI even under post-mortem conditions. Further studies aimed at establishing the sensitivity of the technique are underway.

The results we have obtained encourage us to pursue this approach further, ultimately to be applied to human subjects. Because the availability of whole body scanners is limited, MRI may not necessarily lend itself to routine monitoring. We therefore suggest studies to explore the relationship between urinary chromium levels and the outcome of MRI images, with a view to establishing whether urinary chromium can be used as a predictor of chromium deposited in the lungs. Such work will also have to rely on postmortem analyses of the chromium content of lung tissue in order to obtain a reference point for the "calibration" of MRI. All this could perhaps be achieved if there were records of the chromium levels in the urine and of MRI scans of decedents whose lungs subsequently are analysed postmortem for chromium.

We propose that such studies initially be carried out with human subjects who are exposed solely to chromium(VI), without confounding by other metal compounds. Platers would be an ideal cohort for studies of the toxicokinetics of inhaled chromium(VI). The results obtained with platers may be ultimately applicable to welders and chromium(VI) production workers, but the deposition of ferromagnetic metals such as iron and nickel in the lungs of welders will make the visualisation of chromium(III) by MRI impossible.

10. EFFECT MONITORING IN TISSUES OF THE UPPER RESPIRATORY TRACT

Perhaps the biggest challenge in chromium biomonitoring is to bridge the gap between the monitoring of internalised doses and effect monitoring. Given the problems associated with utilising lymphocytes as a surrogate tissue, effect monitoring ideally should concentrate on cells of the airways. However, this is complicated by the difficulty of access to these tissues. A further problem lies in the small number of cells which can be obtained by biopsy or lavage.

Clearly, what is needed is a technique which allows the determination of DNA damage in single cells. The comet assay, or single cell gel electrophoresis assay, is such a technique. It was first developed by Östling and Johanson in 1984 and later modified and refined by Singh and coworkers and by Olive and her colleagues (see the review by Fairbairn *et al.* 1995). A small number of cells are embedded in agarose on a microscope slide, lysed, electrophoresed and then stained with a fluorescent DNA binding dye. In the electrical field, DNA moves out of the nucleus, such that relaxed or broken regions of the genome migrate further. The objects thus created look strikingly like comets, hence the name of the assay.

Östling and Johanson worked with irradiated cells recovered from biopsy samples of patients receiving radiation therapy. They observed that the extent to which DNA could be liberated from the nucleus during electrophoresis depended on radiation dose. The more damage there was in DNA, the more DNA resided in the "tail" of the comet.

Virtually any eukaryotic cell can be examined using the comet assay and there are well established methods for generating single cell suspensions from biopsy samples. The most commonly used human cells are blood lymphocytes, but cells derived from other tissues including gastric and nasal mucosa have also been used (Pool-Zobel *et al.* 1994). The comets are evaluated by using image analysis. A parameter often referred to as "tail moment" is most frequently determined. "Tail moments" are defined as the product of the length of the comet and the fraction of DNA residing in the tail.

The comet assay is sensitive to DNA damage such as single and double strand breaks and, when alkaline lysis is applied, abasic sites. Strand breaks which are produced by the cell as it initiates DNA repair are also detected. Cross-link type lesions, on the other hand, can only be analysed indirectly. Under standardised conditions, X-rays introduce strand breaks to the DNA of cells, which subsequently will appear as comets with high tail moments. If however cross-links are present the formation of comets will be reduced after pretreatment with radiation. In this way, cross-links formed by nitrogen mustard or *cis*-platinum were readily detected using the comet assay (Fairbairn *et al.* 1995).

We propose to use the comet assay for determinations of DNA damage in cells of the nasal epithelia obtained by using nasal lavage techniques. Before beginning such studies it will be necessary to establish *in vitro* whether the strand breaks and abasic sites induced by chromium(VI) are masked by concurrently formed cross-link type lesions. Whether oxidative DNA damage or cross-link type lesions will dominate in cells exposed to chromium(VI) is likely to depend on factors such as the intracellular concentrations of glutathione and ascorbate anions, with high glutathione levels favouring cross-links. Should cross-links be the major lesions, the protocols developed for cross-linking agents will have to be applied to nasal epithelial cells. However, the work by Pool-Zobel and her colleagues (1994) shows that chromium(VI) induces comets directly, indicating that the modified cross-link protocol is not necessary for detection of chromium(VI)-induced DNA damage. Again, it seems crucial to us that human subjects with single compound exposure, i.e. platers, should be selected for pilot studies.

11. CONCLUSION

The use of blood lymphocytes as surrogate tissue in chromium(VI) biomonitoring has yielded inconclusive results and is flawed in view of the toxicokinetics of inhaled chromium(VI). Urinary chromium, in contrast, is a useful marker of internalised doses of chromium.

The recent advantages in imaging techniques and our own results from MRI studies using an animal model lead us to suggest to extend the monitoring of internalised doses of chromium to the lung. Rather than using MRI as a routine screen, the proposed programme of work should aim to explore the predictive value of urinary chromium in terms of amounts of chromium residing in lung tissue.

We further suggest to explore the usefulness of the comet assay in the monitoring of early biological effects of chromium(VI) in nasal tissue.

ACKNOWLEDGMENTS

We gratefully acknowledge support from Ford Motor Company (U.K.) for work on MRI. Thanks are due to Rob Shayer for providing MRI scans and to Mark Raffray and Montse Casadevall for stimulating discussions.

REFERENCES

Ashby, J. and Richardson, C.R. (1985). Tabulation and assessment of 113 human surveillance cytogenetic studies conducted between 1965 and 1984. Mutat. Res. **154**, 111–133.

Araki, S. and Aono, H. (1989). Effects of water restriction and water loading on daily urinary excretion of heavy metals and organic substances in metal workers. Br. J. Ind. Med. **46**, 389–392.

Angerer, J., Amin, W., Heinrich-Ramm, R., Szadowski, D. and Lehnert, G. (1987). Occupational chronic exposure to metals, I. Chromium exposure of stainless steel welders - biological monitoring. Int. Arch. Occup. Environ. Health **59**, 503–512.

Bigaliev, A.B., Elemesova, M.S. and Turebaev, M.N. (1977). Evaluation of the mutagenic activity of chromium compounds. Gig. Tr. Prof. Zabol. **6**, 37–40.

Bukowski, J.A., Goldstein, M.D., Korn, L.R. and Johnson, B.B. (1991). Biological markers in chromium exposure assessment: confounding variables. Arch. Environ. Health 46, 230–236.

Buttner, B. and Beyersmann, D. (1985). Modification of the erythrocyte anion carrier by chromate. Xenobiotica 15, 735–741.

Choi, Y.J., Kim, Y.W. and Cha, C. W. (1987). A study on sister chromatid exchanges in lymphocytes in some metal plating workers. Korea Univ. Med. J. 24, 249–257.

Fairbairn, D.W., Olive, P.L., O'Neill, K.L. (1995). The comet assay: a compre'_nsive review. Mutat. Res. 339, 37–59.

Gao, M., Levy, L.S., Faux, S.P., Aw, C.T., Braithwaite, R.A. and Brown, S.S. (1994). The use of molecular epidemiological techniques in a pilot study on workers exposed to chromium. Occup. Environ. Medicine 51, 663–668.

Hyodo, K., Suzuki, S., Furuya, N. and Meshizuka, K. (1980). An analysis of chromium, copper and zinc in organs of a chromate worker. Int. Arch. Occup. Environ. Health 46, 141–150.

Husgafvel-Purisainen, K., Kalliomäki, P.L. and Sorsa, M. (1982). A chromosome study among stainless steel welders. J. Occup. Med. 24, 762–766.

IARC (1990). Monographs on the Evaluation of the Carcinogenic Risk of Chemicals to Humans, vol 49, Chromium, Nickel and Welding, International Agency on the Research of Cancer, Lyon.

Jelmert, Ø., Hansteen, I.L. and Langård, S. (1994). Chromosome damage in lymphocytes of stainless steel welders related to past and current exposure to manual metal arc welding fumes, Mutat. Res. 320, 223–233.

Kalliomäki, P.L., Rahkonen, E., Vaaranen, V., Kalliomäki, K. and Aittoniemi, K. (1981). Lung-retained contaminants, urinary chromium and nickel among stainless steel welders. Int. Arch. Occup. Environ. Health 49, 67–75.

Kishi, R., Tarumi, T., Uchino, E. and Miyake, H. (1987). Chromium content of organs of chromate workers with lung cancer. Am. J. Ind. Med. 11, 67–74.

Knudsen, L.E., Boisen, T., Christensen, J.M., Jelnes, J.E., Jensen, G.E., Jensen, J.C., Lundgren, K., Lundsteen, C., Pedersen, B., Wassermann, K., Wilhardt, P., Wulf, H.C. and Zebitz, U. (1992). Biomonitoring of genotoxic exposure among stainless steel welders. Mutat. Res. 279, 129–143.

Kortenkamp, A., Beyersmann, D. and O'Brien, P. (1987) Uptake of chromium(III) complexes by erythrocytes, Toxicol. Environm. Chem. 14, 23–32.

Kortenkamp, A., Casadevall, M., Faux, S., Jenner, A., Shayer, R.O.J., Woodbridge, N. and O'Brien, P. (1996). A role mfor molecular oxygen in the formation of DNA damage during the reduction of the carcinogen chromium(VI) by glutathione. Arch. Biochem. Biophys. 329, 199–207.

Kortenkamp, A., Casadevall, M., da Cruz Fresco, P. and Shayer, R.O.J. (1997). Pathways in the chromium(VI)-mediated formation of DNA lesions: A review. In: Hadjiliadis, N.D. (ed.) "Cytotoxic, mutagenic and carcinogenic potential of heavy metals related to human environment". NATO ASI series, 2. Environment, Vol. 26. Kluwer Academic Publishers, Dordrecht, The Netherlands, pp 15–34.

Kortenkamp, A. (1997). Problems in the biological monitoring of chromium(VI) exposed individuals. Biomarkers 2, 73–80.

Koshi. K., Yagami, T. and Nakanishi, Y. (1984). Cytogenetic analysis of peripheral blood lymphocytes from stainless steel welders. Ind. Health 22, 305–318.

Littorin, M., Hoegstedt, B., Stroembaeck, B., Karlsson, A., Welinder, H., Mitelman, F. and Skerfving, S. (1983). No cytogenetic effects in lymphocytes of stainless steel welders. Scand. J. Work Environ. Health 9, 259–264.

Nagaya, T. (1986). No increase in sister-chromatid exchange frequency in lymphocytes of chromium platers. Mutat. Res 170, 129–132.

Nagaya, T., Ishikawa, N., Hata, H. and Otobe, T. (1991). Sister-chromatid exchanges in lymphocytes from 12 chromium platers: a 5-year follw-up study. Toxicology Letters 58, 329–335.

Paustenbach, D.J., Panko, J.M., Fredrick, M.M., Finley, B.L. and Proctor, D.M. (1997) Urinary chromium as a biological marker of environmental exposure: What are the limitations? Reg. Toxicol. Pharmacol. 26, S23-S34.

Pool-Zobel, B.L., Lotzmann, N., Knoll, M., Kuchenmeister, F., Lambertz, R., Leucht, U., Schroder, H.G. and Schmezer, P. (1994). Detection of genotoxic effects in human gastric and nasal mucosa cells isolated from biopsy samples. Env. Mol. Mutagenesis 24, 23–45.

Popp, W., Vahrenholz, C., Schmieding, W., Krewet, E. and Norpoth, K. (1991). Investigations of the frequency of DNA strand breakage and cross-linking and of sister chromatid exchange in the lymphocytes of electric welders exposed to chromium- and nickel-containing fumes. Int. Arch. Occup. Environ. Health 63, 115–120.

Sarto, F., Cominato, I., Bianchi, V. and Levis, A.G. (1982). Increased incidence of chromosomal aberrations and sister chromatid exchanges in workers exposed to chromic acid in electroplating factories. Carcinogenesis 3, 1011–1016.

Sbrana, I., Caretto, S., Lascialfari, D., Rossi, G., Marchi, M. and Loprieno, N. (1990). Chromosomal monitoring of chromium-exposed workers. Mutat. Res. 242, 305–312.

Stella, M., Montaldi, A., Rossi, R., Rossi, G. and Levis, A.G. (1982). Clastogenic effects of chromium on human lymphocytes in vivo and in vitro. Mutat. Res. 101, 151–164.

Strindsklev, I.C., Hemmingsen, B., Karlsen, J.T., Schaller, K.H., Raithel, H.J. and Langard, S. (1993). Biologic monitoring of chromium and nickel among stainless steel welders using the manual metal arc method. Int. Arch. Occup. Environ. Health 65, 209–219.

Westermann, J. and Pabst, R. (1990). Lymphocyte subsets in the blood: a diagnostic window on the lymphoid system? Immunology Today 11, 406–410.

Wogan, G. (1992). Molecular epidemiology in cancer risk assessment and prevention: Recent progress and avenues for future research. Environ. Health Perspect. 98, 167–178.

Zhitkovich, A., Lukanova, A., Popov, T., Taioli, E., Cohen, H., Costa, M. and Toniolo, P. (1996). DNA-protein crosslinks in peripheral lymphocytes of individuals exposed to hexavalent chromium compounds. Biomarkers 1, 86–93.

Zhitkovich, A. Voitkun, V. and Costa, M. (1996). Formation of the amino acid-DNA complexes by hexavalent and trivalent chromium in vitro: importance of trivalent chromium and the phosphate group. Biochemistry 35, 7275–7282.

6

CHELATE SENSOR METHOD FOR MERCURY

Horacio Kido,[1] Ferenc Szurdoki,[1,2] Mae S. Gustin,[3] and Bruce D. Hammock[1*]

[1]Departments of Entomology and Environmental Toxicology
University of California, Davis
Davis, California 95616
[2]Department of Chemistry, Tufts University
62 Talbot Avenue, Medford, Massachusetts 02155
[3]University Center for Environmental Sciences and Engineering/199
130 Fleischmann Agriculture Building
University of Nevada, Reno, Nevada 89557

1. ABSTRACT

Our chelate sensor method for detecting mercury is based on the high affinity of thiol-containing ligands for the mercuric ion. The method combines the simple ELISA-format with the selective complexation of mercuric ions by dithiocarbamate chelators. The first assay uses a sandwich chelate formed by a ligand immobilized on the well of an ELISA plate, mercuric ion of the analyzed sample, and another ligand bound to the reporter enzyme. Our second assay utilizes competition between the binding of mercuric ions and an organomercury-conjugate to a chelating conjugate. Specifically, it involves a chelator doped on the solid surface and a mercury-containing enzyme-tracer. Both assays were further characterized by testing them under several conditions that might be encountered during some practical applications of the assay. Finally, they were tested with river water and human saliva to yield a good correlation with spike levels.

2. INTRODUCTION

2.1. Toxicological Significance of Mercury

Mercury is one of the most dangerous toxic metals. Before their hazards were better understood, mercury compounds were commonly used in medicine as bactericidal and

---

*To whom correspondence should be addressed.

Metals and Genetics, edited by Sarkar.
Kluwer Academic / Plenum Publishers, New York, 1999.

diuretic agents; they were also widely applied as agricultural fungicides. The toxicity of mercury derivatives gained worldwide attention due to the widespread mercury poisoning of people at the Minamata Bay of Japan from 1953 to 1960. In this case, mercury in industrial waste discharged into the bay bioaccumulated in seafood consumed by humans. This event resulted in a reduction of the use of mercury compounds as agricultural pesticides and in tighter regulation of discharges of mercury to the environment in many countries.

Most of the effects of mercury upon biological systems can be explained in terms of the high affinity of mercurials towards thiols (Zalups and Lash, 1994). The most important target organs of inorganic and organic mercury are the central nervous system and the kidneys. Alkylmercury compounds (e.g., methylmercury and dimethylmercury) are considered to be the most harmful mercury derivatives. Once in the body, mercury accumulates in the liver, brain, and red blood cells. Even moderate poisoning with alkylmercury derivatives can produce typical neurotoxic symptoms (Klaassen et al., 1986). Prenatal exposure to low doses eventually results in mental retardation in children (Suzuki et al., 1991). Severe cases of mercury poisoning may cause paralysis, irreversible or fatal neurological disorders.

The adverse health effects of elementary mercury vapors have also been well known. Hg^0 is rapidly oxidized to Hg^{2+} in erythrocytes or in tissues (Zalups and Lash, 1994, Falnoga et al., 1994); thus, the physiological effects of these two forms of mercury are similar. The initial toxic effects of Hg^0 are observed in the nervous system and later nephrotoxicity develops (Zalups and Lash, 1994). The toxic impact of the Hg^0 vapor released from dental amalgam fillings has recently been assessed (Skerfving et al., 1991; Falnoga et al., 1994; Hultman et al., 1994). A study conducted by Guo et al. (1996) discovered that the mercury content of saliva samples from persons with amalgam fillings ranged from 1 to 300 ppm (5 to 1500 nM).

Mercury pollution is a global problem with serious implications for human health and the environment. A significant fraction of anthropogenic mercury pollution is emitted to the atmosphere. This is generated mostly by fossil fuel combustion, chlorine production, waste incineration, and non-ferrous metal industry and mining. Mercury travels then from the air to lakes and oceans by rain. Methylmercury, an extremely toxic form of mercury, is produced from inorganic mercury contamination by bacteria in aquatic sediments. This mercury derivative accumulates in fish and then enters the human food chain.

2.2. Mercury Analysis

Most common methods of trace analysis and chemical speciation of mercury in environmental and biological samples, e.g., cold vapor atomic absorption spectrometry (Lind et al., 1993), atomic fluorescence spectroscopy (Janjic and Kiurski, 1994), and ICP-MS, require expensive instrumentation and a highly qualified analyst. These techniques are adaptable neither as low-cost, field-portable assays for analyzing large number of samples nor as sensors for continuous monitoring. Wylie et al. (1992) has reported selective ELISAs for mercuric ion (Hg^{2+}) with detection limit of 0.5 ppb. However, this approach is based on monoclonal antibodies, which are expensive to develop.

2.3. Goal

Our goal was to develop simple analytical methods based on the formation of chelates of the metal ion to be detected. We combined the inexpensive and convenient ELISA methodology with the selective and high affinity recognition of mercuric ions by sulfur containing chelators (Szurdoki et al., 1995; Hammock et al., 1995). In other words, we employed sulfur-containing ligands instead of antibodies for the recognition of mercury. Dithiocarbamates were the ligands of choice because their chelates with Hg^{2+} have extremely high thermodynamic stability (Bond and Scholz, 1991). Dithiocarbamates seem to have similar or higher affinity only to several noble metals. In our studies, dithiocarbamates derived from secondary amines were used because these compounds are known to be fairly chemically stable under nonacid conditions in the absence of oxidizing agents (Valentine et al., 1992).

3. MATERIALS AND METHODS

3.1. Reagents and Equipment

The synthesis of the reagents used in the two formats (Figures 1 and 2) has been described by Szurdoki et. al (1995). Microtiter plates were read with a Spectramax 250 microplate reader (Molecular Devices, Menlo Park, CA). Inhibition curves used for data analyses were composed of a 12 point standard curve in quadruplicate and the software package Softmax Pro (Molecular Devices) was used for fitting a sigmoidal curve based on the four-parameter logistic method of Rodbard (1981). Specific features of the mercury assays where indicated below. Aqueous solutions were prepared with nanopure water (16.7 megaohm/cm) obtained from Sybron/Barnstead Nanopure II system (Barnstead Co., Newton, MA). Trace amounts of heavy metals were removed from buffers used in the assay work by Chelex 100 ion exchange resin (100–200 mesh, sodium form, Bio-Rad Laboratories, Richmond, CA) according to the manufacturer's instructions. Glassware and non-disposable plasticware (e.g., containers for buffers, troughs) were kept in 20% nitric acid at room temperature overnight, followed by thorough rinsing with nanopure water (Lansens, 1990; Lau and Ho, 1993). Microtiter plates (96-well Nunc-Immuno Plate, Nunc InterMed, Denmark) and metal-free micro-pipet tips were used without soaking in acid. Between incubation steps, plates were washed four times with sodium/acetate buffer (0.1 M acetate, 0.05% Tween 20, pH 5.5).

3.2. Procedure

The assays were developed by following the principles of solid phase immunoassay (Voller et al., 1976), competitive ELISAs. Prior to inhibition experiments, coating chelator and enzyme tracer dilutions were optimized by using the two dimensional titration method as outlined by Gee et al. (1988).

3.3. Assay Format 1

Assay format 1 was converted to a two-incubation assay from the one-incubation assay described by Szurdoki et al. (1995) in order to avert potential adverse effects upon the activity of the alkaline phosphatase tracer from the sample matrix. Conjugate 1-CONA, dissolved in the coating buffer (sodium carbonate/bicarbonate buffer, 50 mM carbonate/bicarbonate, pH 9.6), was immobilized on the wells of a microplate. The plate was washed. In the first incubation, sample/standard was added to the plate, incubated in 300 mM sodium acetate buffer, 0.05% Tween 20, pH 7 for 60 mins at room temperature. The plate was washed once again. In the second step, tracer 2-AP was added to the plate, incubated in 300 mM sodium acetate buffer, 0.05% Tween 20, pH 7.0 for 60 mins at room temperature. After another wash, the wells were treated with the solution of the enzyme substrate (*p*-nitrophenyl phosphate) and absorbance values were measured.

3.4. Assay Format 2

Just like with assay format 1, assay format 2 was modified to minimize potential matrix effects from environmental samples. The change was the conversion of the assay to a 2-incubation assay from a 1-incubation assay. The wells of a microtiter plate, coated with conjugate 1-CONA as above, were washed and then incubated with the mercury standard or sample in sodium acetate buffer (0.25 M acetate plus 0.05% Tween 20, pH 6). After a washing step, the tracer 3-AP in buffer was added to the plate. A final wash followed and the color reaction was performed and the plate was read as above.

3.5. Analysis of Human Saliva

In the work for the characterization of matrix effects from human saliva upon both assay formats, positive displacement pipettes were used to accurately measure the volume of the viscous saliva samples. These were fortified with different concentrations of mercuric acetate before being diluted with 300 mM sodium acetate buffer (0.05% Tween 20, pH 7.0) or digestion additive. Hydrogen peroxide, nitric acid, trichloroacetic acid, and potassium permanganate were tested as digestion additives to test their effectiveness in reducing matrix interferences from human saliva. After allowing digestions to proceed for 1 hour, the oxidative capacity of hydrogen peroxide and potassium permanganate was eliminated by the addition of sodium bisulfite to the respective digestion mixtures. For the nitric acid and trichloroacetic acid digestion mixtures, the pH was neutralized to 7.0 by the addition of sodium hydroxide. In all cases, the final concentration of saliva in each well was either 10% or 20% v/v.

3.6. Analysis of Water from the Carson River

To evaluate assay format 2 for the analysis of river water, samples were collected from eight sites in the Carson River Drainage Basin of Nevada, which contains pristine and mercury-contaminated waters. The source of the contamination in the drainage basin is a Superfund site consisting of 130 year old mercury-contaminated mill tailings that have been eroded and dispersed along approximately 100 km of the Carson River (Gustin et al., 1994; Miller et al. 1996).

Samples were collected using ultra-clean handling protocols that included wearing polyethylene gloves and utilizing the "clean-hands, dirty hands" technique (Gill and Fitzgerald, 1987; Gill and Bruland, 1990). Samples were collected with Teflon bottles subjected to a rigorous 7-day cleaning cycle that included a 48 hour 50% nitric acid hot bath and two 24 hour cycles of refluxing with optima hydrochloric acid (Keeler et al., 1995). Immediately after sample collection, samples were refrigerated and acidified using 1% v/v optima hydrochloric acid. They were stored in the dark for 3-weeks prior to analysis. This period of time has been shown to be critical for consistent analysis of total mercury (Gill and Bruland, 1990). Half of the samples were analyzed with the rapid assay format 2 (without EDTA) while the other half were analyzed by using cold vapor atomic fluorescence spectrometry (CVAFS).

3.6.1. CVAFS Analysis. Samples were analyzed for total mercury using bromine-monochloride oxidation followed by stannous chloride reduction of ionic mercury to Hg^0 (Bloom and Crecelius, 1983). Mercury was purged from solution using mercury scrubbed ultra-high purity nitrogen. Mercury was collected on gold-coated quartz sand traps analyzed by dual amalgamation and cold vapor atomic fluorescence spectrophotometry (Dumarey et al., 1985; Bloom and Fitzgerald, 1988).

3.6.2. Analysis with Assay Format 2. Because the samples were acidified with hydrochloric acid at the time of collection, they were diluted 1:1 with 0.6 mM sodium acetate, pH 12.6, before analysis with the rapid assay. This dilution increased the pH of the solutions to a range between 6 and 7. Nanopure water that had been scrubbed with Chelex 100 resin was acidified with metal-free hydrochloric acid and diluted 1:1 with 0.6 mM sodium acetate buffer, pH 12.6. This solution was then used to make standard dilutions of mercuric nitrate.

4. RESULTS AND DISCUSSION

4.1. Assay Format 1

This non-competitive assay format works via the formation of a sandwich chelate complex between immobilized chelator protein 1-CONA, a mercuric ion from the sample, and chelator enzyme 2-AP (Figure 1). Even though the IC_{50} of assay format 1 increased from about 3 nM to approximately 30 nM after conversion from a 1-step to a 2-step assay (Figure 2), the intra-assay reproducibility of the results from the modified assay is much better. The linear response of the assay ranges from 10 to 100 nM and the limit of detection is approximately 5 nM (1 ppb). As is the case with antibody-based sandwich immunoassay, at higher concentrations of mercury, the color signal decreases to yield a bell-shaped curve.

Experiments were undertaken to characterize any anion concentration effects from I^-, Cl^-, CO_3^{2-}, PO_4^{3-}, SO_4^{2-}, and Br^-, respectively. These anions were added to the incubation buffer during the first incubation step (with mercury) to get the final concentrations of 10, 1, and 0.1 mM. In each experiment, control solutions that contained only incubation buffer were also analyzed. No effects from Cl^-, CO_3^{2-}, PO_4^{3-}, and SO_4^{2-} were observed. However, as can be seen from Figures 3 and 4, concentration effects from I^- and Br^- are evident. It appears as though I^- complexes with mercury more strongly than with Br^-. Figure 4 sug-

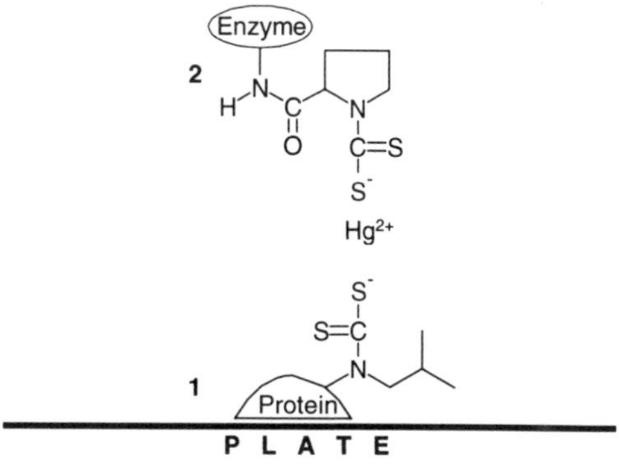

Figure 1. Sandwich assay format 1. This assay offers the advantage of being very rapid and having the concentration of Hg^{2+} directly related to color development up to a certain point. At very high concentrations of Hg^{2+}, color level declines (not shown). Protein is conalbumin (CONA) and enzyme is alkaline phosphatase (AP).

H. Kido *et al.*

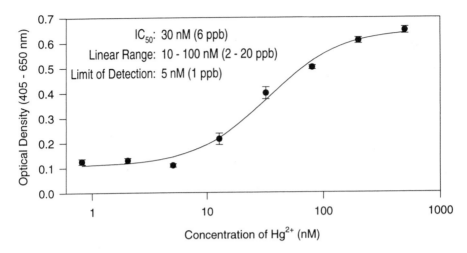

Figure 2. Standard curve for assay format 1.

gests using assay buffer fortified with 10 mM KBr because this treatment may extend the dynamic range of the assay to higher concentrations without considerably diminishing assay performance. The other anions did not interfere with the assay in the concentrations tested.

In order to reduce the cross-reactivity of assay format 1 with other metal ions (Figure 5), ethylenediaminetetraacetic acid (EDTA), imidazole, and 1-(2-pyridylazo)-2-naphthol masking agents were tested. They were added to the incubation buffer during assay incubation with the different metals. The presence of 10 mM EDTA in the metal incubation buffer eliminated cross-reactivity only with Cu^{2+} (Figure 6). Imidazole used at a concentration of 10 mM greatly reduced the interferences by Pd^{2+}, Au^{3+}, and Cu^{2+}, but not by Ag^+ (Figure 7). However, this additive resulted in an increase in the background signal.

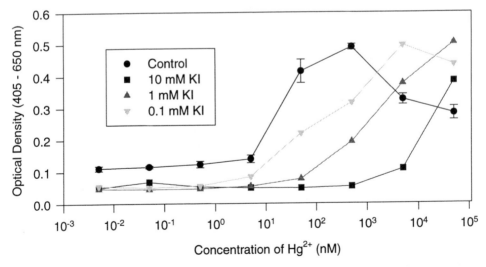

Figure 3. Effect of KI upon assay format 1. Iodide appears to form a stable complex with Hg^{2+} and inhibits the formation of the immobilized sandwich chelate complex with tracer enzyme.

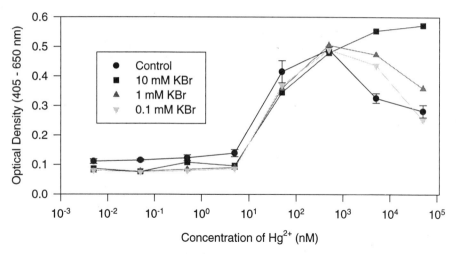

Figure 4. Effect of KBr upon assay format 1. Potassium bromide appears to extend the linear range of the assay. It might be useful to maintain a constant level of bromide at a concentration no less than 10 mM during incubation with standards and samples.

The use of 0.1 mM 1-(2-pyridylazo)-2-naphthol reduced but did not eliminate cross-reactivities with several foreign metal ions (Figure 8). This chelating agent was not used in higher concentrations because of its low solubility.

4.2. Assay Format 2

This assay format works via the competitive inhibition of binding of mercury-linked tracer 3-AP to chelating conjugate 1-CONA by free mercuric ions (Figure 9). After the conversion of assay format 2 to a 2-step assay as previously described, the IC_{50} remained

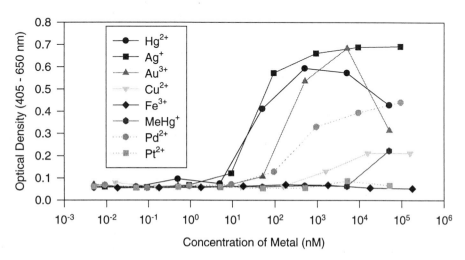

Figure 5. Cross reactivity of assay format 1. The inherent cross reactivity of the assay with silver, gold, and palladium makes it useful for the detection of these precious metals.

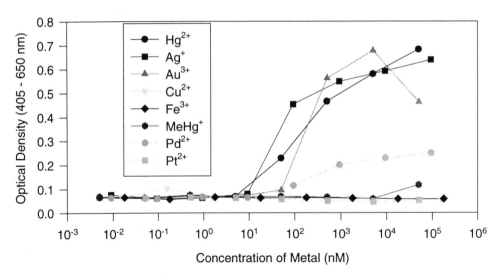

Figure 6. Cross reactivity of assay format 1, 10 mM ethylenediaminetetraacetic acid (EDTA) in metal incubation step. Polyaminocarboxylate chelators such as EDTA have a higher affinity for copper relative to the noble metals. In this example, EDTA effectively masks interference by copper.

at 40 nM (8 ppb), the limit of detection was 1 nM (0.2 ppb), and the cross-reactivity profile with other metals remained the same as described by Szurdoki et. al (1995).

Experiments were conducted to verify optimum pH and ionic strength conditions during the incubation steps. Changes in the ionic strength (50 to 500 mM) of the sodium acetate buffer during incubations with mercury had very little effect upon the performance of the assay (Table 1a). Of the pH values that were tested (5.0, 5.5, 6.0, 6.5, and 7.0), pH 7.0 was the best, resulting in maximal optical density contrast between high and low mercuric ion concentrations (Table 2a). For the incubation with tracer, the optimal concentra-

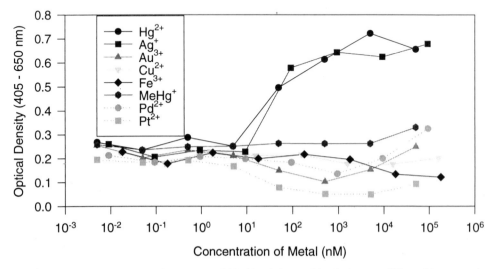

Figure 7. Cross reactivity of assay format 1, 10 mM imidazole in metal incubation step. This masking agent significantly reduced cross reactivity with all metals and concentrations tested except for silver.

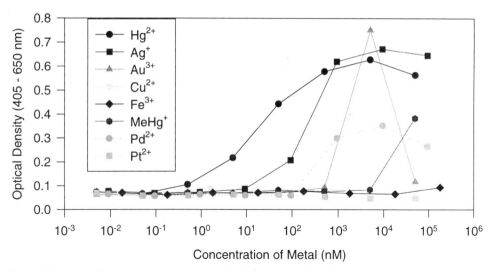

Figure 8. Cross reactivity of assay format 1, 0.1 mM 1-(2-Pyridylazo)-2-naphthol in metal incubation step. Only a slight reduction of interference from copper was achieved with this masking agent.

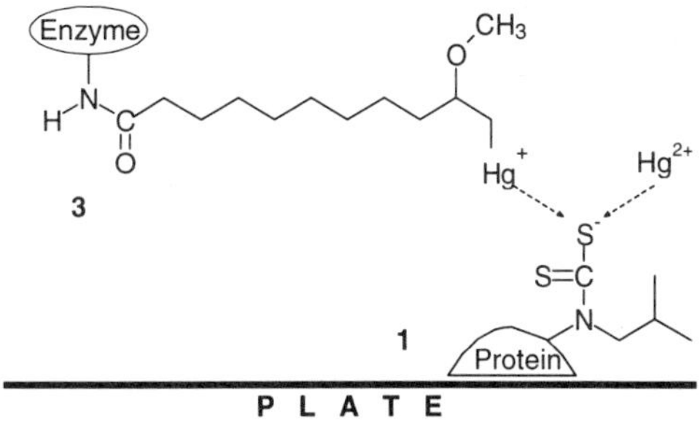

Figure 9. Competitive inhibition assay format 2. Color development is inversely proportional to the concentration of Hg^{2+}. Protein is conalbumin (CONA) and enzyme is alkaline phosphatase (AP).

Table 1a. Optimization of sodium acetate concentration during incubation with mercury

Concentration of sodium acetate (mM) pH 5.5 used in incubation with Mercury[a]	% Optical density contrast*
50	**100**
100	90
200	100
300	90
400	90
500	70

*% Optical density contrast is the difference between optical density values resulting from highest and lowest mercury concentrations tested, and expressed as a percentage of the highest optical density contrast value in the column.

Table 1b. Optimization of sodium acetate concentration during incubation with enzyme tracer

Concentration of sodium acetate (mM) pH 5.5 used in incubation with Tracer[b]	% Optical density contrast*
50	80
100	90
200	90
300	**100**
400	80
500	80

[b]Incubations with mercury were all conducted in 100 mM sodium acetate, pH 5.5

Table 2a. Optimization of pH during incubation with mercury

pH of Buffer (300 mM sodium acetate) used for incubation with Mercury[c]	% Optical density contrast*
5.0	80
5.5	60
6.0	70
6.5	90
7.0	**100**

[c]Incubations with tracer were all conducted in 100 mM sodium acetate, pH 5.5.
*% Optical density contrast is the difference between optical density values resulting from highest and lowest mercury concentrations tested, and expressed as a percentage of the highest optical density contrast value in the column.

Table 2b. Optimization of pH during incubation with tracer

pH of Buffer (300 mM sodium acetate) used for incubation with Tracer[d]	% Optical density contrast*
5.0	40
5.5	40
6.0	70
6.5	**100**
7.0	100

[d]Incubations with mercury were all conducted in 100 mM sodium acetate, pH 5.5.
*% Optical density contrast is the difference between optical density values resulting from highest and lowest mercury concentrations tested, and expressed as a percentage of the highest optical density contrast value in the column.

tion of sodium acetate was 300 mM (Table 1b), and the best pH value was approximately between 6.5 and 7.0 (Table 2b).

The optimized assay included the use of 300 mM sodium acetate, 0.05% Tween 20, pH 7 in both incubation steps. Plate washing was continued with 100 mM sodium acetate at pH 5.5. The IC_{50} was reduced to about 12 nM (2.4 ppb) from 40 nM before. The limit of detection remained at about 1 nM (Figure 10).

Experiments to evaluate the potential use of EDTA to reduce assay cross-reactivity with other metals were undertaken. It was observed that 10 mM EDTA present during the first incubation step was able to eliminate the system's cross-reactivity with copper and iron (Figure 11). However, interference by silver, gold, palladium, and platinum persisted (Figure 12). In addition, EDTA expressed some affinity for mercury, as seen in the higher

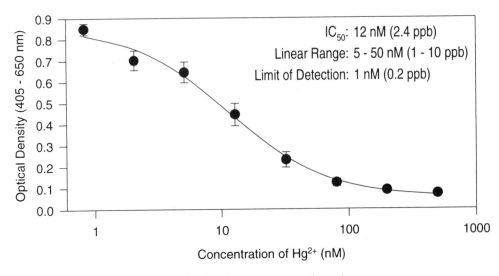

Figure 10. Standard curve for assay format 2.

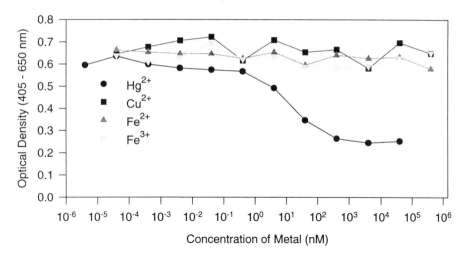

Figure 11. Cross reactivity of assay format 2, 10 mM EDTA in metal incubation step. Cross reactivity with copper and iron was eliminated.

optical density at high mercury concentrations (0.2 optical density units versus 0.1 without EDTA). Both, the IC_{50} and limit of detection remained at 12 nM and 1 nM respectively.

The use of different concentrations of potassium cyanide in the buffer (during incubation with metal) to minimize interference by silver was explored. The dose response of silver was not eliminated by up to 10 mM potassium cyanide.

4.3. Analysis of Water from the Carson River

The standard curve, for the analysis of samples of water from the Carson River, which was generated as previously described had an IC_{50} of approximately 30 nM (6 ppb)

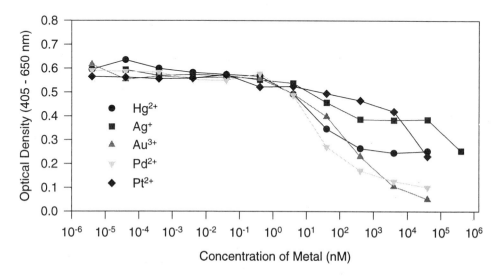

Figure 12. Cross reactivity of assay format 2, 10 mM EDTA in metal incubation step (continued). Cross reactivity with silver, gold, palladium, and platinum persisted.

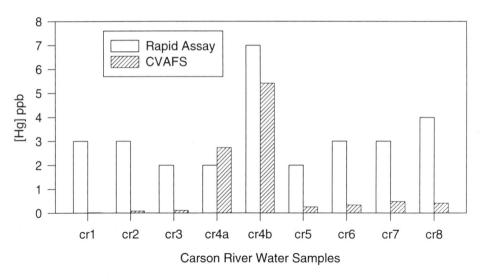

Figure 13. Analysis of Carson River water with assay 2 and atomic fluorescence spectrophotometry. Recoveries of mercury, at low concentrations, by the rapid assay were too high relative to cold vapor atomic fluorescence spectrometry (CVAFS). The correlation is better at a higher concentration of mercury. Interference by a metal other than mercury is probable.

and the effective limit of detection increased to 2 nM (0.4 ppb) due to the 100% dilution of sample. In addition, maximum optical density decreased to about 0.4 units. This is probably due to the increased ionic strength (sodium chloride) in the wells of the plate.

Comparison of results of analysis by both methods suggest that our assay is able to quantify mercury levels in the sample within the linear range of the standard curve from 10 to 100 nM (2 to 20 ppb). However, recoveries from samples containing mercury levels below 2 nM (0.4 ppb) were too high (Figure 13). These might be caused by the presence of foreign metals (Figures 11 and 12) that yielded responses in the assay. The Carson River Drainage Basin is a silver mining area. Masking agents to reduce excessive recoveries were not tested.

4.4. Analysis of Human Saliva

When human saliva was spiked with mercury and tested with both assays, it resulted in excessive signals relative to the control curves with no saliva (Figures 14 and 15). It is likely that some component(s) of saliva adhered to the surface of the plate and induced non-specific binding of the tracer enzymes during the second incubation step.

The above mentioned experiments were repeated except that after being spiked with mercuric acetate, the saliva samples were treated with different digestion additives. Figure 16 shows the results for assay format 1. All of the additives succeeded in reducing the non-specific binding of the enzyme tracer to the plate. However, only hydrogen peroxide resulted in a good dose response curve relative to the control curve containing 300 mM sodium acetate, pH 7.0 instead of saliva and digestion additive. It also appeared to greatly extend the linear range of assay format 1.

Like in the case of assay format 1, all of the digestion additives succeeded in reducing non-specific binding of enzyme tracer to the surface of the plate. The additive that showed the most promise with assay format 2 was trichloroacetic acid (Figure 17).

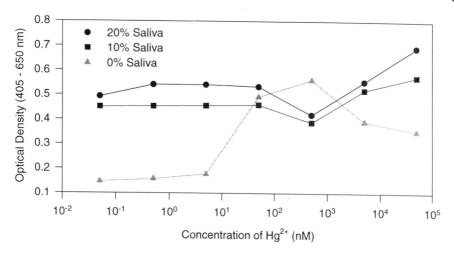

Figure 14. Effect of different concentrations of human saliva upon assay format 1. The saliva may have caused non-specific of the tracer to the microtiter plate.

5. CONCLUSIONS

Overall, the results are promising. Of the anions and their concentrations tested, I^- is the only one that might adversely affect assay format 1 by strongly complexing with mercury. However, a simple chemical treatment of the samples prior to analysis would likely eliminate the interference of iodide ion. As for the cross-reactivity issue, Ag^+ was the only cation for which a potential masking agent has not been found.

The regulatory maximum of the total mercury concentration in environmental water samples is 1 ppb in numerous countries, but the amount of mercury in river and lake water sometimes exceeds this limit due to industrial and urban waste water and mine runoff. Thus, our results hold promise for the monitoring of mercury in environmental water samples.

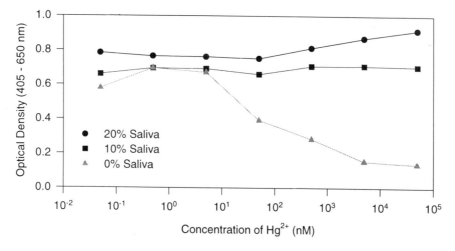

Figure 15. Effect of different concentrations of human saliva upon assay format 2. Non-specific binding of the tracer to the microtiter plate occurred.

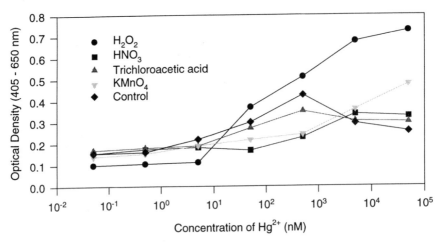

Figure 16. Assay format 1: Test of four digestion additives to reduce matrix effects from saliva spiked with mercuric acetate. Hydrogen peroxide increased the linear range of the assay. The control curve contains buffer instead of saliva and digestion additives.

Proper treatment of human saliva samples with hydrogen peroxide or trichloroacetic acid might enable both assay formats to be used for their analysis. The results are particularly encouraging especially since mercury concentrations normally found in saliva of patients with dental fillings falls within the linear range of the assays.

ACKNOWLEDGMENT

This work was supported in part by the NIEHS Center (ES05707), the NIEHS Superfund Grant (ES04699), NIEHS Training Grant (ES07059), the US EPA (CR814709010), the US EPA Center for Ecological Health Research (CR819658), the NAPIAP (G59520062) and Tektronix Inc. (Beaverton, Oregon).

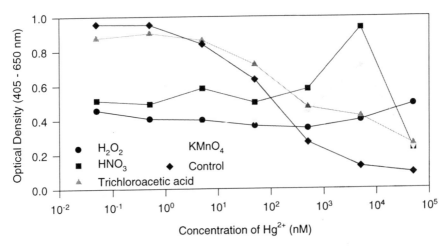

Figure 17. Assay format 2: Test of four digestion additives to reduce matrix effects from saliva spiked with mercuric acetate. Trichloroacetic acid yielded results most similar to the control. The control curve contains buffer instead of saliva and digestion additives.

REFERENCES

Bloom, N. S.; Crecelius, E. A. (1983). Determination of mercury in seawater at subnannogram per liter levels. Mar. Chem. 14, 49–59.

Bloom, N. S.; Fitzgerald, W. F. (1988). Determination of volatile mercury species at the picogram level by low-temperature gas chromatography with cold vapor atomic fluorescence detection. Analitica Chimica Acta, 208, 151–161.

Bond, A. M.; Scholz, F. (1991). Calculation of thermodynamic data from voltammetry of solid lead and mercury dithiocarbamate complexes mechanically attached to a graphite electrode. J. Phys. Chem., 95, 7460–7465.

Dumarey, R.; Temmerman, E.; Dams, T.; Hoste, J. (1985). The accuracy of vapor injection calibration method for determination of mercury by amalgamation/cold vapor atomic fluorescence spectrometry. Analytica Chimica Acta, 170, 337–340.

Falnoga, I.; Mrhar, A.; Karba, R.; Stegnar, P.; Skreblin, M.; Tusek-Znidaric, M. (1994). Mercury toxicokinetics in wistar rats exposed to elemental mercury vapour - modeling and computer simulation. Arch. Toxicol., 68, 406–415.

Gee, S. J.; Miyamoto, T.; Goodrow, M. H.; Buster, D.; Hammock, B. D. (1988). Development of an enzyme-linked immunosorbent assay for the analysis of the thiocarbamate herbicide molinate. J. Agric. Food Chem. 36, 836–870.

Gill, G. A.; Bruland, K. W. (1990). Mercury Speciation in surface freshwater systems in California and other areas. Env. Sci. Tech. 24, 1892–1400.

Gill, G. A.; Fitzgerald, W. F. (1987). Picomolar mercury measurements in seawater and other materials using stannous chloride reduction and two stage-gold amalgamation with gas phase detection. Mar. Chem., 20, 227–243.

Guo, T.; Baasner, J.; Gradl, M.; Kistner, A. (1996). Determination of mercury in saliva with a flow-injection system. Anal. Chim. Acta 320, 171–176.

Gustin, M. S.; Taylor, G. E.; Leonard, T. L. (1994). High levels of mercury contamination in multiple media of the Carson River Drainage Basin of Nevada – Implications for risk assessment. Env. Health Persp., 102, 772–779.

Hammock, B. D.; Szurdoki, F.; Kido, H. (1995). Enzyme amplified, complex linked, competitive and non-competitive assays for the detection of metal ions. U.S. Patent, 5,459,040.

Hultman, P.; Johansson, U.; Turley, S. J.; Lindh, U.; Eneström, S.; Pollard, K. M. (1994). Adverse immunological effects and autoimmunity induced by dental amalgam and allow in mice. FASEB J., 8, 1183–1190.

Janjic, J.; Kiurski, J. (1994). Non-flame atomic fluorescence as a method for mercury traces determination. Wat. Res., 28, 233–235.

Keeler, G.; Glinsorn, G.; Pirrone, N. (1995). Particulate mercury in the atmosphere: Its significance, transport, transformation, and sources. Water Air Soil Poll., 80,159–168.

Klaassen C. S.; Amdur, M. O.; Doull, J. Eds. (1986). Casarett and Doull's Toxicology. The Basic Science of Poisons. 3rd ed.; Macmillan Publishing Co.: New York, NY.

Lansens, P.; Meuleman, C.; Baeyens, W. (1990). Long-term stability of methylmercury standard solutions in distilled, deionized water. Anal. Chim. Acta, 229, 281–285.

Lau, O.-W.; Ho, S.-Y. (1993). Simultaneous determination of traces of iron, cobalt, nickel, copper, mercury and lead in water by energy-dispersive X-ray fluorescence spectrometry after preconcentration as their piperazino-1,4-bis(dithiocarbamate) complexes. Anal. Chim. Acta, 280, 269–277.

Lind, B.; Body, R.; Friberg, L. (1993). Mercury speciation in blood and brain tissue from monkeys - interlaboratory comparison of Magos method with other spectroscopic methods, using alkylation and gas chromatography separation as well as RNAA in combination with Westoo's extraction methods. Fresenius J. Anal. Chem., 345, 314–317.

Miller, J. R.; Rowland, J.; Lechler, P. J.; Desilets, M.; Hsu, L. (1996). Dispersal of mercury contaminated sediments by geomorphic processes, Sixmile Canyon, Nevada, USA: Implications to site characterization and remediation of fluvial environments. Water Air Soil Poll., 86, 373–388.

Rodbard, D. (1981). In Ligand Assay; Langan, J.; Clapp, J. J., Eds.. Masson Publishing: New York; 45–99.

Skerfving, S. (1991). In Advances in Mercury Toxicology; Suzuki, T.; Imura, N.; Clarkson, T. W., Eds.; Rochester Series on Environmental Toxicology; Plenum Press: New York, NY; 411–425.

Suzuki, T.; Imura, N.; Clarkson, T. W., Eds. (1991). Advances in Mercury Toxicology; Rochester Series on Environmental Toxicology; Plenum Press: New York, NY.

Szurdoki, F.; Kido, H.; Hammock, B. D. (1995). Development of rapid mercury assays. Synthesis of sulfur- and mercury-containing conjugates. Bioconjugate Chem., 6, 145–149.

Valentine, W. M.; Amarnath, V.; Graham, D. G.; Anthony, D. C. (1992). Covalent cross-linking of proteins by carbon disulfide. Chem. Res. Toxicol., 5, 254–262.

Voller, A.; Bidwell, D. E.; Bartlett, A. (1976). Enzyme Immunoassays in Diagnostic Medicine: theory and practice. Bull. W.H.O., *53*, 55–64.

Wylie, D. E.; Lu, D.; Carlson, L. D.; Carlson, R.; Babacan, K. F.; Schuster, S. M.; Wagner, F. W. (1992). Monoclonal antibodies specific for mercuric ions. Proc. Natl. Acad. Sci. USA, 89, 4104–4108.

Zalups, R. K; Lash, L. H. (1994). Advances in understanding the renal transport and toxicity of mercury. J. Toxicol. Environ. Health, 42, 1–44.

7

COPPER TRANSPORT AND CERULOPLASMIN DURING LACTATION AND PREGNANCY

Maria C. Linder, Stephanie Donley, David Dominguez, Lisa Wooten,
Farrokh Mehrbod, Philip Cerveza, Roman Shulze, Steven Cotton,
Anjali Sridhar, Norma Lomeli, and Marilyn Hallock

Department of Chemistry and Biochemistry
California State University
Fullerton, California 92843-6866

1. INTRODUCTION

It is well known that several trace elements, and particularly copper and iron, accumulate in the liver of the fetus during the last part of pregnancy (Figure 1) (Linder and Munro, 1974; Linder, 1991). Indeed, at the time of birth, the concentrations of these elements are at much higher levels than at any other time in the normal course of life (Table 1). These trace element accumulations seem to be important for the first part of life during which the newborn's only food is the milk. Concentrations of copper (and particularly iron) in milk are relatively low, in comparison to what is present in other foodstuffs: about 0.3–0.6 ug Cu/ml in cow and human milk versus about 0.7–2.2 ug Cu/g in meats, fish, vegetables and fruits, and much higher levels in whole grains (in the range of 2.8–20 ug Cu/g) (Linder, 1991). Moreover, liver concentrations of copper (as well as iron and zinc) fall dramatically during the suckling period (Figure 1; Table 1), indicating that these prenatal stores are used in support of the growth of the suckling infant.

Despite its relatively low copper levels, the milk is nevertheless a highly important nutritional source of copper during the suckling period. Thus, for example, in rats, the rapid rise in total body copper seen in the fetus at the end of gestation continues, and indeed in this species accelarates, after birth (Figure 2). The data described in this article are consistent with these findings. They emphasize the importance of milk to the copper nutrition of the newborn and suckling infant, at least in rats, and indicate that most of the dietary copper absorbed by the dam is shunted to the mammary gland and milk rather than to the liver and other maternal tissues in this period.

The presence of potentially significant amounts of copper in the form of ceruloplasmin in the milk has only recently begun to receive attention. We and some others have reported that milks from the several mammals examined all contain ceruloplasmin, based

Metals and Genetics, edited by Sarkar.
Kluwer Academic / Plenum Publishers, New York, 1999.

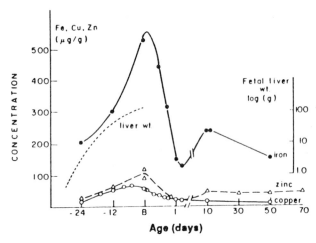

Figure 1. Changes in liver iron, copper, and zinc concentrations during gestation, suckling and later life, in humans. Reprinted with permission from Linder, 1978. Values for liver concentrations of iron (solid dots and dark line), zinc (triangles and dashed line) and zinc (open dots) are indicated and given in ug/g of wet weight. Changes in liver weight before birth (dotted line) are shown as well.

upon the presence of azide-inhibitable oxidase activity with p-phenylene diamine (Linder et al., 1994; Wooten et al., 1996; Nabukhotnyi et al, 1986) or o-dianisidine (Kiyosawa et al., 1995), as well as on immunoassays (Linder et al., 1994; Wooten et al., 1996). Specific mRNA for ceruloplasmin is expressed by the mammary gland (Jaeger et al., 1991; Shulze et al., 1997; Cerveza et al., 1998) and varies in relation to lactational stage in parallel with changes in milk ceruloplasmin concentrations (Shulze et al., 1997; Cerveza et al., 1998). More importantly, we have shown with ^{67}Cu that ceruloplasmin-copper fed to newborn rats is preferentially absorbed over ionic copper (Figure 3). Rat pups were fed 4 ng quantities of radioactive copper in the form of serum ceruloplasmin isolated from a donor adult animal or in the form of ionic Cu(II), both added to small portions of cow's milk. Radioactivity appearing in the liver of the pups (and staying in the digestive tract) was monitored. Uptake of ^{67}Cu from ceruloplasmin was 2–3 times greater than from the ionic form. It is noteworthy that this was no longer the case after weaning (Figure 3). These findings imply that milk ceruloplasmin may serve a special function in the copper nutrition of the newborn.

The studies reported here were designed to further our understanding of the role of the mammary gland and ceruloplasmin in neonatal copper nutrition by (a) following entry of copper into the mammary gland and other tissues over time, in the absence of presence

Table 1. Changes in liver copper concentrations from late gestation, to birth, suckling and adulthood

| Species | Liver copper concentrations (ug/g dry weight) | | | |
	Fetus (3d trimester)	Newborn	Suckling	Adult
Deer	1400	–	–	40
Horse	320	219	–	31
Cow	170	550	–	70
Human	160	300	89	25
Sheep	190	260	180	35
Pig	–	230	300	35
Dog	–	200	300	300
Guinea pig	122	203	–	17
Mouse	64	119	169	20
Rat	45	80	88	18

Data are mean values from Linder, 1991.

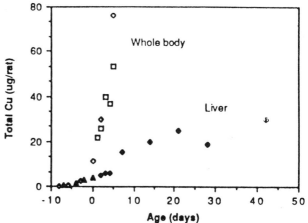

Figure 2. Changes in total body copper and total liver copper before and after birth in rats. Reprinted with permission from Linder, 1991, based upon data from several investigators. Open symbols are for total body copper, closed symbols for totals in liver, all in ug/rat, plotted against days before and after birth.

of lactation, and its appearance in the milk and in milk ceruloplasmin; (b) beginning to examine the nature of milk ceruloplasmin and its expression by the mammary gland in relation ot lactation.

2. MATERIALS AND METHODS

2.1. Milk, Tissue, and Serum Sources

Samples of human breast milk were obtained from healthy human volunteers recruited by Lisa Wooten, RN, as previously described (Wooten et al., 1996), and were at

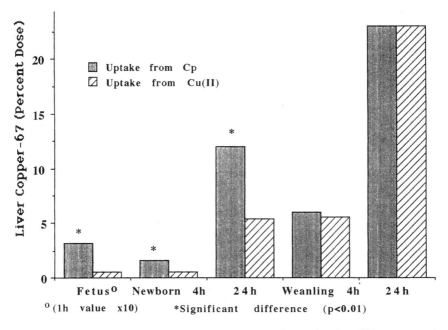

Figure 3. Preferential uptake of copper from ceruloplasmin by the rat fetus and newborn. Values are mean data for radioactivity in liver after uptake of ^{67}Cu from ^{67}Cu-ceruloplasmin (dark bars) or ^{67}Cu(II)-NTA (light bars) at various times after injection into the amniotic sac (Fetus) or feeding to newborn or weanling rats. From Wooten et al., 1996. Starred Means were significantly different (p < 0.01).

various stages of lactation, from 3 days to more than 8 months post partum. Human serum was obtained from volunteers in our laboratory with the assistance of the university's Health Center. All samples of serum and non fat milk (obtained by centrifugation) were stored at -20° or -85°. Pig milk and serum samples, as well as biopsies of mammary gland, were collected from Hampshire x Yorkshire cross sows (200–400 kg) in the Swine Unit of the Department of Animal and Veterinary Science, California Polytechnic University, Pomona, CA, with the collaboration of Professors Steve Wickler, PhD, D.V.M. and Edward Fonda, PhD (Wooten et al., 1996; Cerveza et al., 1998). Samples were taken from lactating sows on days 2–4 and 32–34 days post-partum as well as from adult non-pregnant sows, and those at the end of gestation. Milk was collected from pre-washed teats into sterile 50 ml culture tubes (Corning) after i.v. injection of oxytocin (20 units). To obtain biopsies, sows were minimally restrained and anesthetized locally with an injection of 2% lidocaine into a single teat. The site was prepared surgically, and about 1.5 g of mammary tissue was removed. Biopsies were immediately frozen on dry ice and stored at -85° until RNA was extracted for quantitation of ceruloplasmin mRNA.

2.2. Rat Treatments and Radioisotope Experiments

Radioactive copper [67Cu(II)] was received in dilute HCl from the University of Missouri Research Reactor at Columbia or from Los Alamos National Laboratories (Los Alamos, NM) (Wooten et al., 1996; Lee et al., 1993). Nitrilotriacetate (NTA), in approximately 1:1 molar ratio, was added; the solution was neutralized with NaOH and brought to isotonicity with NaCl solution, for injection into lactating and non-lactating rats. Volumes of up to 0.6 ml, containing 50–300 uCi (20–100 ng Cu), were injected i.p. or i.v. by tail vein (Lee et al., 1993). Tail vein injections, milking, and euthanasia of the rats occurred during ketamine/xylazine anesthesia (87 mg + 13 mg, respectively, per kg body weight). Rats were milked not more than once per 24 h and were kept with their pups inbetween. Euthanasia was by exsanguination after pneumothorax (Lee et al., 1993). Milk, blood and tissue samples, including mammary gland, were collected and counted for radioactivity by gamma counter (Packard Cobra II, Downer's Grove, IL). Milk and plasma samples were kept frozen at -20° until analysed further. Rats (3–4 mo old) were of the Sprague Dawley strain, from Simonson Laboratories (Gilroy, CA). Pregnant rats were shipped on approximately days 12–15 of gestation. Radioactivity with milk ceruloplasmin was determined on phosphate buffered saline-washed immunoprecipitates of known volumes of rat milk (pretreated to remove casein), using specific goat IgG against the 0.6 Rf form of rat ceruloplasmin.

2.3. Assays of Total Protein, Ceruloplasmin, and [67]Cu-Ceruloplasmin

Total protein was determined by the method of Bradford, using reagents and protocols from BioRad (Richmond, CA), with bovine serum albumin standards. Ceruloplasmin concentrations in milk were determined by sandwich ELISA, developed by Dr. David Loeffler (Sinai Hospital, Detroit MI), using the 0.6 Rf form of rat ceruloplasmin prepared in our laboratory (Middleton and Linder, 1993) as a standard and antigen. Specific antibodies were raised in rabbits and a goat. Ceruloplasmin oxidase activity was measured with p-phenylene diamine, as previously described (Wooten et al., 1996).

2.4. SDS-PAGE and Immunoblotting

SDS-Polyacrylamide gel electrophoresis was performed in 1.5 mm thick minigels in a Hoefer Mighty Small unit (Hoefer, San Francisco, CA) according to Laemmle, using the

Hoefer protocols and 7.5% acrylamide in the resolving gel. Prestained and unstained high range molecular weight standards were from BioRad. Gels were stained with Coomassie Blue G250 or were transferred, by a Hoefer semi dry apparatus, to Immobilon PSQ membranes (Millipore), using CAPS buffer (100 mM, pH 10.5). For development of the immunoblot, the membrane was blocked with nonfat dry milk (1%) in TTBS (0.9% NaCl, 0.1% Tween 20, 100 mM Tris HCl, pH 7.5). The primary antibody was rabbit anti human serum ceruloplasmin (Dako, Glostrup, Denmark); the secondary antibody (goat anti rabbit IgG) was conjugated with horse radish peroxidase.

2.5. RNA Extraction, RT-PCR, and Northern Analysis

Total RNA was extracted by standard procedures, using guanidinium thiocyanate, as previously described (Madani and Linder, 1992). Purity was assessed by the ratio of absorbance at 260/280 nm. (Samples analysed further had ratios > 1.8). Northern analysis was of 50–200 ug RNA separated in 1.2% agarose gel electrophoresis, transferred to Zetaprobe nylon membranes (BioRad, Richmond, CA) and probed with a ^{32}P-UTP-labeled single stranded RNA prepared from a human cDNA originally obtained from Barbara Bowman and Funmei Yang (University of Texas, at Austin). This probe corresponded to exons 8–10 of the human ceruloplasmin gene (Harris et al., 1995). It was obtained by RT-PCR and cloned into a pCR-Script SK+ vector (Stratagene, San Diego, CA) in TA OneShot Super Competent Cells (Invitrogen). Quantities of RNA applied were based upon absorbance at 260 nm, assuming an absorbance of 1.0 corresponds to 40ug/ml. In some cases, ceruloplasmin mRNA was quantitated from autoradiographs of slot blots obtained with the BIO-DOT SF blotter system. For Northerns and slot blots, RNA loading was confirmed and quantitated by densitometry of images obtained from ethidium bromide stained 18 S and 28 S RNA on the transfer membranes or from autoradiographs of 18 S rRNA blots hybridized with a ^{32}P-labeled probe for rat 18 S rRNA (Iris Gonzalez (Hahnemann Medical College, Philadelphia, PA). Quantities of ceruloplasmin mRNA were calculated relative to these controls. Autoradiography was with Kodak X-OMAT AR film exposed for 1–5 days at -85°. Densitometry was of images obtained with the Speedlight Imaging System (San Diego, CA) analysed with the NIH Image program.

3. RESULTS

3.1. Uptake of Copper by the Mammary Gland

To assess the avidity of the mammary gland for copper in connection with lactation, rats were injected i.p. with 20–50 ng quantities of ^{67}Cu(II) as the NTA complex during days 2–5 of lactation, and radioactivity in mammary gland, liver and other tissues was monitored from 1 to 24 h and longer. Figure 4 shows that radioactive copper rapidly entered the mammary gland. The highest levels of radioactivity occurred at the earliest time points examined (1–2h), after which radioactivity rapidly declined, initially in parallel with the fall in serum radioactivity (see later). At 1–2 h after administration, the total radioactivity in mammary gland was much higher than that in any of the other tissues examined. It was also much greater than that in the liver and kidney to which most of the copper entering the blood from the diet normally first goes (Linder, 1991; Weiss and Linder, 1985). To make certain that the route of entry of the copper did not affect these re-

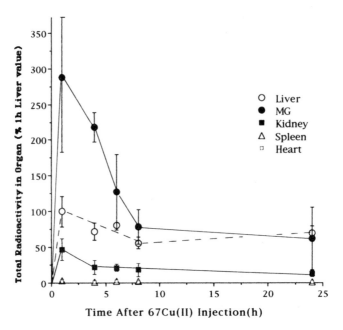

Figure 4. Total radioactivity in mammary gland and other organs of lactating rats at various times after injection of ^{67}Cu(II)-NTA. Mean values (± SD) for groups of 3–5 animals. Values are from several experiments and are given as the percent of the mean radioactivity in liver at 1h.

sults, the same kind of study was done using the intravenous tail vein route. No differences were observed.

Since the mammary gland has to our knowledge not been examined before with regard to its potential avidity for copper, we also examined uptake in non-lactating, virgin rats of the same age and sex. In contrast to what we observed in lactation, almost no ^{67}Cu entered the mammary gland: as seen previously, the liver and kidney had most of the radioactivity 1–4 h after administration, and the total ^{67}Cu in mammary gland was more than 100-fold less than that in the liver. The difference in copper uptake with and without lactation could not be attributed to a difference in weight of the mammary gland, which was also substantial: Mammary weight was 12 ± 3 g (Mean ± SD; N=22) during days 2–5 of lactation, and 2.6 ± 1.2 g (N=12) in virgin rats of comparable age, only a 5-fold difference.

Radioactive copper appeared in the milk almost as fast as it appeared in the mammary gland (Figure 5). Especially early on, but also at the other time points examined, the *concentration* of radioisotope in the milk greatly exceeded that in plasma. The amounts of radioactivity associated with milk ceruloplasmin at different times after treatment were also determined. Ceruloplasmin was precipitated from the casein-free supernatants of known portions of milk with the help of specific antibody against the 0.6 Rf form of rat serum ceruloplasmin (see Methods). The washed pellets were counted for radioactivity and this was compared to the total radioactivity in the original milk samples. Consistently, over all time points examined, about one quarter of the total ^{67}Cu was associated with ceruloplasmin (27 ± 8; N=24), indicating that secretion of ceruloplasmin from the mammary gland into the milk paralleled the secretion of other forms of copper into that fluid. By ELISA, the milk contained 11.0 ug ceruloplasmin per ml (± 3.7; N=4).

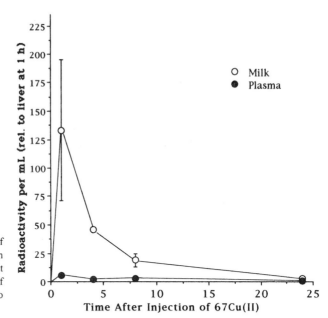

Figure 5. Relative concentrations of [67]Cu radioactivity (Mean ± SD) in milk and plasma of lactating rats at different times after injection of [67]Cu(II)-NTA. Values are relative to those for liver, at 1h.

3.2. The Nature of Ceruloplasmin in Milk

During our initial investigations of the ceruloplasmin in milk (reported in Wooten et al., 1996), an additional preliminary finding was that pig milk ceruloplasmin consistently eluted slightly later than that of pig *serum* in open column size-exclusion chromatography on Sephadex G150 (Figure 6). In these studies, the elution of ceruloplasmin was determined by its oxidase activity. On the same column, using exactly the same procedures, the peak of milk ceruloplasmin elution occurred after that for serum ceruloplasmin. This occurred whether or not the milk sample was applied before that of serum and suggested that the milk protein was somewhat smaller or more compact than the serum protein. On the other hand, analyses of *human* milk and serum by FPLC on Superdex 200 indicated no differences in apparent molecular weight. We thus carried out additional studies, at both the protein and mRNA levels, to ascertain what, if any, major differences there might be between the milk and serum proteins.

Ceruloplasmins were purified from pig and human milks and sera and/or immunoprecipitated from casein-free milk supernatants. Samples were then separated in SDS-polyacrylamide gel electrophoresis, transferred to PVDF membranes, and immunoblotted with specific antibody against human ceruloplasmin. The same kinds of results were obtained for pig and human samples. As shown in Figure 7A, there were no detectable differences in the apparent molecular weights of the milk and serum ceruloplasmins. It is worth noting, however, that two forms of ceruloplasmin were detected in both fluids, one with the typical Mr of about 135 k, the other distinctly larger (Mr about 155 k). Removal of carbohydrate by incubation with endoglycosidase F lowered the Mr of both forms equally (Figure 7B), suggesting that the forms reflect differences in ceruloplasmin polypeptide length rather than carbohydrate.

Potential differences in the size and sequence of liver and mammary gland ceruloplasmins was also examined by Northern analysis and RT-PCR (Cerveza et al., 1998). For this, total RNA was extracted from portions of pig liver and mammary gland, separated in agarose gel electrophoresis, transferred to nylon membranes, and hybridized with a 400 bp

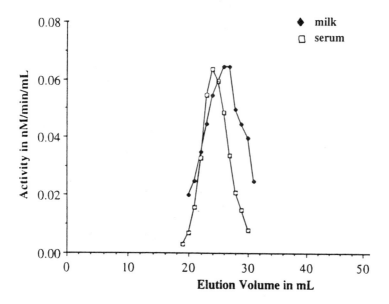

Figure 6. Elution of ceruloplasmin oxidase activity in size exclusion chromatography using a 1 × 50 cm column of Sephadex G150. One ml samples of serum (open squares) or milk (filled diamonds) were applied, and oxidase activity was measured in the 1 ml fractions collected, using p-phenylene diamine.

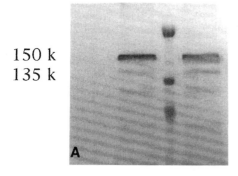

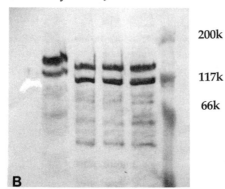

Figure 7. Apparent molecular weights of milk and serum ceruloplasmins determined by immunoblotting, before and after treatment with endoglycosidase F. (A) Western blot of pig milk (Mlk) and serum (Ser) ceruloplasmin in comparison to unstained (USS) and pre-stained (SS) molecular weight standards, using specific antibody against human ceruloplasmin. (B) Western blot of samples incubated in parallel without (-Glyc.) and with (+ Glyc.) various amounts of the glycosidase, in comparison with molecular weight standards that were unstained (Std. unst.) or prestained (Std. stained).

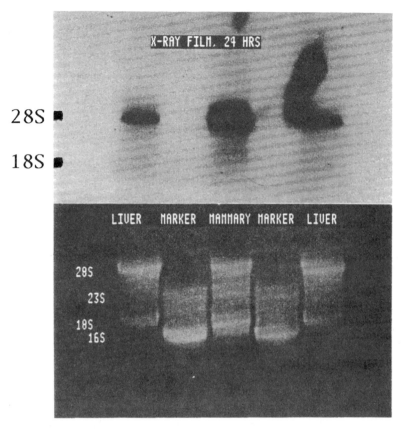

Figure 8. Northern analysis of ceruloplasmin mRNA in pig liver and mammary gland. The positions of the liver and mammary gland RNA samples applied are indicated on the picture of the ethidium bromide stained nylon membrane, after transfer (below), along with the position of bacterial rRNAs used as markers.

probe for pig serum ceruloplasmin. Figure 8 shows that there was no obvious difference in the size of the mRNAs for ceruloplasmin in liver and mammary gland. That in mammary gland was clearly no smaller than that in liver. There also was no evidence of more than one mRNA, although by this method one would not be able to easily distinguish between the presence of one versus two mRNAs differing in size by less than 10–15%.

Different areas of the ceruloplasmin mRNAs in mammary gland and liver were also examined by RT-PCR (Cerveza et al., 1998). The primers used were designed to encompass all the exons encoding the amino acid sequence of human ceruloplasmin. Different combinations of primer pairs (> 10 combinations) were PCR cycled with the cDNAs. Products were examined for potential differences in size by agarose electrophoresis. Examples are shown in Figure 9. No differences in the size of PCR products obtained for ceruloplasmin of mammary gland and liver were evident.

The expression of ceruloplasmin mRNA by the mammary gland at different stages of lactation, as well as in late pregnancy and in the virgin state, was also examined (Cerveza et al., 1998). Values for total RNA per g tissue obtained for biopsies of pig mammary gland differed with physiological state, as might be expected (Figure 10, solid bars). Pregnancy induced a increase from the virgin state, and total RNA per g increased again in

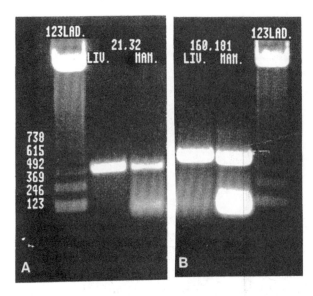

Figure 9. Comparative RT-PCR of ceruloplasmin exons in pig liver and mammary gland: some examples. PCR products were separated in ethidium bromide containing 1.2% agarose gels and sized based on 123 ladders. (A) shows products using markers spanning exons 2 and 3; (B) shows those for exons 16 and 18.

conjunction with birth and lactation. The concentration of ceruloplasmin mRNA as a percentage of total RNA was also analysed. For this, similar amounts of total RNA were slot-blotted onto nylon membranes and probed with anti-sense RNA to pig ceruloplasmin, and with 18S rRNA (to correct for RNA loading). The data indicate that the proportion of total RNA in the mammary gland ascribable to ceruloplasmin mRNA also increased substantially from late pregnancy to the start of lactation. It then declined again, over the next month of lactation. When combined with the data on total tissue RNA content, the data indicate there was a doubling of total ceruloplasmin mRNA in the mammary gland from the virgin state to late pregnancy, and more than a 4-fold fruther rise by early lactation (Figure 10, medium dark bars). This was followed by a 40% decline by day 33 of lactation. The changes in daily milk output of ceruloplasmin paralleled changes in ceruloplasmin mRNA (Figure 10, light bars), further substantiating the linkage between mammary gland ceruloplasmin expression and secretion of this protein as part of the milk.

4. DISCUSSION

We have shown that the mammary gland is extraordinarily active in terms of copper uptake and transport into the milk, at least in the case of the rat. Under normal circumstances, the liver and kidney are the first organs to take up most of the copper entering the blood from the diet or after its injection i.p. or iv. (Linder, 1991; Linder and Hazegh-Azam, 1996; Linder et al., 1998), and we have shown here that the mammary gland takes up very little. Lactation brings about a dramatic change in the mammary gland. Not only does the gland increase about 5-fold in size, but its avidity for uptake of copper from the maternal blood plasma is enhanced more than 100-fold. The overall rate of uptake thus exceeds that of the liver and kidney combined, whether examined on the basis of uptake per g tissue or total uptake per organ.

The kinetics of uptake indicate that copper is absorbed from the exchangeable copper pool of the maternal blood plasma rather than from maternal blood plasma ceruloplasmin. The radioactive tracer attained its highest concentrations in the mammary gland at

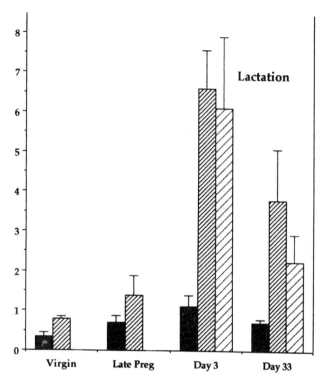

Figure 10. Changes in ceruloplasmin mRNA expression in mammary gland, total RNA and milk ceruloplasmin production, in relation to lactation. From Cerveza et al., 1998. Values are means ± SD for 4 samples from sows at each stage, including shortly before birth, and 3 and 33 days post partum. Total RNA (mg/g; dark bars), total ceruloplasmin mRNA (relative units/g; medium bars) and milk ceruloplasmin production (mg/day × 0.1; light bars) are indicated.

the earliest time points examined, which was one h after administration of the isotope. At this time, virtually all of the copper in the plasma is attached to albumin and transcuprein in the plasma of such rats (Weiss and Linder, 1985; Lee et al., 1993), and little or none of it has as yet appeared in plasma ceruloplasmin (which is being formed in the liver). This contrasts to what occurs in pregnancy. Here, we have shown that maternal plasma ceruloplasmin is the preferred source (or perhaps even the only source) of copper for the placenta and fetus (Linder et al., 1998; Lee et al., 1993).

Our data demonstrate that well over half of the copper entering the dam finds its way rather directly to the mammary gland and milk of the lactating rat. A relatively small proportion is retained by the dam. This suggests that, if deficiency on the part of the dam is to be avoided, the lactating rat must be consuming more dietary copper and/or absorbing a larger proportion of dietary copper than is normally the case. [Non-lactating rats absorb about 40–50% of dietary copper (Linder, 1991).] This possibility still needs to be tested directly. In any event, our findings help to explain how the suckling rat pup is able to accumulate as much copper as it does, as rapidly as it does, in the early part of life when it is growing very fast (Figure 2).

A substantial proportion of the copper in rat milk appears to be associated with ceruloplasmin, and our data indicate the the appearance of ceruloplasmin-copper in the milk parallels that of other forms of copper. This suggests that the secretory process by which milk is produced and released is the rate limiting step for appearance of copper and ceruloplasmin in the milk, and that insertion of copper into ceruloplasmin, which is thought to occur in the trans Golgi network (Murata et al., 1995; Hirasawa et al., 1997) parallels its incorporation into other milk components. As indicated by our previous work in which we fed rat pups [67]Cu-ceruloplasmin or [67]Cu(II), the copper in ceruloplasmin may be the form most available to the pup from the milk. Moreover, in analogy with the findings for milk lactoferrin (Davidson and Lonnerdal, 1988), there may be specific ceruloplasmin receptors on the mucosal surfaces of the intestinal mucosa of newborn mammals that facilitate absorption ceruloplasmin-copper, just as specific receptors on cells of internal organs may facilitate uptake of ceruloplasmin-copper from the blood plasma by internal organs (Linder et al., 1998; Linder, 1991). Indeed, preliminary evidence from our laboratory, using pigs, supports this hypothesis (Linder and Parekh, 1997).

The ceruloplasmin in milk appears to be the same as that in the blood plasma. Using milk and serum from two different species, we found that both have the same apparent molecular weights. Indeed, both fluids appear to have the same two major forms, including one form which is larger than expected from the amino acid sequence of purified human serum ceruloplasmin (Takahashi et al., 1984). In further support of no differences between the milk and serum proteins, Northern analysis indicated that the mRNA(s) for ceruloplasmin in mammary gland and liver are of the same apparent size. There were also no obvious differences in sequence, as suggested by the finding that PCR products obtained after reverse transcription of mammary gland and liver RNA appeared to be identical in size and of the expected sizes. The reason for the two different forms of ceruloplasmin in milk and serum requires further investigation, although it appears that it is not due to differences in total amounts of carbohydrate.

Our studies also show that milk ceruloplasmin does not derive from the maternal blood plasma. As determined by following appearance of its copper, the time course of ^{67}Cu-ceruloplasmin release into the milk was completely different from (and much earlier than) the appearance of ^{67}Cu-ceruloplasmin in plasma. As shown many times in the past (Weiss and Linder, 1985; Vargas et al., 1994), plasma ceruloplasmin only becomes radiolabeled after a considerable lag, during which radioisotope is incorporated into the protein during its synthesis in the liver. Usually, no ^{67}Cu-ceruloplasmin is detectable until 2 h after ^{67}Cu(II) administration. In the studies here described, most of the radiolabeled ceruloplasmin that would appear in the milk was already present 1 h after ^{67}Cu(II) administration, indicating that the mammary gland must have been its source.

Ceruloplasmin mRNA expression in the pig mammary gland varied markedly in relation to lactation, being highest in the first days after birth and lowest in virgin animals. The calculated daily production of milk ceruloplasmin by the pig paralleled mammary gland ceruloplasmin mRNA expression. Thus, although some ceruloplasmin mRNA is expressed even in the absence of lactation, there is a clear parallelism between expression and production of ceruloplasmin for the milk, and there can now be no question about the origin of milk ceruloplasmin. The maternal serum is not the source, and the mammary gland's avidity for copper is unparalleled in lactation and directly relates to release of ceruloplasmin-copper into the milk. We conclude that at least in rats, the milk is the main source of copper for the newborn, and that ceruloplasmin-copper produced by the mammary gland may play a particularly important nutritional role in the transfer of copper from the mother to her offspring.

ACKNOWLEDGMENTS

Supported in part by U.S. PHS Grants R55 HD 30424 and SO6 GM 08258, by NSF Grants REU 5369 and DUE 9352396, and by a Minigrant from California State University, Fullerton.

REFERENCES

Cerveza, P., Cotton, S., Mehrbod, F., Lomeli, N., Sridhar, A., Linder, M.C., Wickler, S., Fonda, E. (1998). Milk ceruloplasmin and mammary gland ceruloplasmin expression in relation to lactation in the pig. Submitted.

Davidson, L.A. and Lonnerdal, B. (1988). Specific binding of lactoferrin to brush border membrane: ontogeny and effect of glycan chain. Am. J. Physiol. 254, G580-G585.

Harris, Z.L., Takahashi, Y., Miyajima, H., Serizawa, M., MacGillivray, R.T.A., and Gitlin, J.D. (1995) Aceruloplasminemia: Molecular characterization of this disorder of iron metabolism. Proc. Natl. Acad. Sci. USA 92, 2539–2543.

Hirasawa, F., Kawarada, Y., Sato, M., Suzuki, S/. Terada, K., Miura, N., Fujii, M., Kato, K., Takisawa, Y., Sugiyama, T. (1997). The effect of silver administration on the biosynthesis and the molecular properties of rat ceruloplasmin. Biochim. Biophys. Acta 1336, 195–201.

Jaeger, J.L., Shimizu, N., Gitlin, J.D. (1991). Tissue-specific ceruloplasmin gene expression in the mammary gland. Bioch. J. 280, 671–677.

Kiyosawa, I., Matsuyama, J., Nyui, S., Fukuda, A. (1995). Ceruloplasmin concentration in human colostrum and mature milk. Biosci. Biotechnol. Biochem. 59, 713–714.

Lee, S.H., Lancey, R.W., Montaser, A., Madani, N., and Linder, M.C. (1993). Transfer of copper from mother to fetus during the latter part of gestation in the rat. Proc. Soc. Exp. Biol. Med. 203, 428–439.

Linder, M.C. (1978). Function and metabolism of trace elements. In: U. Stawe (Ed.) "Perinatal Physiology." Plenum, New York, pp 425–454.

Linder, M.C. (1991). "The Biochemistry of Copper." Plenum, New York.

Linder, M.C., Cerveza, P., Wooten, L., Sun, J., Shulze, R., Wang, T-P. (1994). Ceruloplasmin in milk and amniotic fluid. FASEB J. 8, A711.

Linder, M.C., Hazegh-Azam, M. (1996). Copper biochemistry and molecular biology. Am. J. Clin. Nutr. 63, 797S-811S.

Linder, M.C., Lomeli, N., Wooten, L., Cerveza, P., Cotton, S. and Shulze, R. (1998) Copper transport. Am. J. Clin. Nutr. 67, 965S-971S.

Linder, M.C., Munro, H.N. (1974). Iron and copper metabolism in development. Enzyme 15, 111–138.

Linder, M.C. and Parekh, D. (1997). Receptors for milk ceruloplasmin in the brush border of piglet intestinal mucosa. In: P.W.F. Fischer, M.R., L'Abbe, K.A. Cockrell, R.S. Gibson (Eds.). "Trace Elements in Man and Animal - 9." NRC Research Press, Ottawa, pp 77–78.

Madani, N., Linder, M.C. (1992). Differential effects of iron and inflammation on ferritin synthesis on free and membrane-bound polyribosomes. Arch. Biochem. Biophys. 299, 206–213.

Middleton, R.B., Linder, M.C. (1993). Synthesis and turnover of ceruloplasmin in rats treated with 17-beta-estradiol. Arch. Biochem. Biophys. 302, 362–368

Murata, Y., Yamakawa, E., Iizuka, T., Kodama, H., Abe, T., Seki, Y., Kodama, M. (1995). Failure of copper incorporation into ceruloplasmin in the Golgi apparatus of LEC rat hepatocytes. Biochem. Biophys. Res. Commun. 209, 349–355.

Nabukhotnyi, T.K., Markevich, V.E., Pavlyuk, V.P., Kostyrya, E. (1986). Ceruloplasmin isozymes in human milk. Vopr. Okhr. Materin. Det. 31, 15 (CAS Abstract).

Nakamura K, Endo F, Ueno T, Awata H, Tanoue A, Matsuda I. Excess copper and ceruloplasmin biosynthesis in long-term cultured hepatocytes from Long-Evans cinnamon (LEC) rat, a model of Wilson disease. J Biol Chem 1995; 270: 7656–7660.

Shulze, R.A., Wooten, L., Cerveza, P., Cotton, S., Linder, M.C. (1997). Ceruloplasmin expression by mammary gland and its concentraiton in milk. In: P.W.F. Fischer, M.R., L'Abbe, K.A. Cockrell, R.S. Gibson (Eds.). "Trace Elements in Man and Animal - 9." NRC Research Press, Ottawa, pp 69–70.

Takahashi, N., Ortel, T.L., Putnam, F.W. (1984). Single chain structure of human ceruloplasmin: The complete amino acid sequence of the whole molecule. Proc. Natl. Acad. Sci. USA 84, 4413–4417.

Vargas, E.J., Shoho, A.R., Linder, M.C. (1994). Copper transport in the Nagase analbuminemic rat. Am. J. Physiol. 267, G259-G269.

Weiss, K.C., Linder, M.C. (1985). Copper transport in rats involving a new plasma protein. Am. J. Physiol. 249, E77-E88.

Wooten, L., Shulze, R.A., Lancey, R.W., Lietzow, M., Linder, M.C. (1996). Ceruloplasmin is found in milk and amniotic fluid and may have a nutritional role. J. Nutr. Biochem. 7, 632–639.

8

CAP43

A New Gene Induced by a Rise in Free Intracellular Ca^{2+} Following Ni^{2+} Exposure

Konstantin Salnikow, Daoji Zhou, Tomasz Kluz, Cheng Wang, and Max Costa*

Institute of Environmental Medicine and
Kaplan Comprehensive Cancer Center
New York University Medical Center
57 Old Forge Road, Tuxedo, New York 10987

1. INTRODUCTION

Nickel (Ni) compounds have been implicated in elevating the incidence of lung and nasal cancers in numerous epidemiological studies of nickel refinery workers (IARC, 1990; Doll, 1970; Bennett, 1984; Shen, 1994; Langard, 1994). Many different types of cancer have also been induced at the site of exposure in experimental animal models following inhalation or parenteral administration of various Ni compounds (Ottolenghi, 1975; Sunderman, 1981; Sunderman, 1989). Carcinogenic Ni compounds transform primary human and rodent fibroblasts in tissue cultures (Biedermann and Landolph, 1987; Conway and Costa, 1989; Trott et al., 1995). Ni compounds exhibit low mutagenic activity in gene mutation assays but are clastogenic with a preference for genetically inactive heterochromatin (Conway and Costa, 1989).

Nickel and several other heavy metals have been found to enhance the expression of transcription factors, metallothionein (MT), heat-shock (HS) proteins and acute phase reactant proteins (Karin, et al., 1980; Jin and Ringertz, 1990; Epner and Herschman, 1991; Bauman, et al., 1993; Yiangou and Papaconstantinou, 1993). Recently we have found that thrombospondin I gene expression was inactivated in mouse and hamster Ni-transformed and Ni-treated cells (Salnikow et al., 1997). The inactivation of thrombospondin I expression resulted from an increased level of ATF-1 transcription factor that negatively regu-

_____

*To whom all correspondence should be addressed. Telephone: (914) 351-2368; Fax: (914) 351-2118.

Metals and Genetics, edited by Sarkar.
Kluwer Academic / Plenum Publishers, New York, 1999.

lated thrombospondin I gene expression (Salnikow et al., 1997). In contrast to previous findings, this effect favored Ni since other metals, such as cadmium and cobalt, were inactive. The mechanism for induction of ATF-1 has not been elucidated. In the present study we continue to investigate the molecular and cellular targets of Ni toxicity. Using the differential display technique (Liang and Pardee, 1992), we cloned a gene that was specifically induced by exposure to non-toxic levels of both water-soluble and -insoluble Ni^{2+} compounds in a dose-and time-dependent manner in human bronchoalveolar epithelial A549 cells (Zhou et al., 1998). No other metal compound tested, significantly induced expression of this gene in A549 cells, indicating that this gene was expressed with marked specificity to Ni^{2+} exposure (Zhou et al., 1998).

Using this gene as a marker, we hoped to identify signaling pathways leading to enhanced gene expression. We found that the calcium ionophore A23187 induced *Cap43* expression as efficiently as nickel, and BAPTA-AM, an intracellular calcium chelator, abolished expression of *Cap43* caused by either Ni or calcium ionophore. Direct measurement of intracellular calcium levels showed that Ni treated cells exhibited a sustained elevation of free intracellular Ca^{2+}. Okadaic acid, an inhibitor of serine/threonine phosphatases, but not vanadate, a tyrosine phosphatases inhibitor, also induced *Cap43* gene expression suggesting that serine/threonine phosphorylation was involved in *Cap43* induction. The effect of okadaic acid can also be abolished by BAPTA-AM, suggesting that the phosphorylation pathway was Ca^{2+} dependent. *Cap43* expression is likely to be a useful tool to enhance our understanding of calcium homeostasis in general and how Ni interferes with this process.

2. EXPERIMENTAL PROCEDURES

2.1. Cell Culture

A549 (CCL 185) and HUV-EC-C (CRL 1730) cells were purchased from the American Type Culture Collection (ATCC) (Rockville, MD). All cells were maintained at 37°C as monolayers in a humidified atmosphere containing 5% CO_2.

2.2. Northern Blot Analysis

Total RNA was extracted from cells immediately following exposure using an UL-TRASPEC™ RNA isolation system (Biotecx) and separated by electrophoresis 20 μg total RNA/lane) in 1.2% agarose/formaldehyde gels. *Cap43*, actin, or glyceraldehyde 3 phosphate dehydrogenase probes were labeled with [^{32}P]-a-dCTP using a Random Primed DNA Labeling Kit (Boehringer Mannheim).

2.3. Intracellular Calcium

Intracellular calcium was measured using a long-wavelength fluorescent probe, fluo-3-acetoximethyl ester. Cells were treated for different periods of time with nickel or other pharmacological agents, washed and resuspended in HEPES buffer at 2×10^6 cells/ml and incubated at 37°C with 5 μM Fluo-3AM for 40 min. After incubation, cells were washed three times with HEPES buffer and fluorescence intensity (F) was recorded for 400 s on a Spex Fluorolog (Edison, NJ), with emission settings of 525 nm and excitation of 506 nm. Maximum fluorescence intensity (F_{max}) was measured after addition of 40 mM digitonin, and minimum fluorescence intensity was recorded after addition of 6 mM EGTA. The $[Ca^{2+}]i$ level was calculated according the equation: $[Ca^{2+}]i = 400nm \cdot (F - F_{min})/(F_{max} - F)$.

2.4. Immunofluorescence Microscopy of A549 Cells with Anti-*Cap43* Antibodies

A549 cells were fixed with freshly prepared 3% paraformaldehyde for 3 min, and then treated with 0.5% triton X-100 for 15 min, washed with PBS and overlaid with primary polyclonal antibodies against human *Cap43*. After washing in PBS, the coverslips were incubated in an appropriate diluted FITC-tagged second antibody (Sigma, St. Louis, MO) for 30 min. The coverslips were mounted in 70% glycerol and diluted with PBS. Fluorescent images were recorded using a Zeiss photomicroscope on Kodak T-Max 3200 film and were developed in a Kodak T-Max developer.

2.5. Transfection and Controlled Expression of the *Cap43* Gene

RNA was isolated from Ni-treated A549 cells using TRIzol (Gibco BRL) and converted to single stranded cDNA using SuperScript preamplification system for the first strand cDNA synthesis (Gibco BRL). A 1.3 kbp DNA fragment containing an open reading frame of the *Cap43* gene was obtained by PCR using an upper primer-5' with the following sequence: CTCGCGTTAGGCAGGTGACAG, and a lower primer-5' having the following sequence: TGGCAGGCAGGGGGCGAAAAG, under the following conditions: 95°C for 2 min, and then 25 cycles-denaturation-94°C for 45 sec, annealing 61°C for 45 sec, extention 72°C for 45 sec, final extention 72°C for 10 min. PCR products were cloned into a pCR 3.1 expression vector (Invitrogen, San Diego, CA). Orientation of the *Cap43* reading frame in the isolated clones was verified using a double digestion with Nhe I-Hind III restriction endonucleases. The DNA sequence was confirmed using a Sequenase kit (US Biochemical, Cleveland, OH). Expression vectors of *Cap43* in the sense and antisense orientations were transfected into A549 cells by calcium phosphate precipitation, followed by selection in the presence of 600 μg/ml G418. Northern blot analysis of G418 resistant clones confirmed higher levels of *Cap43* expression for the sense vector host cells, and lower levels of expression with the antisense expression vectors.

3. RESULTS

3.1. The *Cap43* Is a Novel Gene Induced by Ni Compounds

Using differential display we cloned a gene that was induced up to 30-fold by soluble and insoluble Ni compounds in cultured human A549 cells (Zhou et al., 1998). The cDNA sequence of the *Cap43* gene that we obtained had a predicted open reading frame encoding 394 amino acid residues with a deduced molecular weight of 43.4 kd, and an isoelectric point of 5.3 (Figure 1). Northern blot analysis revealed a single band of approximately 3.0 kb. The sequence matched that of other human genes recently cloned from endothelial and colon epithelial cells indicating that they were the same gene product (Kokame et al., 1996; van Belzen et al., 1997). No functions were ascribed to the gene, except for the finding that the gene product was a marker of colon epithelium differentiation and was diminished in its expression in colon cancers (van Belzen et al., 1997). The high content of serines and threonines (16%) suggested that *Cap43* may be a good substrate for protein kinases. Potential phosphorylation sites for protein kinase C (3 sites), casein kinase II (3 sites), and tyrosine kinase (1 site) were found in the predicted sequence of *Cap43* (Figure 1). The *Cap43* gene had no transmembrane domain or zinc finger motif

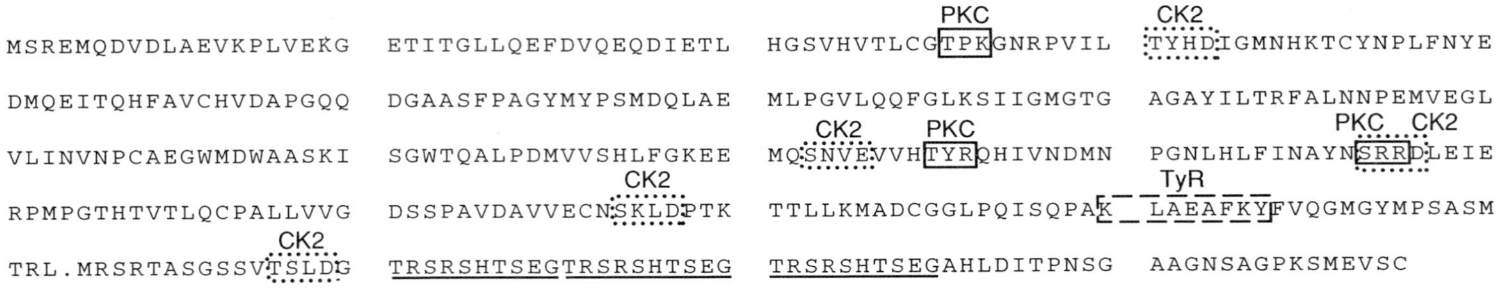

```
                                                      PKC             CK2
MSREMQDVDLAEVKPLVEKG  ETITGLLQEFDVQEQDIETL  HGSVHVTLCGTPKGNRPVIL  TYHDIGMNHKTCYNPLFNYE
                                                                        PKC   CK2
DMQEITQHFAVCHVDAPGQQ  DGAASFPAGYMYPSMDQLAE  MLPGVLQQFGLKSIIGMGTG  AGAYILTRFALNNPEMVEGL
                                         CK2    PKC
VLINVNPCAEGWMDWAASKI  SGWTQALPDMVVSHLFGKEE  MQSNVEJVHTYRQHIVNDMN  PGNLHLFINAYNESRRDLEIE
                            CK2                                          TyR
RPMPGTHTVTLQCPALLVVG  DSSPAVDAVVECNSKLDPTK  TTLLKMADCGGLPQISQPAK  LAEAFKYFVQGMGYMPSASM
                CK2
TRL.MRSRTASGSSVTSLDG  TRSRSHTSEGTRSRSHTSEG  TRSRSHTSEGAHLDITPNSG  AAGNSAGPKSMEVSC
```

Figure 1. Amino acid sequence of *Cap43*. The ten amino acid repeats on the C-terminal part of protein are underlined. Potential phosphorylation sites for protein kinase C (solid boxes), caseine kinase 2 (dotted boxes), and tyrosine kinase (dashed boxes) were also identified using PROSITE pattern search.

Table 1. *Cap43* induction by metal compounds A549 cells

Metals	Maximal induction (folds over basal levels)	Metals	Maximal induction (folds over basal levels)
Ni_3S_2	30	$HgCl_2$	2
$NiCl_2$	30	$NaAsO_2$	1
$ZnCl_2$	1	$MgCl_2$	1
$CoCl_2$	2	$Na_3VO_4^{\bullet}$	1
$CuCl_2$	1	Cisplatin[*]	1
$Pb(AC)_2$	1	K_2CrO_4	1
$CdCl_2$	3	FeS	1

A549 cells were exposed to the above agents for 24 hr at a variety of concentrations spanning from non toxic to toxic levels.
*These agents can neither induce *Cap43* expression nor change the extent of *Cap43* induction by 1 mM of $NiCl_2$.

or metal binding domains, but it had a new motif consisting of 10 amino acids repeated three times in the C terminus of the protein (Figure 1).

3.2. Specificity of *Cap43* Induction by Nickel in A549 Cells, and the Search for Signal Transduction Pathway Involved in Ni-Induced *Cap43* Expression

In order to investigate whether other metal compounds were able to induce the *Cap43* gene, A549 cells were treated with various metal compounds, and total RNA from exposed cells was assessed for *Cap43* expression. Apart from Ni_3S_2 and $NiCl_2$, the metal compounds tested were water-soluble chloride salts of zinc, cobalt, copper (II), cadmium, magnesium and mercury, water-soluble lead acetate, cisplatin, sodium vanadate, sodium arsenite, potassium chromate, and water-insoluble ferrous sulfide. These metal compounds did not significantly induce the expression of the *Cap43* gene (Table 1) at doses that ranged from non-toxic to lethal levels. A representative Northern blot showing the effect of cadmium on *Cap43* gene expression is shown on Figure 2. Apart from Ni, Cd was another metal to produce some induction 3-fold (Table 1). Hg^{2+} and Co^{2+} produced a 2-fold induction but all of these effects were small when compared with the 30-fold induction attributable to Ni^{2+} exposure.

Many toxic metals including Ni compounds are capable of producing oxidative stress in cells. To study whether oxidative stress was involved in the induction of *Cap43* expression, paraquat and hydrogen peroxide were used in a wide range of concentrations. Both failed to induce *Cap43* expression (Table 2) indicating that the observed increased expression of *Cap43* following Ni exposure was not likely due to oxidative stress induced by the Ni compounds themselves. Two mM of 2-mercaptoethanol or heat shock at 42°C for 15 min also failed to induce *Cap43*, suggesting that *Cap43* induction by Ni did not involve an unfolded protein response.

3.3. Phosphorylation Is Involved in *Cap43* Gene Induction

To test whether phosphorylation was involved in a signaling pathway leading to enhanced *Cap43* expression, we used two phosphatase inhibitors, sodium vanadate, a tyrosine phosphatase inhibitor, and okadaic acid, a specific inhibitor of serine/threonine phosphatases. Sodium vanadate, alone or in combination with Ni did not affect *Cap43* expression (Figure 3a), suggesting that tyrosine phosphorylation was not involved, however, okadaic acid (Figure 3b) alone was found to be a good inducer, suggesting that ser-

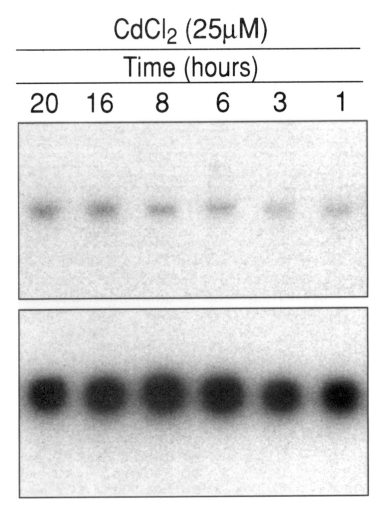

Figure 2. Effect of CdCl₂ on *Cap43* induction in A549 cells. A549 cells were treated with 25 μM of CdCl₂ for the time periods indicated in the Figure. Total RNA was isolated, separated by electrophoresis, transferred to filters and the filter was first hybridized with the *Cap43* probe (top). Then the membrane was stripped and rehybridized with β-actin (bottom) to verify the quantity of RNA loaded in each lane.

Table 2. *Cap43* induction by metal compounds A549 cells

Agent	Dose range tested	Maximal induction (folds of Basal levels)
H₂O₂	0.04-500 μM	1
Paraquat	20-200 μM	1
Dibutyryl cAMP*	250-2000 μM	1
Trifluoperazine*	2.5-20 μM	1
Homocysteine*	74-6000 μM	1
Homocysteine thiolactone	6 mM	1
S-adenosyl-homocysteine	6 mM	1
2-mercaptoethanol	50-2000 μM	1
Amethopterin	0.4-50 μM	1

A549 cells were exposed to the above agents for 24 hr.
*These agents can neither induce *Cap43* expression nor change the extent of *Cap43* induction by 1 mM of NiCl₂.

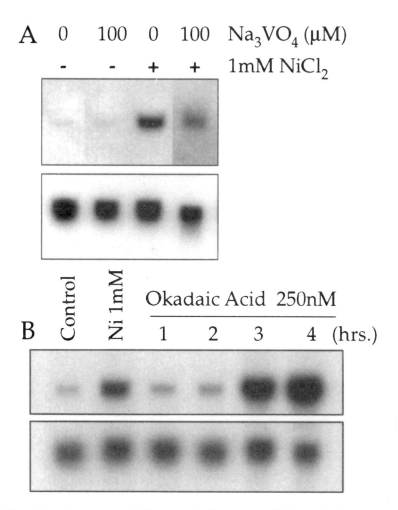

Figure 3. Effect of protein phosphatase inhibitors on *Cap43* expression. A) Effect of sodium vanadate on *Cap43* expression. Cells were treated with 100 µM of sodium vanadate for 20 h. 15 µg of total RNA was subjected to Northern blot analysis as described in the Methods section. Total RNA was isolated and the blot was first hybridized with the *Cap43* probe (top), then the membrane was stripped and rehybridized with β-actin (bottom). B) Effect of okadaic acid on *Cap43* expression. Cells were treated with 250 nM of okadaic acid for different periods of time, or treated with Ni for 8 h. Fifteen µg of total RNA was subjected to Northern blot analysis as described in the Methods section. Total RNA was isolated and the blot was first hybridized with the *Cap43* probe (top), then the membrane was stripped and rehybridized with β-actin (bottom).

ine/threonine phosphorylation was a component of the Ni-induced signaling pathway. To further narrow down the possible serine/threonine phosphorylation pathways we used dibutyryl cyclic AMP, an activator of PKA, at concentrations of up to 2 mM but found no effect on *Cap43* induction. Trifluoperazine (TFP), an inhibitor of calmodulin-dependent phosphorylation also had no effect on Ni-induced *Cap43* expression (not shown). Calphostin C and chelerythrine chloride, specific inhibitors of PKC also did not inhibit *Cap43* gene induction by Ni, suggesting that PKC was not involved in *Cap43* gene expression. We were also not able to inhibit *Cap43* gene expression using wortmannin, an inhibitor of phosphatidylinositol 3-kinase.

3.4. The Role of Intracellular and Extracellular Ca^{2+} in *Cap43* Induction

Ca^{2+} ionophore A23187 was a very efficacious inducer of *Cap43* in A549 cells, suggesting that a rise in free intracellular Ca^{2+} was involved in *Cap43* induction (Figure 4a). Direct measurements of calcium levels in A549 cells showed sustained elevations of free intracellular Ca^{2+} caused by 20 h of treatment of A549 cells with $NiCl_2$, or calcium ionophore A25187 (Figure 4b). The induction of *Cap43* by $NiCl_2$, or calcium ionophore

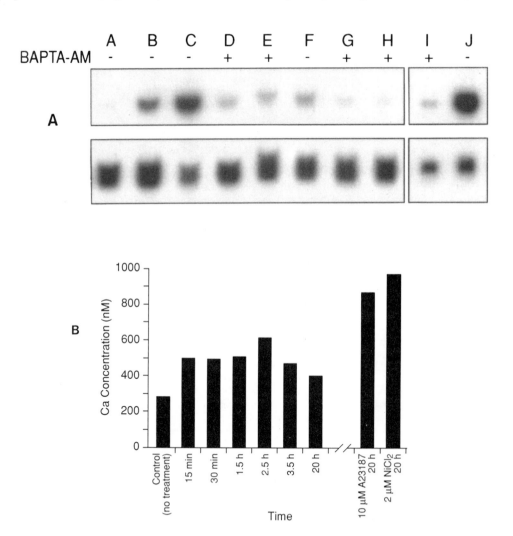

Figure 4. Intracellular calcium activates *Cap43* gene expression. **A)** Effect of BAPTA-AM on expression of *Cap43* mRNA. A549 cells were treated with 1 mM $NiCl_2$ (B and D), 250 nM of okadaic acid (C and E), 0.5 μM of thapsigargin (F and G) and 7 μM of calcium ionophore A23187 (I and J) for 20 h. In lanes D, E, G, and J, cells were simultaneously treated with 100 μM of BAPTA-AM. BAPTA-AM alone did not alter *Cap43* mRNA expression (H). Untreated A549 cells are shown in lane A. Total RNA was isolated and the blot was first hybridized with the *Cap43* probe (top), then the membrane was stripped and rehybridized with β-actin (bottom). **B)** Levels of intracellular free calcium in A549 cells treated with Ni and Calcium ionophore A23187. A549 cells were treated with 1 mM of $NiCl_2$ for different periods of time or with 2 mM of $NiCl_2$ or 10 μM of calcium ionophore A23187 for 20 h. Intracellular calcium levels were measured using the Fluo-3 fluorescence.

A25187 was abolished when intracellular Ca^{2+} was chelated by BAPTA-AM, suggesting that increased levels of Ca^{2+} were essential for *Cap43* gene induction (Figure 4a).

The *Cap43* gene has been independently cloned by another laboratory based upon its induction in endothelial cells (HUV-EC-C) by homocysteine (Kokame et al., 1996). We tested the induction of *Cap43* by 1mM $NiCl_2$ or by 6 mM of homocysteine in different cell lines (Zhou et al., 1998), and found that $NiCl_2$ significantly induced *Cap43* expression in all of the human cell lines tested including HUV-EC-C cells, while homocysteine only induced *Cap43* in HUV-EC-C cells (Zhou et al., 1998). When Ni and homocysteine were added together the level of expression was identical to that attributed to Ni alone. It was of interest to address whether the effect of homocysteine on *Cap43* expression in HUVEC-C cells was mediated by an elevation of intracellular Ca^{2+}. Figure 5 showed that the induction of *Cap43* in HUVEC-C cells was abolished by the intracellular Ca^{2+} chelator BAPTA-AM. The importance of an initial elevation of intracellular Ca^{2+} was confirmed in experiments where BAPTA-AM failed to attenuate *Cap43* induction by Ni when added 4 h after Ni exposure, suggesting that an early rise in cellular Ca^{2+} levels were important for *Cap43* induction (not shown).

Because Ni has previously been shown to be a calcium channel blocker we investigated whether changes in extracellular Ca^{2+} levels would affect *Cap43* gene expression. Direct chelation of calcium with EGTA added to culture media did not enhance gene expression (Figure 6). To further study the effects of extracellular Ca^{2+} levels on *Cap43* expression cells were incubated in phosphate buffer containing various levels of Ca^{2+} up to a

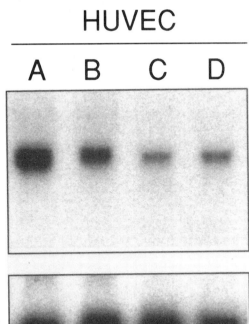

Figure 5. *Cap43* expression in HUVEC-EC cells. *Cap43* was induced by calcium ionophore A23187 and homocysteine in HUVEC cells. The expression was abolished by the addition of 100 μM of intracellular calcium chelator BAPTA-AM. Total RNA was isolated and the blot was first hybridized with the *Cap43* probe (top), then the membrane was stripped and rehybridized with β-actin (bottom).

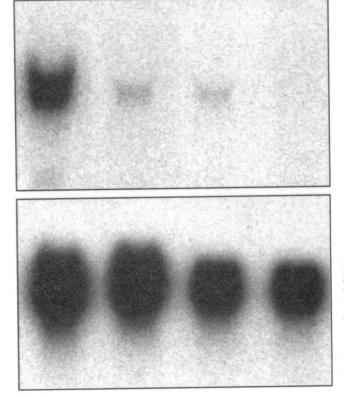

Figure 6. Effect of EGTA on *Cap43* gene expression. A549 cells were treated with 1 mM of $NiCl_2$ (A), 0.5 mM of EGTA (B), 1.0 mM EGTA (C), or not treated (D). Total RNA was isolated and the blot was first hybridized with the *Cap43* probe (top), then the membrane was stripped and rehybridized with β-actin (bottom).

maximum of 7 mM. Changing the extracellular Ca^{2+} levels did not affect *Cap43* gene expression (not shown). At the same time, Ni induced *Cap43* gene expression regardless of the presence or absence of extracellular Ca^{2+}.

3.5. *Cap43* mRNA Induction Results in an Elevation of the *Cap43* Protein

We have prepared antibodies against a peptide synthesized from amino acid 340-353 of *Cap43* and have found a corresponding increase of *Cap43* immunoreactive intracellular protein induced by $NiCl_2$ parallels the rise in the levels of *Cap43* mRNA (Figure 7). *Cap43* protein was not found in the nucleus but was localized to the perinuclear region, and cytoplasm. It was interesting to note that *Cap43* protein elevation induced by Ni was as striking as its mRNA induction (Figure 7).

3.6. *Cap 43* Is a Growth/Differentiation Related Gene

A549 cells have been engineered to overexpress or to underexpress *Cap43* mRNA by transfecting a *Cap43* expression vector oriented in either the sense or antisense direction. Northern analysis confirmed that we had derived several clones with lower levels of *Cap43* expression when the antisense vector was transfected, as well as several clones

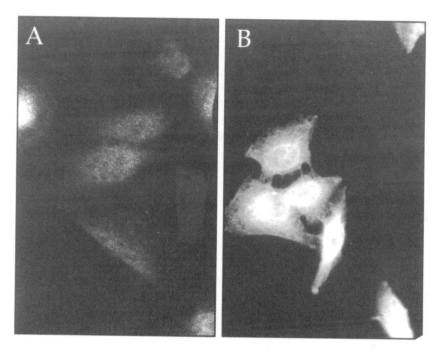

Figure 7. Effect of NiCl₂ on *Cap43* protein induction. Rabbit polyclonal antiserum was prepared against peptide 340–353 of *Cap43* amino acid sequence. (A). A549 cells not treated or treated for 24 h with 1 mM NiCl₂ (B) were fixed and stained for immunoreactive protein as described in the Methods section.

with *Cap43* overexpression levels using the sense vector. Colonies derived from antisense *Cap43* cells were generally smaller and grew poorly as compared with the sense clones. To test whether cell growth was impaired in the antisense clones we measured the daily changes in cell number in cells transfected with the sense or antisense clones. Cell division was profoundly inhibited in cells with lowered *Cap43* expression, whereas cell growth was normal in the sense clones (Figure 8).

4. DISCUSSION

To enhance our understanding of the molecular mechanisms involved in Ni toxicity we have cloned a gene that was specifically induced in human A549 cells by Ni com-

Figure 8. Growth curve of A549 cells transfected with *Cap43* expression vectors. The 4-1-2-cell line contained the *Cap43* sequence in the sense orientation, or -2-1 and 3-2-2- cell lines contained the *Cap43* sequence in the antisense orientation. The levels of expression of endogenous *Cap43* mRNA and the *Cap43* transgene were determined by Northern blot. The endogenous and inducible levels of expression for the *Cap43* mRNA were significantly lower in the 3-2-1 and 3-2-2 cell lines compared to A549 cells or the 4-1-2 cell lines.

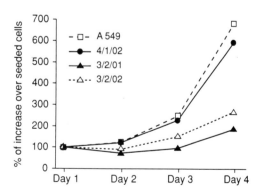

pounds. We found that this gene was also induced by calcium ionophore A23187, and the induction of the gene by Ni or calcium ionophore was abolished when free intracellular calcium was chelated with BAPTA-AM. The direct measurement of free calcium levels confirmed the notion that they were elevated in Ni-treated cells. Therefore, this gene was called *Cap43*, or calcium activated protein with a molecular weight of 43 kd. This is one of the first studies showing that Ni toxicity resulted in sustained elevation of intracellular calcium. The mechanisms leading to the elevation of intracellular calcium in Ni-treated cells are currently not well understood. Ni is known to be a Ca^{2+} channel blocker (Zamponi et al., 1996). It is conceivable that intracellular Ca^{2+} initially dropped in response to Ni treatment but this was followed by a compensatory rise of free Ca^{2+} by its release from intracellular stores. A blockade of calcium entry into the cell by Ni may eventually also elevate free Ca^{2+}. Another possibility is that Ni can interact with a Ca^{2+} sensor or Ca^{2+} receptor on plasma membrane and this interaction activated Ca^{2+} release inside the cell. In our experiments, however, modulations of extracellular calcium levels did not affect *Cap43* expression. We have shown that in the absence of Ca^{2+} outside of cells, or an increase of extracellular Ca^{2+} levels up to 7 mM did not alter *Cap43* induction by 1 mM of $NiCl_2$, indicating that if Ni were interacting with a calcium receptor, it probably was not a Ca^{2+} binding domain. On the other hand, the equal potency of soluble and insoluble Ni compounds in terms of inducing *Cap43* suggested that this may be a membrane related effect since water-soluble $NiCl_2$ produces much lower doses of intracellular soluble Ni^{2+} compared with the more carcinogenic Ni_3S_2 or NiS (Costa, 1991).

The route of divalent Ca^{2+} entry into cytoplasm has not been completely resolved. Even for the well studied calcium ionophore, it is not clear whether it enhances Ca^{2+} influx via direct action at the plasma membrane or by stimulation of store-regulated cation entry (Morgan and Jacob, 1994).

The function of *Cap43* is presently unknown, however, it was independently cloned by another group who showed it to be induced by homocysteine in HUVEC cells (Kokame et al., 1996). Their data is consistent with our study since we also have shown that homocysteine alone induced Cap43 gene expression in HUVEC cells. Moreover we were able to induce *Cap43* gene expression in HUVEC cells using the $NiCl_2$, and calcium ionophore A23187. The induction of *Cap43* in HUVEC cells by homocysteine, calcium ionophore, or Ni was abolished with BAPTA-AM confirming that the elevation of intracellular Ca^{2+} was the common signal for *Cap43* gene induction. The ability of homocysteine to activate Ca^{2+}-dependent processes is restricted to the HUVEC cell because homocysteine mainly acts through N-methyl-D-aspartate (NMDA) receptor, a glutamate-gated calcium channel receptor expressed on a surface of only certain types of cells (Dalton et al., 1997). This explains why homocysteine did not induce *Cap43* in other types of cells such as lung epithelial and fibroblasts. In contrast Ni compounds induced *Cap43* in all human tested cell lines, with a maximum induction being achieved at 24 h after exposure.

Analysis of the amino acid sequence of the *Cap43* protein did not reveal any features that would identify its function in the Ca^{2+} signaling pathway, however, from the antibody staining, we know that *Cap43* was localized in the perinuclear region and was not a nuclear protein. Its localization was consistent with a possible location in the endoplasmic reticulum, the major site of Ca^{2+} storage in the cell.

Elevations of free intracellular Ca^{2+} are known to activate Ca^{2+}-dependent phosphorylation (Rosen et al., 1995). Protein kinase C is one of the kinases activated by Ca^{2+}. Therefore, the involvement of the protein kinase C in Ni-induced signal transduction pathways was examined. However, two specific inhibitors of protein kinase C, calphostin C and chelerythrine chloride, had no effect on *Cap43* induction by Ni or by calcium ionophore. Trifluoperazine, an inhibitor of calmodulin-dependent phosphorylation, also had no effect on Ni-induced Cap43 expression, suggesting that perhaps another calcium-dependent protein kinase may be involved in *Cap43* gene induction.

The number of unrelated genes currently known to be induced by a rise in free intracellular Ca^{2+} includes glucose-regulated genes (Resendez et al., 1985) prolactin (Delidow et al., 1992) as well as transcription factors and chaperone proteins (Resendez et al., 1986; Choi et al., 1994). However, the level of induction of *Cap43* was striking compared with other genes. Since calcium is an important second messenger, *Cap43* may be involved in transducing Ca^{2+}-dependent effects on cell growth. To examine this possibility we cloned the *Cap43* coding region in the sense and antisense orientations into an expression vector and transfected it into A549 cells. Clones that had *Cap43* in the sense orientation had growth rates similar to parental cells, however, clones that had *Cap43* in the antisense orientation grew slower supporting the idea that *Cap43* was a growth-related gene.

Cap43, called the DrgI gene by another investigator, was shown to be a differentiation marker of colon cells, since it was induced approximately 20-fold during differentiation of HT29 cells (van Belzen et al., 1997). In adenomas *Cap43* gene expression was found to be 2-7-fold lower compared to biopsies from normal colon tissue, and in adenocarcinomas, the level of *Cap43* expression was lower by 2-12-fold (van Belzen et al., 1997). This data suggested that the loss of *Cap43* expression can be used as a prognosis marker in colon cancer.

In conclusion our study identifies the signal for *Cap43* induction as a rise in free intracellular Ca^{2+} and we show that *Cap43* was the most induced and persistent gene expression marker that responded to a sustained rise in free intracellular Ca^{2+}. Calcium is an important second messenger that mediates the effect of growth factors. Here we have also demonstrated the importance of expression of *Cap43* in cell growth. It is already clear that *Cap43* is an important gene, however, further work is required to understand the function of *Cap43*.

ACKNOWLEDGMENTS

This work was supported by grant numbers ES05512 and ES00260 from the National Institute of Environmental Health Sciences and grant number CA16087 from the National Cancer Institute.

REFERENCES

Bauman, J.W., Liu, J., and Klaassen, C.D. (1993). Production of metallothionein and heat-shock proteins in response to metals. Fund. Appl. Toxicol. 21, 15–22.

Bennett, B.G. (1984). Environmental nickel pathways to man. In: Sunderman, F.W. Jr. (Ed.) "Nickel in the Human Environment." IARC Scientific Publication No. 53, Lyon, pp 487–495.

Biedermann, K.A. and Landolph, J.R. (1987). Induction of anchorage independence in human diploid foreskin fibroblasts by carcinogenic metal salts. Cancer Res. 47, 3815–3823.

Choi, A.M., Tucker, R.W., Carlson, S.G., Weigand, G., and Holbrook, N.J. (1994). Calcium mediates expression of stress-response genes in prostaglandin A2-induced growth arrest. FASEB J. 8, 1048–1054.

Conway, K. and Costa, M. (1989). Nonrandom chromosomal alterations in nickel-transformed Chinese hamster embryo cells. Cancer Res. 49, 6032–6038.

Costa, M. (1991). Molecular mechanisms of nickel carcinogenesis. Annu. Rev. Pharmacol. Toxicol. 31, 321–327.

Dalton, M.L., Gadson, P.F., Wrenn, R.W., and Rosenquist, T.H. (1997). Homocysteine signal cascade: production of phospholipids, activation of protein kinase C, and the induction of c-*fos* and c-*myb* in smooth muscle cells. FASEB J. 11, 703–711.

Delidow, B.C., Lail-Trecker, M., and White, B.A. (1992). Effects of calcium and calcium ionophores on prolactin gene expression in GH3 and 235–1 rat pituitary tumor cells. Mol. Endocrinol. 6, 1268–1276.

Doll, R., Morgan, L.G., and Speizer, F.E. (1970). Cancers of the lung and nasal sinuses in nickel workers. Br. J. Cancer 24, 623–632.

Epner, D. E. and Herschman, H.R. (1991). Heavy metals induce expression of the TPA-inducible sequence (TIS) genes. J. Cell. Physiol. 148, 68–74.

Jin, P. and Ringertz, N.R. (1990). Cadmium induces transcription of proto-oncogenes c-*jun* and c-*myc* in rat L6 myoblasts. J. Biol. Chem. 265, 14061–14064.

Karin, M., Andersen, R.D., Slater, E., Smith, K., and Herschman, H.R. (1980). Metallothionein mRNA induction in HeLa cells in response to zinc or dexamethasone is a primary induction response. Nature 286, 295–297.

Karin, M., Haslinger, A., Holtgreve, H., Cathala, G., Slater, E., and Baxter, J.D. (1984). Activation of a heterologous promoter in response to dexamethasone and cadmium by metallothionein gene 5'-flanking DNA. Cell 36, 371–379.

Kokame, K., Kato, H., and Miyata, T. (1996). Homocysteine-respondent genes in vascular endothelial cells identified by differential display analysis. J. Biol. Chem. 271, 29659–29665.

Langard, S. (1994). Nickel-related cancer in welders. Sci. Total Environ. 148, 303–309.

Liang, P. and Pardee, A.B. (1992). Differential display of eukaryotic messenger RNA by means of the polymerase chain reaction. Science 257, 967–971.

Morgan, A.J. and Jacob, R. (1994). Ionomycin enhances Ca^{2+} influx by stimulating store-regulated cation entry and not by a direct action at the plasma-membrane. Biochemical J. 300, 665–672.

Ottolenghi, A.D., Haseman, J.K., Payne, W.W., Falk, H.L., and MacFarland, H.N. (1975). Inhalation studies of nickel sulfide in pulmonary carcinogenesis of rats. J. Natl. Cancer Inst. 54, 1165–1172.

Pozzan, T., Rizzuto, R. Volpe, P., and Meldolesi, J. (1994). Molecular and cellular physiology of intracellular calcium stores. Physiol. Rev. 74, 595–636.

Resendez, E. Jr., Attenello, J.W., Grafsky, A., Chang, C.S., and Lee, A.S. (1985). Calcium ionophore A23187 induces expression of glucose-regulated genes and their heterologous fusion genes. Mol. Cell. Biol. 5, 1212–1219.

Resendez, E. Jr., Ting, J., Kim, K.S., Wooden, S.K., and Lee, A.S. (1986). Calcium ionophore A23187 as a regulator of gene expression in mammalian cells. J. Cell. Biol. 103, 2145–2152.

Rosen, L.B., Ginty, D.D., and Greenberg, M.E. (1995). Calcium regulation of gene expression. Adv. Second Messenger Phosphoprotein Res. 30, 225–253.

Salnikow, K., Wang, S., and Costa, M. (1997). Induction of activating transcription factor I by nickel and its role as a negative regulator of thrombospondin I gene expression. Cancer Res. 57, 5060–5066.

Shen, H.M. and Zhang, Q.F. (1994). Risk assessment of nickel carcinogenicity and occupational lung cancer. Environ. Health Perspect. 102, Suppl. 1, 275–282.

Sunderman, F.W. Jr. (1981). Recent research on nickel carcinogenesis. Environ. Health Perspect. 40, 131–141.

Sunderman, F.W. Jr. (1989). Mechanisms of nickel carcinogenesis. Scand. J. Work Environ. Health 15, 1–12.

Trott, D.A., Cuthbert, A.P., Overell, R.W., Russo, I., and Newbold, N.F. (1995). Mechanisms involved in the immortalization of mammalian cells by ionizing radiation and chemical carcinogens. Carcinogenesis 16, 193–204.

van Belzen, N., Dinjens, W.N.M., Diesveld, M.P.G., Groen, N.A., van der Made, A.C.J., Nozawa, Y., Vliestra, R., Trapman, J., and Bosman, F.T. (1997). A novel gene which is up-regulated during colon cell differentiation and down-regulated in colorectal neoplasms. Lab. Invest. 77, 85–92.

Yiangou, M. and Papaconstantinou, J. (1993). The differential induction of alpha1- acid glycoprotein and serum amyloid A genes by heavy metals. Biochim. Biophys. Acta 1174, 123–132.

Zamponi, G.W., Bourinet, E., and Snutch, T. P. (1996). Nickel block of a family of neuronal calcium channels: subtype- and subunit-dependent action at multiple sites. J. Membrane Biol. 151, 77–90.

Zhou, D., Salnikow, K., and Costa, M. (1998). *Cap43*, a novel gene specifically induced by Ni^{2+} compounds. Cancer Res., in press.

9

ZINC AND METALLOTHIONEIN IN MAMMALIAN CELL CYCLE REGULATION

Detmar Beyersmann and Carsten Schmidt

Department of Biology and Chemistry
University of Bremen
D-28334 Bremen, Germany

1. IS ZINC A MITOGENIC FACTOR?

Zinc has been discussed to be an inorganic proliferation factor for eukaryotic tissues since about a quarter of a century. It has been found to stimulate DNA synthesis in chick embryo fibroblasts (Rubin and Koide, 1973) and mouse 3T3 fibroblasts (Chesters et al., 1989) and to activate thymidine kinase in various cell types (Prasad, 1979). It has been postulated to act as a second messenger of mitogenesis with baby hamster kidney cells (Grummt et al., 1986). More recently, zinc has been found to be involved in several signal transducing enzymes critical for the induction of mitogenesis in mammalian cells. Zinc stimulated protein kinase C (Zalewski et al., 1990), tyrosine kinase (Vener and Loeb, 1992) and MAP kinase (Hansson et al., 1996), and inhibited a protein phosphatase (Wang et al., 1992). Furthermore, zinc ions have been shown to be required for the onset of differentiation of myoblasts (Petrie et al., 1991). On the genetic level, zinc is an essential constituent of an important class of DNA binding transcription factors possessing zinc finger structures (Berg and Shi, 1996).

On the other hand, zinc deprivation has been shown to induce apoptosis in lymphoid cells (Martin et al., 1991), in thymocytes (McCabe et al., 1993), in lymphocytes (Zalewski et al., 1993) and in PC12 cells (Villalba et al., 1995). In accordance with these findings, toxic zinc concentrations of 250 μM prevented the dexamethasone-induced apoptosis in leukemic cell lines (Adebodun and Post, 1995), and $ZnCl_2$ in a dose-dependent manner inhibited apoptotic DNA-fragmentation in isolated liver cell nuclei (Lohmann and Beyersmann, 1995). In contrast with these observations, addition of excessive $ZnCl_2$ induced apoptosis in human liver cells (Paramanathan et al., 1979), hepatoma cells (Xu et al., 1996) and thymocytes (Fraker et al., 1995). This type of induction of apoptosis by zinc does not appear to be the rule as in most other systems, excessive zinc caused cell necrosis.

Is Zn^{2+} a good candidate for a second messenger function as it is for Ca^{2+}? Zinc differs from calcium by forming kinetically inert complexes with sulfurs and nitrogens of proteins. Hence Zn^{2+} ions may function as regulators of cellular processes, but unlike cal-

Metals and Genetics, edited by Sarkar.
Kluwer Academic / Plenum Publishers, New York, 1999.

145

cium ions they are not suitable for the control of fast metabolic responses. They seem to be better candidates for more profound changes in life and death of cells as the onset of proliferation and differentiation. If so, Zn^{2+} ions must meet further condiditions, i.e. their intracellular concentration has to be controlled in a homeostatic manner, and the intracellular zinc level has to be modulated by signals promoting growth and differentiation.

2. CELLULAR ZINC UPTAKE AND CONTROL OF INTRACELLULAR ZINC HOMEOSTASIS

A most pertinent question is how cellular zinc homeostasis is maintained. Whereas the homeostasis of intracellular free calcium ions is precisely controlled by the regulation of cellular Ca^{2+} uptake and export, the question of whether and how the cellular uptake of zinc ions is regulated in in response to cellular functions has not been elucidated satisfactorily. The intracellular zinc concentration of mammalian cells seems to be subject to a much wider variation than that of calcium, and the factors confining the cellular zinc ion concentration are not yet fully understood.

The mechanisms of zinc uptake are quite diverse, depending on the type of tissue. With liver cells, two phases of uptake could be distinguished (Pattison and Cousins, 1986; Taylor and Simons, 1994): First, there is a fast-exchanging, labile zinc pool responding to the extracellular zinc concentration, but independent of hormones. Second, there is a slow-exchanging zinc pool stimulated by dexamethasone and other hormones. The latter mechanism was inhibited by actinomycin D and cycloheximide (Steinebach and Wolterbeek,1992). Human fibroblasts use a K^+-dependent mechanism for the basal, hormone-independent zinc uptake and a Ca^{2+}-dependent mechanism for the hormone-dependent route (Ackland and McArdle, 1996). In addition, fibroblasts seem to use endocytosis for a part of their zinc uptake (Grider and Vazquez, 1996). Rabbit kidney cells use two different mechanisms for zinc uptake by apical and basolateral cells, and neither route depends on the cellular metallothionein level (Gachot et al., 1994). Whereas erythrocytes needed ATP for zinc uptake (Schmetterer, 1978), no requirement for ATP has been found for zinc uptake by most other cell types, and also the uptake into rat liver nuclei was not requiring ATP (Hechtenberg and Beyersmann, 1995).

After its uptake, zinc seems to be subject to a non-homogeneous intracellular distribution in hepatocytes, as detected by a zinc-specific fluorescent probe, Zinquin (Coyle et al., 1994). This "starry night" appearance of Zinquin fluorescence either may be caused by interaction of the dye with intracellular membrane constituents or by a vesicular sequestration of Zn^{2+} ions. The concentration of intracellular free Zn^{2+}-ions in biological media is very low due to tight binding of the ion to proteins such as serum albumins in blood and intracellular proteins, especially metallothionein, which acts as a zinc ion buffer (Cousins, 1985). The intracellular free Zn^{2+}-concentrations reported by various authors vary considerably depending on the type of tissue and the methods of analysis (Table I). In several cell types, assays with the chelating fluorescent probes fura-2 and mag-fura-5 as well as with the NMR-probe 5-F-BAPTA yielded free Zn^{2+} concentrations in the lower nanomolar range. However, with each method of analysis, there exist some major drawbacks. The fluorescent probes of the fura-type are useful only in the absence of interfering divalent ions like Ca^{2+} and heavy metal ions which form fluorescent fura-complexes, too. The NMR-technique is very elaborate and time consuming and is relatively unsensitive. However, the NMR method has the great advantage to allow the simultaneous measurement of different intracellular free ion species, because 5-F-BAPTA exhibits different chemical shifts when bound to different ion species. The fluorescent probe with the highest specifity

Table I. Intracellular Zn^{2+} ion concentrations and methods of determination

Cell type	Zinc concentration	Method of analysis	References
Erythrocytes	0.024 nM	^{65}Zn radiolabelling	Simons 1991
Synaptosomes	1.4 nM	^{19}F-NMR of 5-F-BAPTA complex	Denny and Atchison, 1994
Human leukemic cells	1 nM	^{19}F-NMR of 5-F-BAPTA complex	Adebodun and Post, 1995
PC12 cells	0.5 nM	^{19}F-NMR of 5-F-BAPTA complex	Benters, 1997
Stimulated chromaffin cells	0.4–2 nM	Fura-2 fluorescence	Zalewski et al., 1994
Living neurons	35–45 nM	Mag-fura-5 fluorescence	Sensi et al., 1997
Resting splenocytes	20–50 µM	Zinquin fluorescence	Zalewski et al., 1993
Hepatocytes	0.6–2.7 µM	Zinquin fluorescence	Coyle et al., 1994

for zinc is Zinquin, whose fluorescence is confounded only by the presence of Cd^{2+} ions in toxic concentrations. The drawback with Zinquin is that it mobilizes labile Zn^{2+} ions from protein-bound states and rather reports the total mobilizable Zn^{2+} ion level than the concentration of free Zn^{2+} ions. Zinquin in special conditions may even cause depletion of zinc from metallothionein and degradation of metallothionein (Coyle et al., 1994).

3. SUBCELLULAR LOCALIZATION OF ZINC

As discussed above, compared to the extremely fine regulation of the cytoplasmic Ca^{2+} concentration, the intracellular zinc concentration is much more dependent on the extracellular zinc level. Obviously, the uptake of Zn^{2+} through the plasma membrane is not as restrictive as that of Ca^{2+}. After its induction by elevated intracellular zinc, metallothionein plays a major role in the adjustment of the intracellular free Zn^{2+} level. There is also some evidence for the intracellular containment of Zn^{2+} by inclusion into vesicles: The fluorescence of the intracellular Zn^{2+} probe Zinquin is inhomogeneous with denser appearance in structures resembling intracellular vesicles (Zalewski et al., 1993). It has still to be elucidated whether these structures really represent zinc storing vesicles. In brain cells an active uptake and vesicular storage of zinc and mobilization of Zn^{2+} by stimulation of cells by neurotransmitters has been observed (Wenzel et al., 1997; Budde et al., 1997).

Is there a regulation of the Zn^{2+}-concentration in cell nuclei? If zinc ions function as regulators of genetic elements, their intranuclear concentration has to be regulated, too. Dietary zinc is taken up relatively fast into the cell nuclei of liver, kidney and spleen, as estimated from the total nuclear zinc (Cousins and Lee-Ambrose, 1992). The nuclear pore complex does not allow the free flow of cations, and uptake of cations seems to be regulated by the nuclear pores (Mazzanti et al., 1991). Further support of this theory is the observation that the uptake of total zinc by isolated liver cell nuclei follows the medium concentration but the concentration of free intranuclar Zn^{2+}-ions is confined by an unknown mechanism (Hechtenberg and Beyersmann, 1995).

4. REGULATION OF INTRACELLULAR ZINC AND METALLOTHIONEIN IN PROLIFERATION AND DIFFERENTIATION

One important type of molecules regulating the intracellular free Zn^{2+}-concentration are the metallothioneins which act as zinc buffers (Cousins, 1985). The expression of met-

Table II. Metallothionein (MT) induction and intracellular localization

Cell type	Observations	References
Developing Drosophila melanogaster	Induction of MT	Silar et al., 1990
Regenerating rat liver	Nuclear MT highest in S-G2/M phases	Tsujikawa et al., 1994
Rat hepatocytes	Epidermal growth factor and insulin augment nuclear localization of MT	Tsujikawa et al., 1991; Moffat et al., 1994
Fetal and newborn rat and human hepatocytes	Nuclear localization of MT	Templeton et al., 1985; Nartey et al., 1987a
Human tumor cells	Nuclear accumulation of MT in rapidly proliferating cells	Nartey et al., 1987b; Kuo et al., 1994;
Human colon cancer cells	Cytoplasmic MT peak at G1/S transition	Nagel and Vallee, 1995
Human tumor cells	Nuclear localization of MT sensitive to chilling and energy depletion	Woo et al., 1996
Developing rat brain	Changes in metallothionein and zinc	Ono and Cherian, 1998

allothionein genes is induced by hormones that stimulate cell proliferation as insulin and dexamethasone. The induction of metallothionein is coupled to cell cycle phases. With several rapidly growing tumor cell types, a predominant nuclear localization of metallothionein is found, and with several cell lines stimulated for proliferation, metallothionein is accumulated in cell nuclei during S- to G2/M phases (Table II). Metallothioneins are also induced in the development of rat liver (Andrews et al., 1987), mouse brain and Drosophila. Furthermore, the ratio of cytoplasmic to nuclear localization of metallothionein changes in the course of developmental processes. Nuclear metallothionein is also observed in fetal and newborn rat and human hepatocytes, but declines during adolescence. However, it is not clear, whether and how these shifts are related to changes in the intracellular localization of zinc.

5. ZINC AND METALLOTHIONEIN LEVELS AND SUBCELLULAR LOCALIZATION DURING PROLIFERATION AND DIFFERENTIATION OF MURINE 3T3L1 FIBROBLASTS

5.1. Flow Cytometric Analysis of Proliferation and Differentiation of 3T3L1 Cells

Mouse 3T3L1 fibroblasts can be stimulated by hormones to undergo a synchronous process of proliferation and subsequent differentiation to adipocytes. This process proceeds in three stages (Figure 1). For the first eight hours after treatment with a mixture of dexamethasone, insulin and isobutylmethylxanthin, cells are predominantly in the G_1 / G_0 phase of the cell cycle. After 24 h there is an onset of rapid proliferation where more than 70% of the cells are in the proliferative S- and G_2/M-phases. After further 24 h, proliferation declines to nearly the same level as at the start of induction, and the cells are found predominantly (70 to 80%) in the G_1 / G_0 phase. During this time, the cells start to differentiate to the adipocyte-type, i.e. they accumulate intracellular lipid vesicles. This process is completed between 96 and 168 h after induction. After the onset of differentiation 72 h and later after the first stimulation, cells cannot be further stimulated by hormones to undergo proliferation (data not shown).

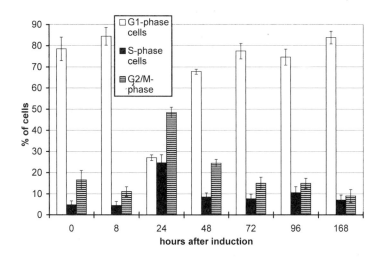

Figure 1. Flow cytometric analysis of proliferating and differentiating of 3T3L1 cells after stimulation by hormones. Cells were induced by a mixture of dexamethasone, insulin and 3-isobutyl-1-methylxanthine, and assayed at the indicated times.

5.2. Induction of Metallothionein and Uptake of Zinc by Stimulated 3T3L1 Cells

Figure 2 shows the concentrations of metallothionein content in 3T3L1 cells in the course of proliferation and differentiation. Paralleling the rise of the fraction of S-phase cells, the cellular levels of total metallothionein and total zinc rise to a peak and decline as the cells differentiate to adipopcytes. In these experiments, it is undiscernible whether zinc or metallothionein changes first and which subsequently. It may be assumed, however, that metallothionein is induced by dexamethasone and insulin in 3T3L1 cells as it is the case in other cell types. If this is the case, the labile intracellular zinc will be bound to the additional metallothionein. The rise in total cellular zinc would have to occur retarded as a consequence of the hormone-induced rise of metallothionein, but this is not detected in this system.

5.3. Subcellular Localization of Metallothionein and Zinc in Stimulated 3T3L1 Cells

Simultaneously with the entry of the induced cells into the S-phase, metallothionein is translocated from the cytoplasm to the nucleus (Figure 3). In non-stimulated, confluent cells, antibodies against metallothionein mainly bind in the cytoplasm. At the stage of rapid proliferation, 24 h after hormonal induction, 70% of the metallothionein is found in the nucleus. During the subsequent process of differentiation to adipocytes, the intracellular localization of metallothionein is reversed. Paralleling the decrease of the proportion of cells in proliferative S- and G_2 / M phases, metallothionein appears again in the cytoplasm. In fully differentiated adipocytes, where the proliferation rate has declined nearly completely, metallothionein is nearly not detectable within cell nuclei.

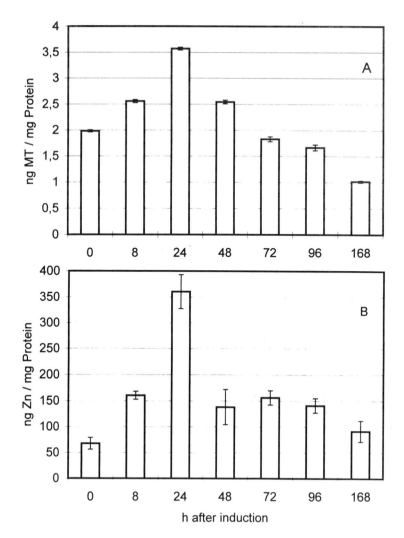

Figure 2. Cellular levels of total metallothionein (A) and total zinc (B) in the course of proliferation and differentiation of 3T3L1 cells after stimulation by hormones. Cells were induced as described in the methods section and analyzed for metallothionein by the silver staining method of Scheuhammer and Cherian (1991) and for total zinc by atomic absorption spectroscopy.

To analyse the subcellular distribution of labile zinc in 3T3L1 cells, the zinc-specific intracellular fluorescent probe Zinquin was employed. We confirmed the finding of Zalewski et al.(1993) who observed a inhomogeneous, "starry" distribution of Zinquin fluorescence in the cytoplasm of mammalian cells, indicating a vesicular storage of zinc. The intracellular distribution of Zinquin fluorescence was predominantly in the cytoplasm, whereas the nuclei appeared void of staining (data not shown). In contrast with the observed translocation of metallothionein to cell nuclei during the proliferative stages of the cell cycle, there was only a slight increase in the fraction of cells with nuclear Zinquin staining. Thus, the intracellular localization of labile zinc does not parallel the nuclear translocation of metallothionein in proliferating 3T3L1 cells.

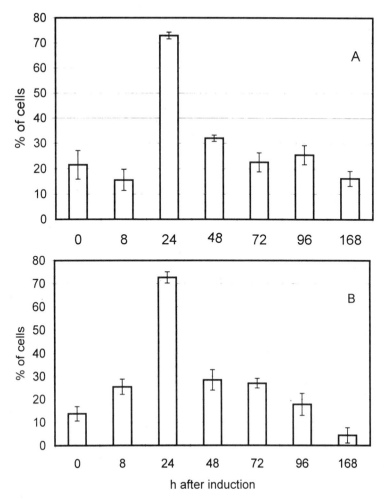

Figure 3. Nuclear localization of metallothionein in the course of proliferation and differentiation of 3T3L1 cells after stimulation by hormones. A, Fraction of cells in the proliferative S- and G2 / M stages; B, Fraction of cells with nuclear localization of metallothionein. Nuclear metallothionein was determined by immunostaining as dscribed in the methods section.

6. TOXIC EFFECTS OF ZINC DEPRIVATION AND OVERLOAD

Zinc is essential for cells to enter the S-phase, since treatment with the membrane-permeating zinc chelator TPEN prevents stimulated 3T3L1 cells from proceeding to the proliferative S and G2 / M stages (Table III). As reported in the introduction, zinc deprivation causes apoptosis of various cell types. This is also true for 3T3L1 cells. After treatment of these cells with 10 μM of the membrane-permeant zinc chelator TPEN, the cells undergo apoptosis as demonstrated by the appearance of small, so-called sub-G1-phase cells in the flow cytometer, which represent apoptotic bodies (Figure 4). The cells were most resistant to zinc deprivation at 24 hours after induction, i.e. during the proliferative phase, whereas the confluent control cells at time zero and the differentiating cells at 72 h after induction were more sensitive.

Table III. Inhibition of proliferation of 3T3L1 cells by the zinc chelator TPEN

Time (h)	TPEN-treated cells in S and G2/M stages	Untreated cells in S and G2/M stages
0	21.88 ± 4.11	21.35 ± 4.48
24	24.40 ± 2.75	72.93 ± 3.88
48	31.72 ± 2.33	32.89 ± 2.75
72	29.95 ± 2.55	22.50 ± 2.89

At the indicated times after stimulation by hormones, cells were treated with TPEN for 24 h, and the fraction of cells in proliferative stages S and G2 / M were estimated by flow cytometry.

On the other hand, excessive zinc is cytotoxic and causes necrosis as measured with the Neutral Red assay (Figure 5). At variance with the relative resistance of confluent cells to zinc deprivation, this cell stage is most sensitive to elevated concentrations of zinc and the cells die when exposed to more than 40 µM $ZnCl_2$ for 24 h, whereas the differentiated cells at 72 h and more tolerate the highest zinc concentrations of up to 100 µM $ZnCl_2$ for 24 h. Proliferating cells at 24 h after induction exhibit an intermediate sensitivity to excess $ZnCl_2$.

Is there an intermediate zinc concentration between deprivation and excess of extracellular zinc which stimulates cell proliferation as it has been observed with some other cell types? To test whether extracellular zinc stimulates cell proliferation in the absence of hormones, cells were incubated with the non-toxic concentration of 20 µM $ZnCl_2$ for 24 h and subjected to a flow cytometric analysis. These cultures did not show a significantly

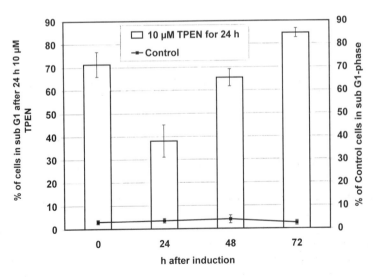

Figure 4. Apoptosis induced by the zinc chelator TPEN in 3T3L1 cells after various times following stimulation of proliferation and differentiation by the zinc chelator TPEN. Bars represent the fraction of cells in the subG₁ state, points represent the values of the untreated control. Cell cultures were supplemented with 10 µM of TPEN at the times indicated and treated for further 24 h. The subG₁ fraction was determined as the fraction of cells appearing as small, apoptotic bodies in the flow cytometer.

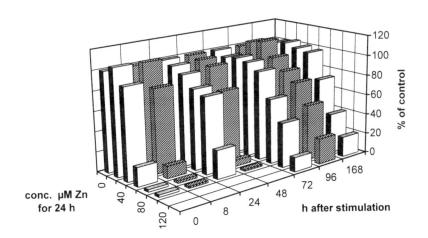

Figure 5. Fraction of 3T3L1 cells surviving the toxic action of $ZnCl_2$ at various stages of proliferation and differentiation. At various times after induction, cell cultures were supplemented with various concentrations of $ZnCl_2$ at and incubated for further 24 h. Surviving cells were stained with Neutral Red.

enhanced fraction of proliferating cells in S- and G2/M-phases as compared to controls without additional zinc supplementation over the zinc content of culture medium constituents (data not shown).

7. A MODEL OF COORDINATED INDUCTION OF ZINC UPTAKE AND METALLOTHIONEIN SYNTHESIS

Figure 6 depicts a model for possible interactions of metallothionein and zinc in hormone-stimulated 3T3L1 cells. At the time of peak proliferation, i.e., 24 h after stimulation by hormones, the cells exhibit a maximum of zinc uptake, of metallothionein synthesis and its translocation to cell nuclei. It has not been possible so far to dissect zinc uptake and metallothionein synthesis from each other, and it has not been feasible to estimate the level of zinc localized in nuclei, because the zinc probe Zinquin reflects only the labile part of Zn^{2+} ions. A simple model would suggest that dexamethasone and insulin induce the expression of the metallothionein gene(s), and that this protein would evoke further cellular Zn^{2+} uptake because of its high affinity for this ion. The zinc-thionein complex would have to be actively transported into the nucleus where it delivered Zn^{2+} to such proteins that bind the metal ion tighter and function in the regulation of early genes of cell proliferation. A major objection against this simple model is that it suggests an initial decrease in cellular zinc at the time of fresh metallothionein synthesis, a change which has not ben detected.

A more refined model would take into account an initial stimulation of zinc uptake by the hormone treatment. It has been observed by several authors that hormones activate a slow exchanging zinc uptake (Pattison and Cousins, 1986; Taylor and Simons, 1994; Ackland and McArdle, 1996). This activation is slow because it depends on synthesis of RNA and protein. In this model, the enhanced intracellular concentration of Zn^{2+} would

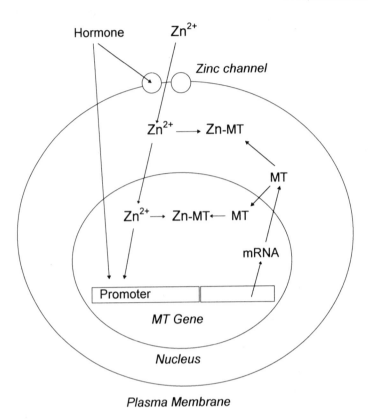

Figure 6. Scheme illustrating possible mechanisms of regulation of intranuclear zinc and metallothionein (MT).

activate metal regulated elements in the promoter of the metallothionein gene and induce its expression. As in the first model, Zn^{2+} and metallothionein would combine either in the cytoplasm with subsequent nuclear translocation of the zinc thionein complex, or combine in the nucleus after translocation of the newly synthesized protein. Because of the high affinity of metallothionein for Zn^{2+} the former alternative seems to be more probable. In both cases, nuclear accumulation of metallothionein would require an active transport through the nuclear pore complex, which has not been demonstrated so far. In any case, after the peak during active proliferation, there is a downregulation of the metallothionein synthesis and of zinc uptake. In a simple way, this would occur by the cessation of hormonal stimulation. This possibility can be excluded, however, because the cells cannot be stimulated any more by hormones after the onset of differentiation. Hence, an active proteolytic degradation of metallothionein and regulated export of is concluded.

What is the role of metallothionein in apoptosis induced by zinc deficiency? If hormones induce metallothionein on conditions of normal zinc supply, cells will respond by taking up additional zinc. At variance, if metallothionein is induced on conditions of zinc deprivation, zinc deficiency will be exacerbated and cells respond by entering into apoptosis. On the other hand, highly excessive zinc will no longer be buffered away by metallothionein and cause misfunction of membrane metal transporters and severe imbalance of intracellular ion homeostasis resulting in necrosis. Differentiated cells are less sensitive to this toxic action of excessive zinc, probably due to a downregulation of metal transport as compared to growing cells.

8. METHODS

8.1. Cell Culture and Stimulation

Murine 3T3L1 fibroblasts provided by Dr. A. Ullrich, Martinsried, were grown at 37 °C in Dulbecco Modified Eagle's Medium (DMEM) containing 5% newborn calf serum and supplemented with Na_2CO_3 and penicillin/streptomycin. Cell proliferation and subsequent differentiation to adipocytes was stimulated by fresh DMEM containing 10% fetal calf serum plus 0.5 mM 3-isobutyl-1-methylxanthine, 0.25 μM dexamethasone and 10 μg/ml insulin. After two days, the medium was changed to DMEM supplemented with fetal calf serum and insulin only, and the medium was changed every second day.

8.2. Flow Cytometry

Cells were trypsinized, washed twice with ice-cold phosphate buffered saline (PBS), fixed with 70% ice-cold ethanol and stored at -20° C.The fixed cells were washed with PBS and stained for 45 min with PBS containing 0.05 mg/ml propidium iodide, 0.1 mg/ml RNase A. Analyses of cellular DNA content and distribution were performed in a Coulter Counter Epics 1.5 Flow Cytometer using Lysis Software.

8.3. Determination of Metallothionein Concentration and Cellular Localization

Total metallothionein was quantified by the silver saturation assay according to Scheuhammer and Cherian (1991). The intracellular localization of metallothionein was assayed by immuno-histochemical staining. Cells were grown on eight-well Lab-Tek chamber slides (Nunc, Denmark). At confluence, cells were rinsed twice with PBS and fixed in 3% paraformaldehyde for 10 min. Cells were permeabilized with 0.1% Triton X-100 in PBS for 3 min and incubated with 5% goat serum to block unspecific antibody binding. After a further wash with PBS, cells were incubated for 1 h at room temperature with a primary E9 mouse monoclonal antibody against MT (Dako, Denmark), diluted 1:30 in 5% goat serum/PBS. After two further washes with PBS, the secondary goat anti mouse-IgG labelled with fluorescein (Sigma, Deisenhofen, diluted 1.350 in 5% goat serum/PBS) was applied for 1 h at room temperature. After three washes with PBS, the chambers were stripped and the cells were mounted with 1,4-diazabicyclo[2,2,2]octane (Fluka) and examined in a fluorescence microscope (Axiovert 50 of Zeiss, Oberkochem). No reactivity was observed with a MT control alone or the secondary antibody alone. To quantify the nuclear/cytoplasmic distribution of MT, 100 to 200 cells were randomly chosen and the predominant localization of MT was determined. Each experiment was performed threefold. Nuclei were delineated by staining fixed cells with 2 μg/ml of Hoechst 33528 for 10 min at room temperature.

8.4. Measurement of Total Cellular Zinc and of Intracellular Labile Zinc

Total zinc was determined by flame atomic absorption spectroscopy in a Perkin Elmer A 2380 atomic absorption spectrophotometer. Intracellular labile Zn^{2+} was assayed with the fluorescent probe Zinquin (Alexis, Grünberg). Cells were transferred to eight-well Lab-Tek chamberslides (Nunc, Denmark) and grown to confluence. Cells were rinsed twice with warm (37°C) Buffer A (25 mM Hepes of pH 7.35 containing 1 mM $MgCl_2$, 1 mM $CaCl_2$, 120 mM Na Cl, 5 mM KCl, 1 mM NaH_2PO_4, 5 mM Glucose, and 0.3% bovine serum albumin). Cells were incubated in the dark at 37°C with 25 μM Zinquin in buffer A. After 30 min, cells were washed twice with icecold PBS, equipped with cover slips and examined in a fluorescence microscope (Olympus) using 364 nm for excitation and reading emission at 485 nm.

REFERENCES

Ackland, M.L., and McArdle, H.J. (1996). Cation-dependent uptake of zinc in human fibroblasts. BioMetals 9, 29–37.

Adebodun, F., and Post, J.F.M. (1995). Role of intracellular free Ca(II) and Zn(II) in dexamethasone-induced apoptosis and dexamethasone resistance in human leukemic CEM cell lines. J. Cell. Physiol. 163, 80–86.

Andrews, M.I., Gallant, K.R., Cherian, M.G. (1987). Regulation of the ontogeny of rat liver metallothionein mRNA by zinc. Eur. J. Biochem. 166, 527–531.

Atar, D., Backx, P.H., Appel, M.M., Gao, W.D., Marban, E. (1995). Excitation-transcription coupling mediated by zinc influx through voltage-dependent calcium channels. J. Biol. Chem. 270, 2473–2477.

Benters, J., Flögel, U., Schäfer, T., Leibfritz, D., Hechtenberg, S., Beyersmann, D. (1997). Study on the interactions of cadmium and zinc ions with cellular calcium homeostasis using 19F-NMR spectroscopy. Biochem. J. 322, 793–799.

Berg, J.M., Shi, Y. (1996). The galvanization of biology: Agrowing appreciation for the roles of zinc. Science 271, 1081–1085.

Budde, T., Minta, A., White, J.A., Kay, A.R. (1997). Imaging free zinc in sysnaptic terminals in live hippocampal slices. Neuroscience 79, 347–358.

Canzoniero, L.M.T., Sensi, S.L., Choi, D.W. (1997). Measurement of intracellular free zinc in living neurons. Neurobiol. Dis. 4, 275–279.

Chesters, J.K., Petrie, L., Vint, H.. (1989). Specifity and timing of the Zn2+ requirement for DNA synthesis by 3T3 cells. Exp. Cell Res. 184, 499–508.

Cousins, R.J. (1985). Absorption, transport, and hepatic metabolism of copper and zinc: special reference to metallothionein and ceruloplasmin. Physiol. Revs. 65, 238–309.

Cousins, R.J., and Lee-Ambrose, L.M. (1992). Nuclear zinc uptake and interactions and metallothionein gene expression are influenced by dietary zinc in rats. J.Nutr. 122, 56–64.

Coyle, B.J., Zalewski, P.D., Philcox, J.C., Forbes, I.J., Ward, A.D., Lincoln, S.F., Mahadevan, I., and Rofe, A.M. (1994). Measurement of zinc in hepatocytes by using a fluorescent probe, Zinquin: relationship to metallothionein and intracellular zinc. Biochem. J. 303, 781–786.

Denny, M.F., Atchison, W.D. (1994). Methylmercury-induced elevations in intrasynaptosomal zinc concentrations: an 19F-NMR study. J. Neurochem. 63, 383–386.

Fraker, P.J., Osati, A.F., Wagner, M.A., King, L.E. (1995). Possible roles for glucocorticoids and apoptosis in the suppression of lymphopoiesis during zinc deficiency: a review. J. Am. Coll. Nutr. 14, 11–17.

Gachot, B., Tauc, M., Wanstoc, F., Morat, L., and Poujol, P. (1994). Zinc transport and metallothionein induction in primary cultures of rabbit kidney proximal cells. Biochim. Biophys. Acta 1191, 291–298.

Grider, A., and Vazquez, F. (1996). Nystatin affects zinc uptake in human fibroblasts. Biol. Trace Elem. Res. 54, 97–104.

Grummt, F. Weinmann-Dorsch, C., Schneider-Schaulies, J., and Lux, A. (1986). Zinc as a messenger of mitogenic ionduction. Exp. Cell Res. 163, 191–200.

Hansson, A. (1996). Extracellular zinc ions induce mitogen-activated protein kinase activity and protein tyrosine phosphorylation in bombesin-sensitive Swiss 3T3 fibroblasts. Arch. Biochem. Biophys. 328, 233–238.

Hechtenberg,S., and Beyersmann, D. (1995).Regulation of nuclear calcium and zinc: interference by toxic metal ions. In: Sarkar, B. (Ed.) "Genetic Response to Metals". Marcel Dekker, New York, Basel, Hongkong, pp. 1–20.

Kuo, S.-M., Kondo, Y., DeFilippo, J.M., Ernstoff, M.S., Bahnson, R.R., Lazo, J.S. (1994). Subcellular localization of metallothionein IIA in human bladder tumor cells using a novel epitope-specific antiserum. Toxicol. Appl. Pharmacol. 125, 104–110.

Lohmann, R.D., and Beyersmann, D. (1993). Cadmium and zinc mediated changes in the Ca2+-dependent endonuclease in apoptosis. Biochem. Biophys. Res. Commun. 190, 1097–1103.

Martin, S.J., Mazdai, G., Strain, J.J., Cotter, T.G., Hannigan, B.M. (1991). Programmed cell death (apoptosis) in lymphoid and myeloid cell lines during zinc deficiency. Clin. Exp. Immunol. 83, 338–343.

Mazzanti, M., DeFelice, L.J., Cohen, J., Malter, H. (1990). Ion channels in the nuclear envelope. Nature 343, 764–767.

McCabe, M.J., Jiang, S.A., Orrenius, S. (1993). Chelation of intracellular zinc triggers apoptosis in mature thymocytes. Lab. Investig. 69, 101–110.

McNall, A.D., Etherton, T.D., Fosmire, G.J. (1995). The impaired growth induced by zinc deficiency in rats is associated with decreased expression of hepatic insulin-like growth factor I and growth receptor genes. J. Nutr. 125, 874–879.

Moffat, P., Lachapelle, M., Plaa, G.L., and Denizeau, F. (1994). Induction and localization of metallothionein in isolated rat hepatocytes exposed to zinc and epidermal growth factor. In: Collery, P., Littlefield, N.A., Etienne, J.C. (Eds.) "Metal Ions in Biology and Medicine". John Libbey Eurotext, Paris, pp. 577–562.

Nagel, W.W., Vallee, B.L. (1995). Cell cycle regulation of metallothionein in human colonic cancer cells. Proc. Natl. Acad. Sci. USA 92, 579–583.

Nartey, N.O., Banerjee, D., Cherian, M.G. (1987a). Immunohistochemical localization of metallothionein in cell nucleus and cytoplasm of fetal human liver and kidney and its changes during development. Pathology 19, 233–248.

Nartey, N.O., Cherian, M.G., Banerjee, D. (1987b). Immunohiszochemical localization of metallothionein in human thyroid tumors. Am. J. Pathol. 129, 177–182.

Ono, S.I., Cherian, M.G. (1998). Changes in rat brain metallothionein and zinc during development in transgenic mice. Biol. Trace Elem. Res. 61, 41–49.

Paramanantham, R., Sit, K.-H., Bay, B.-H. (1997). Adding Zn^{2+} induces DNA fragmentation and cell condensation in cultured human Chang liver cells. Biol. Trace Elem. Res. 58, 135–147.

Pattison, S.E., Cousins, R.J. (1986). Zinc uptake and metabolism by hepatocytes. Federation Proc. 45, 2805–2809.

Petrie, L., Chesters, J.K., Franklin, M. (1991). Inhibition of myoblast differentiation by lack of zinc. Biochem.J. 276, 109–111.

Prasad,A.S. (1979). Clinical, biochemical and pharmacological role of zinc.Annu. Rev. Pharmacol. Toxicol. 20, 393–426.

Rubin, H., Koide, T. (1973). Stimulation of DNA synthesis and 2-deoxy-D-glucose transport in chick embryo cultures by excessive metal concentrations and by a carcinogenic hydrocarbon. J. Cell. Physiol. 81, 387–396.

Scheuhammer, A.M., Cherian, M.G. (1991). Quantification of metallothionein by silver saturation. Methods Enzymol. 205, Part B, 78–83.

Schmetterer, G. (978). ATP dependent uptake of zinc by human erythrocyte ghosts. Z. Naturforsch. 33c, 210–215.

Sensi, S., Canzoniero, L.M.T., Yu, S.P., Ying, H.S., Koh, J.-Y., Kerchner, G.A., Choi, D.W. (1997). Measurement of intracellular free zinc in living cortical neurons: routes of entry. J. Neurosci. 17, 9554–9564.

Silar, P., Theodore, L., Mokdad, R., Errais, N.E., Cadic, A., Wegnez, M. (1990). Metallothionein Mto gene of Drosophila melanogaster: structure and regulation. J. Mol. Biol. 215, 217–224.

Simons, T.J.B. (1993). Measurement of free Zn2+ ion concentration with the fluprescent probe mag-fura-2 (furapta). J. Biochem. Biophys. Methods 27, 25–37.

Steinebach, O.M., Wolterbeek, H.T. (1992). Effects of cadmium, copper and metallothionein synthesis inhibiting and stimulating compounds on zinc uptake and accumulation in rat hepatoma HTC cells. Chem.-Biol. Interact. 84, 199–220.

Taylor, J.A., Simons, T.J.B. (1994). Effect of dexamethasone on zinc transport in rat hepatocytes in primary culture. Biochim. Biophys. Acta 1193, 240–246.

Templeton, D.M., Banerjee, D., Cherian, M.G. (1985). Metallothionein synthesis and localization in relation to metal storage in rat liver during regeneration. Can. J. Biochem. Cell. Biol. 63, 16–22..

Tsujikawa, K., Imai, T., Kakutani, M., Kayomori, Y., Mimura, T., Otaki, N., Kimura, M., Fukuyama, R., Shimizu, N. (1991). Localization of metallothionein in nuclei of growing primary cultured adult rat hepatocytes. FEBS Lett. 283, 239–242.

Tsujikawa, K., Suzuki, N., Sagawa, K., Itoh, M., Sugiyama, T., Kohama, Y., Otaki, N., Kimura, M., Mimura, T. (1994). Induction and subcellular localization of metallothionein in regenerating rat liver. Eur. J. Cell Biol. 63, 240–246.

Vener, A.V., Loeb, J. (1992).Zinc-induced tyrosine phosphorylation of hippocampal p60c-src is catalyzed by another tyrosine kinase. FEBS Lett. 308, 91–93.

Villalba, M., Ferrari, D., Bozza, A., Del Senno, L., Di Virgilio, F. (1995). Ionic regulation of endonuclease activity in PC12 cells. Biochem. J. 311, 1033–1038.

Wang, Y., Pallen, C.J. (1992). Expression and characterization of wild-type, truncated, and mutated forms of the intracellular region of the receptor-like protein tyrosine phosphatase HPTP beta. J. Biol. Chem. 267, 16696–16702.

Wenzel, H.J., Cole, T.B., Born, D.E., Schwartzkroin, P.A., Palmiter, R.D. (1997). Ultrastructural localization of zinc transporter-3 (ZnT-3) to synaptic vesicle membranes within mossy fiber boutons in the hippocampus of mouse and monkey. Proc. Natl. Acad. Sci. USA 94, 12676–12681.

Woo, E.S., Kondo, Y., Watkins, S.C., Hoyt, D.G., Lazo, J.S. (1996). Nucleophilic distribution of metallothionein in human tumor cells. Exp. Cell Res. 224, 365–371.

Xu, J., Xu, Y., Nguyen, Q., Novikoff, P.M., Czaja, M.J. (1996). Induction of hepatoma cell apoptosis by c-myc requires zinc and occurs in the absence of DNA fragmentation. Am. J. Physiol. 270, G60-G70.

Zalewski, P.D., Forbes, I.J., Betts, W.H. (1993).Correlation of apoptosis with change in intracellular labile Zn(II) using Zinquin [(2-methyl-8-p-toluenesulphonamido-6-quinolyloxy)acetic acid], a new specific fluorescent probe for Zn(II). Biochem. J. 296, 403–408.

10

EFFECT OF METAL COMPOUNDS ON THE FUNCTION OF ZINC FINGER PROTEINS INVOLVED IN DNA REPAIR

Andrea Hartwig,[1]* Leon H. F. Mullenders,[2] Monika Asmuß,[1] Maike Benters,[1] and Ines Krüger[1]

[1]University of Karlsruhe
Institute of Food Chemistry
D -76128 Karlsruhe, Germany
[2]University of Leiden
MGC-Department of Radiation Genetics and Chemical Mutagenesis
2333 AL Leiden, The Netherlands

SUMMARY

Compounds of nickel(II), cadmium(II), cobalt(II) and arsenic(III) have been shown previously to disturb DNA repair processes at low, non-cytotoxic concentrations. They inhibit nucleotide excision repair involved in the removal of a broad spectrum of DNA lesions induced by environmental mutagens as well as the repair of oxidative DNA damage generated endogenously due to oxygen metabolism. This raises the question why DNA repair systems are such sensitive intracellular targets of metal-induced toxicity. We found that different steps are affected and different mechanisms account for metal-induced repair inhibition. Nickel(II) and cadmium(II) interfere with the very first step of the repair process by disturbing DNA-protein interactions involved in DNA damage recognition. Since these effects are reversible by the addition of magnesium(II) and zinc(II), respectively, to nuclear extracts derived from metal-treated cells, the inactivations occur at the protein level and not at the level of gene expression and/or translation of the respective DNA repair proteins. Potentially very sensitive targets for toxic metal ions are zinc finger struc-

---

*To whom reprint requests should be addressed: Prof. Dr. Andrea Hartwig, University of Karlsruhe, Institute of Food Chemistry, Postfach 6980, D-76128 Karlsruhe, Germany. Tel.: 49-721-608-2936; Fax: 49-721-608-7254; E-mail: Andrea.Hartwig@chemie.uni-karlsruhe.de

Metals and Genetics, edited by Sarkar.
Kluwer Academic / Plenum Publishers, New York, 1999.

tures in DNA binding motifs, which have been identified in several DNA repair enzymes. They include the mammalian XPA protein essential for the recognition of DNA damage during nucleotide excision repair as well as the bacterial formamidopyrimidine-DNA gly-cosylase (Fpg protein) involved in the removal of oxidative DNA base modifications. By applying both proteins as models for zinc finger DNA repair proteins, we observed no in-hibitory effect on the Fpg protein in the presence of up to 1 mM nickel(II) or cobalt(II); however, 10 μM nickel(II) disturbed XPA-DNA interactions on a cisplatin-damaged oli-gonucleotide. With respect to other metal compounds, 50 μM cadmium(II), 5 μM cop-per(II) and 50 nM mercury(II) lead to pronounced inhibitions of the Fpg protein. The observed inhibitions were partly prevented by simultaneous treatment with zinc(II) in case of cadmium(II) and copper(II), but not mercury(II). Our results indicate that with respect to the Fpg protein, its zinc finger structure is a sensitive target for cadmium(II) and cop-per(II), but not for nickel(II) or mercury(II), providing further evidence that each metal exerts unique mechanisms of repair inhibition.

1. INTRODUCTION

1.1. Mutagenicity and Oxidative DNA Damage Induced by Metal Compounds

Epidemiological and/or animal studies point towards a carcinogenic potential of sev-eral metals including chromium, nickel, cadmium, arsenic, cobalt and lead (IARC, 1980; 1990; 1991; 1993). However, despite their carcinogenic activities, compounds of nickel(II), cadmium(II), cobalt(II) and arsenic(III) are not mutagenic in bacterial test sys-tems and only weakly, or—in the case of arsenic(III)—not mutagenic in mammalian cells in culture. Furthermore, the weak mutagenicity is often restricted to comparatively high, cytotoxic concentrations of the respective compounds (for review see Hartwig, 1995). Nevertheless, due to the redox chemistry of transition metals and their ability to activate oxygen species in the course of redox reactions, one frequently discussed mechanism in metal-induced genotoxicity is the formation of oxidative DNA damage. In this context, compounds of chromium(VI), iron(II)/iron(III), nickel(II) and cobalt(II) have been shown to damage DNA in vitro, for example in the presence of H_2O_2. Furthermore, oxidative DNA damage might occur via the inactivation of cellular defence systems, as has been postulated in the case of cadmium(II) (for reviews see Klein et al., 1991; Sunderman, 1993; Sugiyama, 1994; Kasprzak, 1995). Whether or not the induction of oxidative DNA damage is relevant in intact cells cannot readily be answered from subcellular test sys-tems. The most reactive oxygen species like hydroxyl radicals are rather short-lived and in intact cells defence systems exist to detoxify free oxygen radicals. Therefore, the potential relevance of oxidative damage for intact cells depends on uptake, complex ligands, intra-cellular distribution and reductants, all of which affect the concentrations of metal ions available for redox reactions close to critical targets like the DNA. For example, when in-vestigating the induction of DNA strand breaks and oxidative DNA base modifications in intact HeLa cells, we observed that both nickel(II) and cadmium(II) induce DNA strand breaks in a dose-dependent manner, starting at non-cytotoxic concentrations. In contrast, the induction of specific oxidative DNA base modifications was restricted to high, cyto-toxic concentrations of nickel(II) and no increase was observed after treatment with cad-mium(II) (Dally and Hartwig, 1997).

1.2. Enhancing Effects in Combination with Other DNA Damaging Agents and Interactions with DNA Repair Systems

In spite of the missing mutagenicity of metal compounds in bacterial test systems, arsenic(III), copper(II), nickel(II), cadmium(II) and manganese(II) increased the mutation frequency when combined with UV-light and/or alkylating agents suggesting that an inhibition of DNA repair processes may be the predominant mechanism in metal-induced genotoxicity. Even though these results obtained in bacterial test systems are not directly transferable to eucaryotic cells due to considerable differences in cellular repair systems, most carcinogenic metal compounds have been shown to increase the cytotoxicity, mutagenicity and clastogenicity when combined with different types of DNA damaging agents in mammalian cells (for review see Hartwig, 1995). In the case of arsenic(III), nickel(II), cadmium(II), cobalt(II) and lead(II) these effects were found in a close relationship to interactions with DNA repair processes.

1.2.1. Nucleotide Excision Repair. Nucleotide excision repair is the major repair system involved in the removal of DNA damage induced by UV radiation, environmental and food mutagens as well as DNA lesions generated by some cytostatic drugs like cis-diamminedichloroplatinum(II) (cisplatin). According to the current knowledge it involves at least 30 different proteins and enzymes in mammalian cells, including those which are defective in patients suffering from the DNA repair disorder Xeroderma Pigmentosum (XP) complementation groups A through G. It can be roughly subdivided into three different steps, namely the incision at both sides of the lesion, the repair polymerisation leading to the displacement of the damaged oligonucleotide and finally the ligation of the repair patch. By investigating the effect of metal compounds on the removal of DNA lesions induced by UV radiation and by applying inhibitors of distinct steps of this repair pathway, compounds of nickel(II), cobalt(II), cadmium(II) and arsenic(III) have been shown to inhibit nucleotide excision repair. With respect to all metal compounds investigated, the incision step was affected at the lowest concentrations (Hartwig et al., 1994; Kasten et al., 1997; Hartwig et al., 1996a; Hartwig et al., 1997). Nevertheless, cobalt(II) inhibited additionally the polymerization and—even though at higher concentrations—arsenic(III) as well as nickel(II) the ligation step of the repair process (Li and Rossman, 1989; Lee-Chen et al., 1993; Hartwig et al., 1994; Hartwig et al., 1997).

1.2.2. Base Excision Repair. In contrast to nucleotide excision repair, which removes different types of bulky DNA base damage initiated by the coordinated action of many proteins, some forms of DNA base damage are excised by a specific class of DNA repair enzymes called glycosylases. They catalyse the hydrolysis of the N-glycosidic bonds linking the damaged base to the deoxyribose-phosphate backbone, thereby releasing the damaged moiety as a free base. This process generates sites of base loss called apurinic or apyrimidinic (AP) sites, which are further processed in a multistep process. AP endonucleases produce incisions by hydrolysis of phosphodiester bonds 5' or 3' to each AP site followed by the exonucleolytic excision of the terminal deoxyribose-phosphate residue and finally the polymerisation and ligation of the repair patch (for review see Friedberg et al., 1995). DNA glycosylases act specifically on one or few substrates, and DNA base excision repair is mainly responsible for the removal of different types of endogenous DNA damage, including oxidative DNA base modifications arising as a consequence of oxygen consumption.

In a previous study we elucidated the effect of nickel(II) and cadmium(II) on the repair of oxidative DNA base modifications induced by visible light in HeLa cells. Current evidence suggests that the prevailing oxidative DNA base modification generated by this treatment is 7,8-dihydro-8-oxoguanine (8-hydroxyguanine), a premutagenic DNA base modification, and to a lesser extent DNA strand breaks, presumably due to the reaction of singlet oxygen or excited intracellular photosensitizer molecules with DNA (Schneider et al., 1990; Boiteux et al., 1992; Pflaum et al., 1994). Both cadmium(II) and nickel(II) inhibited the repair of the induced DNA damage at low, non-cytotoxic concentrations. In the case of nickel(II), the closure of DNA strand breaks and the removal of Fpg-sensitive sites were inhibited at concentrations of 100 μM and 50 μM, respectively. While cadmium(II) had no effect on the repair of DNA strand breaks, the repair of the oxidative DNA base modifications was impaired at 0.5 μM and higher (for a more detailed description see Dally and Hartwig, 1997). Therefore, even though oxidative DNA base modifications are not induced by both metal compounds at biologically relevant concentrations, their extent may be enhanced indirectly by nickel(II) and cadmium(II): since oxidative DNA damage is induced continuously in living cells due to oxygen metabolism, an impaired repair of these lesions may result in an increased background frequency of the respective DNA lesions.

1.3. Aim of the Study

As described above, compounds of nickel(II), cadmium(II), cobalt(II) and arsenic(III) have been shown previously to disturb different types of DNA repair processes at low, non-cytotoxic concentrations. This raises the question why DNA repair systems are such sensitive intracellular targets of metal-induced toxicity. In principle, several reasons could account for the repair inhibition. First, the inhibition could be due to a decreased level of repair enzymes, for example due to a decreased expression of the respective genes. Second, DNA structures could be disrupted by the toxic metal compounds, preventing DNA repair enzymes to bind. Third, the inhibition could be due to an inactivation of the repair enyzmes itself. To elucidate these aspects on the molecular level, we applied invitro systems to investigate DNA-protein interactions in the presence of toxic metal ions.

2. RESULTS

2.1. Effect of Metal Compounds on DNA Damage Recognition during Nucleotide Excision Repair

Nucleotide excision repair is a complex event, and even though not fully understood, much progress has been made to resolve the process of nucleotide excision repair on the cellular and molecular level. One important prerequisite for the initiation of repair events is the recognition of the DNA lesions, and some proteins involved in DNA damage recognition have been identified during the last years. With respect to UV-induced DNA damage, XPA is absolutely required for lesion recognition; its binding to damaged DNA is greatly enhanced by the replication protein A (RPA). Additionally, even though not essential for DNA repair in vitro, XPE is thought to serve as an accessory factor in damage recognition (for review see Naegeli, 1995). Finally, other proteins exerting high affinities for UV-irradiated DNA have been described recently, which are distinct from known Xeroderma pigmentosum complementation groups, and which may have yet unknown functions in the cellular response to DNA damage (Wakasugi et al., 1996; Ghosh et al., 1996).

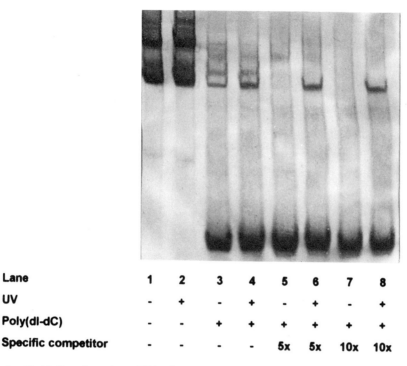

Lane		1	2	3	4	5	6	7	8
UV		-	+	-	+	-	+	-	+
Poly(dI-dC)		-	-	+	+	+	+	+	+
Specific competitor		-	-	-	-	5x	5x	10x	10x

Figure 1. Specific binding of proteins to UV-irradiated DNA. HeLa nuclear cell-free extracts were incubated with a digoxygenin-endlabeled oligonucleotide either unirradiated or irradiated with UV and DNA-protein interactions were analysed by a gel-mobility-shift assay. Poly(dI-dC) and an excess of unlabeled, unirradiated oligonucleotide of the same sequence (specific competitor) were added where indicated (from Hartmann and Hartwig, 1998).

With respect to those metals inhibiting the incision step of nucleotide excision repair, we investigated whether they disturb the actual incision reaction or already the recognition of DNA damage after treatment of intact cells with low, biologically relevant concentrations of the metal compounds. DNA damage recognition requires the preferential binding of one or more proteins to damaged DNA as compared to undamaged DNA. To identify proteins which bind with high affinity to UV-irradiated DNA we conducted a gel-mobility-shift assay by applying a 48 bp digoxygenin-labeled oligonucleotide, either unirradiated or irradiated with 18 kJ/m² UV, and HeLa nuclear cell free extracts. In the absence of competitor DNA, no difference was observed between proteins binding to both oligonucleotides, indicating unspecific DNA binding. However, when poly(dI-dC) was added as unspecific competitor DNA as well as a five- or tenfold excess of the unirradiated, unlabeled oligonucleotide of the same sequence as a specific competitor, a mobility shift of the UV-irradiated oligonucleotide was observed, indicating the binding of one or more proteins to damaged DNA, whereas the unspecific binding to the non-damaged oligonucleotide was diminished completely (Figure 1). The specific binding to the UV-irradiated oligonucleotide was suppressed entirely when adding an excess of UV-irradiated, unlabeled oligonucleotide, adding further evidence that the protein(s) interact(s) specifically with UV-irradiated DNA. These proteins have been characterized with respect to their molecular weight by applying a magnetic DNA affinity purification procedure. Two specific proteins were detected with molecular weights of 34 and 40 kDa; they could correspond to XPA and a subunit of RPA; however, this aspect has to be further elucidated (for further details see Hartmann and Hartwig, 1998).

In a next step we investigated whether the specific binding of the proteins to the UV-irradiated oligonucleotide is disrupted by nickel(II), cadmium(II) or arsenic(III). According to incubation protocols established in previous studies, which insure sufficient uptake of the respective metal compounds, HeLa cells were preincubated with $NiCl_2$ for 24 h at concentrations between 50 and 600 µM, with cadmium(II) for 2 h at concentrations between 0.5 and 5 µM or with 10 µM arsenic(III) for 24 h, which did not reduce the colony forming ability (Hartwig et al., 1994; Dally and Hartwig, 1997; Hartwig et al., 1997). Afterwards, nuclear extracts were prepared, and the gel-mobility shift assay was performed as described above. While 5 and 25 µM $NiCl_2$ had no effect, a diminished protein binding was observed at 50 µM $NiCl_2$ and higher, resulting in 62 % residual binding at 50 µM and 14 % at 600 µM. One possible mechanism of binding inhibition consists in the competition of nickel ions with divalent magnesium ions, and the inhibition of the incision frequency after UV-irradiation has been shown to be partly reversible by the addition of excess magnesium(II) (Hartwig et al., 1994). Therefore, we investigated whether magnesium(II) is able to reverse the inhibitory effect of nickel in our test system. When the concentration of magnesium(II) in the gel-shift buffer was enhanced from 5 mM (standard reaction) to 10 mM, the protein binding of nuclear extracts from nickel-treated cells was restored almost completely; only at 600 µM nickel(II) a reduction of the binding capacity

Table 1. Effect of nickel(II) and cadmium(II) on the specific binding of proteins to UV-irradiated DNA and protective interaction of magnesium(II) and zinc(II)

Metal		Protein binding (% of control)
Ni(II) [µM]	Mg(II) (mM)	
0	5*	100
50	5*	62.3 ± 24.1
50	10	84.5 ± 19.5
250	5*	49.0 ± 11.2
250	10	82.5 ± 7.5
600	5*	14.3 ± 10.3
600	10	27.0 ± 2.0
Cd(II) [µM]	Zn(II) [µM]	
0	0	100
0.5	0	58.0 ± 3.3
0.5	100	96.0 ± 3.5
1	0	46.0 ± 9.0
1	100	89.0 ± 4.5
5	0	11.3 ± 11.4
5	100	76.0 ± 22.5

*5 mM magnesium(II) are required in the standard reaction.
HeLa cells were incubated for 24 h with the respective concentrations of $NiCl_2$ or for 2 h with $CdCl_2$ before the preparation of the nuclear protein extracts and the conduction of the gel-mobility-shift assay. The binding reaction was carried out in the presence of poly(dI-dC) and a fivefold excess of unlabeled, unirradiated oligonucleotide of the same sequence (specific competitor) (data from Hartmann and Hartwig, 1998).

to 27 % of the control was still observed. In the case of cadmium(II), a dose-dependent reduction of DNA-protein interactions was seen at all concentrations applied; at 0.5 μM, the protein-binding to the UV-irradiated oligonucleotide was reduced to 58% and at 5 μM to 11 % compared to untreated control cells. In contrast to the results obtained with nickel(II), an enhanced concentration of magnesium(II) in the gel-shift buffer had no impact on the observed inhibition (data not shown); however, the addition of 100 μM zinc(II) to the binding reaction led to a largely restored binding behaviour at all cadmium-concentrations applied (Table 1). No inhibition of the binding reaction was observed in the presence of arsenic(III), indicating that the damage recognition is not disturbed by this metal.

2.2. Interaction of Carcinogenic Metal Compounds with the Function of Isolated Zinc Finger Proteins

The results presented so far indicate that compounds of nickel(II) and cadmium(II) inactivate nuclear proteins involved in DNA damage recognition. This raises the question whether it is possible to define structural elements which might present particular sensitive targets for toxic metal ions. Potential candidates are zinc finger structures in DNA binding motifs, which have been identified in several DNA repair enzymes, such as the bacterial UvrA protein and the mammalian XPA protein, both essential for DNA damage recognition during nucleotide excision repair, as well as the bacterial formamidopyrimidine-DNA glycosylase (Fpg protein) involved in the removal of oxidative DNA base modifications including the premutagenic 8-hydroxyguanine (Friedberg et al., 1995). Therefore, in the present study we investigated the effect of toxic metal compounds on two DNA repair proteins with zinc finger structures, namely the Fpg protein and the XPA protein.

2.2.1. The Bacterial Fpg Protein. The Fpg protein is a repair glycosylase derived from E. coli which specifically removes 8-hydroxyguanine as well as imidazol ring-opened forms of guanine and adenine. It combines a DNA glycosylase and an AP endonuclease activity, thus it converts Fpg-sensitive DNA lesions into DNA strand breaks, which, in intact cells, are further processed by base excision repair (see above). The Fpg protein contains one zinc ion located in one single zinc finger motif, where zinc is complexed to four cysteines, and which mediates DNA binding (for review see Boiteux, 1993). To test the activity of the Fpg protein, we used isolated PM2 DNA damaged by methylene blue plus visible light, which generates predominantly 8-hydroxyguanine due to the action of endogenous photosensibilisators or singlet oxygen (Boiteux et al., 1992; Schneider et al., 1990). While intact PM2 DNA is mainly supercoiled, the introduction of DNA single strand breaks converts it into the open circular form; if DNA single strand breaks occur in close proximity on opposite DNA strands, PM2 DNA is finally converted into the linear form. All three forms can be separated by agarose gel electrophoresis and quantified densitometrically to calculate the frequency of DNA strand breaks as a measure of the catalytic activity of the Fpg protein.

With respect to carcinogenic metal compounds, we investigated the effect of nickel(II), cadmium(II), copper(II) and cobalt(II) on the activity of the Fpg protein and compared it to the highly toxic mercury(II), which is, however, not considered to be a human carcinogen (IARC, 1993) (Table 2). While nickel(II) and cobalt(II) had no effect at concentrations up to 1 mM, all other compounds showed pronounced inhibitions. Cadmium(II) inhibited the Fpg protein at concentrations of 10 μM and higher in a dose-dependent manner. This effect was largely reversible by the simultaneous incubation with

Table 2. Effect of metal compounds on the activity of the bacterial Fpg protein

Metal compound	Concentration range tested	Lowest inhibitory concentration	Effect of zinc(II)[1]
Nickel(II)	1–1000 µM	no inhibition	–
Cobalt(II)	1–1000 µM	no inhibition	–
Cadmium(II)	0.1–1000 µM	50 µM	inhibition partially reversible
Copper(II)	0.1–1000 µM	5 µM	inhibition partially reversible
Mercury(II)	0.01–1000 µM	0.05 µM	no effect

[1]Simultaneous incubation with other metal compounds
Isolated supercoiled PM2 DNA was treated with methylene blue plus visible light, incubated with the Fpg protein, which had previously been incubated with the respective metal compounds, and the resulting DNA strand breaks were quantified by agarose gel electrophoresis.

equimolar concentrations of zinc, indicating that the displacement of zinc ions by cadmium may be the underlying mechanism of enzyme inactivation by cadmium(II). Similar effects were observed by copper(II): concentrations of 5 µM and higher led to a pronounced inhibition of the Fpg protein. However, this effect was only partly reversible by simultaneous incubation with tenfold higher concentrations of zinc(II). With respect to mercury(II), the activity of the Fpg protein was inhibited at concentrations as low as 0.05 µM, leading to a complete inactivation at 0.1 µM. In contrast to the results obtained with cadmium(II) and copper(II), the addition of zinc(II) had no impact on the inhibitory effect, indicating that the zinc finger structure is not the main target of mercury. Nevertheless, the inhibition was reversible by the subsequent incubation with reduced glutathione, pointing towards an inactivation of essential SH-groups outside the zinc finger structure.

2.2.2. The Mammalian XPA Protein. The ability of the mammalian XPA protein to recognize and bind to damaged DNA has been studied in the absence and presence of nickel(II). We applied an digoxygenin-labeled oligonucleotide damaged by cisplatin, and the binding capacity of the XPA protein was investigated by a gel-mobility-shift assay. When adding nickel(II) to the gel-shift reaction, the binding of XPA was inhibited at 10 µM and higher (Table 3), indicating that the XPA protein provides a sensitive target for nickel-induced toxicity. Further studies will have to clarify whether the inactivation occurs by the disturbance of the zinc finger structure or by other mechanisms.

Table 3. Effect of nickel(II) on the binding of XPA to cisplatin-damaged DNA

Nickel(II) (µM)	Protein binding (% of control)
0	100
10	67.9
20	57.1
50	53.0
75	12.5

The XPA protein was incubated with up to 75 µM NiCl$_2$ where indicated before the conduction of the gel-mobility-shift assay.

3. CONCLUSIONS AND PERSPECTIVES

What do these experiments tell us? First, the results provide new insights into the mechanisms leading to the inhibition of nucleotide excision repair. They demonstrate that compounds of nickel(II) and cadmium(II) act on the very first step of the repair process, namely the recognition of the DNA damage induced by UVC radiation. These DNA-protein interactions are disturbed after treatment of the HeLa cells with low, non-cytotoxic concentrations of the respective metals. In the case of nickel(II), 50 µM decreased the protein binding to 62 % compared to the control; this is about 15-fold below the concentration, where the colony forming ability starts to decline under these treatment conditions (Hartwig et al., 1994). Similarily, the incubation of the cells with as little as 0.5 µM cadmium(II) reduced the protein binding to 58 %, while the colony forming ability was affected only at concentrations higher than 20 µM (Dally and Hartwig, 1997). As stated above, several reasons could account for the observed disturbance of DNA-protein interactions, including a reduced transcription and/or translation of the respective proteins, the distortion of DNA structures or the inactivation of the proteins due to metal binding. Our results support the last alternative. Since the nuclear protein fractions are derived from metal-treated cells, changes in DNA structures in these cells are not expected to have any impact on the outcome of the experiments. The nearly complete restoration of the DNA-protein interactions after the addition of excess magnesium(II) or zinc(II), respectively, to the nuclear extracts excludes a reduction in the content of the specific proteins in the metal-treated cells. Therefore, the experiments presented in this study point towards an inactivation of nuclear proteins by nickel(II) and cadmium(II), where the competition with essential metal ions seems to play a predominant role (Figure 2).

Our results indicate that the zinc finger structure of the Fpg protein is a sensitive target for cadmium(II) and copper(II), but not for cobalt(II), nickel(II) or mercury(II). Nevertheless, nickel(II) inactivates the XPA protein, the mechanism of which is still unclear.

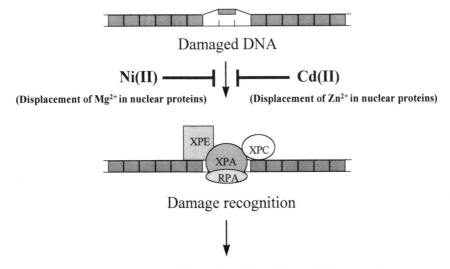

Figure 2. Schematic presentation of the effects of nickel(II) and cadmium(II) on DNA damage recognition during nucleotide excison repair and proposed mechanism of action. Shown is a current model of proteins involved in damage recognition, some of which are defective in Xeroderma pigmentosum (XP) patients belonging to different complementation groups; additionally, the replication protein A (RPA) is needed for damage recognition.

The results are distinct from those reported by Sarkar (1995) when applying a steroid hormone receptor, where the DNA binding activity is mediated by a zinc finger structure as well. When the apoprotein was reconstituted with cadmium(II) or cobalt(II), the DNA binding behaviour was restored completely by cadmium(II), but only partially by cobalt(II). Even though the two approaches are not directly comparable, this discrepancy suggests that despite the fact that zinc finger structures are common motifs in many DNA binding proteins, each member of this family appears to have its own structural features and sensitivities towards toxic metal compounds. With respect to the inhibition of DNA repair, our findings provide further evidence that even though DNA repair processes are inactivated by all carcinogenic and/or toxic metal compounds investigated, each metal exerts unique mechanisms of repair inhibition.

What do these findings mean for the risk assessment associated with the exposure towards these metal compounds? With respect to nickel(II), our results contradict a current model presented by Oller et al. (1997), according to which high concentrations of soluble nickel(II) would be needed for cellular uptake, and, once nickel(II) ions were inside the cell, they would bind to cytoplasmic ligands and would not reach the cell nucleus in concentrations high enough to have a genetic effect. Since in our study nuclear extracts were prepared from nickel-treated cells, the results clearly show that soluble nickel(II) is also able to reach nuclear targets in amounts high enough to inactivate nuclear proteins.

The relevance of the observations is closely related to the relevance of DNA repair processes. DNA is continuously damaged by endogenous and environmental mutagens. For example, a measurable steady-state level of oxidative DNA damage exists in all aerobic cells due to reactive oxygen species arising in the course of oxygen consumption (Ames et al., 1993; Hartwig et al., 1996; Pflaum et al., 1997). Other factors which contribute to the instability of the DNA include UV radiation, spontaneous demethylation of the DNA bases, hydrolysis of the glycosidic bond as well as a broad spectrum of environmental and food mutagens (Lindahl, 1993). If left uncorrected, DNA damage interferes with many essential transactions including transcription and DNA replication, which may lead to cell death. If cells survive, there is an increased probability of errors occurring during replication of a damaged template. Therefore, DNA repair systems act as key components of cellular defense to minimize the mutagenic consequences of DNA damage associated with endogenous processes and with mutagenic exposures at workplaces and in the general environment. Regarding the effects of metal compounds, our results show that even low concentrations lead to a partially repair deficient state in exposed cells, which may increase the risk of cancer.

ACKNOWLEDGMENTS

The authors would like to thank Dr. Serge Boiteux, Commisariat Energie Atomique, Fontanay aux Roses, France, and Dr. A. Eker, Rotterdam, Netherlands, for kindly providing the Fpg protein and the XPA protein, respectively. This work was supported by the Deutsche Forschungsgemeinschaft, grant no. Ha 2372/1–1.

REFERENCES

Ames, B. N., Shigenaga, M.K. and Hagen, T.M.(1993) Oxidants, antioxidants, and the degenerative diseases of aging. Proceedings of the National Academy of Sciences in the United States of America 90, 7915–7922.

Boiteux, S., Gajewski, E., Laval, J. and Dizdaroglu, M. (1992) Substrate specificity of the *Escherichia coli* Fpg protein (Formamidopyrimidine-DNA glycosylase): Excision of purine lesions in DNA produced by ionizing radiation or photosensitization. Biochemistry 31, 106–110.

Dally, H. and Hartwig, A. (1997) Induction and repair inhibition of oxidative DNA damage by nickel(II) and cadmium(II) in mammalian cells. Carcinogenesis 18, 1021–1026.

Friedberg, E.C., G.C. Walker and W. Siede (1995) "DNA Repair and Mutagenesis." ASM Press, Washington D.C.

Ghosh, R., Peng, C.H. and Mitchell, D.L. (1996) Evidence for a novel DNA damage binding protein in human cells. Proc. Natl. Acad. Sci. USA 93, 6918–6923.

Hartmann, M. and Hartwig, A. (1998) Disturbance of DNA damage recognition after UV-irradiation by nickel(II) and cadmium(II) in mammalian cells. Carcinogenesis 19, 617–621.

Hartwig, A., Mullenders, L.H.F., Schlepegrell, R., Kasten, U. and Beyersmann, D. (1994) Nickel(II) interferes with the incision step in nucleotide excision repair. Cancer Res. 54, 4045–4051.

Hartwig, A. (1995) Current aspects in metal genotoxicity. BioMetals 8, 3–11.

Hartwig, A., Schlepegrell, R., Dally, H. and Hartmann, M. (1996a) Interaction of carcinogenic metal compounds with DNA repair processes. Annals of Clinical and Laboratory Science 26, 31–38.

Hartwig, A., Dally, H. and Schlepegrell, R. (1996b) Sensitive analysis of oxidative DNA damage in mammalian cells: Use of the bacterial Fpg protein in combination with alkaline unwinding. Toxicology Letters 88, 85 - 90.

Hartwig, A., Gröblinghoff, U.D., Beyersmann, D., Natarajan, A.T., Filon, R. and Mullenders, L.H.F. (1997) Interaction of arsenic(III) with nucleotide excision repair in UV-irradiated human fibroblasts. Carcinogenesis 18, 399–405.

IARC (1980) Monographs on the Evaluation of the Carcinogenic Risk of Chemicals to Humans, Vol. 23: Some Metals and Metallic Compounds. IARC, Lyon.

IARC (1990) Monographs on the Evaluation of Carcinogenic Risks to Humans, Vol. 49: Chromium, Nickel and Welding. IARC, Lyon.

IARC (1991) Monographs on the Evaluation of Carcinogenic Risks to Humans, Vol. 52: Chlorinated Drinking Water; Chlorination By-Products; some other Haloginated Compounds; Cobalt and Cobalt Compounds. IARC, Lyon.

IARC (1993) Monographs on the Evaluation of Carcinogenic Risks to Humans, Vol. 58: Beryllium, Cadmium, Mercury, and Exposures in the Glass Manufacturing Industry. IARC, Lyon.

Kasprzak, K.S. (1995) Possible role of oxidative damage in metal-induced carcinogenesis. Cancer Invest. 13, 411–430.

Kasten, U., Mullenders, L.H.F. and Hartwig, A. (1997) Cobalt(II) inhibits the incision and the polymerization step of nucleotide excision repair in human fibroblasts. Mutation Res. 383, 81–89.

Klein, C.B., Frenkel, K. and Costa, M. (1991) The role of oxidative processes in metal carcinogenesis. Chem. Res. Toxicol. 4, 592–604.

Lee-Chen, S.F., Wang, M.C., Yu, C.T., Wu, D.R. and Jan, K.Y. (1993) Nickel chloride inhibits the DNA repair of UV-treated but not methyl methanesulfonate-treated Chinese hamster ovary cells. Biol. Trace Elem. Res. 37, 39–50.

Li, J.H. and Rossman, T.G. (1989) Inhibition of DNA ligase activity by arsenite: A possible mechanism of its comutagenesis. Mol. Toxicol. 2, 1–9.

Lindahl, T. (1993) Instability and decay of the primary structure of DNA. Nature 362, 709–715.

Naegeli, H. (1995) Mechanisms of DNA damage recognition in mammalian nucleotide excision repair. FASEB J. 9, 1043–1050.

Oller, A.R., Costa, M. and Oberdörster, G. (1997) Carcinogenicity assessment of selected nickel compounds. Toxicol. Appl. Pharmacol. 143, 152–166.

Pflaum, M., Boiteux, S. and Epe, B. (1994) Visible light generates oxidative DNA base modifications in high excess of strand breaks in mammalian cells. Carcinogenesis 15, 297–300.

Pflaum, M., Will, O. and Epe, B. (1997) Determination of steady-state levels of oxidative DNA base modifications in mammalian cells by means of repair endonucleases. Carcinogenesis 18, 2225–2231.

Sarkar, B. (1995) DNA recognition by stereoid hormone receptor zinc fingers: Effects of metal replacement and protein-protein dimerization interface. In: Sarkar, B. (Ed.) Genetic response to metals, pp. 237–253.

Schneider, J.E., Price, S., Maidt, M.L., Gutteridge, J.M.C. and Floyd, M.A. (1990) Methylene blue plus light mediates 8-hydroxy-2´-deoxyguanosine formation in DNA preferentially over strand breakage. Nucleic Acids Res. 18, 631–635.

Sugiyama M. (1994) Role of antioxidants in metal-induced damage. Cell Biology and Toxicology 10, 1–22.

Sunderman, F. W. Jr. (1993) Search for molecular mechanisms in the genotoxicity of nickel. Scand. J. Work Environ. Health 19, Suppl. 1, 75–80.

Wakasugi, M., Abe, Y., Yoshida, Y, Matsunaga, T. and Nikaido, O. (1996) Purification of a novel UV-damaged-DNA binding protein highly specific for (6–4) photoproduct. Nucleic Acids Research 24, 1099–1104.

11

CHROMIUM(III) AS A MALE PRECONCEPTION CARCINOGEN IN MICE

Lucy M. Anderson,[1] Marek A. Sipowicz,[1] Wei Yu,[1] Bhalchandra A. Diwan,[2] Lisa Birely,[2] Diana C. Haines,[2] Charles W. Riggs,[3] and Kazimierz S. Kasprzak[1]

[1]Laboratory of Comparative Carcinogenesis
Division of Basic Sciences
National Cancer Institute
[2]SAIC-Frederick, Inc.
[3]Data Management Services, Inc.
Frederick Cancer Research and Development Center
Frederick, Maryland 21702

1. INTRODUCTION

1.1. Paternal Exposures and Risk of Cancer in Offspring

Numerous epidemiological studies have suggested significant correlation between paternal exposure to chemical carcinogens or radiation and incidence of childhood cancers, including brain tumor, Wilm's tumor of kidney, leukemia, hepatoblastoma, retinoblastoma and sarcoma in soft tissues (Savitz and Chen, 1990; Olsen et al., 1991; Bunin et al., 1992; Tomatis, 1994). Interpretations of some of these data have been controversial, particularly with regard to radiation effects. However, recently several very large studies of paternal smoking, conducted in different parts of the world, have established associations between paternal preconception smoking levels and risk of cancer in children, with a high degree of statistical significance (Sorahan et al., 1997a and b; Ji et al., 1997). These results provide strong confirmation that preconception carcinogenesis does occur in humans.

One of the associations frequently uncovered by epidemiological investigations is that of paternal occupational exposures to metals with childhood cancers of a variety of types (Olsen et al., 1991; Bunin et al., 1992; Wilkins and Wellage, 1996; Knight and Marrett, 1997; Cordier et al., 1997). In addition to the childhood cancers listed above, testicular cancer, which has possibly been increasing in young men, has recently been linked to paternal occupation in the year before conception, with odds ratios of 3.3 for metal workers and 5.8 for men employed in metal products industries (Knight and Marrett, 1997).

Metals and Genetics, edited by Sarkar.
Kluwer Academic / Plenum Publishers, New York, 1999.

1.2. Preconception Carcinogenesis in Animal Models

Treatment of male rodents, either *in utero* or as adults, with chemical carcinogens or X- or neutron irradiation, can cause increased incidence of tumors in their offspring as well as in later generations (for reviews see Tomatis et al., 1992; Yamasaki et al., 1992; Tomatis, 1994; also Mohr et al., 1995; Watanabe et al., 1996). Many types of tissue have been shown to be targeted, including the nervous system and uterus of rats; lung, lymphoid tissue, ovary, uterus, liver, intestine, and skin of mice; and neuroendocrine cells, forestomach, and uterus of hamsters. The kinds of chemical carcinogen found to have this effect are also diverse and include urethane, N-ethyl-N-nitrosourea, N-nitrosodiethylamine, 4-nitroquinoline N-oxide, diethylstilbestrol, and cyclophosphamide.

In view of the frequency with which metals exposures have been implicated in possible human preconception carcinogenesis (see above), we wished to determine whether metals could have a preconception carcinogenic effect in rodents. In a preliminary trial (Anderson et al., 1994) we exposed male Swiss mice to each of the metals common in welding fumes, two weeks before mating with untreated female mice. Offspring were killed at twenty weeks of age and examined for tumors. This model was designed to utilize the murine primary lung tumor as the endpoint, as it has a short latency for development, and has repeatedly been shown to be an indicator of preconception carcinogenic effects of both organic chemical carcinogens and radiation. Among the offspring, the incidence of mice bearing lung tumors was 0.9% in the controls and 0.7% in the offspring of males treated with sodium fluoride, which is prominent in welding fumes. Lung tumor incidence in the offspring after paternal treatment with chromium(III) chloride was 7%, a significant increase, and a well developed lymphoblastic lymphoma was also found. Paternal exposure to iron(II) or iron(III) chloride, manganese(II) chloride, nickel(II) chloride, cobalt(II) chloride, or sodium chromate was associated with offspring lung tumor incidences of 2.5–3%, suggestive differences but not of statistical significance.

These preliminary findings indicated that metals may indeed be preconception carcinogens. We have now confirmed the preconception effect of chromium(III) chloride, with observations of the offspring extended through old age.

1.3. Toxic Effects of Chromium

Humans are exposed to both hexavalent chromium, Cr(VI), and trivalent chromium, Cr(III). Occupational exposure generally involves both species, and has been associated with increased risk of respiratory and possibly gastrointestinal cancer (reviewed in Alcedo and Wetterhahn, 1990; Cohen et al., 1993; Costa, 1997). Ulceration of skin and nasal tissues and liver, kidney, and gastrointestinal damage may occur. These effects have been ascribed largely to Cr(VI), which is quite toxic both to cells in culture and in whole animals. It is carcinogenic in animals, for the respiratory tract and subcutaneous tissues (reviewed in Alcedo and Wetterhahn, 1990; Cohen et al., 1993).

These toxic effects of Cr(VI) are thought to be due to reactive intracellular reduction products, including the terminal valence state, Cr(III). In spite of this, administered Cr(III) is minimally toxic, has not been associated with cancer risk in humans, and has been negative for carcinogenicity in numerous rodent trials (see Alcedo and Wetterhahn, 1990; and Cohen et al., 1993). This lack of biological activity in most *in vivo* settings appears to be due to inability of Cr(III) to cross cell membranes. With *in vitro* cell-free systems, Cr(III) binds to DNA (Hneihen et al., 1993), causing monoadducts and cross-links, and resulting in single-base substitution and deletion mutations (Tsou et al., 1997). It stimulates formation of amino acid-DNA and protein-DNA complexes, involving especially tyrosine, cysteine, histidine, methionine, and threonine (Salnikow et al., 1992). Also, Cr(III) alters DNA-polymerase interactions such that fidelity of replication is compromised (Snow, 1991). In addition, Cr(III) can catalyze the formation of reactive oxygen species and oxidative base damage in DNA, in the presence of hydrogen peroxide (Shi et al., 1993; Tsou et al., 1996).

These capabilities presumably pertain to Cr(III) generated intracellularly by reduction of Cr(VI), although much of the trivalent species is sequestered by interaction with certain amino acids in peptides (Hneihen et al., 1993). Treatment of intact cells with Cr(VI) has been shown to result in stable complexes of DNA with Cr(III) and glutathione or amino acids (Zhitkovich et al., 1995). The intracellular mutagenicity of Cr(III) probably is determined in part by the ligand environment (Warren et al., 1981).

The impermeability of the cell membrane to Cr(III) is not absolute. Insoluble forms of Cr(III), including chromium oxide and chromium chloride, may be taken up into cells in vacuoles, and have been shown to cause mutations and transformation in mammalian cells (Elias et al., 1986; Biederman and Landolph, 1990). Increased production of reactive oxygen species occurred in peritoneal macrophages and hepatic organelles after treatment of rats with either Cr(III) or Cr(VI) (Bagchi et al., 1995).

There is some evidence that Cr may be particularly toxic to the male reproductive system. Cr(III) was reported to be more toxic to the testes of rabbits than Cr(VI), causing inhibition of several enzymes and histological degeneration (Behari et al., 1978). Cr(III) was found to accumulate in the interstitial tissue of hamster testes (Danielsson et al., 1984). In rat testes, no effects were seen with Cr(III) (Ernst, 1990), but Cr(VI) has been repeatedly found to be toxic, causing testicular atrophy and reduction in sperm (Ernst, 1990), leakage of Sertoli-cell tight junctions and mitochondrial damage in late stage spermatids (Murthy et al., 1991), and reduction in Leydig cells and testosterone production (Chowdhury, 1995). Even a relatively low dose caused a decrease in serum testosterone (Ernst and Bond, 1992). In mice both Cr(III) chloride and Cr(VI) as potassium dichromate, given in the drinking water, reduced male fertility and caused a decrease in the weights of seminal vesicles and preputial glands, but an increase in testes weights (Elbetieha and Al-Hamood, 1997), indicating possible disturbance of the pituitary-gonadal endocrine axis.

2. EXPERIMENTAL ASSESSMENT OF PRECONCEPTION EFFECTS OF CHROMIUM(III)

2.1. Outline of Experiment

Swiss Cr:NIH(s) mice were obtained from the Animal Production Area of the Frederick Cancer Research and Development Center and maintained under standardized pathogen-free conditions. Male mice were treated once i.p. with 1 mmol/kg of $CrCl_3 \cdot 6H_2O$ in sterile acidified water at 6 weeks of age, or water only, and mated two weeks later with five 8-week old females. This timing insured that sperm utilized in fertilization would have been exposed post-meiotically, a stage of high sensitivity to preconception carcinogenic effects (Nomura, 1982, 1984). There were eleven fathers in each group with surviving offspring, with no differences in numbers of females impregnated, or numbers of babies born or weaned. Some of the male offspring were killed at intervals (data not shown). All other offspring were held without further treatment until natural morbidity. Offspring analyzed at natural death included, for the chromium group, 71 males from 20 litters and 72 females from 21 litters, and for the controls, 71 females from 23 litters and 48 males from 14 litters. Average survival times were similar between groups (average 743 and 730 days for the chromium and control males, and 848 and 857 days for the chromium and control females, respectively).

All mice were subjected to complete necropsy, tissues fixed in formalin, and masses and lesions embedded in paraffin, sectioned at 5 μm, and stained with hematoxylin and eosin for pathological diagnosis. Statistical tests included the Fisher exact test and the chi square test of homogeneity for incidence comparisons; Kruskal-Wallis and Wilcoxon rank-sum methods for multiplicities; and the correlated binomal C(alpha) test for homogeneity of the proportions of tumor bearers per litter and per sire. Except where noted, one-sided tests were utilized, since changes were in the direction predicted by the hypothesis.

The most significant changes occurring in the offspring of treated fathers, as compared with controls, are summarized graphically in this chapter. Full presentation of the numerical data will be published elsewhere (Yu et al., submitted).

2.2. Tumors in Secretory Organs

Adrenal pheochromocytomas occurred in offspring of chromium-treated males of both sexes, whereas none were found in controls (Figure 1), a highly significant difference by statistical test (P = 0.01). Thyroid follicular cell adenomas occurred in control female offspring, but had a significantly higher incidence in the male plus female offspring of chromium-treated males, including 2 carcinomas in male offspring (Figure 1). A significant increase in Harderian gland tumors, mainly adenomas, was found in the male offspring of the treated fathers (Figure 1).

2.3. Female Secretory Tissues

Lung tumors, which in the mouse arise from secretory alveolar type 2 and/or Clara cells, showed an increased incidence in the female offspring, only, of chromium-treated fathers (Figure 2). The multiplicity of these tumors was also increased, from 1.1 ± 0.2 in controls to 2.0 ± 0.3 in the chromium group females (P = 0.02, two-sided Mann-Whitney test). Non-neoplastic changes showing a treatment-related increase were ovarian cysts and uterine cystic endometrial hyperplasia.

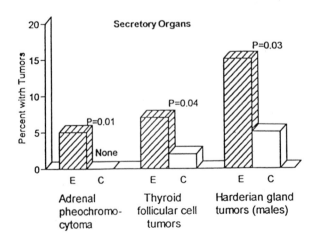

Figure 1. Tumors in secretory organs of offspring of chromium(III)-treated fathers (hatched bars) and control offspring (open bars). Data for both sexes were combined for adrenal pheochromocytomas and thyroid tumors, since incidences did not differ significantly between sexes. P values are from one-sided Fisher Exact Tests.

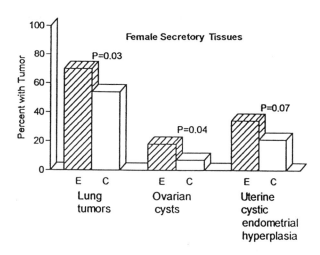

Figure 2. Tumors or abnormalities in secretory tissues of female offspring of chromium(III)-treated fathers (hatched bars) and controls (open bars). Lung tumors were adenomas and carcinomas of the peripheral lung, derived from secretory alveolar type 2 or Clara cells. All tumors ≥ 0.1 mm in largest dimension were counted. P values are from one-sided Fisher Exact Tests.

2.4. Male Reproductive Glands

Tumors in seminal vesicles were mainly sarcomas, with one granular cell tumor and one histiocytic sarcoma in chromium group offspring and a hemangiosarcoma in a control (Figure 3). These may have been increased in incidence in the chromium group (borderline significance). Changes in Cowper's gland included one polyp and two cystadenomas in the chromium group offspring, plus dilation in this tissue in both groups. Incidences of any change in Cowper's gland, and of tumors in seminal vesicles or Cowper's gland, were significantly greater in the chromium group male offspring compared with controls.

2.5. Non-Neoplastic Changes in Male Kidneys and Livers

Various abnormalities were observed in male kidneys, significantly more frequently in those sired by the chromium-treated fathers (Figure 4). The most common changes were nephropathy and cysts. Other alterations included hydronephrosis, hypertrophy, lymphocytic infiltrates, and hyaline droplets, and the incidence of these together was also signifi-

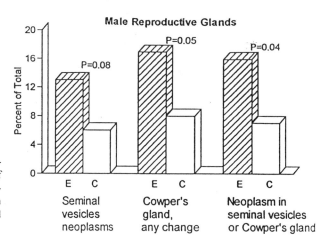

Figure 3. Neoplasms and abnormalities in male reproductive glands of offspring of chromium(III)-treated fathers (hatched bars) and controls (open bars). P values are from one-sided Fisher Exact Tests.

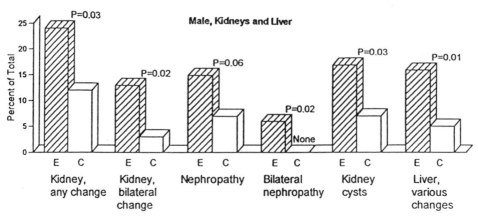

Figure 4. Non-neoplastic abnormalities in male kidneys and livers in offspring of chromium(III)-treated fathers (hatched bars) and controls (open bars). The various changes in the livers included all abnormalities as listed in text, except neoplasms and preneoplastic foci (4% in treated, 1% in controls) and inflammation (5% in treated, 3% in controls). P values are from one-sided Fisher Exact Tests.

cantly greater in the chromium group males compared with controls (13% vs. 5%, P = 0.04). Bilateral alterations were also more common in this group.

Hepatic abnormalities included hepatocellular hypertrophy and necrosis, coagulative necrosis, pigmented macrophages, lymphocytic infiltrates, extramedullary hematopoiesis, and cysts. These together were about three times more common among the male offspring of the chromium-treated fathers (Figure 4).

2.6. Frequency of Effects of Preconception Chromium Exposure

Offspring of ten of the eleven chromium-treated males presented one or more of the significant treatment-related effects described above. Data for some specific tissues are presented in Figure 5, showing that 60–80% of these fathers were involved. For lesions occurring in the control offspring, fewer fathers contributed, and this difference was of statistical significance, or nearly so, for pheochromocytomas, multiple lung tumors (females), and bilateral kidney changes (males), in spite of the small number of fathers.

The percentages of total offspring showing one or more of these three changes were 21% for females and 14% for males, compared with 7% and 3% of control females and males, respectively (Figure 6), differences of statistical significance. If tumors of thyroid and Harderian gland (males) and abnormalities of male liver are included, the percentage of total offspring affected by paternal chromium was 23% for females and 39% for males, compared with 10% and 20% of control females and males, respectively, also differences of statistical significance.

We tested whether any of the significant changes, or any other pathological alteration, were more likely to occur in the same animal than expected by chance. The only significant correlations were for seminal vesicle neoplasm with any kidney change (P = 0.014), pheochromocytoma with pituitary tumor (females) (P = 0.026), three or more lung tumors with ovarian cysts (P = 0.019), and Harderian gland tumors with lymphoma (females) (P = 0.053) (all P values are 2-sided). Statistical tests did not reveal any father-specific or litter-specific actions.

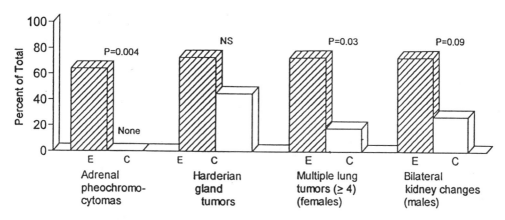

Figure 5. Percentage of fathers with affected offspring after preconception chromium(III) (hatched bars) compared with control fathers (open bars). Endpoints selected for test were those with the greatest apparent effect of the preconception exposure (see preceding figures), since the numbers of fathers was small (eleven in each group). P values are from two-sided Fisher Exact Tests.

3. DISCUSSION

These results confirm our previous observation (Anderson et al., 1994), that chromium(III) is a preconception carcinogen in mice. Although the effects were not large in magnitude, they occurred frequently, in at least 10–15% of the offspring, and most of the fathers produced affected offspring. Numerous tissues were targeted, and both neoplastic and nonneoplastic lesions were increased. The appearance of pheochromocytomas was especially notable, as the adrenal medulla has not been reported to be a target for preconception carcinogenesis previously, and these are rare spontaneous or induced tumors in mice (Tischler and Sheldon, 1996). The increased incidence of thyroid follicular cell and Harderian gland tumors, and the significant association of pituitary tumors with pheochromocytomas in the females, suggests involvement of the endocrine axis. Endocrine involvement is also indicated by the increase in ovarian cysts, which may be caused by administration of steroid hormones (Maekawa and Yoshida, 1996), and by cysts and hydronephrosis in kidneys, which also occur after treatment with steroid and xenobiotics

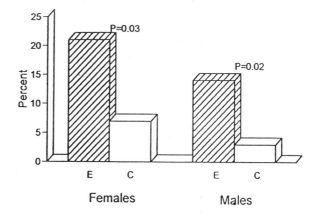

Figure 6. Percent of all offspring showing one of the major effects of preconception chromium(III) carcinogenesis: adrenal pheochromocytomas, high multiplicity of lung tumors (females), and/or bilateral kidney abnormalities (males); hatched bars, from treated fathers, open bars, controls. P values are from two-sided Fisher Exact Tests.

with hormone-like properties (Wolf and Hard, 1996). These renal conditions may be congenital, especially if bilateral; bilateral kidney disease was more common in the chromium group male offspring.

Our previous observation (Anderson et al., 1994), of an increase in lung tumors in offspring of chromium-exposed fathers, was confirmed in the present study, with the effect being limited to female offspring. Most of the tumors were found in females in the earlier experiment also. Although the effect in lung was of statistical significance, it was not as large as that reported for another strain of mouse after preconception X-ray, urethane or 4-nitroquinoline-N-oxide (Nomura, 1982). More recently, paternal X-ray treatment of BALB/c mice was reported not to alter the incidence of lung tumors in progeny (Cattanach et al., 1995). It seems likely that the outcome of preconception exposures in mice is highly strain and tissue specific, as well as oncogen specific. In the present experimental series, a separate group of male mice were treated with urethane before mating, with offspring again held until natural death. While some of the urethane effects were similar to those of the chromium treatment (increase in pheochromocytomas, thyroid tumors, Harderian gland tumors, and hepatic changes), the chromium-associated increases in male reproductive organ neoplasms and in kidney abnormalities were lacking after urethane. Furthermore preconception urethane caused increases in neoplasms of the glandular stomach in males and in lymphomas in females, and decreases in histiocytic sarcomas in females, changes which were not seen after chromium (Yu et al., submitted). It is clear that the effects observed cannot be ascribed to nonspecific consequences of general chemical insult.

The time at which the chromium exerted its effects on the offspring is uncertain. In a parallel study of the distribution and clearance of Cr(III) after intraperitoneal treatment as in the preconception carcinogenesis protocol, we found that whole testis was relatively effective in taking up and retaining chromium, with levels that were about 10% those in liver and 30% those in pancreas and spleen (Sipowicz et al., 1997). The amounts in testes did not change significantly over a two week time period. Levels in sperm were not determined. If chromium was present in the sperm at the time of fertilization, then effects throughout the period of embryogenesis would have been possible. However, the many preconception effects of urethane (see above) suggest direct effects on the sperm, since this chemical is cleared from the body within one day.

The clear positive action of chromium(III) as a preconception carcinogen in mice is somewhat surprising, as this ion is non-toxic and has been described as minimally carcinogenic (see Introduction). However, as noted above, the biological ineffectiveness of chromium(III) is related to its inability to cross the membranes of many cells; insoluble complexes may be taken up by phagocytosis. In the testis, Cr(III) may enter cells more effectively than in other tissues. In an autoradiographic study with hamsters, chromium, especially Cr(III) from $^{51}CrCl_3$ (dose unspecified), was strongly accumulated in the testicular interstitial tissues, but not in blood vessels or the spermatogenic epithelium (Danielsson et al., 1984). An effect on the Leydig cells and hence on hormone production was postulated.

Cr (III) *in vitro*, and presumably within the cell, reacts readily with DNA in a variety of ways and may also generate reactive oxygen species. To test for formation of reactive oxygen species in testes, we assayed DNA from whole testes for levels of 8-hydroxydeoxyguanosine, at intervals of one to thirty-five days after administration of chromium(III) chloride at the same dose used in the preconception carcinogenesis study. No differences were found (data not shown). Increases in 8-hydroxydeoxyguanosine could still have occurred in germ cells or other specific cell types within the testis; this will require further investigation.

The frequency of the preconception effect of Cr(III), affecting most of the fathers and at least 10–15% of the offspring (after correction for control incidences), is much greater than occurs at known structural gene loci after treatment of fathers with mutagens (Nomura, 1984). For example, in a recent study of effects of ethylnitrosourea (ENU) administered 8–21 days before mating, the maximum effect on dominant cataract mutations plus recessive mutations at seven loci was 7/4617 offspring, or 0.15% (Favor et al., 1990). Use of transgenic reporter genes to measure rates of male germ cell mutagenicity by ENU or x-rays has confirmed these low rates (Provost and Short, 1994; Katoh et al., 1994). Paternal exposure to x-rays or ENU caused various fetal mouse developmental abnormalities in an approximate ten-fold higher percentage of offspring, about 2% (Nomura, 1982, 1984; Kirk and Lyon, 1984; Nagao and Fujikawa, 1990), a value still about ten-fold lower than the effect in our study. Less well defined toxicological outcomes of paternal exposures, such as dominant lethality and respiratory distress, have ten to thirty percent incidences, similar to the frequency of pathological alterations in our current study (Paschin, 1990; Nomura et al., 1990). Possibly the largest preconception carcinogenic effect reported thus far was a fourteen-fold increase in liver tumors in male offspring of C3H male mice treated with ^{252}Cf two weeks before mating with C57BL/6 females, from 3% in controls to 43% in treated mice (Takahashi et al., 1992; Watanabe et al., 1996).

Thus complex outcomes of preconception exposures, including neoplasia, can occur with high frequency. This may imply that mutation in any one of numerous genes can result in the observed outcomes, or that hypermutable sequences, such as microsatellites, are altered. Alternatively, the mechanism of the preconception effects may be novel and indirect, and involve an effect on gene expression, for example, by disregulation of the gene imprinting process (Anderson et al., 1994). The localization of Cr(III) in the interstitial tissue, rather than the germinal epithelium, of hamster testes (Danielsson et al., 1984), and decreases in hormones after chromium exposure (Ernst and Bond, 1992; Chowdhury, 1995), suggest an indirect effect. Evidence of hormonal dysregulation has also been presented for mice (Elbetieha and Al-Hamood, 1997). In the case of Cr(III), an excessive dose of this essential dietary ion (Mertz and Schwartz, 1959) could cause abnormality in the metabolic processes it regulates.

4. SUMMARY AND CONCLUSIONS

Occupational exposures of fathers to metals has been linked epidemiologically to increased risk of cancer in their children. Preconception carcinogenesis, as a result of exposures of fathers to organic carcinogens or radiation, has been described in mice, rats, and hamsters. A previous preliminary study had shown that, of the metals prominent in welding fumes, chromium(III) appeared to be a preconception carcinogen for mice. In a confirmation of this effect, we treated male Swiss mice with chromium(III) chloride (1 mmol/kg) two weeks before mating. Offspring were held without further treatment until natural death.

The preconception exposure to chromium(III) chloride resulted in the appearance of pheochromocytomas of the adrenal medulla. None occurred in controls, and the difference was significant by statistical test. Other effects of statistical significance involved an increase in lesions found at low incidence in the controls and included follicular cell tumors of thyroid, Harderian gland tumors (males), lung tumors (females), tumors in seminal vesicles or Cowper's gland, ovarian cysts, uterine cystic endometrial hyperplasia, abnormalities of male kidney including nephropathy and cysts, and a variety of abnormalities of male liver. The effect was of high frequency, with offspring of ten of the eleven exposed males showing at least one of the significant effects, and with at least 15% of females and 10% of males being affected.

The mechanism of this remarkable biological activity of Cr(III) is unknown, but could involve mutations at multiple sites or at hypermutable loci, and/or novel effects on gene expression, possibly by derangement of imprinting events or via disruption of the metabolic processes normally influenced by this metal ion.

REFERENCES

Alcedo, J.A., Wetterhahn, K.E. (1990). Chromium toxicity and carcinogenesis. Int. Rev. Exp. Path. 31, 85–108.

Anderson, L.M., Kasprzak, K.S., Rice, J.M. (1994). Preconception exposure of males and neoplasia in their progeny: Effects of metals and consideration of mechanisms. In: Mattison, D.R., Olshan, A.F. (Eds.) "Proc. Int. Symp. Male-Mediated Developmental Toxicity." Plenum, New York, pp 129–140.

Bagchi, D., Hassoun, E.A., Bachi, M., Stohs, S.J. (1995). Chromium-induced excretion of urinary lipid metabolites, DNA damage, nitric oxide production, and generation of reactive oxygen species in Sprague-Dawley rats. Comp. Biochem. Physiol. 110C, 177–187.

Behari, J., Chandra, S.V., Tandon, S.K. (1978). Comparative toxicity of trivalent and hexavalent chromium in rabbits. III. Biochemical and histological changes in testicular tissues. Acta Biol. Med. Germ. 37, 463–468.

Biedermann, K.A., Landolph, J.R. (1990) Role of valence state and solubility of chromium compounds on induction of cytotoxicity, mutagenesis, and anchorage independence in diploid human fibroblasts. Cancer Res. 50, 7835–7842.

Bunin, G.R, Rose, P.G., Noller, K.L., Smith, E. (1992). Carcinogenesis. In: Paul, M. (Ed.) "Occupational and Environmental Reproductive Hazards: a Guide for Clinicians." Williams & Wilkins, Baltimore, pp 76–88.

Cattanach, B.M., Patrick, G., Papworth, D., Goodhead, D.T., Hacker, T., Cobb, L., Whitehill, E. (1995). Investigation of lung tumour induction in BALB/cJ mice following paternal X-irradiation Int. J. Radiat. Biol. 67, 607–615.

Chowdhury, A.R. (1995) Spermatogenic and steroidogenic impairment after chromium treatment in rats. Indian J. Exp. Biol. 33, 480–484.

Cohen, M.D., Kargacin, B., Klein, C.B., Costa, M. (1993). Mechanisms of chromium carcinogenicity and toxicity. Crit. Rev. Toxicol. 23, 255–281.

Cordier, S., Lefeuvre, B., Filippini, G., Peris-Bonet, R., Farinotti, M., Lovicu, G., Mandereau, L. (1997). Parental occupation, occupational exposure to solvents and polycyclic aromatic hydrocarbons and risk of childhood brain tumors (Italy, France, Spain). Cancer Causes Control 8, 688–697.

Costa, M. (1997). Toxicity and carcinogenicity of Cr(VI) in animal models and humans. Crit. Rev. Toxicol. 27, 431–442, 1997.

Danielsson, B.R.G., Denckèr, L., Lindgren, A., Tjälve, H. (1984). Accumulation of toxic metals in male reproduction organs. Arch. Toxicol., Suppl. 7, 177–180.

Elbetieha, A., Al-Hamood, M.H. (1997). Long-term exposure of male and female mice to trivalent and hexavalent chromium compounds: effect on fertility. Toxicology 116, 39–47.

Elias, Z., Poirot, O., Schneider, O., Danière, M.C., Terzetti, F., Guedenet, J.C., Cavelier, C. (1986). Cellular uptake, cytotoxic and mutagenic effects of insoluble chromic oxide in V79 Chinese hamster cells. Mutat. Res. 169, 159–170.

Ernst, E. (1990). Testicular toxicity following short-term exposure to tri- and hexavalent chromium: an experimental study in the rat. Toxicol. Lett. 51, 269–275.

Ernst, E., Bonde, J.P. (1992). Sex hormones and epididymal sperm parameters in rats following sub-chronic treatment with hexavalent chromium. Hum. Exp. Toxicol. 11, 255–258.

Favor, J., Neuhäuser-Klaus, A., Ehling, U.H. (1990). The frequency of dominant cataract and recessive specific-locus mutations and mutation mosaics in F_1 mice derived from post-spermatogonial treatment with ethylnitrosourea. Mutat. Res. 229, 105–114.

Hneihen, A.S., Standeven, A.M., Wetterhahn, K.E. (1993). Differential binding of chromium(VI) and chromium(III) complexes to salmon sperm nuclei and nuclear DNA and isolated calf thymus DNA. Carcinogenesis 14, 1795–1803.

Ji, B-T., Shu, X-O., Linet, M.S., Zheng, W., Wacholder, S., Gao, Y-T., Ying, D-M., Jin, F. (1997). Paternal cigarette smoking and the risk of childhood cancer among offspring of nonsmoking mothers. J. Natl. Cancer Inst. 89, 238–244.

Katoh, M., Inomata, T., Horiya, N., Suzuki, F., Shida, T., Ishioka, K., Shibuya, T. (1994). Studies on mutations in male germ cells of transgenic mice following exposure to isopropyl methanesulfonate, ethylnitrosourea or X-ray. Mutat. Res. 341, 17–28.

Kirk, K.M., Lyon, M.F. (1984). Induction of congenital malformations in the offspring of male mice treated with X-rays at pre-meiotic and post-meiotic stages. Mutat. Res. 125, 75–85.

Knight, J.A., Marrett, L.D. (1997). Parental occupational exposure and the risk of testicular cancer in Ontario. JOEM 39, 333–338.

Maekawa, A., Yoshida, A. (1996). Susceptibility of the female genital system to toxic substances. In: Mohr, U., Dungworth, D.L., Capen, C.C., Carlton, W.W., Sundberg, J.P., Ward, J.M. (Eds.) "Pathobiology of the Aging Mouse." Vol. 1. ILSI Press, Washington, D. C., pp 481–493.

Mertz, W., Schwartz, K. (1959). Chromium as a trace metal and essential nutrient. Am. J. Physiol. 196, 614–618.

Mohr, U., Emura, M., Kamino, K., Steinmann, J., Kohler, M., Morawietz, G., Dasenbrock, C., Tomatis, L. (1995). Increased risk of cancer in the descendants of Syrian hamsters exposed prenatally to diethylnitrosamine (DEN). Int. J. Cancer 63, 86–91.

Murthy, R.C., Saxena, D.K., Gupta, S.K., Chandra, S.V. (1991). Ultrastructural observations in testicular tissue of chromium-treated rats. Reprod. Toxicol. 5, 443–447.

Nagao, T., Fujikawa, K. (1990). Genotoxic potency in mouse spermatogonial stem cells of triethylenemelamine, mitomycin C, ethylnitrosourea, procarbazine, and propyl methansulfonate as measured by F_1 congenital defects. Mutat. Res. 229, 123–128.

Nomura, T. (1982). Parental exposure to X rays and chemicals induces heritable tumours and anomalies in mice. Nature 296, 575–577.

Nomura, T. (1984). Quantitative studies on mutagenesis, teratogenesis and carcinogenesis in mice. In: Tazima, Y., Kondo, S., Kuroda, T. (Eds.) "Problems of Threshhold in Chemical Mutagenesis." The Environmental Mutagen Society of Japan, Shizouka, pp 27–34.

Nomura, T., Gotoh, H., and Namba, T. (1990). An examination of respiratory distress and chromosomal abnormalities in the offspring of male mice treated with ethylnitrosourea. Mutat. Res. 229, 115–122.

Olsen, J.H., Brown, P.deN., Schulgen, G., Jensen, O.M. (1991). Parental employment at time of conception and risk of cancer in offspring. Eur. J. Cancer 27, 958–965.

Paschin, Y.V. (1990). Presumed gene mutations detected in F_1 progeny of mouse males treated with ethyl methanesulfonate. Mutat. Res. 229, 185–187.

Provost, G.S., Short, J.M. (1994). Characterization of mutations induced by ethylnitrosourea in seminiferous tubule germ cells of transgenic B6C3F₁ mice. Proc. Natl. Acad. Sci. USA 91, 6564–6568.

Salnikow, K., Zhitkovich, A., Costa, M. (1992). Analysis of the binding sites of chromium to DNA and protein *in vitro* and in intact cells. Carcinogenesis 13, 2341–2346.

Savitz, D.A., Chen, J. (1990). Parental occupation and childhood cancer: Review of epidemiologic studies. Environ. Health Perspect. 88, 325–337.

Shi, X., Dalal, N.S., Kasprzak, K.S. (1993). Generation of free radicals from hydrogen peroxide and lipid hydroperoxides in the presence of Cr(III). Arch. Biochem. Biophys. 302, 294–299.

Sipowicz, M.A., Anderson, L.M., Utermahlen, W.E., Issaq, H.J., Kasprzak, K.S. (1997). Uptake and tissue distribution of chromium(III) in mice after a single intraperitoneal or subcutaneous administration. Toxicol. Lett 93, 9–14.

Snow, E.T. (1991). A possible role for chromium(III) in genotoxicity. Environ. Health Perspect. 92, 75–81.

Sorahan, T., Lancashire, R.J., Hultén, M.A., Peck, I., Stewart, A.M. (1997). Childhood cancer and parental use of tobacco: deaths from 1953 to 1955. Br. J. Cancer 75, 134–138.

Sorahan,T., Prior, P., Lancashire, R.J., Faux, S.P., Hultén, M.A., Peck, I.M., Stewart, A.M. (1997). Childhood cancer and parental use of tobacco: deaths from 1971 to 1976. Br. J. Cancer 76, 1525–1531.

Takahashi, T., Watanabe, H., Dohi, K., Ito, A. (1992). ^{252}Cf relative biological effectiveness and inheritable effect of fission neutrons in mouse liver tumorigenesis. Cancer Res. 52, 1948–1953.

Tischler, A.S., Sheldon, W. (1996). Adrenal medulla. In: Mohr, U., Dungworth, D.L., Capen, C.C., Carlton, W.W., Sundberg, J.P., Ward, J.M. (Eds.). "Pathobiology of the Aging Mouse." Vol. 2. ILSI Press, Washington, D.C., pp 135–151.

Tomatis, L. (1994). Transgeneration carcinogenesis: A review of the experimental and epidemiological evidence. Jpn. J. Cancer Res. 85, 443–454.

Tomatis, L., Narod, S., Yamasaki, H. (1992). Transgeneration transmission of carcinogenic risk. Carcinogenesis 13, 145–151.

Tsou, T-C., Chen, C-L., Liu, T-Y., Yang, J-L. (1996). Induction of 8-hydroxydeoxyguanosine in DNA by chromium(III) plus hydrogen peroxide and its prevention by scavengers. Carcinogenesis 17, 103–108.

Tsou, T-C., Lin, R-J., Yang, J-L. (1997). Mutational spectrum induced by chromium(III) in shuttle vectors replicated in human cells: Relationship to Cr(III)--DNA interactions. Chem. Res. Toxicol. 10, 962–970.

Warren, G., Schultz, P., Bancroft, D., Bennett, K., Abbott, E.H., Rogers, S. (1981). Mutagenicity of a series of hexacoordinate chromium(III) compounds. Mutat. Res. 90, 111–118.

Watanabe, H., Takahashi, T., Lee, J-Y., Ohtaki, M., Roy, G., Ando, Y., Yamada, K., Gotah, T., Kurisu, K., Fujimoto, N., Satow, Y., Ito, A. (1996). Influence of paternal [252]Cf neutron exposure on abnormal sperm, embryonal lethality, and liver tumorigenesis in the F_1 offspring of mice. Jpn. J. Cancer Res. 87, 51–57.

Wilkins, J.R., 3rd, Wellage, L.C. (1996). Brain tumor risk in offspring of men occupationally exposed to electric and magnetic fields. Scand. J. Work Environ. Health 22, 339–345.

Wolf, D.C., Hard, G.C. (1996). Pathology of the kidneys. In: Mohr, U., Dungworth, D.L., Capen, C.C., Carlton, W.W., Sundberg, J.P., Ward, J.M. (Eds.). "Pathobiology of the Aging Mouse." Vol. 1. ILSI Press, Washington, D.C., pp 331–344.

Yamasaki, H., Loktionov, A., Tomatis, L. (1992). Perinatal and multigenerational effect of carcinogens: Possible contribution to determination of cancer susceptibility. Environ. Health Perspect. 98, 39–43.

Yu, W., Sipowicz, M.A., Haines, D.C., Birely, L., Diwan, B.A., Riggs, C.W., Kasprzak, K.S., and Anderson, L.M. (1998). Preconception urethane or chromium(III) treatment of male mice: multiple neoplastic and nonneoplastic changes in offspring. Submitted.

Zhitkovich, A., Voitkun, V., Costa, M. (1995). Glutathione and free amino acids form stable complexes with DNA following exposure of intact mammalian cells to chromate. Carcinogenesis 16, 907–913.

12

DNA CLEAVAGE VS. CROSS-LINKING USING NICKEL PEPTIDES

Mechanistic Aspects

Robyn P. Hickerson,[1] Victor Duarte,[1] J. David Van Horn,[1] Ronelito J. Perez,[1] James G. Muller,[1] Steven E. Rokita,[1] and Cynthia J. Burrows[2]

[1]Department of Chemistry and Biochemistry
University of Maryland
College Park, Maryland 20742
[2]Department of Chemistry
University of Utah
315 S. 1400 East
Salt Lake City, Utah 84112

1. INTRODUCTION

Rarely do the late transition metal ions Ni^{2+} and Cu^{2+} exist as aquated ions *in vivo*; rather they find protein residues as effective chelating agents. In serum, albumin is one of the principal scavengers of these ions, in part via the N-terminus which in the human protein begins with the sequence Asp-Ala-His (DAH) (Harford and Sarkar, 1997). Since the characterization of this site, the N-terminal XXH motif has been identified in a number of other peptides and proteins, including human spermine protamine HP2a, histatin 3, and the neuromedins C and K (Bal et al., 1997). Nickel has long been known as a carcinogenic metal, especially in the form of nickel ore particulates which more readily enter cells by phagocytosis (Costa et al., 1994). Once inside the cell, complexation to nuclear proteins, especially histones, could be one important mechanism for toxicity in which redox active nickel species are capable of inflicting oxidative damage on proteins and DNA (Kasprzak, 1996; Kasprzak, 1995). Both DNA strand breaks and DNA-protein cross-links have been identified as lesions resulting from exposure to excess nickel (Ciccarelli et al., 1981; Ciccarelli and Wetterhahn, 1982). Meanwhile, bioconjugates containing the N-terminal XXH motif have been designed for the express purpose of protein or DNA modification. In order to understand the molecular mechanisms underlying nickel toxicity as well as to provide a foundation for the design of DNA-targeting agents, we have carried out a series of

Metals and Genetics, edited by Sarkar.
Kluwer Academic / Plenum Publishers, New York, 1999.

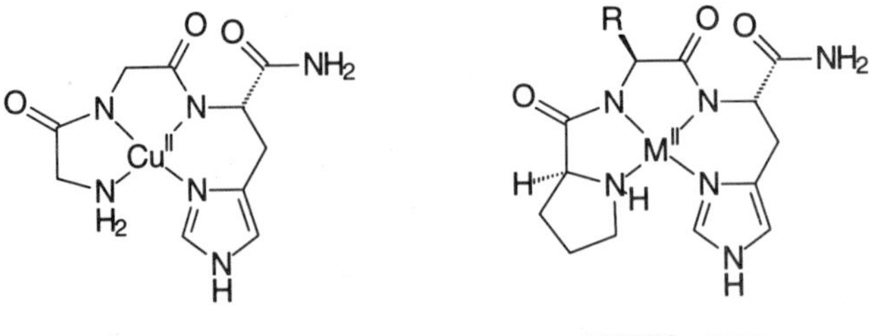

CuGGH-CONH₂   MPXH-CONH₂

Drawing 1.

mechanistic investigations pertaining to nickel peptide mediated DNA oxidation. Here we review our current understanding of these processes.

2. THE N-TERMINAL XXH PEPTIDE MOTIF

Nickel(II) and copper(II) form stable, square-planar complexes with the N-terminal XXH peptide motif (also termed "ATCUN" (Harford and Sarkar, 1997)) through ligation to the amino terminus, two deprotonated amide nitrogens, and the histidine imidazole group as shown in the crystallographically characterized CuGGH-CONH₂ complex shown below (Camerman et al., 1976). The first and second amino acids "X" may be any α-amino acid residue as long as the amide bonds are formed from primary amines, *i.e.* the second "X" may not be proline. A number of such peptides have been synthesized and characterized, and the motif has been recently reviewed (Harford and Sarkar, 1997).

2.1. GGH-COR

The simplest nickel(II) or copper(II) binding unit is the GGH-COR sequence bearing either a carboxylate (R = OH) or a carboxamide (R = NH₂) C-terminus. Most studies are carried out with the carboxamide terminus since these metal complexes are stable to air, while the carboxylate-terminal metal peptides undergo autoxidative decarboxylation (see section 4.1 below). Tripeptides may be synthesized by standard solid-phase techniques, and formation of the square-planar complexes in aqueous, neutral solution upon addition of nickel(II) or copper(II) is readily apparent by their yellow and pink-red colors, respectively. Coordination in this strong in-plane ligand field environment also leads to a substantial stabilization of the +III oxidation state for both of these metals. Characteristic spectral and cyclic voltammetric data are given in Table 1.

2.2. PXH-CONH₂

In order to verify that proline could occupy the first position of the XXH motif,(Harford and Sarkar, 1997) we synthesized the tripeptides PKH-CO₂H and PKH-CONH₂ and prepared their copper(II) and nickel(II) complexes. The data in Table 1 indicate that these PXH-type peptides do form square-planar complexes according to their electronic spectra; however, the incorporation of a secondary amine in place of a primary amine in the N-terminal position results in about a 150 mV destabilization of the nickel(III) state since $E_{1/2}$ values are raised from an average of about 700±30 mV to about 850 mV. Similar changes

Table 1. Spectral and electrochemical data for Ni(II) and Cu(II) XXH complexes

	λ_{max} in nm (ϵ in $M^{-1}cm^{-1}$)	$E_{1/2}^{III/II}$ (mV vs. SCE)
CuGGH-CO_2H	525 (103)	685
NiGGH-CO_2H	425 (~ 100)	720
CuPKH-CO_2H	520 (107)	803 (E_{ox}, irrev.)
CuPKH-$CONH_2$	520 (112)	791
NiPKH-CO_2H	424 (102)	845
NiPKH-$CONH_2$	424 (117)	860
NiKGH-CO_2H	420 (73)	686
NiKGH-$CONH_2$	422 (77)	731
NiRGH-CO_2H	420 (84)	671
NiRRH-CO_2H	420 (94)	720

are seen for the copper(III/II) couple. In each case the cyclic voltammograms are quasi-reversible, at best, with typical peak separations of 100–200 mV. On the other hand, the presence of a carboxylate vs. carboxamide terminus has little effect on the $E_{1/2}$. These studies indicate that proline can occupy the first position of the XXH motif and can be used to tune the redox potential of the resulting metal complex.

2.3. KGH-$CONH_2$, RGH-$CONH_2$, etc.

Substantial work has focused on the incorporation of positively charged amino acid residues into the XXH motif, principally lysine (K) and arginine (R), in order to increase the affinity of small peptides towards DNA (Liang et al., 1995). Data presented in Table 1 show that the identity of the side chain—lysine vs. arginine vs. glycine—has little effect on the coordination chemistry and redox behavior of the nickel(II) complexes. A related motif involving ornithine at the first position yields similar results (Liang et al., 1995). Incorporation of lysine or ornithine at position 1 of the XXH motif allows synthetic extension of the N terminus through the δ or ϵ-amino group of ornithine or lysine, respectively.

3. DNA OXIDATION CATALYZED BY Ni(II)-XXH-$CONH_2$

The N-terminal XXH motif has now been incorporated into a number of DNA-binding protein fragments including a three-helix bundle of Hin recombinase, (Mack and Dervan, 1992) a zinc-finger portion of Sp1, (Nagaoka et al., 1994) a Fos/Jun hybrid, (Harford et al., 1996) a PNA-DNA hybrid, (Footer et al., 1996) and (naturally-occurring) in human protamine HP2a. In each case, coordination of nickel(II) or copper(II) to the XXH motif followed by treatment with a peracid oxidant or H_2O_2 leads to oxidative damage at nearby DNA residues. In some cases, direct strand scission is observed, implying ribose chemistry. In others, an alkaline work-up enhances the cleavage which could point to nucleobase damage.

3.1. Ribose vs. Nucleobase Oxidation

In order to elucidate the key features of metal-peptide-catalyzed DNA oxidation, two research groups, ours (Muller et al., 1997) and Prof. E. Long's (Liang et al., 1998) have launched mechanistic investigations. The results so far indicate that two different pathways operate—one via direct strand scission following hydrogen atom abstraction from the C4' position of the *deoxyribose* (Liang et al., 1998), akin to Fe-bleomycin chemistry (Scheme 1, part A), and the other via one-electron oxidation of the *guanine* nucleobase leading to piperidine-labile strand scission (Scheme 1, part B) (Burrows and Muller, 1998; Muller et al., 1997). The extent to which one pathway or the other is followed appears to depend on (*i*) ionic strength, (*ii*) the identity and stoichiometry of the oxidant, and (*iii*) the presence of delivery agents in the form of bioconjugates or naturally occurring proteins.

As a model target for oxidative DNA modification we chose to study the EcoRI/RsaI 167-base pair restriction fragment from the plasmid pBR322. This double-stranded DNA presents several types of sequence contexts including AT tracts and guanines in several environments, notably a 5'-GG-3' sequence. In G_n sequences, the redox potential of Gs located on the 5' end (or in the center when n>2) is markedly lower than the 3'-G,(Sugiyama and Saito, 1996) and thus oxidation occurring by a one-electron mechanism displays a clear signature of 5'-*G*G-3' >> 5'-*G*A-3' > 5'-*G*Py-3'. Since oxidative chemistry on nucleobases does not generally lead to direct strand scission, it is important to use other methods to learn

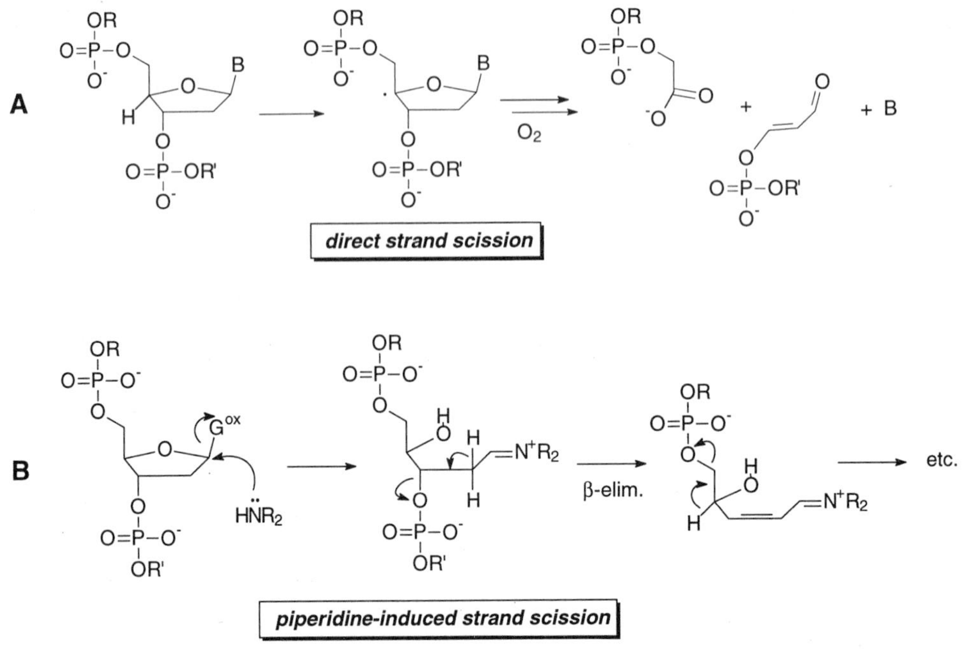

Scheme 1.

about these oxidative lesions (Burrows and Muller, 1998). For example, many oxidative modifications to nucleobases lead to strand scission after a standard piperidine treatment involving 0.2 N piperidine at 90°C for 30 minutes. A complementary method involves examination of the ability of DNA polymerase enzymes to extend a primer opposite nucleobase damage (see section 3.4 below). The effect of piperidine treatment (*i.e.* sugar vs. guanine oxidation) is shown in a set of experiments using NiKGH-$CONH_2$ as the catalyst and hydrogen peroxide, monoperoxyphthalate (MMPP) or monoperoxysulfate ($KHSO_5$) as oxidant under conditions of low (10 mM) vs. normal (110 mM) salt concentration (Figure 1).

3.2. Ionic Strength Effect

The intracellular and extracellular concentrations of monovalent cations are approximately 150 mM in human fluids, with K^+ being the predominant cation inside the cell and Na^+ outside the cell. The ionic strength of an aqueous solution has a dramatic effect on the structure and stability of duplex DNA, as well as on the strength of interaction of ionic and hydrophobic small molecules with DNA. At low ionic strength, the duplex is destabilized due to both increased repulsion between phosphodiester anions and a diminished driving force for base stacking and hydrogen bonding. A higher ionic strength stabilizes the duplex resulting in higher melting temperatures (T_m's) and higher affinity to hydrophobic molecules but lower affinity with charged species.

The extent of strand scission before and after piperidine treatment was analyzed for different concentrations of 3 oxidants at low and normal ionic strength. Raw data, in the form of a phosphorimagery scan, are shown in Figure 1, and selected comparisons are shown quantitatively in Figure 2. It is clear from a comparison of graphs A and B in Figure 2 that much less oxidative damage occurs under normal ionic strength conditions (A), and that the vast majority of chemistry is due to guanine oxidation (graph A, top vs. bottom), particularly at a 5'-*G*G-3' site. On the other hand, low ionic strength conditions (Figure 2B) greatly favor reaction at AT tracts as reported by Liang, et al. (Liang et al., 1998) In contrast to their report, however, we find that piperidine does enhance the extent of cleavage by factors of 3 to 10-fold, both at AT tracts and at guanines (Figure 2B, top vs. bottom).

The switch from one-electron oxidation of guanine nucleobases under normal conditions to deoxyribose oxidation with AT specificity under low ionic strength conditions could be due to either of two phenomena. One possibility is that NiKGH-$CONH_2$ has a preference for binding in the minor groove of the narrower, more hydrophobic AT tracts under low salt conditions, as suggested by Long and coworkers (Liang et al., 1995). On the other hand, AT tracts are also the most readily denatured, or "melted", regions under low salt conditions, and if melting improves the accessibility of the deoxyribose to oxidation reagents, these sites will be preferentially oxidized. To test this hypothesis, it would be useful to compete, intramolecularly, duplex vs. single-stranded AT tracts; but this experiment, to our knowledge, had not yet been done.

3.3. Role of Oxidant

Recent experiments by Liang, et al. (Liang et al., 1995) demonstrate that the nature of the oxidant responsible for hydrogen atom abstraction from the deoxyribose group is different from that which we had proposed responsible for guanine oxidation (Muller et al., 1997). Furthermore, guanine chemistry seems to be markedly enhanced by use of a stoichiometric excess of peracid oxidants, $KHSO_5$ or MMPP. To confirm this latter effect, we examined the difference in the extent of oxidation using excess $KHSO_5$ (50 μM) and

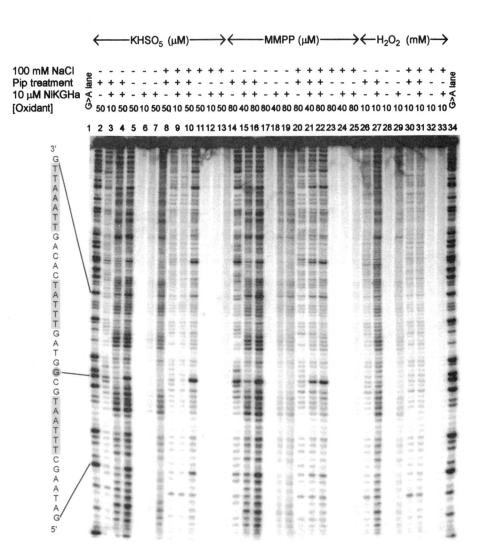

Figure 1. Phosphorimagery (Molecular Dynamics Storm 840) scan of denaturing polyacrylamide gel indicating cleavage sites of the 5'-end-labeled 167-base pair fragment of pBR322 (*Eco*RI/*Rsa*I) mediated by NiKGH-CONH$_2$ and various oxidants, KHSO$_5$ (2 min, lanes 2–13), magnesium monoperoxyphthalate (15 min, lanes 14–25), and hydrogen peroxide (30 min, lanes 26–33) using concentrations indicated above the lanes. All reactions were carried out in 10 mM cacodylate buffer (pH 7.1) with or without 100 mM NaCl added. The Ni complex was pre-incubated with the DNA for 10 min in buffer prior to addition of the oxidant. After oxidation, reactions were quenched with EDTA/HEPES at pH 7. Piperidine-treated lanes were allowed to react with 0.2 M piperidine for 30 min at 90°C. The portion of the sequence analyzed is shown at left with AT tracts and a highly reactive G residue highlighted.

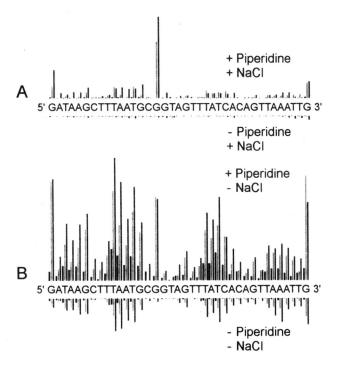

Figure 2. Portions of the gel shown in Fig. 1 were quantified using ImageQuaNT software. Appropriate control lanes were subtracted in all cases. A. Comparison of piperidine (upper bars) and non-piperidine-treated (lower bars) lanes under normal (110 mM) salt conditions. At each nucleotide site, the first bar represents data obtained with 10 mM H_2O_2 (black, lane 31 vs. 33), the second is 80 µM MMPP (light gray, lane 22 vs. 25), and the third, 50 µM $KHSO_5$ (dark gray, lane 10 vs. 13). **B.** Same as above under low salt (10 mM) conditions: 10 mM H_2O_2 (lane 27 vs. 29), 80 µM MMPP (light gray, lane 16 vs. 19), and 50 µM $KHSO_5$ (dark gray, lane 4 vs. 7).

stoichiometric $KHSO_5$ (10 µM) compared to NiKGH-$CONH_2$. These results are quantified in Figure 3. As expected, the extent of strand scission after piperidine treatment (normal salt conditions) increases at every site with higher oxidant concentration, and in agreement with the previous report,(Liang et al., 1998) guanine oxidation is particularly enhanced. These data suggest a special role for peracid oxidants in guanine oxidation such that there is a higher dependency on the concentration of HSO_5^- or MMPP for formation of $G^{\bullet+}$ (or for its further reaction) than for C4' hydrogen atom abstraction.

3.3.1. Comparison of Monoperoxysulfate, Monoperoxyphthalate, and Hydrogen Peroxide. To further elucidate the role of oxidant in nickel peptide-mediated DNA oxidation, comparisons were made as shown in Fig. 1 and quantified in Fig. 2. At each nucleobase site, the graphs in Fig. 2 show the extent of reaction for H_2O_2 (left bar), MMPP (center bar), and $KHSO_5$ (right bar). In all cases, the peracid oxidants MMPP and $KHSO_5$ show approximately the same pattern of reactivity, although MMPP is generally used in higher concentration (80 vs. 50 µM) in order to achieve the same result. Hydrogen perox-

Figure 3. The enhancement of G oxidation by higher concentrations of $KHSO_5$ is seen by subtracting the amount of piperidine-induced cleavage observed with 10 µM $KHSO_5$ (Fig. 1, lane 9) from data obtained with 50 µM $KHSO_5$ (Fig. 1, lane 10). Oxidation is enhanced at all sites, but particularly at a highly reactive 5'-GG-3' site.

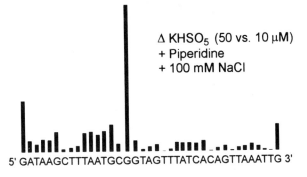

ide is much less reactive and never shows significant guanine enhancement upon piperidine treatment, although it does show a slight preference for AT tracts under low salt conditions.

3.3.2. In Situ Formation of Monoperoxysulfate via Sulfite Autoxidation.

Considerable mechanistic insight into guanine oxidation was gained by the discovery that nickel(II) peptides such as NiKGH-CONH$_2$ could catalyze the in situ formation of HSO$_5^-$ from aerobic oxidation of sulfite/bisulfite (Muller et al., 1997). This likely occurs by a four-step mechanism related to that proposed for Mn(III) and other ions, (Brandt and van Eldik, 1995) *i.e.*:

$$Ni(III) + SO_3^{2-} \rightarrow Ni(II) + SO_3^{-\bullet} \tag{1}$$

$$SO_3^{-\bullet} + O_2 \rightarrow SO_5^{-\bullet} \quad \text{(diffusion controlled)} \tag{2}$$

$$Ni(II) + SO_5^{-\bullet} \rightarrow Ni(III) + SO_5^{2-} \tag{3}$$

$$SO_5^{2-} + H^+ \rightarrow HSO_5^- \quad (pKa = 9.4) \tag{4}$$

Steps 1 and 3 of this mechanism are readily catalyzed by NiKGH-CONH$_2$ because its redox potential (731 mV vs. SCE) is sufficiently high to oxidize SO$_3^{2-}$ (E = 390 mV vs. SCE) and sufficiently low to reduce SO$_5^{-\bullet}$ (E = 860 mV vs. SCE) to monoperoxysulfate (Neta et al., 1988). Thus, while NiKGH-CONH$_2$ does not react alone with dioxygen, it will react in the presence of a sacrificial reductant (sulfite) to ultimately generate a much stronger oxidant, monoperoxysulfate, which in turn is decomposed by nickel complexes to produce a DNA-damaging oxidant. Our previous studies with nickel peptides showed that use of 100 μM Na$_2$SO$_3$ in the presence of air gave nearly identical DNA chemistry to the use of preformed, 100 μM KHSO$_5$ (Figure 4) (Muller et al., 1997) That a freely diffusible SO$_4^{-\bullet}$ was not formed in the subsequent decomposition of HSO$_5^-$ was clearly demonstrated by alcohol quenching studies. However, sulfate radical does lead to guanine chemistry, and we have proposed that a metal-ligated sulfate radical may be responsible for the ultimate DNA damage (Scheme 2) (Muller et al., 1996). Related studies have suggested either HO$^\bullet$ or SO$_3^{-\bullet}$ as reactive species from nickel/sulfite chemistry (Shi et al., 1994).

If a high-valent nickel-sulfate intermediate is formed upon reaction of either HSO$_5^-$ or SO$_3^{2-}$/O$_2$ (via HSO$_5^-$ formation *in situ*), one might expect it to react like SO$_4^{-\bullet}$. Sulfate

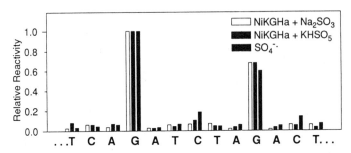

Figure 4. Comparison of the reactivity of a single-stranded oligodeoxynucleotide, 5'-d(ATATCAGATCTAGAC-TAT)-3', with 10 μM NiKGH-CONH$_2$ + 100 μM Na$_2$SO$_3$/O$_2$ (white bars) vs. 10 μM NiKGH-CONH$_2$ + 100 μM KHSO$_5$ (gray bars) vs. SO$_4^{-\bullet}$ generated from photolysis of S$_2$O$_8^{2-}$ (black bars). Cleavages were observed by PAGE after standard piperidine treatment.

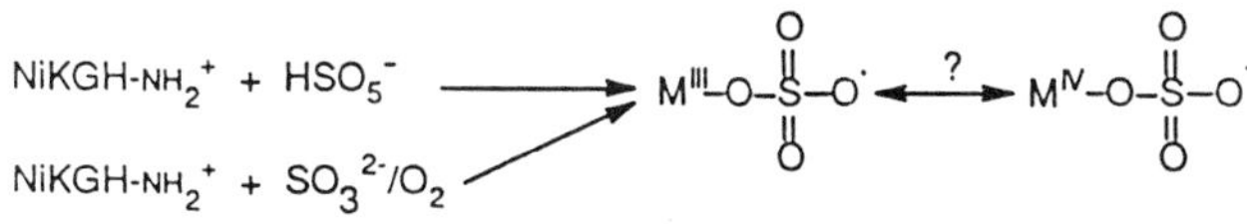

$$NiKGH\text{-}NH_2^+ + HSO_5^- \longrightarrow$$

$$NiKGH\text{-}NH_2^+ + SO_3^{2-}/O_2 \longrightarrow$$

$$M^{III}\text{-}O\text{-}\overset{\displaystyle O}{\underset{\displaystyle O}{\overset{\|}{\underset{\|}{S}}}}\text{-}O\cdot \quad \overset{?}{\longleftrightarrow} \quad M^{IV}\text{-}O\text{-}\overset{\displaystyle O}{\underset{\displaystyle O}{\overset{\|}{\underset{\|}{S}}}}\text{-}O^-$$

Scheme 2.

radical is known to be a good one-electron oxidant with about a 100-fold preference for abstraction of an electron from guanine compared to abstraction of a hydrogen atom from ribose. In contrast, $HO^\bullet$ and high-valent metal hydroxide species carry out both hydrogen atom abstraction and one-electron oxidation. For example, a nickel(III) hydroxide complex with a peptide ligand (~700 mV vs. SCE) would probably have too low a potential to mediate guanine oxidation (~1 V vs. SCE). Thus, we agree with the recent conclusions of Long and coworkers that nickel peptide-mediated oxidation of deoxyribose vs. guanine likely arises from different reactive intermediates, potentially $Ni^{III}\text{-}OH$ vs. $Ni^{IV}\text{-}OSO_3^-$.

3.4. Single-Stranded vs. Double-Stranded Guanine Reactivity

In order to investigate the difference in reactivity between single-stranded and double-stranded DNA targets, we utilized the 7.2 kb single-stranded plasmid M13mp18. Analysis was performed by DNA polymerase extension of a radiolabeled primer complementary to a specific region of the target. Primer extension analysis should identify any sites of direct strand scission as well as major sites of nucleobase modification. The results of NiKGH-CONH$_2$-mediated oxidation of the plasmid with either HSO_5^- or SO_3^{2-}/O_2 are shown in Figure 5. Control studies (lanes 1 and 2) verified that complete extension of the primer was observed when the nickel complex was absent. On the other hand, both oxidants led to specific polymerase "stops" when NiKGH-CONH$_2$ was present. The majority of these stop points were located one base before a G on the target strand indicating that the site of damage was usually guanine. The product of guanine oxidation has not yet been firmly identified, however it is likely to be an imidazolone or oxazolone derivative. From other work, we know that 7,8-dihydro-8-oxoguanine is not a stop point for most DNA polymerases using PCR conditions, nor is it particularly piperidine sensitive,(Muller et al., 1998) so this lesion is not likely to be responsible for the stop points (or piperidine-labile cleavages) observed in this work.

From these experiments, it is clear that guanine is the principal target of oxidation with single-stranded DNA under normal salt conditions using a stoichiometric excess of oxidant. We have previously proposed a role for binding of a high-valent nickel intermediate to guanine's N7 to explain the high selectivity of certain nickel complexes for exposure of this site (Burrows and Rokita, 1996; Burrows and Rokita, 1994).

3.5. Implications for Bioconjugate and Metal Toxicity Studies

Studies with nickel(II) complexes of model tripeptides bearing a carboxamide side-chain lead to the following general conclusions. First, nickel(II) can bind to the N-terminal XXH motif of peptides and proteins with similar coordination geometry and relatively similar redox properties. The presence of a proline residue at position 1 of the XXH motif is tolerated, although the ligand field strength is presumably somewhat weaker since the $E_{1/2}$ is about 150 mV higher. Secondly, the nickel(II) complexes are relatively stable in the presence of dioxygen, although some degradation is observed over a period of days with aque-

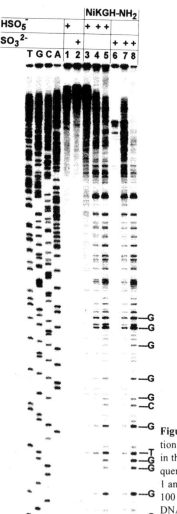

Figure 5. PAGE analysis of an extended primer (5'-end-labeled) after oxidation of a single-stranded plasmid template, M13mp18, using NiKGH-CONH$_2$ in the presence of KHSO$_5$ or Na$_2$SO$_3$/O$_2$. Lanes T, G, C, and A are Sanger sequencing lanes using ddATP, ddCTP, ddGTP, and ddTTP, respectively. Lanes 1 and 2 are control lanes without nickel peptide but with 50 μM KHSO$_5$ and 100 μM Na$_2$SO$_3$, respectively. Lanes 3–5: primer extension after oxidation of DNA by 0.1, 1.0 and 10 μM NiKGH-CONH$_2$ + 50 μM KHSO$_5$. Lanes 6–8: primer extension after oxidation of DNA by 0.1, 1.0, and 10 μM NiKGH-CONH$_2$ + 100 μM Na$_2$SO$_3$ in air.

ous solutions of nickel peptide complexes. On the other hand, the peptide complexes are quite reactive with peracid oxidants, and to a lesser extent with H$_2$O$_2$. Pertinent to nickel toxicity is the interesting observation that nickel peptides will catalyze the formation of a strong peracid oxidant (HSO$_5^-$) from physiologically available species—sulfite and dioxygen—under conditions that are possible inside a cell. In this vein, the ability of other cellular reductants to interact with nickel peptides and dioxygen will be important to explore.

Third, the intrinsic reactivity of small nickel peptides and peracid oxidants under normal salt conditions involves one-electron oxidation of the guanine nucleobase. Bioconjugates presumably direct the chemistry to specific sites when the nickel peptide complex is tethered to a specific DNA-binding unit. Thus, studies carried out under low ionic strength conditions may mimic the minor groove binding and sugar chemistry observed with bioconjugates. Finally, more studies will need to be performed in order to fully explore the differences between the two pathways—sugar vs. nucleobase oxidation—and to completely characterize the products and gain insight into reactive intermediates.

4. AUTOXIDATIVE DECARBOXYLATION OF Ni(II)-XXH-CO₂H

An early study by Ciccarelli and Wetterhahn of the molecular basis of nickel carcinogenesis identified DNA-protein cross-links as one of the lesions potentially responsible for toxicity (Ciccarelli et al., 1981). Given the avidity of nickel for protein binding, a likely mechanism would involve nickel-mediated protein oxidation leading to an electrophilic species that could be trapped by a DNA nucleophile. In order to investigate this possibility, the O_2 reactivity of the NiXXH motif was examined, based on previous reports of metal-mediated peptide autoxidation (Bossu et al., 1978; Sakurai and Nakahara, 1979).

4.1. Reaction with Dioxygen

Nickel(II) tripeptide complexes with *carboxamide* C-termini including NiPKH-CONH₂ and NiKGH-CONH₂ were stable in air-saturated aqueous solutions at pH 7–8, although some decomposition could be discerned over a period of days. In contrast, tripeptide complexes of the NiXXH-CO₂H type with *carboxylate* C-termini were reactive with dioxygen in aqueous solution at pH 7.5. The oxidation process was monitored by the growth of a nickel(III) band at 310 nm over a period of hours. The oxidations of the nickel(II) complexes of carboxy-terminal PKH, KGH, RGH and RRH peptides were observed to be very similar in behavior.

The final product of NiGGH-CO₂H autoxidation has been crystallographically characterized as a decarboxylated peptide with addition of a hydroxyl group in place of the carboxylate (Bal et al., 1994). A mechanism for the formation of this species has been proposed (see Figure 6) (Perez et al., 1998).

4.2. DNA Chemistry

The mechanism shown in Figure 6 suggests that there are at least two types of peptide intermediates that could undergo reaction with nucleic acids. First, a free radical species such as the carbon-centered radical **A** (or alternatively, a peroxyl radical derived from O_2 reaction with this radical) could be responsible for oxidative damage to DNA. Indeed, a study of NiKGH-CO₂H (1 mM) and NiRRH-CO₂H with closed circular plasmid DNA (pBR322) showed plasmid nicking (form I to form II conversion) after incubation in air for 60 min. at 23°C, pH 7.5 (Perez et al., 1998).

A second potential intermediate of interest is the N-acylimine **B** (Figure 6). Such a species should be highly reactive toward nucleophiles. In fact, the product **C** ultimately observed by Bal et al. during autoxidation of NiGGH-CO₂H could have arisen from addition of H_2O to a imine precursor, forming the hemiaminal.

In the presence of a DNA nucleophile such as G-N7, G-N^2, A-N1, A-N3, A-N^6, T-N3 or C-N^4, a covalent adduct of a nucleobase cross-linked to the peptide could be formed. Preliminary evidence for such a species has recently been obtained, although a high resolution structure is still lacking.

4.3. Implications for Bioconjugate and Metal Toxicity Studies

The autoxidative decarboxylation of a NiXXH-CO₂H complex can lead to oxidative DNA damage and a DNA-peptide covalent cross-link, both of which are lesions observed

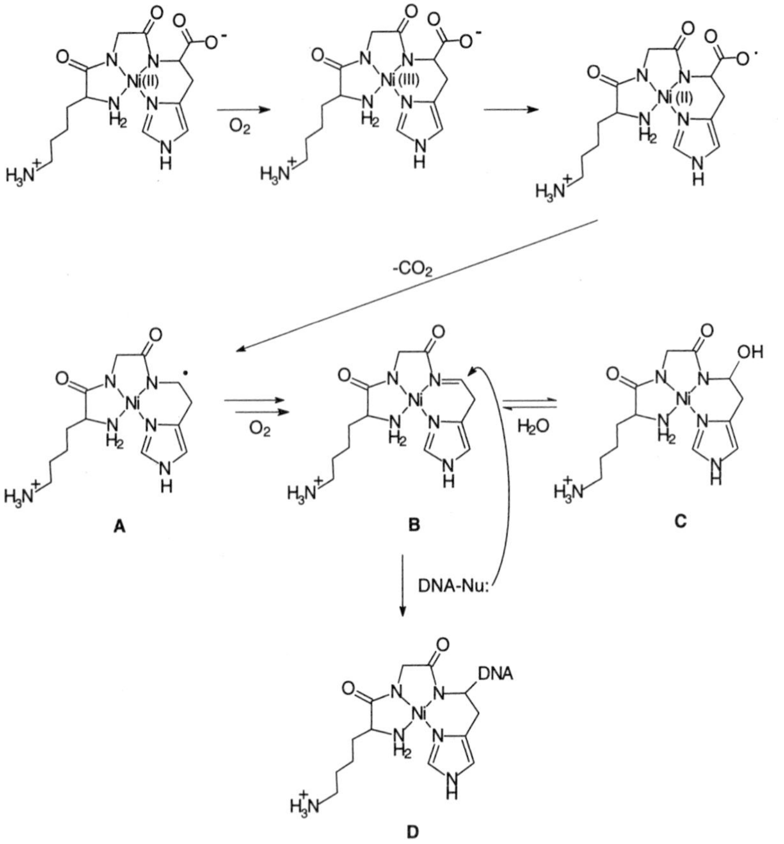

Figure 6. Proposed mechanistic pathway for autoxidative decarboxylation of NiKGH-CO$_2$H.

in the exposure of cells to excess nickel. However, the requirement of the N-terminus for effective binding to the metal, and of the C-terminal carboxylate for O$_2$ reaction, limits this particular mechanism to the study of tripeptides. We are currently exploring the use of alternative amino acids to replace both of these requirements; *i.e.*, an alternative ligand to replace the N-terminal amine and a redox active amino acid to replace the carboxylate. Success in this endeavor would provide futher evidence for the hypothesis of nickel-mediated protein-DNA cross-links under physiological conditions.

For bioconjugate work, the same is true. It is already possible to imagine the incorporation of the NiXXH-CO$_2$H motif into the C-terminus of a bioconjugate by connection via the ε-amino group of lysine. Thus, this motif should enjoy application to systems where a DNA adduct is desired rather than DNA cleavage.

5. SUMMARY

The current interest and application of the nickel and copper-binding XXH peptide motif has mushroomed since its initial studies by Sarkar and coworkers three decades ago (Sarkar and Wigfield, 1968). This versatile metal complex, based on naturally occurring

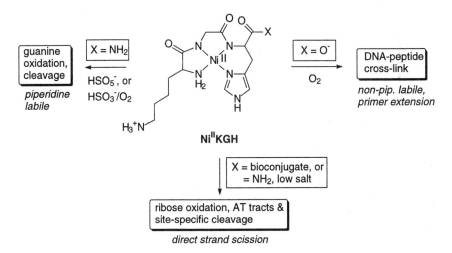

Figure 7. The versatility of the amino-terminal nickel-binding peptide motif, NiXXH, is seen in the three different types of DNA chemistry manifested. With peracid oxidants, the intrinsic reactivity of this complex is toward guanine in duplex or single-stranded DNA. Under low salt conditions, or through delivery with a DNA-binding bioconjugate, direct strand scission at either AT tracts or specific recognition sites can be realized. In the specific case of a carboxylate terminus, the peptide complex is reactive with O_2, and an electrophilic intermediate capable of forming DNA-peptide cross-links is obtained.

peptide sequences, can be tuned to participate in nucleobase oxidation, direct strand scission via deoxyribose chemistry, or DNA-peptide cross-linking (Figure 7). Given the redox activity of these metal complexes and the fact that their chemistry can be selected by judicious choice of amino acid components, it is surprising that Nature has not exploited this motif in metalloenzymes…at least, not to our present knowledge.

ACKNOWLEDGMENTS

The authors thank the National Institute of General Medical Sciences for a grant (GM-49860) in support of this work. RPH is an NIH-MARC predoctoral fellow (GM-18403). Travel funds for VD were provided by a US-France cooperative research grant from NSF (INT-9603313).

REFERENCES

Bal, W., Djuran, M. I., Margerum, D. W., Gray, E. T., Jr., Mazid, M. A., Tom, R. T., Nieboer, E., and Sadler, P. J. (1994). Dioxygen-induced decarboxylation and hydroxylation of [Ni^{II}(glycyl-glycyl-L-histidine)] occurs via Ni^{III}: X-ray crystal structure of [Ni^{II}(glycyl-glycyl-alpha-hydroxy-D,L-histamine)].3H$_2$O. J. Chem. Soc., Chem. Commun., 1889–1890.

Bal, W., Jezowska-Bojczuk, M., and Kasprzak, K. S. (1997). Binding of nickel(II) and copper(II) to the N-terminal sequence of human protamine HP2. Chem. Res. Toxicol. *10*, 906–914.

Bossu, F. P., Paniago, E. B., Margerum, D. W., and Kirksey, J. L., Jr. (1978). Trivalent nickel catalysis of the autoxidation of nickel(II) tetraglycine. Inorg. Chem. *17*, 1034–1042.

Brandt, C., and van Eldik, R. (1995). Transition metal-catalyzed oxidation of sulfur(IV) oxides: Atmospheric-relevant processes and mechanisms. Chem. Rev. *95*, 119–190.

Burrows, C. J., and Muller, J. G. (1998). Oxidative nucleobase modifications leading to strand scission. Chem. Rev. *98*, 1109–1151.

Burrows, C. J., and Rokita, S. E. (1996). Nickel complexes as probes of guanine sites in nucleic acid folding. In Metal Ions In Biological Systems, A. Sigel and H. Sigel, eds. (M. Dekker: New York), pp. 537–559.

Burrows, C. J., and Rokita, S. E. (1994). Recognition of guanine structure in nucleic acids by nickel complexes. Acc. Chem. Res. *27*, 295–301.

Camerman, N., Camerman, A., and Sarkar, B. (1976). Molecular design to mimic the copper(II) transport site of human albumin. The crystal and molecular structure of copper(II)-glycylgycyl-L-histidine-N-methyl amide monoaquo complex. Can. J. Chem. *54*, 1309–1316.

Ciccarelli, R. B., Hampton, T. H., and Jennette, K. W. (1981). Nickel carbonate induces DNA-protein cross-links and DNA strand breaks in rat kidneys. Cancer Lett. *12*, 349–354.

Ciccarelli, R. B., and Wetterhahn, K. E. (1982). Nickel distribution and DNA lesions induced in Rat tissues by the carcinogen nickel carbonate. Cancer Res. *42*, 3544–3549.

Costa, M., Salnikov, K., Cosentino, S., Klein, C. B., Huang, Z., and Zhuang, Z. (1994). Molecular mechanisms of nickel carcinogenesis. Environ. Health Perspect. *102, suppl. 3*, 127–130.

Footer, M., Egholm, M., Kron, S., Coull, J. M., and Matsudaira, P. (1996). Biochemical evidence that a D-loop is part of a four-stranded PNA-DNA bundle. Nickel-mediated cleavage of duplex DNA by a Gly-Gly-His Bis-PNA. Biochemistry 35, 10673–10679.

Harford, C., Narindrasorasak, S., and Sarkar, B. (1996). The designed protein M(II)-Gly-Lys-His-Fos(138–211) specifically cleaves the AP-1 binding site containing DNA. Biochemistry 35, 4271–4278.

Harford, C., and Sarkar, B. (1997). Amino terminal Cu(II)- and Ni(II)-binding (ATCUN) motif of proteins and peptides: Metal binding, DNA cleavage, and other properties. Acc. Chem. Res. *30*, 123–130.

Kasprzak, K. S. (1996). Oxidative DNA damage in metal-induced carcinogenesis. In Toxicology of Metals, L. W. Chang, L. Magos and T. Suzuki, eds. (Boca Raton: CRC Press), pp. 299–320.

Kasprzak, K. S. (1995). Possible role of oxidative damage in metal-induced carcinogenesis. Cancer Investigation *13*, 411–430.

Liang, Q., Ananias, D. C., and Long, E. C. (1998). Ni(II).Xaa-Xaa-His induced DNA cleavage: Deoxyribose modification by a common "activated" intermediate derived from KHSO₅, MMPP, or H₂O₂. J. Am. Chem. Soc. *120*, 248–257.

Liang, Q., Eason, P. D., and Long, E. C. (1995). Metallopeptide-DNA interactions: Site-selectivity based on amino acid composition and chirality. J. Am. Chem. Soc. *117*, 9625–9631.

Mack, D. P., and Dervan, P. B. (1992). Sequence-specific oxidative cleavage of DNA by a designed metalloprotein, Ni(II)GGH(Hin139–190). Biochemistry *31*, 9399–9405.

Muller, J. G., Duarte, V., Hickerson, R. P., and Burrows, C. J. (1998). Gel electrophoretic detection of 7,8-dihydro-8-oxoguanine and 7,8-dihydro-8-oxoadenine via oxidation by Ir(IV). Nucleic Acids Res. *26*, 2247–2249.

Muller, J. G., Hickerson, R. P., Perez, R. J., and Burrows, C. J. (1997). DNA damage from sulfite autoxidation catalyzed by a nickel(II) peptide. J. Am. Chem. Soc. *119*, 1501–1506.

Muller, J. G., Zheng, P., Rokita, S. E., and Burrows, C. J. (1996). DNA and RNA modification promoted by [Co(H₂O)₆]Cl₂: Guanine selectivity, temperature dependence and mechanism. J. Am. Chem. Soc. *118*, 2320–2325.

Nagaoka, M., Hagihara, M., Kuwahara, J., and Sugiura, Y. (1994). A novel zinc finger-based DNA cutter: Biosynthetic design and highly selective DNA cleavage. J. Am. Chem. Soc. *116*, 4085–4086.

Neta, P., Huie, R. E., and Ross, A. B. (1988). Rate constants for reactions of inorganic radicals in aqueous solution. J. Phys. Chem. Ref. Data *17*, 1027–1247.

Perez, R. J., Muller, J. G., Rokita, S. E., and Burrows, C. J. (1998). Oxidative DNA damage mediated by metal-peptide complexes. Pure Appl. Chem., *in press*.

Sakurai, T., and Nakahara, A. (1979). Reaction of nickel(II)-glycylglycyl-L-histidine complex with molecular oxygen and formation of decarboxylated species. Inorg. Chim. Acta *34*, L243-L244.

Sarkar, B., and Wigfield, Y. (1968). Evidence for albumin-Cu(II)-amino acid ternary complex. Can. J. Biochem. *46*, 601–607.

Shi, X., Dalal, N., and Kasprzak, K. S. (1994). Enhanced generation of hydroxyl radical and sulfur trioxide anion radical from oxidation of sodium sulfite, nickel(II) sulfite, and nickel subsulfide in the presence of nickel(II) complexes. Environ. Health Perspect. *102 (Suppl. 3)*, 209–217.

Sugiyama, H., and Saito, I. (1996). Theoretical studies of GG-specific photocleavage of DNA via electron transfer: Signficant lowering of ionization potential and 5'-localization of HOMO of GG bases in B-form DNA. J. Am. Chem. Soc. *118*, 7063–7068.

13

COPPER-DEPENDENT DNA SCISSION

David S. Sigman,[*] Lisa Milne, Yue Xu, James Gallagher, David M. Perrin, and Clark Pan

Molecular Biology Institute
University of California, Los Angeles
Los Angeles, California 90095-1570

1. INTRODUCTION

The 1,10-phenanthroline-cuprous complex is a chemical nuclease which cleaves DNA in the presence of the coreactant H_2O_2 either as the unlinked 2:1 1,10-phenanthroline-copper complex $((OP)_2Cu^+)$ or as the 1:1 complex (OP-Cu) linked to a targeting ligand with high affinity for a specific DNA sequence or structure (Sigman, et al., 1993; Sigman, 1996). Since our discovery of this reaction in the late "70"s (Sigman et al., 1979), our laboratory has been interested in identifying the detailed course of the scission chemistry and in exploiting its unique features to study ligand-DNA interactions. In this summary, I would like to outline recent progress made in understanding the chemical mechanism of the reaction. Then I would like to review a new approach to devising gene specific inhibitors which relies on the reactivity of this synthetic nuclease activity to define a hybridization site within the catalytically competent open-complex formed by RNA polymerase and a promoter. Finally, I would like to demonstrate that the protein targeted nuclease activity of OP-Cu provides a novel approach for investigating protein-DNA interactions.

2. DISCOVERY OF THE NUCLEASE ACTIVITY

Early work led to the recognition that the universal inhibition of DNA and RNA polymerases by 1,10-phenanthroline (OP) did not reflect the coordination of an essential metal at the active site of these enzymes as proposed by Loeb, Mildvan and colleagues (Slater et al., 1971; Springgate et al., 1973). The dependence of the reaction on thiol con-

---

[*] Address correspondence to: David S. Sigman: Phone: 310-825-8903; Fax: 310-206-7286; e-mail: sigman@mbi.ucla.edu

Metals and Genetics, edited by Sarkar.
Kluwer Academic / Plenum Publishers, New York, 1999.

centrations and trace levels of copper suggested rather that OP scavenged trace amounts of copper from the buffer to form a cupric complex. This complex was then reduced to the cuprous complex, and oxidation of the cuprous complex led to the production of hydrogen peroxide, an essential coreactant in the overall chemistry (D'Aurora *et al.,* 1977; D'Aurora *et al.,* 1978). These two coreactants cleaved the primer/template DNA, and, in turn, generated products which are directly responsible for the observed inhibition by OP (Sigman *et al.,* 1979). Relief of inhibition by hydrolysis of the products with exonuclease III and alkaline phosphatase suggested that the chemical identities of these products were 3'-phosphates and 3'-phosphoglycolates (Reich *et al.,* 1981). It was well known that micrococcal nuclease, which generates 3'- phosphates and 5'-hydroxyl groups, produced effective inhibitors of the enzyme (Pope *et al.,* 1982).

3. KINETIC MECHANISM

The efficiency of the chemical nuclease (eq. 1) relies on the reversible binding of the tetrahedral chelate to the minor groove of B-DNA.

$$(OP)_2Cu^+ + DNA \rightleftharpoons (OP)_2Cu^+\text{--- } DNA \xrightarrow{\quad H_2O_2 \quad} \text{scission products} \qquad (1)$$

This noncovalent intermediate is also responsible for the specificity of this reaction for the distinct secondary structures of DNA (Pope and Sigman, 1984). B-DNA is the preferred structure while A- DNA is cleaved roughly 7-fold less efficiently. Z-DNA is not cut nor is single-stranded DNA which is prevented from forming hairpin structures by base specific modification reagents. Neither Z-DNA nor single-stranded DNA possess a recognizable minor groove where the coordination complex can bind near the oxidatively sensitive C-1' hydrogen. Reaction of a heteroduplex composed of single strands of DNA and RNA supports the conclusion that binding rather than covalent chemistry dictates rates of scission (Pope *et al.,* 1984). Assaying cleavage of both strands within the heteroduplex reveals that both strands are cut at equal rates which are both slower than that observed with B-DNA. Subsequent work showed that the cleavage of RNA proceeds more sluggishly than that of DNA most likely because RNA is composed of less reactive single stranded regions and the chemical nuclease resistant A-duplex of RNA (Murakawa *et al.,* 1989).

4. CHEMICAL MECHANISM OF THE SCISSION REACTION

Our work has focused on the mechanism of scission of duplex DNA by $(OP)_2Cu^+$. The goal was to determine the chemical structure of the products formed. These studies have demonstrated that the stable oxidative products are 5-methylene furanone, 5'- and 3'-phosphomonoesters, and free pyrimidines and purines (Sigman, 1986; Kuwabara, *et al.,* 1986). These products are fully consistent with oxidative attack in the minor groove and the 3'-stagger of the reaction products of the complementary strands (Sigman, 1986; Kuwabara, *et al.,* 1986; Yoon *et al.,* 1988; Yoon *et al.,* 1990). The source of the carbonyl oxygen in 5-methylene furanone was investigated by cleaving calf thymus DNA in the presence of $H_2{}^{18}O$, $H_2{}^{18}O_2$ and $^{18}O_2$. The goal was to determine if the cleavage reaction

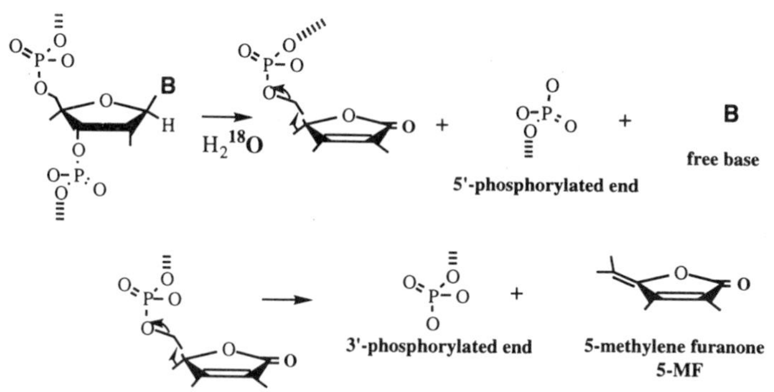

Figure 1. Products of scission by $(OP)_2Cu^+$. Stable products formed in the scission reaction. The C-1 oxygen is derived from the water oxygen.

proceeded by an oxygen rebound mechanism as has been observed with the manganese complex of the cationic porphyrin derivative, meso-tetrakis(4-N-methylpyridiniumyl)porphyrin (Pitie *et al.,* 1995; Pratviel *et al.,* 1995). Our results clearly indicated that $H_2^{18}O$ was the sole source of the carbonyl oxygen of C-1 of 5-methylene furanone (Figure 1) (Meijler *et al.,* 1997). Oxygen from either $H_2^{18}O_2$ or $^{18}O_2$ did not appear in the 5-methylene furanone. Although we favored the prior elimination of the 5'-phosphomonester terminus because it readily took into account the instability of the postulated intermediate under our reaction conditions, Greenberg and colleagues have noted that the chelate itself catalyzes the scission of the phosphodiester backbone of a model oligonucleotide at rates that might be sufficient to account for the observed reaction (Chen and Greenberg, 1998).

5. $(OP)_2Cu^+$ AS A PROBE FOR TRANSCRIPTION INITIATION

The use of the chemical nuclease as a footprinting reagent was one of the earliest applications of $(OP)_2Cu^+$. Although it accurately located the binding site of the *lac* repressor on *lac* UV5 DNA (Kuwabara and Sigman, 1987), we found that an unusual pattern of bands appeared when the *E. coli* wt, Ps and UV5 *lac* promoters were footprinted in the presence RNA polymerase (Spassky and Sigman, 1985). In addition to a pronounced of region protection, the cuprous complex of OP and some of its derivatives produced strong cleavage sites within the region covered by the polymerase, which strongly indicated that the enzyme was inducing structural changes in the DNA template (Thederahn *et al.,* 1990; Spassky *et al.,* 1985; Spassky, 1986). These results suggested that the tetrahedral hydrophobic cuprous chelates of OP bind productively to the open-complex and lead to the scission of the template strand at positions -6 to -3. Corresponding scission was not observed on the nontemplate strand. For the chelate prepared from 5-phenyl-OP, scission is also observed in the template strand at position +2 and +3.

Support for this hypothesis came from the demonstration that the redox inert copper chelates prepared with 2,9-dimethyl-1,10-phenanthroline (neocuproine = NC) were effective inhibitors not only of procaryotic promoters but also of eucaryotic promoters as well (Mazumder *et al.,* 1993; Mazumder *et al.,* 1994; Perrin *et al.,* 1994; Perrin *et al.,* 1996). Inhibition of transcription from *lac* UV5 by the cuprous complex of 5-phenyl- 2,9-dimethyl-1,10-

phenanthroline occurs with an I_{50} of approximately 5 μM, which could result from the chelate binding to the open-complex and/or to other initiation complexes formed at the start of RNA synthesis (Perrin *et al.,* 1996). Indeed, the binding affinity to one of these intermediates corresponds best to the observed I_{50} for runoff transcription (Perrin *et al.,* 1996). Previous reports on the cellular toxicity of these complexes indicate that, despite the simple structures, these chelates are remarkably potent cytotoxins. For example, they affect an array of cells ranging from *S. typhymirum* (Feig *et al.,* 1988) to mycoplasma (Antic *et al.,* 1977) to L1210 cells (Mohindru *et al.,* 1983) at lower concentrations than they inhibit transcription.

The hyperreactivity of $(OP)_2Cu^+$ within the open-complex has prompted us to systematically mutate the DNA sequence of the *lac* UV5 open complex at positions -6 to -4 which are highly variable in *E. coli* promoters and are the primary sites of hyperreactivity (Gallagher *et al.,* 1996). The goal of these studies was to determine if there were base specific interactions that were responsible for the observed cleavage sites which are dramatically more impressive than those observed with other oxidative chemical nucleases. Surprisingly, these changes had no effect on the catalytic activity, the affinity of the chelates for the open-complex, nor the sites of scission. Transforming this sequence of bases to those characteristic of the *trp* EDCBA complex similarly had no effect on the cleavage. These results indicate that upstream regions of the promoter must be principally responsible for the observed cleavage patterns.

6. GENE-SPECIFIC INHIBITORS

The toxicity of the tetrahedral 2,9-dimethyl-1,10-phenanthroline (neocuproine=NC) complexes of cuprous ion indicates that they should not be used in a general approach for the development of transcription inhibitors. However, the single stranded DNA formed at the start of transcription in these procaryotic promoters, and presumably in eucaryotic promoters, would be a most appropriate target for the design of highly selective gene specific inhibitors. Each gene would be anticipated to have a characteristic DNA sequence which could be targeted by a complementary oligonucleotide. Relatively short oligonucleotides could be used, because the transcriptional machinery will have melted the DNA upon activation of the gene so that it would not be necessary to unwind the DNA from its nontemplate complement. Oligonucleotides and their analogs would be mechanism-based inhibitors because they would depend on the normal biochemical action of RNA polymerase, the enzyme essential for RNA synthesis. According to this design, OP itself would not be crucial for increasing the affinity of the complementary oligonucleotide for its target. However, it could play the essential role of providing an experimental approach for demonstrating the binding orientation of the inhibitory oligonucleotide analog to the unwound DNA strand at the initiation of transcription (Perrin *et al.,* 1994) [Figure 2].

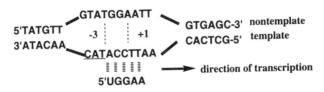

Figure 2. Hybridization of complementary oligonucleotide analogs to *lac* UV5 open-complex.

This new approach to the development of gene specific inhibitors contrasts with other methods which have been proposed that might be appropriate in a pharmacological context. These include: a) antisense deoxyoligonucleotides (Matteucci and Wagner, 1996); b) ribozymes targeted to specific mRNA (Milligan *et al.,* 1993); c) the inhibition of gene expression by formation of triple-helices at regulatory and coding sequences (Helene and Toulme, 1990); and d) minor groove specific polyamide derivatives (Gottesfeld *et al.,* 1997). Although previous workers have shown that short dinucleotides could prime DNA synthesis (Grachev *et al.,* 1984), our work is the first which seeks to exploit systematically the reactivity and accessibility of single stranded DNA formed in activated transcription units.

To date, we have explored the viability of this approach with two *E. coli* transcription units *in vitro,* the *lac* UV5 and the *trp* EDCBA. These two promoters have been chosen because they are both relatively strong promoters, do not require additional activation proteins for efficient transcription, and form open-complexes reactive to $(OP)_2Cu^+$. Our initial experiments demonstrated hybridization to the open-complex of *lac* UV5 by gel retardation, kinetics, and most importantly targeted scission in which OP was directed to the hybridization site by covalent attachment to 5'-UGGAA-3' corresponding to sequence positions -3 to +2 (Perrin *et al.,* 1994) . Our results demonstrated targeted scission at precisely the sites that we anticipated based on our studies.

A crucial test for the success of this approach is whether stringent inhibition can be achieved. This feature of the inhibition mechanism was tested by comparing the inhibition and targeted scission to both the *trp* and *lac* promoters. The former has the sequence 5'-CGCAA-3' from positions -3 to +2 while the latter has the sequence 5'-UGGAA-3' as noted above. In each case, we synthesized 2'-O-methyl oligoribonucleotides which terminated at the 3'-ends with a 3'-deoxyribose. We discovered two salient characteristics at the hybridization site of these two promoters. The first is that it is possible to demonstrate preferential *inhibition* of transcription by the cognate oligonucleotide (Perrin *et al.,* 1997). For example, 5'-CGCAA-3' inhibited only transcription from the *trp* EDCBA promoter and 5'-UGGAA-3' inhibited only the *lac* promoter [Figure 3].

Perhaps, more significantly we have shown the precise targeting of *scission* of the template strand exclusively by the congruent oligonucleotide (Perrin *et al.,* 1997). Only the inhibitory oligonucleotides target scission. Even with a pentamer, the targeted scission is stringent! The precise site of hybridization can be defined further by preparing oligonucleotides labeled on the 3'-end with OP. When the oligonucleotide analog 5'-GUGGA-OP-3' is prepared, cleavage is observed at positions -2,-1, and +1 [Figure 4].

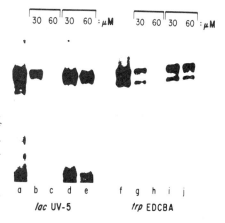

Figure 3. Selective inhibition of the *lac* UV5 And *trp* EDCBA promoters 5'-UGGAA-3' and 5'-CGCAA-3'. Concentrations of inhibitory pentanucleotides are indicated. Transcription directed by *lac* UV5 lanes a-e; lane a: control; lanes b,c: UGGAA synthesized with propynyl U; lanes d,e: CGCAA. Transcription directed by *trp* EDCBA lanes f-j; lane f: control; lanes g,h: CGCAA; lanes i,j: UGGAA synthesized with propynyl U.

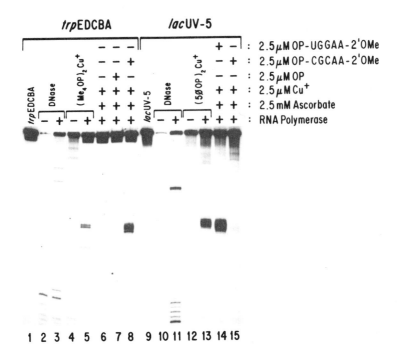

Figure 4. OP-rCGCAA-directed oxidative scission of the *trp* EDCBA open complex demonstrates gene-specific association with the *trp* EDCBA promoter and not the *lac UV5* Promoter. Lane 1: *trp* EDCBA promoter fragment labeled on template strand. Lanes 2 and 3: DNase footprint of the promoter and open-complex respectively. Lanes 4 and 5: 3,4,7,8-tetramethyl-OP-copper footprint of the *trp* EDCBA promoter and open-complex respectively. Lane 6: Copper control (2 µM), Lane 7: OP-Copper control (2.5 µM-2 µM), Lane 8: Site-specific scission directed by 2.5 µM OP-rCGCAA-(2′OMe)$_{2-5}$ to *trp* EDCBA. Lane 9: *lac UV5* promoter labeled on the template strand. Lanes 10 and 11: DNase footprinting of the *lac UV5* promoter and open-complex respectively. Lanes 12 and 13: 5ϕOP-copper footprinting of the *lac UV5* promoter and open-complex respectively. Lane 14: (positive control) Site-specific scission directed by 2.5 µM OP-rUGGAA-(2′OMe)$_{2-5}$ to the *lac UV5* promoter. Lane 15: (negative control) Site-specific scission directed by 2.5 µM OP-rCGCAA-(2′OMe)$_{2-5}$ to the *lac UV5* promoter. Scission in lanes 6–8, 14 and 15 was initiated by addition of sodium ascorbate and continued for 25 minutes at 37°C.

Subsequent studies have focused on the oligonucleotides which hybridize most efficiently to the open-complex. A range of pentamers has been prepared. We find that the most stable hybridization is observed with 5'-GUGGA-3'(-4 to +1) and 5'-UGGAA-3' (-3 to +2). Unless the pentameric oligonucleotide overlaps the initiation site of RNA synthesis at positions +1,+2, or +3, neither inhibition nor oligonucleotide targeted scission is observed.

To investigate the length dependence of the hybridizing oligonucleotides, we used the protection of open-complex scission by 5'-OP-UGGAA-3' as an assay. We have found that oligonucleotides terminating at position +2 of length 5, 6, and 7 nts are effective blockers of scission while oligonucleotide analogs of chain length 8 to 11 are not. Oligonucleotides complementary to the top strand neither inhibit RNA polymerase activity nor target scission. Our findings underscore that the hybridization site formed in the open-complex is well defined and stringent. Its tolerance for abiological nucleotides with higher T_m's due to modification of the bases and ribose will help determine if these single stranded regions will be appropriate targets for the design of potent gene specific inhibitors.

7. PROTEIN-TARGETED DNA SCISSION

In addition to providing a method to independently monitor the hybridization of oligonucleotides to complementary sites, the discovery that the 1:1 OP-Cu complex could be targeted to specific sites also suggested that it would be possible to transform DNA binding proteins into site specific scission reagents (Chen and Sigman, 1987; Bruice *et al.*, 1991; Sigman *et al.*, 1993; Sutton *et al.*, 1993). Rather than depending on the relatively serendipitous affinities of a hydrophobic tetrahedral chelate, targeted scission provides an approach for generating cleavage reagents of preselected specificity. The general strategy which we employed depended on examination of a high resolution structure of a DNA binding protein generally obtained from an x-ray structure, and then inserting a cysteine residue at a site which was accessible to the minor groove yet unlikely to block the high affinity binding of the protein to its site (Pan *et al.*, 1994). Access to the minor groove is essential because it is the site of the oxidatively sensitive C-1' hydrogen which presumably is attacked both by the

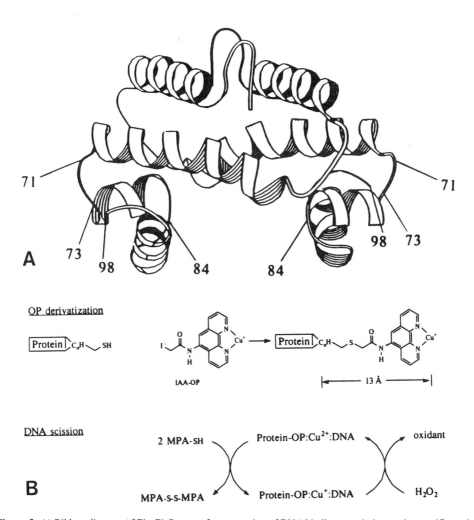

Figure 5. A) Ribbon diagram of Fis. B) Strategy for conversion of DNA binding protein into a site-specific nuclease.

2:1 and 1:1 chelates. Experimental evidence that minor groove reactivity is essential for product scission is the 3'-stagger observed in protein targeted scission as well as in the relatively nonspecific cleavage observed with (OP)₂Cu⁺ (Pan *et al.*, 1994).

Sequence specific binding proteins which have been converted in our laboratory to cleavage reagents include λ phage cro protein (Bruice *et al.*, 1991), *E. Coli* trp repressor (Sutton *et al.*, 1993; Pfau *et al.*, 1994; Landgraf *et al.*, 1996) and factor for inversion stimulation (Fis protein) (Pan *et al.*, 1994; Pan *et al.*, 1996; Pan *et al.*, 1996), and the Drosophila engrailed protein (Pan *et al.*, 1995). Others have transformed the *E. coli* cyclic AMP binding protein (Pendergrast *et al.*, 1994) and the TBP binding protein from yeast (Cox *et al.*, 1997). Most recently, we have transformed the single-strand specific gene V protein into an efficient cleavage reagent in a protocol which accomplishes sequence specific scission of single stranded targets generated by R-loop formation (Chen *et al.*, 1998).

Our interest in converting the Fis protein into a nuclease was prompted by structural studies on this protein by UCLA colleagues Dickerson, Johnson and their associates (Yuan *et al.*, 1991). These workers solved the x-ray structure of the protein yet were unable to obtain cocrystals of it with any of its recognition sequences. The difficulty in obtaining cocrystal probably arises from its highly degenerate consensus recognition sequence.

Nevertheless, the structural model of their protein structure indicated that for Fis to achieve sequence specific binding, it would have to bend its DNA recognition sequence between 40–90°. In a joint project with the Johnson laboratory, we tested this hypothesis by preparing a series of chimeric Fis-OP derivatives. These were generated from the mu-

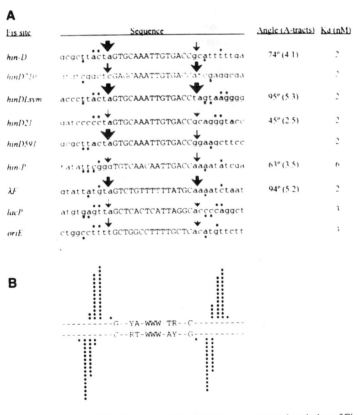

Figure 6. A) Sequences of binding sites of Fis. B) 3'-Stagger apparent in scission of Fis sites.

tants R71C, N73C, N84C and N98C which were alkylated with iodoacetamido derivatives of OP's (Pan *et al.,* 1994; Pan *et al.,* 1996; Pan *et al.,* 1996). The DNA scission patterns of two of the Fis-OP derivatives on four different DNA binding sites allowed us to confirm certain aspects of this working model and infer several new features. One of the most important conclusions is that the overall mode of Fis interaction within target sites containing highly divergent DNA sequences is very similar. The relative locations of the DNA cleavages induced by each Fis-OP derivative at the nine binding sites are virtually identical. These four sites, which are found in different biological contexts had only 4 out of the 15 core nucleotides in common with each other if one considers both orientations. The striking characteristic in all the scission reactions, most commonly with the Fis N73C-OP conjugate, was a 3'- stagger apparent at the sites of scission. This characteristic profile reflects scission within the minor groove which earlier work with the tetrahedral complex had indicated was the site of oxidative scission (Pan *et al.,* 1996).

The similarity in the cleavage patterns by Fis-OP chimeras encouraged us to use targeted scission as an approach to identify high affinity sites capable of bending the DNA near pendant OP derivatives. We examined the cleavage of pUC18 with particular care. Based on the published consensus sequence, 177 possible sites are apparent in the 3 kB plasmid with 13 perfect matches and 164 single mismatches (Pan *et al.,* 1996). However, only three sites of Fis binding were apparent using targeted scission as the assay. These are within the *lac* P, oriE and β-lactamase locus. Novel sites have also been detected in λ phage.

While the location of the Fis dimer relative to the 15 bp core sequence is similar in the different Fis-DNA complexes, analysis of the cleavage efficiencies by Fis N73C-OP within the flanking DNAs suggest significant structural variations. Of the eight half-sites initially investigated, the left half of the Hin enhancer distal site yields by far the strongest cleavage by this derivative. The right half of this same site is the least efficiently cleaved. This implies that the flanking sequence on the left side of the Hin enhancer distal site tends to be in close contact with Fis while the flanking sequence on the right side is not. Since well-characterized Fis sites such as the *hin-D* illustrate this behavior, our results indicate that one reason for the nonstringent consensus core sequence may relate to the bendability of the flanking sequences not typically considered part of the essential canonical site (Pan *et al.,* 1996). Possibly, certain of the various functions of Fis require different degrees of bendability. Activation of transcription such as observed with Fis mutants at R71 may be more sensitive to changes at this site than Hin catalyzed inversion. The interaction of the protein with the flanking sequence rather than with the core sequence may be the crucial determinant in this regard.

8. CONCLUSION

The chemical nuclease of 1,10-phenanthroline-copper provides a novel approach to study structure/function relationships. As the tetrahedral chelate, $(OP)_2Cu^+$ is a footprinting reagent with unanticipated affinity for transcription start sites. It also detects sequence dependent conformational variability of DNA. By covalently tethering a cleavage competent coordination complex in close proximity to an oxidatively sensitive site, this chemical probe is helping to define a new class of transcription inhibitors and identify the binding sites of regulatory proteins within genomic DNA.

ACKNOWLEDGMENTS

Research support was provided by USPHS GM 21199.

REFERENCES

Antic, B. M., van der Goot, H., Nauta, W. T., Balt, S., De Bolster, W. G., Stouthamer, A. H., Verheul, H., and Vis, R. D. (1977). The influence of copper ions on the growth-inhibitory effect of 2,2'-bipyridyl and related compounds on mycoplasmas. Eur. J. Med. Chem. *12*, 573–575.

Bruice, T. W., Wise, J., Rosser, D. S. E., and Sigman, D. S. (1991). Conversion of lambda phage cro into an operator-specific nuclease. J. Am. Chem. Soc. *113*, 5446–5447.

Chen, C.-h. B., and Sigman, D. S. (1987). Chemical conversion of a DNA-binding protein into a site-specific nuclease. Science *237*, 1197–1201.

Chen, C.-h., B., Landgraf, R., Walts, A., Chan, L., Schlonk, P. M., Terwilliger, T. C., and Sigman, D. S. (1998). Scission of DNA at a preselected sequence using a single-strand specific chemical nuclease. Chem. Biol. *5* 283–292.

Chen, T.Q. and Greenberg, M. M. (1998). Model studies indicate that copper phenanthroline induces direct strand breaks via beta-elimination of the 2'-deoxyribonolactone intermediate observed in enediyne mediated DNA damage. J. Am. Chem. Soc. *120* 3815–16.

Cox, J. M., Hayward, M. M., Sanchez, J. F., Gegnas, L. D., van der Zee, S., Dennis, J. H., Sigler, P. B., and Schepartz, A. (1997). Bidirectional binding of the TATA box binding protein to the TATA box. Proc. Nat. Acad. Sci. USA *94*, 13475–80.

D'Aurora, V., Stern, A. M., and Sigman, D. S. (1977). Inhibition of *E. coli* DNA polymerase I by 1,10-phenanthroline. Biochem. Biophys. Res. Comm. *78*, 170–176.

D'Aurora, V., Stern, A. M., and Sigman, D. S. (1978). 1,10-Phenanthroline-cuprous ion complex, a potent inhibitor of DNA and RNA polymerases. Biochem. Biophys. Res. Comm. *80*, 1025–1032.

Feig, A. L., Thederahn, T., and Sigman, D. S. (1988). Mutagenicity of the nuclease activity of 1,10-phenanthroline-copper ion. Biochem. Biophys. Res. Comm. *155*, 338–343.

Gallagher, J., Perrin, D. M., Chan, L., Kwong, E., and Sigman, D. S. (1996). Molecular recognition of tetrahedral 1,10-phenanthroline-cuprous chelates by transcriptionally active open complexes. Dependence on nucleotide sequence of the promoter. Chem. Biol. *3* 739–746.

Gottesfeld, J. M., Neely, L., Trauger, J. W., Baird, E. E., and Dervan, P. B. (1997). Regulation of gene expression by small molecules. Nature *387*, 202–5.

Grachev, M. A., Zaychikov, E. F., Ivanova, E. M., Komarova, N. I., Kutyavin, I. V., Sidelnikova, N. P., and Frolova, I. P. (1984). Oligonucleotides complementary to a promoter over the region -8 to +2 as transcription primers for *E. coli* RNA polymerase. Nucleic Acids Res. *12*, 8509–8524.

Helene, C., and Toulme, J.-J. (1990). Specific regulation of gene expression by antisense, sense and antigene nucleic acids. Biochem. Biophys. Acta *1049*, 99–125.

Kuwabara, M., Yoon, C., Goyne, T. E., Thederahn, T., and Sigman, D. S. (1986). Nuclease activity of 1,10-phenanthroline-copper ion: Reaction with CGCGAATTCGCG and its complexes with netropsin and *Eco*RI. Biochemistry *25*, 7401–7408.

Kuwabara, M. D., and Sigman, D. S. (1987). Footprinting DNA-protein complexes *in situ* following gel retardation assays using 1,10-phenanthroline-copper ion: *E.coli* RNA polymerase-*lac* promoter complexes. Biochemistry *26*, 7234–7238.

Landgraf, R., Pan, C., Sutton, C., Pearson, L., and Sigman, D. S. (1996). Engineering of DNA binding proteins into site-specific cutter: reactivity of Trp repressor-1,10-phenanthroline chimera. Protein Engineering *9*, 603–610.34.

Matteucci, M. D., and Wagner, R. W. (1996). In pursuit of antisense. Nature *384*, 20–22.

Mazumder, A., Perrin, D. M., Watson, K. J., and Sigman, D. S. (1993). A transcription inhibitor specific for unwound DNA in RNA polymerase promoter open complexes. Proc. Nat. Acad. Sci USA *90*, 8140–8144.

Mazumder, A., Perrin, D. M., McMillin, D., and Sigman, D. S. (1994). Interactions of transcription inhibitors with the *E. coli* RNApolymerase lacUV5 promoter open complex. Biochemistry *33*, 2262–2268.

Meijler, M., Zelenko, O., and Sigman, D. S. (1997). Chemical mechanism of DNA scission by 1,10-phenanthroline-copper. Carbonyl oxygen of 5-methylene furanone is derived from water. J. Am. Chem. Soc. *119*, 1135–1136.

Milligan, J. F., Matteucci, M. D., and Martin, J. C. (1993). Current concepts in antisense drug design. J. Med. Chem. *36*, 1923–1937.

Mohindru, A., Fisher, J. M., and Rabinovitz, M. (1983). 2,9-Dimethyl-1,10-phenanthroline (neocuproine): a potent, copper-dependent cytotoxin with anti-tumor activity. Biochem. Pharmacol. *32*, 3627–3632.

Murakawa, G. J., Chen, C.-h. B., Kuwabara, M. D., Nierlich, D., and Sigman, D. S. (1989). Scission of RNA by the chemical nuclease 1,10-phenanthroline-copper ion. Preference for single-stranded loops. Nucleic Acids Res. *17*, 5361–5369.

Pan, C. Q., Feng, J., Finkel, S. E., Landgraf, R., Johnson, R., and Sigman, D. S. (1994). Structure of the Escherichia coli fis protein-DNA complex probed by protein conjugated with 1,10-phenanthroline copper(I) complex. Proc. Natl. Acad. Sci. USA *91*, 1721–1725.

Pan, C. Q., Landgraf, R., and Sigman, D. S. (1995). Drosophila engrailed-1,10-phenanthroline chimeras as probes of homeodomain-DNA complexes. Protein Science *4*, 2279–2288.

Pan, C. Q., Finkel, S. E., Cramton, S. E., Feng, J. A., Sigman, D. S., and Johnson, R. C. (1996). Variable structures of fis-DNA complexes determined by flanking DNA-protein contacts. J. Mol. Biol. *264*, 675–695.

Pan, C. Q., Johnson, R. C., and Sigman, D. S. (1996). Identification of new fis binding sites by DNA scission with fis-1,10-phenanthroline-copper(I) chimeras. Biochemistry *35*, 4326–4333.

Pendergrast, P. S., Ebright, Y. W., and Ebright, R. H. (1994). High-specificity DNA cleavage agent: design and application to kilobase and megabase DNA substrates. Science *265* 959–962.

Perrin, D. M., Mazumder, A., Sadeghi, F., and Sigman, D. S. (1994). Hybridization of a complementary riboolionucleotide to the transcription start site of the *lac*UV-5-*Escherichia coli* RNA polymerase open complex. Potential for gene-specific inactivation reagents. Biochemistry *33*, 3848–3854.

Perrin, D. M., Pearson, L., Mazumder, A., and Sigman, D. S. (1994). Inhibition of prokaryotic and eukaryotic transcription by the 2:1 2,9-dimethyl-1,10-phenanthroline-cuprous complex, a ligand specific for open complexes. Gene *149*, 173–178.

Perrin, D. M., Hoan, V. M., Xu, Y., Mazumder, A., and Sigman, D. S. (1996). Inhibitors of E. coli RNA polymerase specific for the single stranded DNA of transcription intermediates. Tetrahedral cuprous chelates of 1,10-phenanthrolines. Biochemistry *35*, 5318–5326.

Perrin, D. M., Chen, C.-h. B., Xu, Y., Pearson, L, and Sigman, D. S. (1997). Gene specific transcription inhibitors. Oligonucleotides complementary to the template strand of transcription start sites. J. Am. Chem. Soc. *119*, 5746–5747.

Pfau, J., Arvidson, D. N., Youderian, P., Pearson, L. L., and Sigman, D. S. (1994). A site-specific endonuclease derived from a mutant trp repressor with altered DNA-binding specificity. Biochemistry *33*, 11391–11403.

Pitie, M., Bernadou, J., and Meunier, B. (1995). Oxidation At carbon-1' Of DNA deoxyriboses by the Mn-Tmpyp/KHSO$_5$ system results from a cytochrome P-450-type hydroxylation reaction. J. Am. Chem. Soc. *117*, 2935–2936.

Pope, L. M., Reich, K. A., Graham, D. R., and Sigman, D. S. (1982). Products of DNA cleavage by the 1,10-phenanthroline-copper complex. Inhibitors of *E. coli* DNA polymerase I. J. Biol. Chem. *257*, 12121–12128.

Pope, L. E., and Sigman, D. S. (1984). Secondary structure specificity of the nuclease activity of the 1,10-phenanthroline-copper complex. Proc. Natl. Acad. Sci. USA *81*, 3–7.

Pratviel, G., Bernadou, J., and Meunier, B. (1995). Carbon-hydrogen bonds of DNA sugar units as targets for chemical nucleases and drugs. Angew. Chem.-Int. Ed. in English *34*, 746–769.

Reich, K. A., Marshall, L. E., Graham, D. R., and Sigman, D. S. (1981). Cleavage of DNA by the 1,10-phenanthroline-copper ion complex. Superoxide mediates the reaction dependent on NADH and hydrogen peroxide. J. Am. Chem. Soc. *103*, 3582–3584.

Sigman, D. S., Graham, D. R., D'Aurora, V., and Stern, A. M. (1979). Oxygen-dependent cleavage of DNA by the 1,10-phenanthroline-cuprous complex. Inhibition of *E.coli* DNA polymerase I. J. Biol. Chem. *254*, 12269–12272.

Sigman, D. S. (1986). Nuclease activity of 1,10-phenanthroline-copper ion. Accts. Chem. Res. *19*, 180–186.

Sigman, D. S., Mazumder, A., and Perrin, D. M. (1993). Chemical nucleases. Chem. Rev. *93*, 2295–2316.

Sigman, D. S., Bruice, T. W., Mazumder, A., and Sutton, C. L. (1993). Targeted chemical nucleases. Acc. Chem. Res. *26*, 98–104.

Sigman, D. S. "Site specific oxidative scission of nucleic acids and proteins," in *DNA and RNA Cleavers and Chemotherapy of Cancer and Viral Diseases*, vol. Series C-Vol. 479. B. Meunier, Ed. (Kluwer Academic Publishers, Dordrecht/Boston/London, 1996), pp. 119–132.

Slater, J. P., Mildvan, A. S., and Loeb, L. A. (1971). Zinc in DNA polymerase. Biochem. Biophys. Res. Comm.. *44*, 37–43.

Spassky, A., Kirkegaard, K., and Buc, H. (1985). Changes in the DNA structure of the *lac* UV5 promoter during formation of an open complex with *E.coli* RNA polymerase. Biochemistry *24*, 2723–2731.

Spassky, A., and Sigman, D. S. (1985). Nuclease activity of 1,10-phenanthroline-copper ion. Conformational analysis and footprinting of the *lac* operon. Biochemistry *24*, 8050–8056.

Spassky, A. (1986). Following the progression of the transcription bubble. J. Mol. Biol. *188*, 99–103.

Springgate, C. F., Mildvan, A. S., Abramson, S., Engle, J. L., and Loeb, L. A. (1973). *E. coli* deoxyribonucleic acid I, a zinc metalloenzyme. J. Biol. Chem. *248*, 5987–5993.

Sutton, C. L., Mazumder, A., Chen, C. H. B., and Sigman, D. S. (1993). Transforming the *E. coli* trp-repressor into a site-specific nuclease. Biochemistry *32*, 4225–4230.

Thederahn, T., Spassky, A., Kuwabara, M. D., and Sigman, D. S. (1990). Chemical nuclease activity of 5-phenyl-1,10-phenanthroline-copper ion detect intermediates in transcription initiation by E. coli RNA polymerase. Biochem. Biophys. Res. Comm. *168*, 756–762.

Yoon, C., Kuwabara, M. D., Law, R., Wall, R., and Sigman, D. S. (1988). Sequence-dependent variability of DNA structure. J. Biol. Chem. *263*, 8458–8463.

Yoon, C., Kuwabara, M. D., Spassky, A., and Sigman, D. S. (1990). Sequence specificity of the deoxyribonuclease activity of 1,10-phenanthroline-copper ion. Biochemistry *29*, 2116–2121.

Yuan, H. S., Finkel, S. E., Feng, J.-A., Kaczor-Grzeskowiak, M., Johnson, R. C., and Dickerson, R. E. (1991). The molecular structure of wild-type and a mutant Fis protein: Relationship between mutational changes and recombinational enhancer function or DNA binding. Proc. Natl. Acad. Sci. USA *88*, 9558–9562.

14

MODULATION OF GLUCOCORTICOID SIGNALING AND LEUKOCYTE ACTIVATION BY TRANSITION METALS

J. Koropatnick,[*] J. M. DeMoor, and O. M. Collins

Department of Oncology
University of Western Ontario
London Regional Cancer Centre
790 Commissioners Road East
London, Ontario, Canada N6A 4L6

1. METALLOTHIONEINS AND METAL REGULATION

Metallothioneins (MTs) are low molecular weight (typically less than 10,000 daltons), cysteine-rich, metal-binding proteins that are found in all eukaryotes, and transcriptionally regulated by a variety of metals, hormones, and developmental signals. They most certainly play a role in resistance to heavy metal toxicity, and have been postulated to mediate several roles important in metal homeostasis in eukaryotes (Kagi and Schaffer, 1988). They are encoded by a family of genes in mammals. Humans possess 16 MT genes linked on chromosome 16 (including MT-1A, MT1B, MT-1E, MT-1F, MT-1G, MT-2A, MT-3, and MT-4)(West *et al.*, 1990). Rodents have only four functional genes (MT-1, MT-2, MT-3 and MT-4) on chromosome 8 (Quaife *et al.*, 1994). Gene sequences are similar among isoforms, and even between species. This allows cross-hybridization of specific probes among species (Hamer, 1986; Koropatnick *et al.*, 1988). MTs are classically induced by a variety of transition metals, but MTs are expressed without metal induction in many different tissues and cell types. Intriguingly, some isoforms are restricted to specific tissues. MT-3 is expressed predominantly in brain and is not metal-inducible (Uchida *et al.*, 1991). MT-4 is restricted to cornified, stratified, squamous epithelium (which provides a protective surface on skin, tongue, the upper part of the alimentary tract, and the vagina) without metal induction (Quaife *et al.*, 1994). MTs are also expressed during retinoic acid-

[*] Tel: (519) 685-8654; Fax: (519) 685-8646; jkoropat@julian.uwo.ca

Metals and Genetics, edited by Sarkar.
Kluwer Academic / Plenum Publishers, New York, 1999.

induced differentiation of teratocarcinoma cells *in* vitro (Hamer, 1986, and Fraser and Koropatnick, manuscript submitted)), during growth and proliferation in a human prostate stem cell line (Koropatnick *et al.*, 1995), in rat kidney undergoing compensatory hypertrophy (Zalups *et al.*, 1995), in proliferating human colonic cancer cells (Nagel and Vallee, 1995), in cells on the proliferative edge of epithelial tumours *in situ* (Kontozoglou *et al.*, 1989), the proliferating cell compartment of human breast and colon tumours (Meskel *et al.*, 1993), and in human monocytes undergoing a respiratory burst (Leibbrandt and Koropatnick, 1994; Leibbrandt *et al.*, 1994). In rodents and humans, MT expression in visceral yolk sac and liver is developmentally-regulated (Hamer, 1986; Ouellette, 1982; Koropatnick and Duerksen, 1987; Slotkin and Cherian, 1989). Subpopulations of mouse ova and preimplantation embryo cells express high MT levels (Andrews *et al.*, 1991). The intracellular location of MT is dependent upon the stage of tissue development: MT is located in the nuclei of parenchymal cells in fetal and newborn liver, and is primarily cytoplasmic in adult liver (Panemangalore *et al.*, 1983; Nartey *et al.*, 1987; Cherian, 1994). MT isoforms change during differentiation in mice. As inner layer keratinocytes in stratified epithelium differentiate into the outer stratum spinosum, and switch keratin isoform production, they also switch from MT-1 to MT-4 production (Quaife *et al.*, 1994). Cultured embryocarcinoma cells, induced to differentiate by retinoic acid, express MT-1 mRNA and MT protein at increasing levels during differentiation to visceral endoderm (Andrews *et al.*, 1984). Overall, MT is clearly associated with proliferation and hypertrophy, and differentiation and cell activation.

The multiple circumstances under which MTs are expressed has fuelled speculation about their function(s). A role in metabolism and detoxication of metals is strongly suggested by the ability of MT to bind to, and be induced by, heavy metal ions. Other possible functions include control of intracellular redox potential, regulation of thiol metabolism, essential zinc and copper homeostasis, and detoxication of activated oxygen and reactive toxic chemical intermediates. A role for MT in supplying essential zinc and/or copper metalloproteins (including transcription factors, hormone receptors, metalloproteinases, superoxide dismutase, and catalase, among many others), or in sequestering zinc to inactivate metal-dependent molecules has been suggested (Zeng *et al.*, 1991), and MT has been shown to transfer some of its complexed zinc ions to a number of apoproteins (Udom and Brady, 1980). The metal-deficient form of MT (apothionein) has also been demonstrated to block the capacity of zinc-dependent transcription factors (*e.g*, TFIIIA, Sp1, and estrogen receptor) by sequestering zinc (Zeng *et al.*, 1991a,b; Cano-Gauci and Sarkar, 1996). Therefore, MT may act both as a donor or acceptor of zinc. Recent evidence suggests that the capacity of MT to switch between these two roles is coupled to energy metabolism and the cellular redox state (Jiang *et al.*, 1998; Jacob *et al.*, 1998; Fischer and Davie, 1998). Taken together, these data indicate that MT may be a homeostatic molecule that plays a critical role in regulating zinc-requiring proteins, including those that influence gene expression. We describe here the effect of downregulating MT expression in cells responsive to extracellular signals (bacterial lipopolysaccharide [LPS] to induce the respiratory burst in a human monocyte cell line, and the artificial glucocorticoid dexamethasone [DEX] to induce activation of the glucocorticoid receptor in a mouse mammary tumour cell line) using antisense nucleic acids, followed by treatment with LPS or DEX. We also upregulated MT expression by pre-treating cells with zinc or cadmium prior to treatment with PMA or DEX. The results, described below, suggest that MT plays a role in signal transduction and gene transcription events mediated by some, but not all, zinc-requiring transcription factors.

2. ZINC AND METALLOTHIONEIN IN MONOCYTE ACTIVATION

Mononuclear phagocytes (including monocytes and their precursors, and monocyte-derived macrophages of various types found in different sites within the body), and polymorphonuclear neutrophils (PMN) are metabolically active cells with measurable phagocytic capacity and hydrolytic enzyme activity. Monocytes and other cells responding to extracellular signals signals express soluble or membrane-associated proteins, encoded by germ-line genes, to recognize carbohydrate structures generally associated with bacteria. Macrophages, for example, have a receptor for bacterial lipopolysaccharide (LPS). LPS binds to LPS-binding protein (LBP) outside the cells, soluble glycoprotein CD14 (sCD14) accepts LPS from the LPS-LBP complex, and the sCD14-LPS complex interacts with an as yet undefined receptor to "prime" cells for activation. In macrophages, LPS also induces cytokines (primarily TNFα, but including IL-1β), that mediate additional biological responses comprising activation events (Beutler and Cerami, 1989). Activation results in a general increase in oxidative metabolism, enhanced mobility, adherence, phagocytosis, lysosomal enzyme activity, tumour cytotoxicity, and microbicidal activity (Guthrie et al., 1984; Cohn, 1978; Johnston, 1988). Increased secretion of neutral proteases (collagenase, for example), combined with increased motility, permits monocytes to extravasate in response to inflammation and chemotactic stimuli. Cells increase their production of lysosomal hydrolases, colony stimulating factor (CSF), secretion of metalloproteinases, IL-1, IF-γ, and various arachidonic acid metabolites. They also enhance their release of toxic oxygen metabolites. Activated monocytes are stimulated, without exception, to differentiate rapidly into the macrophage line, with increased adhesion and loss of relatively monocyte-specific expression CD14.

Zinc is necessary for the function of signals involved in monocyte/macrophage activation of LPS and other antigen-independent, receptor-mediated signals. In response to these agents, activation proceeds through a G-protein-linked pathway to activate phospholipases C, A2, and D. Phosphorylation of several protein kinases (including tyrosine kinase and mitogen-activated [MAP] kinase) also occurs, and contributes to a cascade of events contributing to cytokine production and arachidonate release. Phospholipase C activation stimulates release of diacylglycerol to activate protein kinase C (PKC), and inositol 1,4,5-trisphosphate (IP$_3$) to increase intracellular calcium levels. PKC activation stimulates release of nuclear factor κB (NF-κB) from its cytoplasmic inhibitor (Iκ-Bα). Released NF-κB is translocated to the nucleus, where it acts as a transcription factor to participate in activation-associated events, including the activation of genes associated with the respiratory burst and IL-1α, IL-1β, and TNFα expression (Hancock et al., 1995; Meier et al., 1995). Zinc is required for the function of PKC (Szallasi et al., 1996) and virtually all of the transcription factors activated by these various signal transduction pathways, including NF-κB (Otsuka et al., 1995). Thus, proteins that transduce activation signals, and the transcription factors that regulate activation-induced gene expression, require zinc for their activity.

We employed a human monocytic leukemia cell line (THP-1) to assess metallothionein expression after activation with bacterial LPS, phorbol myristate acetate, or GM-CSF (Leibbrandt and Koropatnick, 1994). Metallothionein protein (as measured by radioimmunoassay) and mRNA (for MT-2, measured by reverse transcription/polymerase chain reaction) are induced by LPS at the same time as activation (measured by increased production of reactive oxygen intermediates) (Figure 1). LPS did not induce MT expression *in vitro* in a non-monocyte human melanoma cell line that does not undergo activation, nor in other non-activatable human and rodent cell lines, indicating that increased

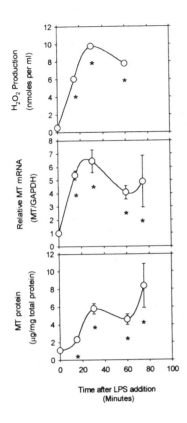

Figure 1. Activation of THP-1 cells with LPS leads to increased MT expression. The effect of 10 μg/ml LPS treatment alone on induction of the respiratory burst (top panel, measured by H_2O_2 production from superoxide radicals), relative MT mRNA accumulation (middle panel, MT-1 plus MT-2 mRNA/glyceraldehyde phosphate dehydrogenase (GAPDH) mRNA, measured by densitometry of northern blots), and MT protein (bottom panel, MT-1 plus MT-2, measured by competitive ELISA). Symbols represent the mean ± SE (n=3). MT mRNA and protein levels measured in cells untreated with LPS were assigned a value of 1.0, and levels in LPS-treated cells were normalized to that value. SE smaller than the symbols are not shown. Asterisks (*) indicate values significantly different from those measured in cells untreated with LPS (p<0.05).

MT expression in response to LPS was an activation-associated phenomenon. Primary human moncytes freshly isolated from healthy males also underwent an LPS-induced respiratory burst (activation) and MT-2 mRNA induction (Leibbrandt and Koropatnick, 1994). The concurrence of MT induction and activation in both primary human monocytes, and a monocyte-derived human tumour cell line, suggests a functional role for MT in monocyte/macrophage activation.

Further experiments were performed to determine whether MT played a causal role in aspects of monocyte behaviour associated with activated cells, or with resting cells untreated with activating agents. We used an antisense RNA strategy to downregulate MT expression by transient transfection of a eukaryotic expression vector (p164/7) directing production of a 372 bp RNA complementary to mouse MT-1, under the control of the cytokine-responsive H-2K^b promoter. The antisense mouse MT-1 sequence was sufficiently complementary to human MT-1 and MT-2 mRNA that it was expected to efficiently associate with and inhibit the biological effectiveness of human MT mRNAs. We found that cationic liposome-mediated transient transfection of this vector did, in fact, effectively suppress MT-2 mRNA induction, and basal and induced total MT protein (including both MT-1 and MT-2 isoforms, which are both bound by the detecting antibody), after treatment with bacterial LPS or cadmium (Leibbrandt *et al.*, 1994). Basal MT protein levels were diminished approximately 25% within 48 hours of transfection. Within 60 minutes of induction with LPS, the level was diminished by more than 80%, in accord with increased antisense RNA accumulation due to the cytokine-responsiveness of the promoter driving antisense MT RNA expression (Figure 2). MT-2 mRNA accumulation induced by subsequent induction with LPS or cadmium was also significantly diminished (Leibbrandt *et*

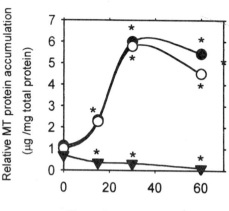

Figure 2. MT protein levels in untransfected THP-1 cells (●), cells transfected with empty control vector (○), and cells transfected with antisense MT vector (?). LPS was added at time 0 to induce the respiratory burst. Values for MT are expressed relative to the amount measured in control vector-transfected cells, untreated with LPS. The mean of 3 samples ± SE, assayed in triplicate by competitive ELISA. Standard errors of means were smaller than the symbols. Asterisks (*) indicate values significantly different from unactivated cells transfected with control vector (p<0.05).

al., 1994). Transfection of vector expressing antisense MT RNA did not significantly decrease THP-1 viability, indicating that the effect was not due to toxicity. Therefore, antisense MT RNA effectively diminished both basal and induced MT protein levels in human monocytes. The physiological consequences of MT downregulation were twofold: 1) sensitivity of monocytes to the toxic effects of cadmium was dramatically increased, in accord with the well-characterized role of MT in detoxifying metals, and 2) the ability of the cells to undergo activation (as assessed by production of reactive oxygen species in response to bacterial LPS) was abolished (Figure 3). In addition, other activation-associated events were affected: the ability of THP-1 cells to adhere to a reconstituted basement membrane or tissue culture plastic (indicative of LPS-induced differentiation into the macrophage line), and to invade into a reconstituted basement membrane, were significantly altered (Leibbrandt *et al.*, 1994). These experiments demonstrate that metallothionein (the primary species associated with transition metals in eukaryotic cells) is a

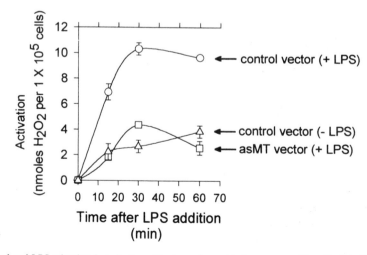

Figure 3. Basal and LPS-stimulated production of hydrogen peroxide from superoxide radicals indicative of activation and respiratory burst in THP-1 cells. LPS (10 μg/ml) was added at time zero to activate cells. The symbols denote the mean of two samples assayed in triplicate ± SE. Lack of error bars indicates that they were smaller than the symbol.

key player in the capacity of human monocytes to undergo activation in response to extracellular signals. Those signals were transduced through a series of zinc-requiring proteins. To explore this further, we assessed how altered MT expression affected the activity of a zinc-requiring transcription factor/signal transduction protein, the glucocorticoid receptor (GR).

3. ZINC AND METALLOTHIONEIN IN HORMONE RESPONSIVENESS

To investigate the role of MT in mediating the activity of a well-characterized transcription factor, we used a derivative of the C127 mouse mammary tumour cell line (denoted 2305) that harbours a stable episomal bovine papilloma virus (BPV)-based vector containing a chloramphenicol acetyltransferase (CAT) reporter gene under the control of the mouse mammary tumour virus (MMTV) promoter (Charron *et al.*, 1989)(kindly supplied by Dr. Trevor Archer, Dept. of Biochemistry, Univ. of Western Ontario, Canada). This promoter responds to dexamethasone signals transduced to the nucleus by the glucocorticoid receptor (GR), which is a zinc finger DNA-binding protein. Nuclear GR interacts with enhancer elements in the MMTV promoter to induce CAT mRNA, protein, and enzyme activity (Lapointe and Baxter, 1989). We transiently transfected into these cells a vector containing a full length mouse MT-1 cDNA designed to express RNA complementary to mouse MT-1 mRNA (pRC/asMT). Orientation and identity of the MT cDNA insert was confirmed by DNA sequence analysis. Control cells were transfected with vector that did not contain MT-1 DNA sequences. Transfected pRC/asMT vector expressed asMT-1 RNA and decreased MT-1 mRNA levels by approximately 60%, as assessed by RT-PCR 48 hours after transfection (Figure 4). MT protein in control cells transfected with pRC/CMV was 2–3 ng per μg total protein (measured by ELISA) and was not detectable in cells transfected with the pRC/asMT vector.

4. ANTISENSE DOWNREGULATION OF MT EXPRESSION INHIBITS GLUCOCORTICOID-INDUCED CAT ACTIVITY

We assessed DEX responsiveness in mouse 2305 cells under conditions where sufficient zinc was available for normal cell growth, and where zinc was depleted. We surmised that endogenous zinc (approximately 4 μM in fetal bovine serum used to supplement nutrient medium)(Palmiter, 1995) might be an alternative source of zinc and diminish the importance of MT in delivering zinc to macromolecules. Therefore, we grew 2305 cells in medium containing fetal bovine serum that had been depleted of zinc by Chelex treatment, and in medium containing unchelated serum. Zinc-depleted medium did not decrease cell growth over a 48 h period when compared to cells grown in zinc-sufficient medium (data not shown). CAT expression in response to dexamethasone induction was significantly reduced (by 10–15%) in asMT-transfected cells in zinc-sufficient medium. This reduction in hormone-responsiveness, although signficant, was relatively minor. However, when cells were maintained in zinc-depleted medum, DEX responsiveness was reduced by 30% in asMT-transfected cells compared to 2305 cells transfected with control vector (Figure 5). This suggested that MT participated in glucocorticoid respon-

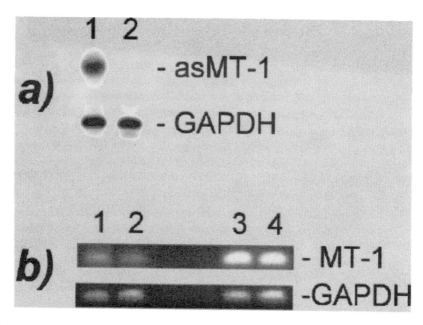

Figure 4. Inhibition of MT-1 mRNA expression by antisense MT-1 RNA in zinc-deficient mouse 2305 cells. A 322 bp mouse MT-1 cDNA was generated by polymerase chain reaction (PCR) from a full-length 400 bp mouse MT-1 cDNA template. The 5' and 3' primers were completely complementary to 21 base sequences within MT-1 cDNA, but also contained non-complementary Not1 restriction enzyme recognition sequences to generate Not1 sites in the PCR product, which was cloned into the Not1 polylinker site of the eukaryotic expression vector pRC/CMV (Invitrogen). Restriction analysis verified that the orientation of the MT-1 cDNA insert in the resulting vector (pRC/asMT) was appropriate for production of antisense MT RNA. The vector was transfected for 6 h in the presence of Lipofectamine (Gibco/BRL) into mouse C127 mammary carcinoma cells containing a GR-responsive CAT expression vector. Cells were allowed to recover for 18–20 h in medium containing zinc (approximately 4 μM), then transferred to medium supplemented with fetal bovine serum that had been depleted of zinc by chelation treatment with Chelex 100 resin (BioRad). Dexamethasone (1×10^{-7} M) was added 24 hours after transfer to zinc-depleted medium, and the cells were incubated for a further 24 h in low zinc conditions in the presence of DEX. Cells were then collected and frozen for CAT assay and mRNA analysis. **a)** Expression of antisense MT-1 RNA in cells transiently transfected with the pRC/asMT expression vector was assessed by specific RT-PCR after dexamethasone treatment. cDNA was generated from antisense MT-1 RNA present in one μg of total cellular RNA isolated from cells transfected with pRC/asMT (Lane 1) or control pRC/CMV vector (Lane 2). The reverse transcription reaction was primed by a specific oligonucleotide that would produce cDNA from antisense MT-1 RNA, but not MT-1 mRNA. A specific primer to allow production of glyceraldehyde phosphate dehydrogenase (GAPDH) cDNA from GAPDH mRNA was included in the reverse transcriptase reaction mix. PCR of the resulting antisense MT- derived cDNA and GAPDH cDNA yielded products of 344 and 752 base pairs, respectively, which were separated by non-denaturing agarose gel electrophoresis, Southern-blotted and hybridized to radiolabeled MT-1 and GAPDH cDNA probes. Hybridization was visualized using a Molecular Dynamics phosphorimager. The intensity of bands is proportional to the amount of asMT-1 RNA and GAPDH mRNA in total cellular RNA: Lane 1: Cells transfected with pRC/asMT-1 vector. Lane 2: Cells transfected with control pRC/CMV vector. **b)** Downregulation of MT-1 mRNA expression by pRC/asMT-1. cDNA from GAPDH mRNA and MT-1 mRNA (and not antisense RNA) was generated by reverse transcription of one μg of total cellular RNA isolated from cells transfected with pRC/asMT-1 (Lanes 1 and 2) or control pRC/CMV (lanes 3 and 4). In this case, the reverse transcription reaction was primed by specific oligonucleotides that would produce cDNA from GAPDH mRNA and MT-1 mRNA, but not antisense MT-1 RNA. The ethidium bromide-stained MT-1 cDNA (344 bp) and GAPDH cDNA (752 bp) bands are shown. Products from two separate RT-PCR reactions from the same pool of zinc-depleted cells transfected with pRC/asMT-1 vector (lanes 1 and 2) or control pRC/CMV vector (lanes 3 and 4) are shown, indicating an approximately 60% decrease in MT-1 mRNA accumulation in cells transfected with the antisense MT-1 RNA expression vector.

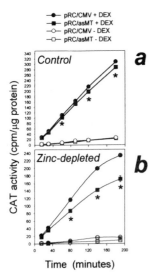

Figure 5. CAT reporter gene expression in response to dexamethasone is inhibited by antisense MT RNA. **a)** *Zinc-sufficient medium:* 2305 cells transfected with control (pRC/CMV) or asMT (pRC/asMT) were grown at all times in medium containing approximately 4 μM zinc. They were treated with or without DEX according to the protocol outlined in the legend to Figure 4 and CAT enzyme activity assessed by a phase extraction assay (a modification of that described by Seed and Sheen, 1988). CAT activity per μg protein in the cell extract is plotted as a function of time. Asterisks (*) indicate significant differences in CAT expression between cells transfected with control (●) or antisense MT-1 (■) expression vectors (p<0.05). **b)** *Zinc-deficient medium:* Differences in DEX-induced CAT expression between 2305 cells transfected with control (●) or antisense MT-1 (■) expression vectors, and maintained in zinc-deficient medium (less than 0.3 μM Zn) for 48 h prior to assay. Asterisks (*) indicate values significantly different from control (p<0.05). Antisense downregulation of MT has an increased effect on glucocorticoid responsiveness when zinc supplies are limited.

siveness to a minor degree under conditions where zinc was freely available, but played a much greater role under conditions where zinc supplies were limited.

5. INDUCTION OF MT BY ZINC ENHANCES GLUCOCORTICOID RESPONSIVENESS

2305 cells grown under normal conditions (without zinc depletion and without transfected vectors) were treated with 20, 40, 80 or 100 μM zinc chloride for 48 h. Treatment with 20 or 40 μM zinc chloride did not induce MT expression, but treatment with 80 or 100 μM did significantly enhance MT protein levels (assessed by ELISA) (Figure 6B). Cells were then assessed for CAT expression in response to induction with DEX. DEX-induced CAT expression was enhanced by zinc in a stepwise manner in correlation with the amount of MT induced by zinc (Figure 6A, B).

6. SUMMARY

Expression of antisense RNA results in a significant decrease in MT-1 mRNA levels within cells. The status of MT-2 mRNA has not been assessed, but the high degree of similarity between mouse MT-1 and mouse MT-2 (greater than 85%), and the fact that expression of RNA antisense to mouse MT-1 has the capacity to downregulate human MT-2 mRNA and protein in a human monocytic leukemia cell line (Leibbrandt *et al.*, 1994) suggests that MT-2 might be similarly downregulated. Reduced MT expression results in reduced capacity of a transfected reporter gene to respond to induction by glucocorticoid, particularly when zinc supplies are limited. The suggested role for MT in regulating zinc supplies within cells (Koropatnick and Leibbrandt, 1995), and the requirement of the glucocorticoid receptor for zinc, suggests that MT may regulate the availability of zinc under low zinc conditions. Although these data indicate only that MT plays a role in some aspect of cellular response to glucocorticoids (GR binding to hormone response elements in the MMTV promoter, the ability of GR to enhance transcription of CAT mRNA, the transla-

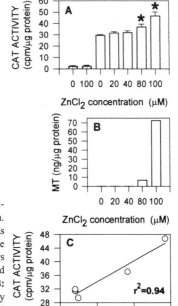

Figure 6. Zinc treatment induces enhanced responsiveness to glucocorticoid induction. 2305 cells were induced with 0–100 μM $ZnCl_2$ for 24 h. **A:** $ZnCl_2$ treatment alone without DEX treatment, at concentrations as high as 100 μM, did not induce CAT activity (the two shaded bars to the left of the graph). However, dexamethasone induction following 24 hours 40 or 80 μM zinc treatment, but not 20 or 40 μM zinc treatment, resulted in significant enhancement of DEX-induced CAT activity (open bars). **B:** Only 40 and 80 μM zinc induced significant MT protein, as measured by a competitive solid phase ELISA. **C:** Zinc-induced enhancement of CAT activity correlated with induced MT expression. Asterisks (*) indicate values significantly different from controls ($p<0.05$).

tion of CAT mRNA, and the capacity of CAT to catalyze the transfer of acetyl groups) evidence to support a direct role in mediating zinc availiability for MT protein is lacking. We have observed no obvious changes in GR protein levels under conditions where MT is downregulated by antisense MT RNA, as measured by western blot analysis (data not shown), suggesting that lowered MT levels do not diminish hormone responsiveness by decreasing GR levels. Our observations are consistent with reports that insertional inactivation (gene knockout) of MT-1 and MT-2 genes in mice is not lethal under normal conditions, but leads to increased sensitivity to the toxic effects of cadmium (Michalska and Choo, 1993; Masters *et al.*, 1994), and to cadmium and a wide range of toxic agents in cultured MT-1/MT-2 gene knockout fibroblasts *in vitro* (Kondo *et al.*, 1995). Interestingly, mice lacking MT-1 and MT-2 are sensitive to both lack of zinc and excess zinc (Kelly *et al.*, 1996), and transgenic mice overexpressing MT-1 resist dietary deficiency (Dalton *et al.*, 1996), suggesting that it plays a role in maintaining essential zinc stores in tissues *in situ*. We suggest that cells lacking MT grow and function normally under conditions where zinc is freely available, both in culture and in whole animals. However, under stress conditions where the requirement for zinc is high or where zinc supplies are limited, some zinc-requiring proteins (including glucocorticoid receptors) would depend more heavily upon metallothionein to supply them with zinc. The fact that metallothionein is produced under conditions of stress (including glucocorticoid induction), and is associated with resistance to stress-inducing toxins (including reactive oxygen intermediates and chemotherapeutic agents) is consistent with this hypothesis.

Induction of MT expression by zinc correlates with enhanced response to glucocorticoid. It is not clear whether enhancement of response is due to increased MT, other gene products induced by zinc. The presence of zinc itself is unlikely to be the cause of enhanced response: addition of 20 μM Zn had no effect on glucocorticoid responsiveness, even though it increased endogenous zinc levels by approximately 5-fold. Only zinc levels that significantly induced MT (40 and 100 μM) zinc appeared to significantly enhance responsiveness to dexamethasone. Priior treatment with metals that induce MT expression may have a complex effect on signalling, depending on cell type and the nature of the signal. Although zinc pretreatment induces MT in both monocytes and in 2395 cells, it diminished the capacity of monocytes to respond to an activation signal (Leibbrandt and Koropatnick, 1994; Koropatnick and Zalups, 1997), but enhanced the capacity of 2305 cells to respond to a glucocorticoid signal. Thus, increased MT expression may have either enhancing or suppressing effects on signalling events. Further experiments in cells lacking functional MT genes (MT-1/MT-2 knockout cells) are under way to explore the mechanism by which zinc alters signal responsiveness.

This novel role for MT in signal transduction and induced gene expression is being further explored by measuring glucocorticoid signaling in MT-1/MT-2 gene knockout fibroblasts in comparison with parental cells with competent MT-1/MT-2 genes, in both normal and zinc-depleted nutrient medium. We are also assessing the effect of zinc induction on glucocorticoid signaling in MT knockout fibroblasts. Finally, further investigation of altered physiological consequences of glucocortoid signaling after addition or depletion of zinc, or by antisense downregulation of metallothionein (for example, glucocorticoid-induced apoptosis in T lymphocytes) are warranted to explore the role of MT in influencing cellular responses to extracellular signals.

REFERENCES

Andrews, G.K., Adamson, E.D., and Gedamu, L. (1984). The ontogeny of expression of murine metallothionein: comparison with the alpha-fetoprotein gene. Dev Biol 103, 294–303.

Andrews, G.K., Huet-Hudson, Y.M., Paria, B.C., McMaster, M.T., De, S.K., and Dey, S.K. (1991). Metallothionein gene expression and metal regulation during preimplantation mouse embryo development (Mt mRNA during early development). Dev Biol 145, 13–27.

Beutler, B., and Cerami, A. (1989). The biology of cachectin/TNF—a primary mediator of the host response. Annu Rev Immunol 7, 625–655.

Cano-Gauci DF and Sarkar B. (1996), May 13. Reversible zinc exchange between metallothionein and the estrogen receptor zinc finger. FEBS Lett 386(1), 1–4.

Charron, J., Richard-Foy, H., Berard, D.S., Hager, G.L., and Drouin, J. (1989). Independent glucocorticoid induction and repression of two contiguous responsive genes. Mol Cell Biol 9, 3127–3131.

Cherian, M.G. (1994). The significance of the nuclear and cytoplasmic localization of metallothionein in human liver and tumor cells. Env Health Persp 102, 131–135.

Cohn, Z.A. (1978). Activation of mononuclear phagocytes: fact, fancy, and future. J Immunol 121, 813–816.

Dalton, T, Fu, K., Palmiter, R.D., and Andrews, G.K. (1996). Transgenic mice that overexpress metallothionein-1 resist dietary zinc deficiency. J Nutr 126, 825–833.

Fischer, E.H. and Davie, E.W. (1998). Recent excitement regarding metallothionein. Proc Natl Acad Sci U S A 95, 3333–3334.

Goyer, A. (1991). Toxicology of metals. In: Toxicology, "The basic science of poisons" (Fourth Ed.). M.O. Amdur, J. Doull, and C.D. Klaassen (eds.), Pergamon Press.

Guthrie, L.A., McPhail, L.C., Henson, P.M., and Johnston, R.B. Jr. (1984). Priming of neutrophils for enhanced release of oxygen metabolites by bacterial lipopolysaccharide. Evidence for increased activity of the superoxide-producing enzyme. J Exp Med 160, 1656–1671.

Hamer, D.H. (1986). Metallothionein. Annu Rev Biochem 55, 913–951.

Hancock, W.W., Grey, S.T., Hau, L., Akalin, E., Orthner, C., Sayegh, M.H., and Salem, H.H. (1995). Binding of activated protein C to a specific receptor on human mononuclear phagocytes inhibits intracellular calcium signaling and monocyte-dependent proliferative response. Transplantation 60, 1525–32.

Jacob, C., Maret, W., and Vallee, B.L. (1998). Control of zinc transfer between thionein, metallothionein, and zinc proteins. Proc Natl Acad Sci U S A 95, 3489–3494.

Jiang L-J, Maret, W., and Vallee, B.L. (1998). The glutathione redox couple modulates zinc transfer from metallothionein to zinc-depleted sorbitol dehydrogenase. Proc Natl Acad Sci U S A 95, 3483–3488, 1998.

Johnston, R.B., Jr. (1988). Current concepts: immunology. Monocytes and macrophages. N Eng J Med 318, 747–752.

Kagi J.H. and Schaffer, A. (1988). Biochemistry of metallothionein. Biochemistry 27, 8509–8515.

Kontozoglou, T.E., Banerjee, D., and Cherian, M.G. (1989). Immunohistochemical localization of metallothionein in human testicular embryonal carcinoma cells. Virchows Archiv A Pathol Anat 415, 545–549.

Kelly E.J., Quaife, C.J., Froelick, G.J., and Palmiter, R.D. (1996). Metallothionein 1 and II protect against zinc deficiency and zinc toxicity in mice. J Nutr 126, 1782–1790.

Kondo, Y., Woo, E.S., Michalska, A.E., Choo, K.H., and Lazo, J.S. (1995). Metallothionein null cells have increased sensitivity to anticancer drugs. Cancer Res 55, 2021–2023.

Koropatnick, J. and Duerksen, J.D. (1987). Nuclease sensitivity of alfafetoprotein, metallothionein-1, and immunoglobulin gene sequences in mouse during development. Dev Biol 122, 1–10.

Koropatnick, J., Pearson, J., and Harris, J.F. (1988). Extensive loss of human DNA accompanies loss of antibody production in heteromyeloma hybridoma cells. Mol Biol Med 5, 69–83.

Koropatnick, J., Kloth, D.M., Kadhim, S., Chin, J.L., and Cherian, M.G. (1995). Metallothionein expression and resistance to cisplatin in a human germ cell tumor cell line. J Pharmaceut Exper Therap 275, 1681–1687.

Koropatnick, J., and Zalups, R.K. (1997). Effect of non-toxic mercury, zinc or cadmium pretreatment on the capacity of human monocytes to undergo lipopolysaccharide-induced activation. Br J Pharmacol 120, 797–806.

Koropatnick, J. and Leibbrandt, M.E.I. (1995). Effects of Metals on Gene Expression, in: Handbook of Experimental Pharmacology (Toxicology of Metals - Biochemical Aspects), vol. 115, R.A. Goyer and M.G. Cherian, eds., Springer-Verlag, pp. 93–120.

Lapointe, M.C. and Baxter, J.D. (1989). in: Anti-inflammatory Steroid Action, Academic Press, San Diego, pp. 3–23.

Leibbrandt, M.E.I., Khokha, R., and Koropatnick, J. (1994). Antisense down-regulation of metallothionein in a human monocytic cell line alters adherence, invasion and the respiratory burst. Cell Growth Differ 5, 17–25.

Leibbrandt, M.E.I. and Koropatnick, J. (1994). Activation of human monocytes with endotoxin induces metallothionein expression and is diminished by zinc. Toxicol Appl Pharmacol 124, 72–81.

Masters, B.A., Kelly E.J., Quaife, C.J., Brinster, R.L., and Palmiter, R.D. (1994). Targeted disruption of metallothionein 1 and 2 genes increases sensitivity to cadmium. Proc Natl Acad Sci U S A 91, 584–588.

Michalska, A.E., and Choo, K.H. (1993). Targeting and germ-line transmission of a null mutation at the metallothionein 1 and 2 loci in mouse. Proc Natl Acad Sci U S A 90, 8088–8092.

Meier, R.W., Niklaus, G., Dewald, B., Fey, M.F., and Tobler, A. (1995). Inhibition of the arachidonic acid pathway prevents induction of IL-8 mRNA by phorbol ester and changes the release of IL-8 from HL 60 cells: differential inhibition of induced expressionn of IL-8, TNFa, IL-1a and IL-1b. J Cell Physiol 165, 62–70.

Meskel, H.H., Cherian, M.G., Martinez, V.J., Veinot, L.A., and Frei, J.V. (1993). Metallothionein as an epithelial proliferative compartment marker for DNA flow cytometry. Mod Pathol 6, 756–760.

Nagel, W.W., and Vallee, B.L. (1995). Cell cycle regulation of metallothionein in human colonic cancer cells. Proc Natl Acad Sci U S A 92, 579–583.

Nartey, N.O., Banerjee, D., and Cherian, M.G. (1987). Immunohistochemical localization of metallothionein in cell nucleus and cytoplasm of fetal human liver and kidney and its changes during development. Pathology 19, 233–238.

Otsuka, M., Fujita, M., Aoki, T., Ishii, S., Sugiura, Y., Yamamoto, T., and Inoue, J. (1995). Novel zinc chelators with dual activity in the inhibition of the kappa B site-binding proteins HIV-IP1 and NF-kB. J Med Chem 38, 3264–70.

Ouellette, A.J. (1982). Metallothionein mRNA expression in fetal mouse organs. Dev Biol 92, 240–246.

Palmiter, R.D. (1995). Constitutive expression of metallothionein-3 (MT-3), but not MT-1, inhibits growth when cells become zinc-deficient. Tox Appl Pharmacol 135, 139–146.

Panemangalore, M., Banerjee, D., Onosaka, S., and Cherian, M.G. (1983). Changes in the intracellular accumulation and distribution of metallothionein in rat liver and kidney during postnatal development. Dev Biol 97, 95–102.

Quaife, C.J., Findley, S.D., Erickson, J.C., Froelick, G.J., Kelly E.J., Zambrowicz, B.P., and Palmiter, R.D. (1994). Induction of a new metallothionein isoform (MT-4) occurs during differentiationn of stratified squamous epithelia. Biochemistry 33, 7250–7259.

Szallasi, Z., Bogi, K., Gohari, S., Biro, T., Acs, P., and Blumberg, P.M. (1996). Non-equuivalent roles for the first and second zinc fingers of protein kinase Cd. Effect of their mutation on phorbol ester-induced translocation in NIH 3T3 cells. J Biol Chem 271, 18299–301.

Seed, B. and Sheen, J.-Y. (1988). A simple phase extraction assay for chloramphenicol acetyl-transferase activity. Gene 67, 271–277.

Slotkin, S.H. and Cherian, M.G. (1989). Elevated metallothionein expression in human fetal liver. Paed Res 24, 326–329.

Stahl, P.D. (1992). The mannose receptor and other macrophage lectins. Curr Opin Immunol 4, 49.

Uchida, Y., Takio, K., Titani, K, Ihara, Y., and Tomonaga, M. (1991). The growth inhibitory factor that is deficient in the Alzheimerís disease brain is a 68 amino acid metallothionein-like protein. Neuron 7, 337–347.

Udom, A.O. and Brady, F.O. (1980). Reactivation in vitro of zinc-requiring apo-enzymes by rat liver zinc-thionein. Biochem J 187, 329–335.

West, A.K., Stallings, R., Hildebrand, C.E., Chiu, R., Karin, M., and Richards, R.I. (1990). Human metal-lothionein genes: structure of the functional locus at 16q13. Genomics 8, 513–518.

Zalups, R.K., Fraser, J, and Koropatnick, J. (1995). Enhanced transcription of metallothionein genes in rat kidney: effect of uninephrectomy and compensatory renal growth. Am J Physiol (Renal) 37, F643-F650.

Zeng, J., Heuchel, R., Schaffner, W, and Kagi, J.H. (1991a). Thionein (apometallothionein) can modulate DNA binding and transcription activation by zinc finger containing factor Sp1. FEBS Lett 279, 310–312.

Zeng, J., Vallee B.L., and Kagi, J.H. (1991b). Zinc transfer from transcription factor IIIA fingers to thionein clusters. Proc Natl Acad Sci U S A 88, 9984–9988.

15

REACTIONS OF COPPER-ZINC SUPEROXIDE DISMUTASES WITH HYDROGEN PEROXIDE

Possible Involvement in Familial Amyotrophic Lateral Sclerosis

Joan Selverstone Valentine

Department of Chemistry and Biochemistry
UCLA
Los Angeles, California 90095-1569

1. COPPER-ZINC SUPEROXIDE DISMUTASE

1.1. Reaction of CuZnSOD with Superoxide

Copper-zinc superoxide dismutase (CuZnSOD) is a highly abundant and ubiquitous cytosolic 32 kDa homodimeric eukaryotic enzyme that catalyzes the disproportionation of superoxide to give dioxygen and hydrogen peroxide (reactions 1–3) (Fridovich, 1997).

$$O_2^- + Cu(II)ZnSOD \xrightarrow{k_1} O_2 + Cu(I)ZnSOD \tag{1}$$

$$O_2^- + 2H^+ + Cu(I)ZnSOD \xrightarrow{k_2} H_2O_2 + Cu(II)ZnSOD \tag{2}$$

$$\text{Sum: } 2O_2^- + 2H^+ \xrightarrow{CuZnSOD} O_2 + H_2O_2 \tag{3}$$

Each of the two subunits of the enzyme contains one zinc ion (Zn^{2+}) and one copper ion in close proximity (Cu^+ in the reduced and Cu^{2+} in the oxidized form of the enzyme). In the case of the oxidized enzyme, the copper(II) ion and the zinc ion are linked by a bridging imidazolate ligand from a histidyl side chain (His 63 in human CuZnSOD). When the copper center is reduced, the bond between copper and the bridging imidazolate is broken, and the imidazolate ring of that histidyl residue picks up a proton to become imidazole (Bertini et al., 1998).

Metals and Genetics, edited by Sarkar.
Kluwer Academic / Plenum Publishers, New York, 1999.

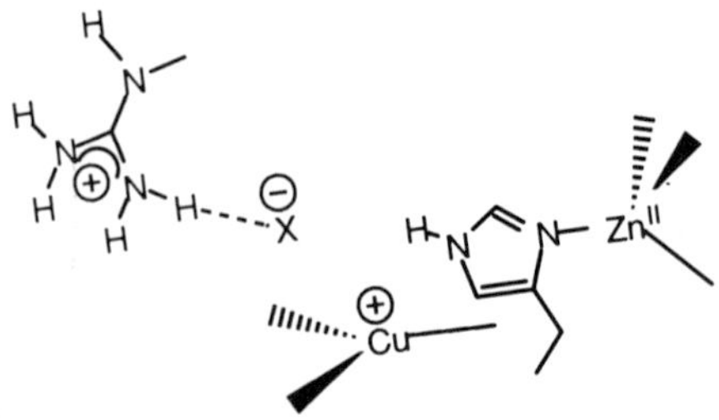

Scheme 1.

1.2. Reaction of CuZnSOD with Other Anions

The copper ion and the guanidinium side chain of a conserved arginine (Arg 143 in human CuZnSOD) in each subunit are located next to each another at the bottom of a funnel-shaped channel that narrows from a shallow depression approximately 24 Å across at the surface of the molecule to a deeper channel about 10 Å wide and finally to an opening of less than 4 Å just above the copper ion. This channel excludes most ligands from binding to the copper center, the exceptions being small anions such as fluoride, cyanide, thiocyanate, azide, and hydroxide which bind to the copper(II) form of the enzyme (Bertini et al., 1998). The copper(I) form of the enzyme also binds some of these anions, probably not directly to the copper ion but instead close to it and to Arg 143 (Leone et al., 1998).

The propensity of CuZnSOD to bind only small anions, rather than other larger ligands that often bind to copper centers, presumably accounts for its remarkable selective for superoxide both as a reducing agent of the oxidized Cu(II) form and as an oxidant of the reduced Cu(I) form of the enzyme. Accordingly, oxidized Cu(II)ZnSOD is reduced exceedingly slowly by ascorbate, normally a highly effective reducing agent for Cu^{2+} in metalloproteins. Reduced Cu(I)ZnSOD is also selectively and rapidly oxidized by superoxide and is oxidized only slowly by dioxygen (Cabelli et al., 1998).

1.3. Reaction of CuZnSOD with Hydrogen Peroxide

CuZnSOD also reacts with hydrogen peroxide. In the case of the wild type bovine enzyme, the reaction occurs exclusively via reaction with HO_2^-, presumably once again because of the selectivity of the enzyme for reaction with small anions (Cabelli et al., 1989). The reaction of HO_2^- with Cu(II)ZnSOD results in reduction of the enzyme (reaction 4). Subsequent reaction of reduced Cu(I)ZnSOD with hydrogen peroxide, in the ab-

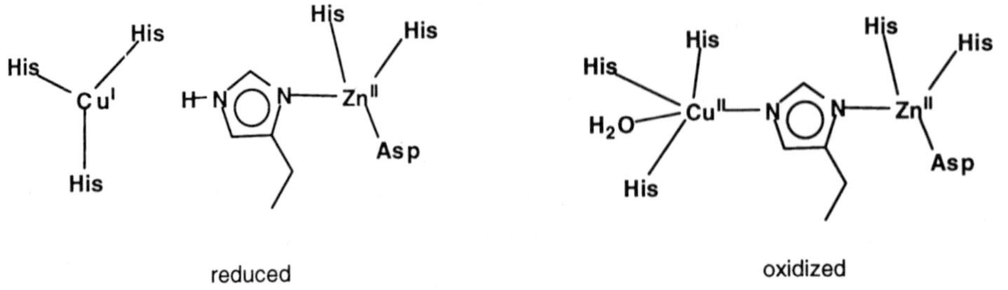

reduced   oxidized

Scheme 2.

sence of added substrates, results in enzyme inactivation (reactions 5 and 6), which has been attributed to oxidative destruction of one or more of the histidyl ligands to the copper ion (Hodgson and Fridovich, 1975; Hodgson and Fridovich, 1975). Alternatively, in the presence of certain substrates, such as formate, xanthine, urate, or certain spin traps, catalytic oxidation of substrates has been observed (reaction 6). The presence of such substrates has, in some cases, been demonstrated to slow down the inactivation process (Hodgson and Fridovich, 1975; Hodgson and Fridovich, 1975).

$$H_2O_2 + Cu(II)ZnSOD \rightarrow O_2^- + 2H^+ + Cu(I)ZnSOD \quad (4)$$

$$H_2O_2 + Cu(I)ZnSOD \rightarrow OH^- + (\bullet OH)Cu(II)ZnSOD \quad (5)$$

$$(\bullet OH)Cu(II)ZnSOD \rightarrow inactive \ CuZnSOD \quad (6)$$

$$(\bullet OH)Cu(II)ZnSOD + substrate \rightarrow Cu(II)ZnSOD + oxidized \ substrate \quad (7)$$

The mechanism proposed in reactions 5–7 for reactions of Cu(I)ZnSOD with hydrogen peroxide are examples of the Fenton reaction, i.e., reaction of a reduced metal center with hydrogen peroxide to generate the oxidized metal and hydroxyl radical, •OH, which then reacts further with a substrate. An aspect of this type of reaction that has been controversial is the degree of freedom of the hydroxyl radical when it is formed, i.e., whether or not it ever breaks free of the metal prior to reaction with substrate (reactions 8–10 versus reactions 8 plus 11).

$$H_2O_2 + M^{n+} \rightarrow OH^- + (\bullet OH)M^{(n-1)+} \quad (8)$$

$$(\bullet OH)M^{(n-1)+} \rightarrow \bullet OH + M^{(n-1)+} \quad (9)$$

$$\bullet OH + substrate \rightarrow oxidized \ substrate \quad (10)$$

$$(\bullet OH)M^{(n-1)+} + substrate \rightarrow M^{(n-1)+} + oxidized \ substrate \quad (11)$$

In the case of wild type CuZnSOD, substrates capable of oxidation by this mechanism are apparently restricted to those small enough to penetrate the channel. Such substrates appear to be reacting with hydroxyl radical either while it is still bound to copper or shortly after its release since such substrates are capable of intercepting hydroxyl radical before it reacts with imidazole rings bound to the copper center and thus inactivates the enzyme. In the case of FALS mutant CuZnSODs (and also possibly metal-substituted derivatives of wild type CuZnSOD with copper ions in the zinc sites) the protein may be more flexible and substrates may enter the substrate access channel more readily. These points are discussed further in section 3 below.

2. SOD-ASSOCIATED FAMILIAL AMYOTROPHIC LATERAL SCLEROSIS

Amyotrophic lateral sclerosis (ALS, Lou Gehrig's disease) is a neurodegenerative disease, with a mean age of onset of 55 years, that is characterized by the progressive loss of motor neurons in the spinal cord. The disease starts with weakness and leads relentlessly to paralysis and, ultimately, to death, usually within two to five years of the onset of symptoms. The disease is inherited in approximately 10 % of cases (familial ALS or FALS), and approximately one-fifth of FALS cases are associated with dominantly inherited, single-site mutations in *SOD1*, the gene that encodes human CuZnSOD (Brown, 1997).

The FALS mutations in CuZnSOD are dominant, and experimental results from several different laboratories have given strong support to a model in which they exert their effects by a gain of function. The strongest evidence came from transgenic mice overexpressing FALS mutant CuZnSODs which developed a motor neuron degenerative syndrome, despite normal or supranormal SOD enzymatic activities. By contrast, transgenic mice overexpressing wild type human as well as mice that lacked CuZnSOD did not develop similar symptoms (Gurney, 1997).

Oxidative damage mediated by copper is implicated in the deleterious effects of FALS mutant CuZnSODs based on the observation of beneficial effects of copper chelators both in a cell culture model (Wiedau-Pazos et al., 1996) and in the FALS SOD-expressing transgenic mice (Hottinger et al., 1997) and on an increasing body of evidence of oxidative damage to tissues in ALS. For example, markers of oxidative damage to nucleic acids, proteins, and lipids have been reported to be elevated in tissue samples from both sporadic ALS (SALS) and SOD-associated FALS patients (Ferrante et al., 1997), and studies of the transgenic mice expressing FALS mutant CuZnSODs have shown beneficial effects of the lipid antioxidant vitamin E (Gurney, 1997; Gurney et al., 1996). In addition, fibroblasts from patients with SOD-associated FALS were found to be more sensitive to hydrogen peroxide than SALS fibroblasts and than fibroblasts of controls (Aguirre et al., 1998).

3. NON-SPECIFIC PEROXIDATIVE ACTIVITY OF MUTANT COPPER-ZINC SUPEROXIDE DISMUTASE AS A POSSIBLE MECHANISM FOR FALS

3.1. EPR Spin Trapping as a Measure of Peroxidative Reactivity

Shortly after the publication of the discovery of the connection between mutations in CuZnSOD and FALS in 1993 (Rosen et al., 1993), the ALS Association sponsored a workshop in Boston whose purpose was to brainstorm the problem, looking for possible connections between CuZnSOD and ALS. During the course of the discussion, two of the CuZnSOD investigators at that workshop—Dr. Earl R. Stadtman, who had earlier pioneered the use of spin traps to probe the peroxidative activity of wild type CuZnSOD (Yim et al., 1990; Yim et al., 1993), and the present author—suggested the possibility that the FALS mutations in human CuZnSOD might result in enhanced peroxidative activity similar to that observed earlier in Dr. Stadtman's laboratory for the wild type bovine enzyme. Thus the hypothesis arose that cellular damage leading ultimately to the death of motor neurons might either directly or indirectly result from oxidation by hydrogen peroxide catalyzed by the copper ions bound to ALS-mutant CuZnSODs. The mechanism proposed is given in reactions 12–14.

$$X_{red} + Cu(II)ZnSOD \rightarrow X_{ox} + Cu(I)ZnSOD$$
$$(X_{red} = \text{cellular reducing agents, including superoxide}) \tag{12}$$

$$H_2O_2 + Cu(I)ZnSOD \rightarrow OH^- + Cu(II)ZnSOD + \bullet OH \text{ (free or copper-bound)} \tag{13}$$

$$\bullet OH \text{ (free or copper-bound)} + \text{substrate} \rightarrow \text{oxidized substrate} \qquad (14)$$

The original demonstration of spin trapping of hydroxyl radical in the reaction of wild type CuZnSOD with hydrogen peroxide was carried out using the bovine enzyme and the spin trap DMPO (5,5'-dimethyl-1-pyrroline N-oxide), and the product was demonstrated by EPR to be DMPO-OH. These experiments were repeated using FALS mutant human CuZnSODs both in Dr. Stadtman's laboratory (Yim et al., 1996; Yim et al., 1997) and our own (Wiedau-Pazos et al., 1996), and enhanced levels of DMPO-OH were observed in both cases.

In Dr. Stadtman's laboratory, the human CuZnSOD mutant enzymes A4V and G93A were overexpressed in Sf9 insect cells infected with recombinant baculovirus (Yim et al., 1996; Yim et al., 1997). The purified mutant CuZnSOD proteins were shown to be fully or nearly fully metallated with copper and zinc ions, and their SOD activities were shown to be identical to each other and to that of the wild type enzyme. However, the peroxidative activities of the mutant and wild type enzymes, as measured by the DMPO EPR spin trapping method, were not identical but fell in the order A4V > G93A > wild type. The enhancement in the case of the FALS mutant enzymes was found to be due to lower Km values for hydrogen peroxide (A4V (13 mM) < G93A (25 mM) < wild-type (44 mM)).

In our laboratory, we have also expressed and purified several FALS mutant human CuZnSODs and are likewise characterizing them in an effort to learn what new properties they have acquired as a consequence of the mutations that might explain their pathogenicity. We are particularly focusing on possible alterations in their metal binding abilities as well as on modifications of their catalytic properties that may have occurred as a consequence of the ALS-causing mutations. In our case, the expression system used for the FALS mutant human CuZnSOD proteins is *S. cerevisiae* that lack the native CuZnSOD (*sod1Δ*). Human wild type and FALS mutant CuZnSOD expressed in this system has been to observed to be properly N-acetylated despite the fact that wild-type yeast CuZnSOD is not similarly modified (Hallewell et al., 1987; Hart et al., 1998; Wiedau-Pazos et al., 1996) (Incidentally, human CuZnSOD expressed in *E. coli* is not N-acetylated.)

Our approach differed from that of Dr. Stadtman's laboratory in that the expressed wild type and FALS mutant human CuZnSOD proteins were converted to apoproteins and then remetallated with Cu^{2+} and Zn^{2+} ions prior to carrying out the EPR experiments with DMPO and hydrogen peroxide (Wiedau-Pazos et al., 1996). The remetallation procedures used were well known to reinsert copper and zinc ions into their proper locations in the wild type mammalian CuZnSOD apoproteins, but we now know that they are not necessarily reliable for the FALS mutant CuZnSOD apoproteins and can result in incorporation of copper in zinc sites and vice versa (Valentine, Goto, and Cabelli, unpublished results). The high SOD activities that were determined for remetallated A4V and G93A in that study provide evidence that the remetallated proteins contained properly remetallated copper sites, but we now know that there may have been significant occupancy of zinc sites by copper ions as well (Valentine, Goto, and Cabelli, unpublished results). Contamination of the zinc sites by copper ions would not have been readily apparent from the spectroscopic measurements that were carried out at that time. It is our current hypothesis that the higher signals of DMPO-OH observed in our experiments relative to those observed in Dr. Stadtman's laboratory may be due to contamination of the zinc sites of our mutant proteins with copper ions resulting in enhancement of the peroxidative activity of the copper ion in the copper site (see below for further discussion). (The possibility that our remetallation procedures caused the difference in the magnitude of the DMPO-OH signals in our laboratory and that of Dr. Stadtman was originally suggested as a note at the end of Yim et al., 1996.)

3.2. Enzyme Inactivation as a Measure of Peroxidative Reactivity

Another method to estimate the relative reactivities of CuZnSOD proteins with hydrogen peroxide is to determine the rate of SOD enzyme inactivation in the presence of hydrogen peroxide (reactions 4–6) (Hodgson and Fridovich, 1975; Hodgson and Fridovich, 1975). This approach has the advantage that the protein itself is the substrate for the oxidation reaction, uncomplicated by the question of substrate accessibility to the active site. Thus the relative rates of inactivation of CuZnSOD and its mutants by hydrogen peroxide are expected to reflect directly the relative ability of these enzymes to catalyze peroxidative reactions of hydrogen peroxide.

Professor Fridovich and co-workers recently reported the results of their studies of reactions of hydrogen peroxide with human FALS mutant CuZnSOD enzymes E100G, G93A, and G93R (which had been expressed in *E. coli*) and reported that no difference was found in the rates of SOD inactivation relative to the wild type enzyme (Liochev et al., 1998). We have also studied the hydrogen peroxide induced inactivation of wild type and G93A FALS mutant G93A enzymes, and we likewise observed that the rates of loss of activity in the presence of hydrogen peroxide were identical for the FALS mutant and the WT enzymes when they were in their CuZn forms (Goto et al., 1998). In addition, however, we carried out the same experiments using the remetallated CuCu derivatives of WT CuZnSOD and G93A, i.e., derivatives in which copper was present both in the copper and in the zinc sites of the enzymes. In these latter cases, the rates of inactivation by hydrogen peroxide were found to be substantially greater for the CuCu derivatives than in the case of the CuZn derivatives of either wild type or G93A (Goto et al., 1998).

In order to determine if the increase in deactivation rate constants for CuCuSOD relative to CuZnSOD was due to changes in the reactivity of copper at the copper site, we studied the stoichiometric reaction of hydrogen peroxide with wild type CuCuSOD and we found that the copper(II) ion in zinc site does not react rapidly enough with hydrogen peroxide to be the source of the enhanced peroxidative activity observed for wild type Cu-CuSOD relative to wild type CuZnSOD (Goto et al., 1998).

3.3. Conclusions

The difference between the EPR spin trap and SOD enzyme inactivation experiments is the difference between an using exogenous scavenger of hydroxyl radical and using the active site itself as a scavenger. It is important to note that an increase in DMPO-OH is not necessarily an increase in the rate constant for •OH radical production but may represent increased accessibility to the active site channel within the mutant enzymes. The increased accessibility may result from the mutant having a more flexible and thereby open active site channel as compared to the wild type protein. The substrate accessibility factor is eliminated when there is no exogenous substrate since the oxidized substrate in the inactivation studies is the SOD enzyme itself.

If abnormally peroxidative activity is indeed the culprit in SOD-associated FALS, oxidative damage to motor neurons may be a direct or indirect consequence of increased reactivity of the FALS mutant CuZnSOD enzymes with hydrogen peroxide. The enhanced level of DMPO-OH observed for the FALS mutant CuZnSOD proteins in their CuZnSOD forms suggests that the FALS mutant proteins provide more access for DMPO (and presumably other substrates) to enter the substrate access channel. This conclusion is also supported by the observation of evidence for increased flexibility of the protein structure in the crystal structure of G37R CuZnSOD (Hart et al., 1998). In addition, increased formation of •OH radical might be a consequence of the presence of some copper rather than zinc in the zinc site, leading to a sub-population of CuCuSOD which is more reactive. An intriguing additional possibility is that mutant SOD enzymes damaged by reaction with hydrogen peroxide could slowly release copper ions that when freed could catalyze deleterious Fenton-type oxidation reactions at or near the sites where they are released.

ACKNOWLEDGMENT

This work was supported by a grant from the National Institute of General Medical Sciences (GM28222).

REFERENCES

Aguirre, T., Van den Bosch, L., Goetschalckx, K., Tilkin, P., Mathijs, G., Cassiman, J. J., and Robberecht, W. (1998). Increased sensitivity of fibroblasts from amyotrophic lateral sclerosis patients to oxidative stress. Ann. Neurol. 43, 452–457.

Bertini, I., Mangar, S., and Viezzoli, M. S. (1998). Structure and properties of copper-zinc superoxide dismutases. Adv. Inorg. Chem. 45, 127 - 250.

Brown, R. H., Jr. (1997). Amyotrophic lateral sclerosis: Insights from genetics. Arch. Neurol. 54, 1246–1250.

Cabelli, D. E., Allen, D., Bielski, B. H., and Holcman, J. (1989). The interaction between Cu(I) superoxide dismutase and hydrogen peroxide. J. Biol. Chem. 264, 9967–9971.

Cabelli, D. E., Riley, D., Rodriguez, J. A., Valentine, J. S., and Zhu, H. (1998). Models of Superoxide Dismutases. In "Biomimetic Oxidations", B. Meunier, ed. (in press.)

Ferrante, R. J., Browne, S. E., Shinobu, L. A., Bowling, A. C., Baik, M. J., MacGarvey, U., Kowall, N. W., Brown, R. H., Jr., and Beal, M. F. (1997). Evidence of increased oxidative damage in both sporadic and familial amyotrophic lateral sclerosis. J. Neurochem. 69, 2064–2074.

Fridovich, I. (1997). Superoxide anion radical (O_2^{-}), superoxide dismutases, and related matters. J. Biol. Chem. 272, 18515–18517.

Goto, J. J., Gralla, E. B., Valentine, J. S., and Cabelli, D. E. (1998). Reactions of hydrogen peroxide with FALS mutant human copper-zinc superoxide dismutases studied by pulse radiolysis, submitted.

Gurney, M. E. (1997). Transgenic animal models of familial amyotrophic lateral sclerosis. J. Neurol. 244 Suppl 2, S15–20.

Gurney, M. E., Cutting, F. B., Zhai, P., Doble, A., Taylor, C. P., Andrus, P. K., and Hall, E. D. (1996). Benefit of vitamin E, riluzole, and gabapentin in a transgenic model of familial amyotrophic lateral sclerosis. Ann. Neurol. 39, 147–157.

Hallewell, R. A., Mills, R., Tekamp-Olson, P., Blacher, R., Rosenberg, S., Otting, F., Masiarz, F. R., and Scandella, C. J. (1987). Amino terminal acylation of authentic human Cu,Zn superoxide dismutase produced in yeast. Bio/Technology 5, 363–366.

Hart, P. J., Liu, H., Pellegrini, M., Nersissian, A. M., Gralla, E. B., Valentine, J. S., and Eisenberg, D. (1998). Subunit asymmetry in the three-dimensional structure of a human CuZnSOD mutant found in familial amyotrophic lateral sclerosis. Protein Sci. 7, 545–555.

Hodgson, E. K., and Fridovich, I. (1975). The interaction of bovine erythrocyte superoxide dismutase with hydrogen peroxide: Inactivation of the enzyme. Biochemistry 14, 5294–5299.

Hodgson, E. K., and Fridovich, I. (1975). The interaction of bovine erythrocyte superoxide dismutase with hydrogen peroxide: Chemiluminescence and peroxidation. Biochemistry 14, 5299–5303.

Hottinger, A. F., Fine, E. G., Gurney, M. E., Zurn, A. D., and Aebischer, P. (1997). The copper chelator d-penicillamine delays onset of disease and extends survival in a transgenic mouse model of familial amyotrophic lateral sclerosis. Eur. J. Neurosci. 9, 1548–1551.

Leone, M., Cupane, A., Militello, V., Stroppolo, M. E., and Desideri, A. (1998). Fourier transform infrared analysis of the interaction of azide with the active site of oxidized and reduced bovine Cu,Zn superoxide dismutase. Biochemistry 37, 4459–4464.

Liochev, S. I., Chen, L. L., Hallewell, R. A., and Fridovich, I. (1998). The familial amyotrophic lateral sclerosis-associated amino acid substitutions E100G, G93A, and G93R do not influence the rate of inactivation of copper- and zinc-containing superoxide dismutase by H_2O_2. Arch. Biochem. Biophys. 352, 237–239.

Rosen, D. R., Siddique, T., Patterson, D., Figlewicz, D. A., Sapp, P., Hentati, A., Donaldson, D., Goto, J., O'Regan, J. P., Deng, H.-X., Rahmani, Z., Krizus, A., McKenna-Yasek, D., Cayabyab, A., Gaston, S. M., Berger, R., Tanzi, R. E., Halperin, J. H., Herzfeldt, B., van den Bergh, R., Hung, W.-Y., Bird, T., Deng, G., Mulder, D. W., Smyth, C., Laing, N. G., Soriano, E., Pericak-Vance, M. A., Haines, J., Rouleau, G. A., Gusella, J. S., Horvitz, H. R., and Brown, R. H. , Jr. (1993). Mutations in Cu/Zn superoxide dismutase gene are associated with familial amyotrophic lateral sclerosis. Nature *362*, 59–62.

Wiedau-Pazos, M., Goto, J. J., Rabizadeh, S., Gralla, E. B., Roe, J. A., Lee, M. K., Valentine, J. S., and Bredesen, D. E. (1996). Altered reactivity of superoxide dismutase in familial amyotrophic lateral sclerosis. Science *271*, 515–518.

Yim, M. B., Chock, P. B., and Stadtman, E. R. (1990). Copper, zinc superoxide dismutase catalyzes hydroxyl radical production from hydrogen peroxide. Proc. Natl. Acad. Sci. USA *87*, 5006–5010.

Yim, M. B., Chock, P. B., and Stadtman, E. R. (1993). Enzyme function of copper, zinc superoxide dismutase as a free radical generator. J. Biol. Chem. *268*, 4099–4105.

Yim, M. B., Kang, J.-H., Yim, H.-S., Kwak, H.-S., Chock, P. B., and Stadtman, E. R. (1996). A gain-of-function of an amyotrophic lateral sclerosis-associated Cu,Zn-superoxide dismutase mutant: An enhancement of free radical formation due to a decrease in Km for hydrogen peroxide. Proc. Natl. Acad. Sci. USA *93*, 5709–5714.

Yim, H.-S., Kang, J.-H., Chock, P. B., Stadtman, E. R., and Yim, M. B. (1997). A familial amyotrophic lateral sclerosis-associated A4V Cu, Zn-superoxide dismutase mutant has a lower K_m for hydrogen peroxide. Correlation between clinical severity and the K_m value. J. Biol. Chem. *272*, 8861–8863.

16

YEAST MITOCHONDRIAL IRON METABOLISM AND THE PATHOPHYSIOLOGY OF FRIEDREICH'S ATAXIA

Jerry Kaplan, Derek Radisky, and Michael Babcock

Department of Pathology School of Medicine
University of Utah
Salt Lake City Utah, 84132

Iron is an essential element required for both redox reactions and for the synthesis of heme. The same properties of iron that render it so useful biochemically- the ease at which it can gain or lose electrons- also makes iron potentially toxic. The toxicity of iron results from its ability to reduce molecular oxygen. A complete reduction of molecular oxygen results in the formation of H_2O which is not dangerous. Incomplete reduction of oxygen leads to the formation of either superoxide anion ($O2^-$) or the hydroxyl radical (OH), both of which are extremely toxic (Spiro, 1974). Cells employ numerous strategies to protect themselves from iron induced oxidant damage. Enzymes, such as superoxide dismutase and catalase, can deactivate oxygen radicals. Cells can store intracellular iron in a relatively safe form within ferritin. The major mechanism, however, that cells employ to protect themselves against iron mediated damage is to regulate iron accumulation. Because iron is so important most cells have multiple transport systems capable of mediating transmembrane iron transport. All iron transport systems studied to date are under tight regulation that is responsive to intracellular iron demand. Studies in E. coli, yeast, and humans have resulted in a detailed understanding of the iron dependent regulation at the molecular level. Based on these studies the pathophysiology of defects resulting from malregulation of transport, either deficiency or excess are readily understandable.

Our knowledge of the regulation of accumulation at the level of the plasma membrane is so exquisite that it exacerbates how little we know about intracellular iron transport. Iron containing proteins are distributed throughout the cell and may be found as integral endoplasmic reticulum proteins, lysosomal proteins and soluble enzymes. We have little knowledge regarding how iron is transported to the sites of usage and what factors regulate that transport. This is particularly true for the mitochondria, the intracellular organelle that requires the most iron. The identity and regulation of mitochondrial iron transporters is unkown. Regulation of iron entry into the mitochondria must be tightly

Metals and Genetics, edited by Sarkar.
Kluwer Academic / Plenum Publishers, New York, 1999.

controlled for several reasons. The first of which is that the mitochondria is the major site of oxygen reduction in cells as oxygen is the terminal acceptor of electrons generated during the tricarboxylic acid cycle. It has been calculated that up to 10% of the consumed oxygen molecules is reduced to either superoxide anion or H_2O_2. Both oxygenated species can, particularly in the presence of iron, be transformed into OH radicals.

The second reason that mitochondrial iron transport must be controlled is that iron, along with protoprophyrin IX, is a substrate for heme. The initial and final steps in the porphryin biosynthetic pathway are localized within the mitochondria while a number of intermediate steps occur in the cytosol (Conder et al., 1991; Jordan, 1994). The insertion of iron into the protoporphyrin IX macronucleus occurs in the mitochondria and is mediated by the enzyme ferrochelatase. This enzyme is embedded in the mitochondrial matrix and relies on an iron sulfur cluster for its enzymatic activity (Dailey et al., 1994; Ferreira et al., 1994). Thus, the final step in heme synthesis requires iron transport into the mitochondria, which also is the site of oxygen radical generation. That iron transport occurs in this potentially dangerous environment implies that mitochondrial iron accumulation must be a highly regulated process. Studies on plasma membrane iron transport, another highly regulated process, have shown that defects in regulation results in pathophysiological states. Either iron excess or deprivation results in disease states. It might be expected that a similar circumstances would apply to mitochondria; iron excess or depletion may result in mitochondrial dysfunction leading to disease. It is also expected that mitochondrial iron transport must be regulated in response to demand. The rate of heme biosynthesis varies widely among cell types. As the synthesis of porphyrins is highly regulated, usually by regulating substrate flow, it is expected that the rate of iron transport should be regulated in response to porphyrin synthesis.

Until recently our knowledge of mitochondrial iron metabolism was restricted to an identification of human disease states that was manifested by an increase in mitochondrial iron deposition. The sideroblastic anemia's show excessive deposition of iron within the mitochondria of developing erythrocytes (Fitzsimons and May, 1996). Other tissues do not show the phenotype of iron deposition. The biochemical basis of two disorders of sideroblastic anemia is known. The genetic form is due to a defect in the erythrocyte specific amino levulinic acid (ALA) synthase (Fitzsimons and May, 1996). The rate of protoporphyrin IX synthesis can not match the rate of iron accumulation resulting in the deposition of iron with mitochondria. A sporadic form of the disorder has been identified as a defect in cytochrome C oxidase (Gattermann et al., 1997). Although it is not clear how this defect results in iron accumulation. Based on analysis of the genetic form of sideroblastic anemia and on studies that utilize inhibitors of heme synthesis it was suggested that iron can only exit mitochondria as heme. Thus, an inhibition of heme synthesis would result in iron deposition due to a lack of iron transport out of mitochondria (Ponka, 1997). This conclusion implies that iron transport was regulated at the site of iron uptake into mitochondria. The explanation for the restriction of the phenotype of sideroblasts to reticulocyte mitochondria was that the rate of mitochondrial iron uptake was largest in this cell type. Recent studies in a simple eucaryote, the budding yeast *Saccharomyces cerevisiae*, have suggested a different view of mitochondrial iron metabolism. Mitochondrial iron transport may be regulated at the level of iron exit, and that defects in iron egress can lead to mitochondrial iron accumulation and human disease (Babcock et al., 1997).

Saccharomyces cerevisiae has long been a useful model system to study biological process as diverse as metabolism and genetics. In the past few years it has been recognized that *S. cerevisiae* is also a model system to study the genetics and biochemistry of human disease. This realization resulted from the analysis of both the complete sequence of the yeast genome and of the sequence of human expressed sequence tags (EST). Depending upon the level of stringency, up to 34 percent of yeast genes have human homologues (Botstein et al., 1997). In some instances the human homologues are identical to the yeast gene and can functionally replace the yeast gene. In instances where the genes are not identical often there is enough sequence similarity between the two that the function of human genes can be inferred from analysis of the yeast gene. The homology between yeast and human is highest in basic biochemical process that are common to all species.

Our laboratory and others have exploited the molecular biology and genetics of yeast to identify genes involved in iron metabolism. Over the past few years studies have elucidated in great detail the process of plasma membrane transport. These studies have not only identified plasma membrane transporters but have also identified genes that are involved in the assembly of plasma membrane transporters. The basis of the approach is to devise a selection system to identify mutants defective in iron metabolism (Klausner and Dancis, 1994) (Askwith et al., 1996). The genes that are responsible for the mutants can be cloned by complementation of the phenotype using a genomic library. One of the advantages of yeast is that most yeast genes do not have introns so that a genomic library can be used for complementation studies. Genes can complement because they are either allelic to the mutated gene or they act as high copy suppressers. High copy suppressers are usually found when the genomic library is made using a high copy plasmid as opposed to an integrative or a centromeric plasmid, both of which are in low copy. There are rigorous genetic tests which can distinguish between a complementing gene functioning as a high copy suppresser. High copy suppressers may be informative as they reveal process related to but not specific to the original defect.

Insight into yeast mitochondrial iron metabolism was recently derived from analysis of a high copy suppresser. Our laboratory had identified a collection of mutants which were defective for growth on low but not high iron medium (Askwith et al., 1996). Most of the mutants were defective in either the high affinity iron transport system or in genes involved in the assembly of the high affinity iron transport system (Askwith and Kaplan, 1998). Mutants were also identified in genes that encode iron containing enzymes (Li and Kaplan, 1996). Using this screen we identified mutants and subsequently genes whose function still remains unknown. One such gene is referred to as BM-8. The original mutant was identified because it showed poor growth on low iron medium. The gene responsible for the defect was identified through complementation of the phenotype using a genomic library. The gene was proven to be allelic by genetic criteria and a yeast strain with a deletion was shown to be non-viable indicating that the gene product was essential. The sequence of the gene gave no clues as to its function and we still do not know what the gene does.

The same genomic library, however, also contained a plasmid that could weakly complement the phenotype. This complementation was only seen when a high copy plasmid was used, as the same genomic insert on a low copy plasmid gave no complementation. This and other observations indicated that the gene responsible for the complementation was not allelic to the defect and was acting as a high copy suppresser (Babcock et al., 1997). Since the entire yeast genome was sequenced, just a small amount of sequence of the genomic insert was required to identify the genes carried by the plasmid. Subcloning resulted in the identification of the complementing open reading frame. Only one sequence was discovered when the reading frame was used to probe the data base of gene sequences looking for homologues genes. The yeast gene was discovered to be highly homologous to a recently identified gene responsible for the lethal disease Friedreich's Ataxia.

Friedreich's Ataxia (FA) is the most common of the genetic ataxias. The clinical manifestation of ataxia is loss of gait and motor coordination due to the absence of sensory neurons. It is a lethal disorder with an age of onset usually in the early teens. The mean age at death is 37. While the disease presents as an ataxia the cause of death is most invariably a cardiac myopathy resulting in heart failure. The disease gene was identified by a strict positional cloning approach (Campuzano et al., 1996). Most of the mutations were found to be a result of a triplet, GAA, expansion. What differentiates FA from other neurological diseases due to triplet expansions is that most expansions occur in exons while in FA the expansion occurs in an intron. The size of the expansion is correlated with the severity of the disorder (Campuzano et al., 1997). Apparently the expansion inhibits transcription resulting in a reduced amount of mRNA. The greater the expansion the lower the concentration of functional mRNA. All individuals showing the expansions have some functional mRNA, suggesting that a null mutation would be lethal. The highest expression of the gene occurs in those tissues which are most affected, spinal cord and heart. While the sequence of the gene revealed much about the molecular basis of the defect, it provided little information about the normal function of the gene or of the pathophysiology of the disorder. Because we had identified the yeast homologue of the gene, termed YFH1 (*Yeast Homologue of Frataxin*) we examined its effect on cellular function particularly in regards to iron metabolism.

A strain with a deletion in the *YFH1* gene was generated by homologous recombination (Babcock et al., 1997). The strain showed poor growth in glucose containing medium and no growth in medium which contained glycerol/ethanol as carbon sources. Yeast requires an intact respiratory system to grow on glycerol/ethanol, whereas growth on glucose can occur simply by glycolysis. The lack of growth on glycerol indicates that the deletion strain (Δ*yfh*) had defective mitochondrial function. A diploid between wild type cells (*YFH1*) and the deletion strain (Δ*yfh1*) could grow on glycerol ethanol. When the strain was sporulated (the diploid results in four haploids) two of the four haploids were unable to grow on glycerol ethanol. This result suggested that the defective gene acted as a Mendelian recessive. Transformation of the deletion stain with a plasmid containing the normal gene resulted in an improvement of growth on glucose medium but did not result in growth in glycerol ethanol. This result suggested that *YFH1* was responsible for some facet of mitochondrial integrity. The absence of the gene product resulted in the loss of respiratory competency due to a mutations in the mitochondrial genome. A diploid between a deletion strain and a wild type strain could survive for two reasons. The wild type strain not only provided a good copy of the nuclear *YFH1* gene, but also provided a good copy of the mitochondrial genome.

Two genetic experiments confirmed this secenerio. First, a diploid of Δ*yfh1* and a mitochondrial tester strain could not grow on glycerol ethanol. Mitochondrial tester strains are cells that have an intact nuclear genome but are missing a mitochondrial genome. The second genetic experiment involved transforming a diploid that carried a deleted *YFH1* gene on one chromosome, and a functional *YFH1* gene on the other chromosome, with a plasmid that contained a functional *YFH1* gene. When the diploid was sporulated all of the haploids were able to grow on glycerol ethanol. Half of the haploids could grow because they contained a functional chromosomal copy of *YFH1* gene. The other half could grow because they contained a functional YFH1 gene carried on the plasmid. When cells containing the chromosomal deletion were forced to lose the plasmid they were no longer able to grow on glycerol ethanol. Cells that contained the chromosomal copy of the gene could still grow on that medium. This result formally proved that *YFH1* was a nuclear gene that when defective resulted in loss of mitochondrial function. Further studies demonstrated deletions in the yeast mitochondrial genome in strains with a deleted *YFH1* (Koutnikova et al., 1997; Wilson and Roof, 1997).

Evidence that *YFH1* had a function related to mitochondria was strengthened by examination of the subcellular distribution of the expressed protein. A fusion protein was engineered between *YFH1* and the Green Fluorescent protein (GFP). The cDNA for the GFP was fused to the 3' end of the *YFH1* gene which generated a fusion protein that had the amino terminus of *YFH1* and the carboxyl terminus of GFP. This protein was localized to mitochondrial by fluorescence microscopy using an antibody to the mitochondrial protein porin as a reference(Babcock et al., 1997). When GFP is expressed by itself it is uniformly distributed in the cytosol. Further expression of GFP as the amino terminus of the fusion protein also resulted in a cytosolic distribution of the protein. These observations demonstrate that Yfh1p is a mitochondrial protein whose amino terminus is required for mitochondrial localization.

Since overexpressed *YFH1* had an effect on iron metabolism- high copy suppression of an iron dependent gene- we examined iron metabolism in the Δ*yfh1* strain. Atomic absorption spectroscopy revealed that Δ*yfh1* cells demonstrated a two fold increase in cellular iron content relative to wild type cells. When the cells were subfractionated it was discovered that mitochondria showed a 10–15 fold increase in iron content relative to wild type mitochondria. Measurements of cellular iron transport using ^{59}Fe demonstrated that the Δ*yfh1* strain showed a constitutively high rate of iron transport. The high affinity iron transport system is comprised of an oxidase and a permease, the products of the *FET3* and *FTR1* genes. Transcription of these genes is highly iron regulated. In the presence of iron, as is found in iron rich medium, transcription is repressed. Whereas iron starvation results in the transcription of these two genes. Wild type cells show little expression of the high affinity iron transport system in rich medium, whereas mutant cells show both high levels of transcription and high levels of iron transport. The transcriptional regulation of *FET3* and *FTR1* is mediated through the product of the *AFT1* gene (Yamaguchi-Iwai et al., 1995). Deletion of this gene results in the absence of transcription. A gain of function mutant of *AFT1* was identified that resulted in constitutive transcription of the high affinity iron transport system. This mutant *AFT1*^{up} leads to cellular iron levels that are as high as in the Δ*yfh1* strain. Mitochondria isolated from the iron-loaded *AFT1*^{up} cells shows no increased iron content. This result demonstrates that simply increasing cellular iron content does not result in increased mitochondrial iron. The increase in mitochondrial iron must result from the absence of specific processes.

The presence of high mitochondrial iron suggests a mechanism for pathogenicity, as iron in the presence of oxygen can be toxic. The potential toxicity was demonstrated by examining cell growth in H_2O_2 containing medium. Cells with a deletion in the *YFH1* gene were incapable of growth in 0.02% H_2O_2, while both wild type cells and *AFT1*^{up} cells grew perfectly well. Only at higher concentration of H_2O_2 did *AFT1*^{up} cells show a growth deficit. This result suggests that increased mitochondrial iron can generate toxic radicals from H_2O_2. That the increase in mitochondrial iron lead to the growth deficit was shown by use of genetic constructs. A diploid, in which one chromosome contained a deleted *YFH1* gene was transformed with a plasmid that had the *YFH1* gene under the control of the methionine (met) promoter, which confers inducibilty on the genes that it regulates. In the absence of methionine the gene is transcribed, and transcription is repressed in the presence of methionine. When the diploids are sporulated in the absence of methionine, the haploids containing the deleted chromosomal copy of *YFH1* can grow on glycerol ethanol because of expression of the plasmid copy of *YFH1*. As long as *YFH1* is expressed cells are capable of growing on glycerol ethanol regardless of medium iron content. In the presence of methionine *YFH1* gene transcription is inhibited and within two hours there is an induction of the high affinity iron transport system resulting in increased rates of iron uptake. There is also a time dependent loss or respiratory competence that is dependent on the concentration of medium iron. The higher the iron content the greater the percentage of cells that are respiratory defective. This result demonstrates that mitochondrial iron is causal to the respiratory defect rather than a secondary response to defective respiration.

The observation that a deletion of *YFH1* results in an increase in mitochondrial iron suggests that the gene product either must regulate iron entry into mitochondria or regulate iron export. These two alternatives were distinguished experimentally through use of the methionine regulated gene construct. In the presence of methionine iron accumulates in mitochondria, when methionine is removed the accumulated iron is rapidly lost. The presence of the gene results in the *loss* of iron, suggesting that the gene product is involved in mitochondrial iron export. Sequence analysis of the *YFH1* gene indicates that it encodes a small soluble protein. It does not show transmembrane motifs indicating that it is not a transporter, and thus it probably regulates a transporter. A potential candidate for a transporter may be the *ATM1* gene that encodes a mitochondrial ABC transporter (Kispal et al., 1997). This protein shows 12 transmembrane domains and an ATP binding site. Strains which show a deletion in *ATM1* also show increased mitochondrial iron accumulation which may even exceed the levels seen in $_\Delta$YFh1 cells. While *ATM1* may be an iron exporter, genetic or biochemical experiments have not yet demonstrated an interaction between the two gene products.

These studies suggest an explanation for the phenotype seen when *YFH1* is deleted in yeast cells. The *YFH1* gene regulates mitochondria iron export. In the absence of the gene product iron accumulates in mitochondria because it can not be exported. The accumulation of iron occurs at the expense of cytosolic iron, lowering the concentration of cytosolic iron. This decrease in cytosolic iron results in the loss of iron from the transcription factor Aft1p resulting in transcription of the high affinity iron transport genes *FET3* and *FTR1*. The increased concentration of these gene products results in increased iron transport bringing more iron into the cell. This iron however is transported into mitochondria where it accumulates thus never increasing cytosolic iron levels. Iron within the mitochondria interacts with H_2O_2 formed as a result of respiratory activity and generates toxic radicals. The radicals oxidize proteins, lipids, and DNA leading to the generation of mutations in the mitochondrial genome. Yeast can dispense with respiration and survive solely on glycolysis. Yeast with mutations in respiration can grow because the lack of respiration will prevent further generation of toxic radicals even in the face of an excessive mitochondrial iron accumulation.

The explanation for the $\Delta yfh1$ phenotype in yeast can be extended to human cells that show deficits in Frataxin, with one notable exception. Mammalian cells can not survive solely on glycolysis as they require some level of respiration. The tissues that show the greatest clinical effect of the Frataxin defect are those that have the highest requirement for respiration. Greater than 70% of cardiac ATP levels result from the mitochondrial oxidation of fatty acids. In mammalian systems defective mitochondrial respiration results in apoptosis and cell death. This increased cell death may explain the loss of neurons and the cardiac hypertrophy seen in FA.

There are several pieces of data that suggest that Frataxin plays a role in humans similar to the role of Yfh1 in yeast. First, Frataxin encodes a mitochondrial protein (Babcock et al., 1997; Campuzano et al., 1997). The mitochondrial localization of Frataxin has been demonstrated using both antibodies and GFP fusion proteins. Second, heart biopsies from patients with FA show lowered activities for mitochondrial enzymes that rely on iron-sulfur clusters, as iron-sulfur cluster enzymes are highly susceptible to oxidative damage (Rotig et al., 1997). Third, heart biopsies show increased iron content as revealed by the cytochemical Pearl stain (Lamarche et al., 1980). In total, these results support the view that FA is a mitochondrial defect. A critical experiment to prove that the pathophysiology in FA is similar to that described for yeast model requires the demonstration of excessive iron within mitochondria isolated from FA patients. In the absence of that demonstration the model while strong is still speculative. The model, however, is robust as it provides for definitive experiments that can test its validity. The studies on *YFH1* demonstrate that the transport of iron into subcellular organelles is similar to iron transport across the plasma membrane. Both processes are regulated and malregulation results in dysfunction and in human disease.

REFERENCES

Askwith, C., and Kaplan, J. (1998). Iron and copper transport in yeast and its relevance to human disease [In Process Citation]. Trends Biochem Sci *23*, 135–8.

Askwith, C. C., de Silva, D., and Kaplan, J. (1996). Molecular biology of iron acquisition in Saccharomyces cerevisiae. Mol Microbiol *20*, 27–34.

Babcock, M., de Silva, D., Oaks, R., Davis-Kaplan, S., Jiralerspong, S., Montermini, L., Pandolfo, M., and Kaplan, J. (1997). Regulation of mitochondrial iron accumulation by Yfh1p, a putative homolog of frataxin. Science *276*, 1709–12.

Botstein, D., Chervitz, S. A., and Cherry, J. M. (1997). Yeast as a model organism [comment]. Science *277*, 1259–60.

Campuzano, V., Montermini, L., Lutz, Y., Cova, L., Hindelang, C., Jiralerspong, S., Trottier, Y., Kish, S. J., Faucheux, B., Trouillas, P., Authier, F. J., Durr, A., Mandel, J. L., Vescovi, A., Pandolfo, M., and Koenig, M. (1997). Frataxin is reduced in Friedreich ataxia patients and is associated with mitochondrial membranes. Hum Mol Genet *6*, 1771–80.

Campuzano, V., Montermini, L., Molto, M. D., Pianese, L., Cossee, M., Cavalcanti, F., Monros, E., Rodius, F., Duclos, F., Monticelli, A., and et al. (1996). Friedreich's ataxia: autosomal recessive disease caused by an intronic GAA triplet repeat expansion [see comments]. Science *271*, 1423–7.

Conder, L. H., Woodard, S. I., and Dailey, H. A. (1991). Multiple mechanisms for the regulation of haem synthesis during erythroid cell differentiation. Possible role for coproporphyrinogen oxidase. Biochem J *275*, 321–6.

Dailey, H. A., Finnegan, M. G., and Johnson, M. K. (1994). Human ferrochelatase is an iron-sulfur protein. Biochemistry *33*, 403–7.

Ferreira, G. C., Franco, R., Lloyd, S. G., Pereira, A. S., Moura, I., Moura, J. J., and Huynh, B. H. (1994). Mammalian ferrochelatase, a new addition to the metalloenzyme family. J Biol Chem *269*, 7062–5.

Fitzsimons, E. J., and May, A. (1996). The molecular basis of the sideroblastic anemias. Curr Opin Hematol *3*, 167–72.

Gattermann, N., Retzlaff, S., Wang, Y. L., Hofhaus, G., Heinisch, J., Aul, C., and Schneider, W. (1997). Heteroplasmic point mutations of mitochondrial DNA affecting subunit I of cytochrome c oxidase in two patients with acquired idiopathic sideroblastic anemia. Blood *90*, 4961–72.

Jordan, P. M. (1994). Highlights in haem biosynthesis. Curr Opin Struct Biol *4*, 902–11.

Kispal, G., Csere, P., Guiard, B., and Lill, R. (1997). The ABC transporter Atm1p is required for mitochondrial iron homeostasis. FEBS Lett *418*, 346–50.

Klausner, R. D., and Dancis, A. (1994). A genetic approach to elucidating eukaryotic iron metabolism. FEBS Lett *355*, 109–13.

Koutnikova, H., Campuzano, V., Foury, F., Dolle, P., Cazzalini, O., and Koenig, M. (1997). Studies of human, mouse and yeast homologues indicate a mitochondrial function for frataxin. Nat Genet *16*, 345–51.

Lamarche, J. B., Cote, M., and Lemieux, B. (1980). The cardiomyopathy of Friedreich's ataxia morphological observations in 3 cases. Can J Neurol Sci *7*, 389–96.

Li, L., and Kaplan, J. (1996). Characterization of yeast methyl sterol oxidase (ERG25) and identification of a human homologue. J Biol Chem *271*, 16927–33.

Ponka, P. (1997). Tissue-specific regulation of iron metabolism and heme synthesis: distinct control mechanisms in erythroid cells [see comments]. Blood *89*, 1–25.

Rotig, A., de Lonlay, P., Chretien, D., Foury, F., Koenig, M., Sidi, D., Munnich, A., and Rustin, P. (1997). Aconitase and mitochondrial iron-sulphur protein deficiency in Friedreich ataxia. Nat Genet *17*, 215–7.

Spiro, T. G. and. P. Saltman. (1974). Inorganic Chemistry. In Iron in Biochemistry and Medicine, M. A. Jacobs and Worwood, ed. (New York: Academic Press), pp. 1–28.

Wilson, R. B., and Roof, D. M. (1997). Respiratory deficiency due to loss of mitochondrial DNA in yeast lacking the frataxin homologue. Nat Genet *16*, 352–7.

Yamaguchi-Iwai, Y., Dancis, A., and Klausner, R. D. (1995). AFT1: a mediator of iron regulated transcriptional control in Saccharomyces cerevisiae. Embo J *14*, 1231–9.

17

GETTING COPPER INTO MITOCHONDRIA[*]

D. M. Glerum,[1] J. Beers,[2] A. Tzagoloff,[2] F. Punter,[1] and D. Adams[1]

[1]Department of Medical Genetics
University of Alberta
Edmonton, Alberta, T6G 2H7, Canada
[2]Department of Biological Sciences
Columbia University
New York, New York, 10027

1. INTRODUCTION

Defects in mitochondrial function have been implicated in an ever-widening array of neurodegenerative diseases. In particular, the increasing number of disorders in which defects in oxidative phosphorylation (OXPHOS) have been identified suggests that these may form one of the most commonly encountered classes of degenerative diseases (Shoffner and Wallace, 1995). OXPHOS couples the energy generated from electron transport through the respiratory chain to the generation of adenosine triphosphate (ATP). Disorders with specific deficiencies in cytochrome oxidase (COX) account for the largest proportion of defects associated with the respiratory chain (DiMauro *et al.*, 1988; Glerum *et al.*, 1989; Rahman *et al.*, 1996). The existence of COX deficiencies is well-established, with the first reports occurring well over a decade ago (DiMauro *et al.*, 1983; Miyabayashi *et al.*, 1983; Robinson *et al.*, 1987). Despite intensive efforts, however, the causative mutations underlying these diseases have remained elusive.

In our efforts to understand the molecular bases for COX deficiencies, we have been able to exploit the homologies that exist between the cytochrome oxidases in the yeast, *Saccharomyces cerevisiae*, and in humans. In so doing, we have not only begun to answer key questions regarding the biogenesis of cytochrome oxidase, but have also identified unique components of the intracellular copper transport pathway. Our efforts should also prove fruitful in furthering our understanding of human COX deficiencies.

---

[*]Abbreviations used: COX: cytochrome oxidase; imm: inner mitochondrial membrane; ims: mitochondrial intermembrane space; kDa: kilodalton; MnSOD: mangano-superoxide dismutase; mtDNA: mitochondrial DNA; omm: outer mitochondrial membrane; OXPHOS: oxidative phosphorylation; ROS: reactive oxygen species.

Metals and Genetics, edited by Sarkar.
Kluwer Academic / Plenum Publishers, New York, 1999.

1.1. Cytochrome Oxidase: Structure and Function

Cytochrome oxidase is the only component internal to mitochondria that is known to contain copper. This enzyme, which is essential for aerobic life, acts at the last step in the pathway which couples oxidation and phosphorylation to generate cellular energy. It catalyzes the reduction of molecular oxygen to water and concomitantly pumps protons across the inner mitochondrial membrane, thus contributing to the membrane potential which is used to drive the synthesis of ATP (Capaldi, 1990). The eukaryotic enzyme consists of 13 subunits (LaMarche *et al.*, 1992; Taanman and Capaldi, 1992, 1993). The three largest are encoded in the mtDNA and form the catalytic core of the enzyme. Subunits 1 and 2 contain all the redox-active prosthetic groups (Saraste, 1990; Tsukihara *et al.*, 1995), which consist of two heme A molecules and three copper atoms. The remaining 10 subunits are encoded in the nucleus, translated on cytoplasmic ribosomes and imported into mitochondria.

The three mitochondrially encoded subunits are homologous to the three subunits of the enzyme found in most aerobic prokaryotes. Models for cytochrome oxidase function have been largely based on experiments carried out with the prokaryotic enzyme (Calhoun *et al.*, 1994; Cooper, 1990). X-ray crystallographic studies have now yielded detailed structural information about both the bacterial and eukaryotic forms of the enzyme (Iwata *et al.*, 1995; Tsukihara *et al.*, 1996). The three copper atoms, which are essential for assembly and function of the enzyme in both prokaryotes and eukaryotes, are found at two distinct locations. Two copper atoms form a binuclear centre close to the molecular surface of subunit 2 (Tsukihara *et al.*, 1995), facing into the intermembrane space of the mitochondria. This site is in close proximity to the cytochrome *c* binding site. The third copper atom is found deep inside cytochrome oxidase, on subunit 1, where, together with one molecule of heme A, it forms the oxygen binding and reduction site (Tsukihara *et al.*, 1995).

In spite of significant advances in our understanding of the structure and function of cytochrome oxidase, the process by which this multimeric enzyme is assembled remains poorly understood. The complexity of cytochrome oxidase biogenesis stems from several sources. Firstly, the process requires products from both the nuclear and mitochondrial genomes. Secondly, there are a large number of nuclear gene products involved in this process that are not part of the enzyme's final structure. Thirdly, the synthesis and recruitment of the requisite cofactors is inexorably linked to the assembly of the enzyme. In the face of such complexity, a simplified model system to study the assembly process is highly desirable. The yeast *Saccharomyces cerevisiae* presents an ideal model for a number of reasons. It is amenable to genetic and biochemical manipulation, and it is a facultative anaerobe, which means it can grow on fermentable carbon sources either in the absence of oxygen or in the event of a mutation which renders it respiration deficient. Another important advantage in working with *S. cerevisiae* is that its entire genome has been sequenced, which expedites functional analyses of genes and their products. Lastly, but of paramount importance from the perspective of inherited human COX deficiencies, the cytochrome oxidases from yeast and humans are highly homologous, both in terms of subunit structure and function.

1.2. Cytochrome Oxidase: Assembly in Yeast

Studies in yeast have estimated that as many as 40 different genes are involved in the synthesis and assembly of cytochrome oxidase (McEwen *et al.*, 1986; Tzagoloff and Dieckmann, 1990). Table 1 provides a list of the *COX* genes identified thus far. All of the genes on this list have been cloned and their protein products characterized. The nomen-

Table 1. Yeast cytochrome oxidase mutants

Phenotype	Gene	Function	Ref
Missing structural subunit (5)	COX4	Subunit 4	Maarse et al.
			Dowhan et al.
	COX5a	Subunit 5	Koerner et al.
			Koerner et al.
	COX6	Subunit 6	Wright et al.
	COX7	Subunit 7	Aggeler and Capaldi
	COX9	Subunit 7a	Wright et al.
Missing subunits 2 and 3 (10)	PET111	Translation of subunit 2	Poutre and Fox
			Mulero and Fox
	PET54	Translation of subunit 3	Costanzo and Fox
			Brown et al.
	PET122	"	"
	PET494	"	"
Missing subunit 1 (18)	PET309	Subunit 1 mRNA translation	Manthey and McEwen
	MSS18	Processing of subunit 1 pre-mRNA	Seraphin et al.
	MSS51	"	Decoster et al.
	MSS116	"	Seraphin et al.
Heme O deficiency (1)	COX10	Farnesylation of protoheme	Nobrega et al.
			Tzagoloff et al.
Copper metabolism (2)	COX17	Copper delivery to mitochondria	Glerum et al.
	SCO1	Copper transfer	Schulze and Roedel
			Glerum et al.
Assembly arrest (12)	COX11	?	Tzagoloff et al.
	COX14	?	Glerum et al.
	COX15	?	Glerum et al.
	PET100	?	Church et al.
	PET117	?	McEwen et al.
	PET191	?	"

clature for these proteins has not been unified and remains confusing: the genes that encode ancillary proteins required for enzyme assembly, though their gene products are not part of the final structure, are variously referred to as *PET*, *MSS*, *COX* or *SCO* genes. Characterization of mutant alleles has shown that, irrespective of the functional diversities of the gene products, most of the mutant phenotypes are indistinguishable from one another at a biochemical level. Interestingly, they are similar to the phenotype seen in mutants that are missing a structural subunit. The mutants all lack the characteristic cytochrome aa_3 absorption band at 605 nm, have no detectable cytochrome oxidase activity and severely reduced steady-state levels of the mitochondrially-encoded subunits (Glerum et al., 1995, 1997). The loss of the mitochondrial subunit proteins is the result of degradation by intramitochondrial proteases (Arlt et al., 1996; Rep and Grivell, 1996; Rep et al., 1996). The nuclear-encoded subunits of the enzyme are imported into mitochondria where they appear to reside as monomers in the inner membrane (Glerum and Tzagoloff, 1998). These criteria are used to determine the state of assembly of the enzyme and the above phenotype is referred to as assembly arrested. It is remarkable that mutations in a COX-specific translational activator elicit the same phenotype as mutants that fail to synthesize or provide the necessary prosthetic groups. The fact that a wide variety of mutations all result in a failure to assemble cytochrome oxidase strongly suggests that assembly is an all-or-nothing phenomenon.

The technical problems to be surmounted in delineating the steps in such an all-or-nothing reaction represent one of our greatest current challenges. In order to understand

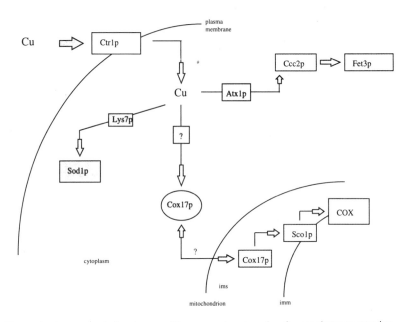

Figure 1. Copper metabolism in yeast. The arrows show transfer of copper between proteins.

the molecular mechanisms underlying human COX deficiencies, we will need to unravel the chronological order of events in the cytochrome oxidase assembly pathway.

1.3. Cytochrome Oxidase: Acquisition of Copper

The insertion of copper into cytochrome oxidase is an essential requirement for both its catalytic activity and its assembly (Capaldi, 1990; Glerum *et al.*, 1996a). Copper, an essential trace metal, is involved in a wide variety of cellular processes, yet we are only now developing an understanding of its transport and metabolism, both in yeast and humans (Cox, 1995; Danks, 1995). Figure 1 presents a schematic overview of our current understanding of the pathways involved in intracellular copper distribution in yeast. Extracellular copper is transported into the cell by Ctr1p, a plasma membrane protein involved in high-affinity copper uptake (Dancis *et al.*, 1994a, 1994b). Once inside the cell, numerous small, hydrophilic proteins carry copper to its internal target sites.

In a unique pathway that links iron and copper transport, Ccc2p accepts copper from Atx1p (Lin *et al.*, 1997) and transfers it to Fet3p (Yuan *et al.*, 1996). The *ATX1* gene was originally identified as a multi-copy suppressor of mutations affecting antioxidant function (Lin and Culotta, 1995) and Ccc2p is a P-type copper transporting ATPase, similar to the proteins defective in Wilson and Menkes disease (Bull *et al.*, 1993; Chelly *et al.*, 1993). The end point for this pathway is Fet3p, the yeast homologue of ceruloplasmin (Askwith *et al.*, 1994). Copper functions as an essential prosthetic group for this protein, which is involved in high affinity iron uptake. In a completely separate pathway, the *LYS7* gene product delivers copper to Sod1p, the cytosolic copper-zinc superoxide dismutase (Culotta et al, 1997).

Recently, a unique pathway for transporting copper to mitochondria and/or cytochrome oxidase was also identified. The *COX17* gene was isolated by complementation of a cytochrome oxidase deficient yeast mutant. The encoded protein is a small, cysteine-rich molecule (Glerum *et al.*, 1996a). Since the respiration-deficient phenotype of *cox17* mutants was rescued by the addition of high levels of copper to the medium, Cox17p was proposed to function in recruiting copper to mitochondria. At the same time as we isolated *COX17*, *SCO1* and *SCO2* were isolated as multi-copy suppressors of the *cox17* mutation (Glerum *et al.*, 1996b). Sco1p was the first cytochrome oxidase assembly factor to be identified (Schulze and Roedel, 1989), and *SCO2* is a *SCO1* homologue identified through the sequencing of the yeast genome (Smits *et al.*, 1994). On the basis of our findings, we proposed that Sco1p acts downstream of Cox17p in providing cytochrome oxidase with its copper prosthetic groups. Mutations in *SCO1* elicit the same respiration-deficient phenotype as seen with *cox17* mutants, but cannot be rescued by supplementation of the growth medium with copper, presumably because Sco1p acts further downstream from Cox17p and is spatially separated from cytosolic copper. It is also clear that the cytochrome oxidase apoenzyme cannot directly acquire copper from the intermembrane space milieu. Further work also demonstrated that *SCO2* is not essential for cytochrome oxidase assembly and that there is only a partial functional overlap between Sco1p and Sco2p.

The *COX17* and *SCO* proteins are also intriguing because they do not share any of the motifs commonly associated with copper transporting proteins, such as the Cys-Pro-Cys and Gly-X-Thr-Cys-X-X-Cys motifs proposed to function in cation transduction and heavy metal binding (Odermatt *et al.*, 1993). However, after the human *COX17* cDNA was cloned (Amaravadi et al; 1997), it became evident that there was a cysteine-rich motif shared by the Cox17ps and the mammalian metallothioneins (upper panel, Figure 2). It was also noted

Figure 2. Putative copper binding motifs of yeast proteins involved in mitochondrial copper metabolism. Dark boxes, identities; light boxes, similarities. Asterisks indicate the amino acids in subunit 2 known to make contact with copper (Tsukihara *et al.*, 1995).

that the Sco proteins share a domain homologous to a part of subunit 2 of cytochrome oxidase (Cox2p) that is involved in binding copper (lower panel, Fig. 2). This homologous domain is conserved in a series of bacterial homologues, some of which are encoded within bacterial copper resistance operons, but for which the precise functions remain unknown.

1.4. Cytochrome Oxidase Deficiencies: Inherited Disorders in Humans

Given that all the OXPHOS enzymes consist of multi-subunit complexes, the large number of respiratory chain deficiencies that have been identified is hardly surprising. Moreover, the complexity of COX biogenesis as outlined above further explains why COX deficiencies form the largest subset of OXPHOS disorders. In addition to all the factors required for correct assembly of cytochrome oxidase in yeast, an added complication in humans is the presence of tissue-specific isoforms for several of the nuclear-encoded subunits (Bonne *et al.*, 1993; Kadenbach *et al.*, 1986). COX deficiencies manifest themselves in a variety of neurodegenerative diseases such as Leigh syndrome (Leigh, 1951), but have also been implicated in late-onset Alzheimer disease (Davis *et al.*, 1997). The most common clinical presentation for COX defects is Leigh syndrome. This neurological disorder is characterized by psychomotor retardation and developmental delay. Other clinical manifestations include ataxia, ophthalmoplegia, hypotonia, and lactic acidosis. The hallmark of the disease is the necrotic lesions found in the brain stem and basal ganglia, the two most oxidative areas of the brain. Leigh syndrome patients have widely varying clinical courses, with most presenting early in life (often the neonatal period) but with death occuring anywhere from the first year up to 6–7 years of age (Robinson *et al.*, 1987). COX deficiencies not associated with Leigh syndrome present with similar clinical phenotypes, but lack the characteristic brain pathology.

The genetic defects for COX deficiencies can have one of two origins (Luft, 1994; Wallace, 1992). Those defects stemming from mutations in the mtDNA are either maternally-inherited or sporadic, whereas defects encoded in the nuclear DNA will result in either an autosomal-dominant or -recessive mode of inheritance. Large scale deletions in mtDNA were the first molecular defects to be identified in patients with mitochondrial myopathies (Holt *et al.*, 1988). Since then, deletions and duplications of mtDNA have been widely reported, both associated specifically with COX deficiency (Keightley *et al.*, 1996; Mita *et al.*, 1989) and with a more general decrease in other respiratory chain enzymes (Poulton *et al.*, 1989; Schon *et al.*, 1989). There are also a variety of point mutations which give rise to specific disorders with established patterns of maternal inheritance (Goto *et al.*, 1990; Shoffner *et al.*, 1990; Wallace *et al.*, 1988). The phenotypes for most of these syndromes are associated with deficiencies in several of the respiratory chain enzyme complexes, including cytochrome oxidase.

While we are starting to understand the molecular bases for the COX defects encoded in mtDNA, those COX deficiencies resulting from mutations in nuclear genes are still poorly understood. Initial studies on COX deficiencies where the defect was suspected to reside in a nuclear gene (based on family history) suggested that defects in subunit constituents might be responsible for the observed defects. A benign, reversible COX deficiency (DiMauro *et al.*, 1983) was thought to stem from a defect in a developmentally-regulated subunit. Studies with the bovine heart and liver enzymes have shown that there are several COX subunits which have tissue-specific isoforms (Capaldi, 1990; Kadenbach *et al.*, 1986), and subsequent studies have demonstrated that there is a developmentally controlled switch in the expression of several of these tissue-specific isoforms (Bonne *et al.*, 1993). In a series of COX deficient patients from Quebec, the biochemical phenotype indicated that the defect was far more severe in liver than in heart and skeletal muscle (Merante *et al.,* 1993), and yet the tissue-specific subunit cDNAs revealed no mutations. Likewise, the cDNA sequencing of all the nuclear-encloded subunits from a large number of COX-deficient patients has failed to reveal any mutations (Adams *et al.,* 1997; DiMauro and DeVivo, 1995). Early biochemical analyses of COX defects suggested that the deficiencies might stem from either a decreased stability of, or a failure to complete assembly of, the holoenzyme (Glerum *et al.,* 1988; Lombes *et al.,* 1991; Robinson *et al.,* 1990).

A mutation in a COX assembly factor gene seems likely based on the evidence presented above. Clearly, studies on COX assembly in yeast have important implications for our understanding of the biochemical nature of human COX mutants. The isolation of yeast COX assembly genes has allowed us to clone the human cDNA homologues for two of these yeast genes (*COX10* and *COX17*) by functional complementation (Amaravadi *et al.,* 1997; Glerum and Tzagoloff, 1994). Indeed, it is only because of the studies in yeast that we have identified any human COX genes thus far. In addition to providing important diagnostic tools for COX deficiencies, these cDNAs will allow identification of the genes and chromosomal loci for these human COX assembly proteins.

Mapping of the COX assembly genes will be important, since these genes may also be involved in other disease phenotypes. The human *COX10* gene was recently mapped to the 1.5-Mb region of chromosome 17 that is duplicated in Charcot-Marie-Tooth disease type 1A (CMT1A; Reiter *et al.,* 1997) and deleted in hereditary neuropathy with liability to pressure palsies (HNPP; Murakami *et al.,* 1997). The deletion found in HNPP would result in a null allele for *COX10*, and combined with the presence of a mutation in the other allele would give rise to clinical COX deficiency. Together, CMT1A and HNPP are the most common peripheral neuropathies, with an estimated frequency of 1:2500 individuals (Ballabio and Zoghbi, 1995). The symptoms associated with HNPP, muscle weakness and atrophy, are common with a subset of those found in COX deficiencies. This overlap of symptoms may partly be the result of the HNPP deletion, given the frequency of its presence in the general population.

The large number of COX assembly genes suggest that there may be more of these genes associated with other genetic diseases. This will clearly be an area of increased activity in the study of inherited cytochrome oxidase deficiencies.

2. DELINEATION OF THE COPPER TRANSPORT PATHWAY INTO MITOCHONDRIA

The mechanism by which Cox17p and Sco1p recruit copper to mitochondria and/or cytochrome oxidase is of immense interest, not only in helping us to understand the assembly pathway for cytochrome oxidase, but also for a more general understanding of intracellular metal transport reactions. In addition, either *COX17* or *SCO1* may be candidates for lesions involved in human COX deficiencies.

2.1. Role of Cox17p in Recruiting Copper to Mitochondria

In the model shown in Figure 1, Cox17p is acting as a copper chaperone for Sco1p. It carries copper from the cytosol to mitochondria. Though our knowledge of this process is far from complete, recent work has started to delineate parts of this copper transport pathway in more detail. Our initial model for Cox17p action did not invoke a direct physical interaction between Cox17p and Sco1p, because Cox17p appeared to be a cytosolic protein (Glerum *et al.,* 1996a) and Sco1p was known to be a component of the mitochondrial inner membrane (Buchwald *et al.,* 1991). However, as shown below, Cox17p actually appears to be partitioned between the cytosol and the mitochondria (Beers *et al.,* 1997).

We previously localized Cox17p to the cytosol using a biotinylated fusion protein, in which the *COX17* sequence was fused at its 3' end to a 300 bp sequence encoding a biotinylation signal (Murtif *et al.,* 1985). The subsequent biotinylated fusion protein was about 8 kDa larger than the expected size of Cox17p, thus doubled in size from the native protein. Even when expressed from a single copy in the genome, this construct completely complemented a *COX17* knock-out strain and the biotinylated protein was found exclusively in the cytosolic fraction (Glerum *et al.,* 1996a). Because we wanted to purify Cox17p to gain a better understanding of its structural and functional properties, we generated an antibody to a carboxyl-terminal peptide of Cox17p. The peptide comprised the terminal 20 amino acids of Cox17p, a region that includes a critical Cys residue (Glerum *et al.,* 1996a). The existence of an antibody allowed us to determine the subcellular localization for the native Cox17p. As shown in Figure 3, Western blot analysis of the subcellular fractionation of a wild-type strain (W303–1B) reveals that a substantial amount of Cox17p is found in the mitochondria. When the amounts of protein in the two fractions are determined, and corrected for the total protein yield, it appears that about 60% of the total Cox17p in the cell is associated with the mitochondria and 40% with the cytosol (Beers *et al.,* 1997). Control experiments with cytochrome b_2 have demonstrated that the Cox17p found in the cytosol is not due to leakage from mitochondria broken during the fractionation experiment. A similar fractionation experiment carried out with a strain that overexpresses Cox17p (ΔCOX17/ST31) demonstrates that Cox17p is again found in both the mitochondria and the post-mitochondrial supernatant. The distributions are altered, with relatively more Cox17p found in the cytosolic fraction, suggesting a limit to the amount of Cox17p that can be taken up by the mitochondria. The existence of Cox17p in

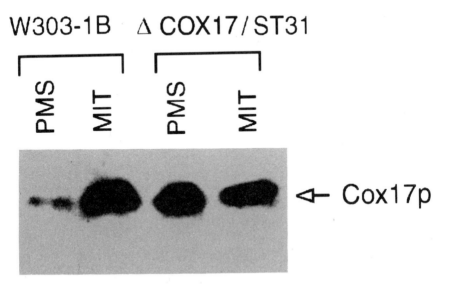

Figure 3. Distribution of Cox17p between the cytosol and the mitochondria. Mitochondria were prepared from the wild-type strain W303–1B and from a strain overexpressing Cox17p (ΔCOX17/ST31). Ten times more total protein was loaded for the mitochondrial (MIT) and post-mitochondrial supernatant (PMS) fractions from the wild-type strain than for the overexpressing strain. Western blot analysis was carried out with an antibody against a C-terminal peptide of Cox17p, as described in Beers *et al.,* (1997). (Reprinted with permission from The Journal of Biological Chemistry.)

two distinct intracellular pools strongly supports the concept that it is acting as a copper shuttle, travelling from the cytosol, across the outer mitochondrial membrane and into mitochondria. However, in order to make this model plausible, the Cox17p that is found in mitochondria must actually be located in the intermembrane space.

We have demonstrated that, indeed, the bulk of the Cox17p that is associated with mitochondria in a subcellular fractionation is localized to the intermembrane space. As shown in Figure 4A, when mitochondria are lysed by osmotic shock to generate mitoplasts, almost

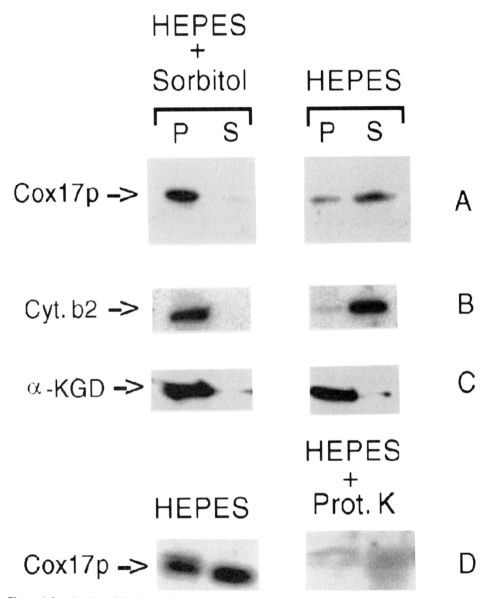

Figure 4. Localization of Cox17p to the mitochondrial intermembrane space. Preparation of mitochondria and mitoplasts, Proteinase K digestions, and Western blotting were all carried out as described in Beers *et al.,* (1997). (Reprinted with permission from The Journal of Biological Chemistry.)

all of the Cox17p becomes associated with the supernatant fraction when the mitoplasts are pelleted by centrifugation. The behaviour of Cox17p mimics that of cytochrome b_2, a known marker of the mitochondrial intermembrane space (Figure 4B). This is in contrast to the pattern seen with α-ketoglutarate dehydrogenase, a known marker of the mitochondrial matrix, which pellets with the mitoplasts (Figure 4C). Further experiments have also shown that the Cox17p found in the intermembrane space is fully accessible to externally added Proteinase K (Figure 4D) when mitochondria are osmotically lysed.

As well as determining the mechanism by which Cox17p can transport copper into mitochondria, it will also be necessary to determine how and where Cox17p becomes loaded with copper in the cytosol. Since it it unlikely that copper exists as a free metal in cells (Cox, 1995; Danks, 1995), there are two possible ways in which Cox17p can obtain its copper: 1) Cox17p can receive the metal from a cytosolic protein which is also acting as a copper shuttle or chaperone; or 2) Cox17p itself is the shuttle and obtains copper directly from Ctr1p at the plasma membrane. Our current experiments are directed at answering these questions using a number of different experimental avenues, including the two-hybrid approach (Fields and Song, 1989; Chien *et al.*, 1991) and searching for extragenic suppressors or synthetically lethal alleles (reviewed in Guarente, 1993). All of the above methods are powerful techniques for identifying protein-protein interactions, which presumably should occur in the case of a direct transfer of copper from one protein/carrier to another.

In addition to determining the donors and acceptors for the copper carried by Cox17p, which will identify the components of this copper transport pathway, there are several other important and intriguing questions raised by our studies. The first is the mechanism by which Cox17p actually enters and exits the mitochondrion. Though the outer mitochondrial membrane is known to be porous to large substances, this is the first example of a molecule that appears to shuttle between the mitochondria and the extra-mitochondrial environment. The other question to be answered is the mechanism by which Cox17p actually transfers copper. We have shown that Cox17p binds copper, albeit with a low stoichiometry. Purified Cox17p contains 0.3 moles of copper per mole of protein (Beers *et al.*, 1997). The protein can bind up to 1.8 moles of copper, as Cu(I), upon incubation in a molar excess of cuprous chloride. These values are in the same range as those recently presented for Atx1p (Pufahl *et al.*, 1997), which also binds copper in the Cu(I) state. Detailed structural studies of Cox17p are needed to elucidate its copper transfer mechanism, which is likely to be unique, particularly in light of the fact that it lacks any homology to the previously studied copper transport proteins or chaperones (Glerum *et al.*, 1996a).

2.2. Roles of Sco1p and Sco2p in Copper Transport into Mitochondria

The only other components of the mitochondrial copper transport pathway that we have identified thus far are the Sco proteins. Because we identified both *SCO1* and *SCO2* as multicopy suppressors of a *cox17* mutation, we proposed that both of these gene products were involved in taking up copper into mitochondria (Glerum *et al.*, 1996b). However, the actual role of Sco2p remains unresolved. It is clearly not required for assembly of cytochrome oxidase, as a *SCO2*-disrupted strain does not have a respiration-deficient phenotype (Glerum *et al.*, 1996b). It was also evident that though highly homologous, there was only a partial overlap of functions between Sco1p and Sco2p.

In proposing a model for the function of Sco1p in providing cytochrome oxidase with copper, there are two possible roles for Sco1p: it can either a) transport copper into the mitochondrial matrix or it can b) transfer copper in the intermembrane space directly onto the cytochrome oxidase apoenzyme. For a number of reasons, some of which are presented below, we favour the second model. A priori, the lack of homology between Sco1p and any of the other known copper transporting P-type ATPases suggests that Sco1p is not involved in transporting a cation through a membrane - it contains neither the Thr-Gly-Glu motif involved in energy transduction nor the Cys-Pro-Cys motif of a putative cation channel (Cox, 1995; Odermatt *et al.*, 1993). Both of these motifs are conserved through a wide variety of prokaryotes, and lower and higher eukaryotes, and would appear to be indispensible for transporting cations across otherwise impermeable membrane barriers. In addition, none of the copper-bearing sites of cytochrome oxidase are found on the matrix side of the inner membrane. The fact that the Cu_A site of the enzyme is completely on the intermembrane space side of the inner membrane (Tsukihara *et al.*, 1996), together with the fact that there are no other known requirements for copper in the mitochondrial matrix, makes the concept of Sco1p as a copper transferase appealing.

    In order for Sco1p to act as a copper transferase, the bulk of the protein would have to be on the intermembrane space side of the membrane. Earlier work on Sco1p, by Buchwald *et al.*, (1991), had shown that the single transmembrane domain in the protein was indispensible for its function and that a construct lacking this domain could not rescue a *SCO1*-disrupted strain—the protein expressed from this construct appeared to be trapped in the mitochondrial matrix (Buchwald *et al.*, 1991). In more recent work, we provided evidence that the bulk of Sco1p is found in the intermembrane space (Beers *et al.*, 1997). Mitochondria from a wild-type yeast strain were either suspended in an isotonic buffer or osmotically lysed to make mitoplasts and then incubated either in the absence or presence of Proteinase K. As shown in Figure 5, Sco1p is protected from digestion by Proteinase K in intact mitochondria, but not in mitoplasts, which means that the bulk of the protein is in a compartment that is exposed to the protease. By comparison, subunit 5 of cytochrome oxidase (Cox5p), which is found in the inner mitochondrial membrane as part of the cytochrome oxidase holoenzyme, is not sensitive to protease digestion, even in mitoplasts. This demonstrates that the largest portion of Sco1p is found in the intermembrane space

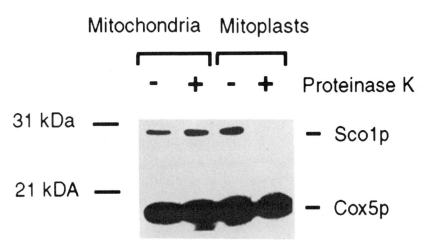

Figure 5. Sco1p is found in the mitochondrial intermembrane space. Mitochondria and mitoplasts were prepared from the wild-type strain W303–1B, and the Proteinase K digestions carried out, as described in Beers *et al.*, (1997). Western blot analysis was performed with antibodies to Sco1p and to subunit 5 of cytochrome oxidase (Cox5p). (Reprinted from The Journal of Biological Chemistry.)

and fits well in a model in which Sco1p accepts copper from Cox17p and transfers it directly onto the maturing cytochrome oxidase apoenzyme.

Though most of our work has focused on the role of Sco1p in providing cytochrome oxidase with copper, the role of Sco2p (if any) in this process must also be elucidated. Our initial studies suggested that perhaps it might be found in a dimer with Sco1p. Sucrose gradient centrifugation analysis of wild-type mitochondria indicated that both native Sco1p and Sco2p migrated at a position equivalent to about twice their known size (data not shown). However, Sco2p continued to migrate at this position even in a strain carrying a *SCO1* disruption. Furthermore, affinity purification of a biotinylated Sco1p fusion protein failed to demonstrate co-purification of Sco2p. All of these results argue against the involvement of Sco1p and Sco2p in a complex. However, the fact that Sco2p is synthesized *in vivo* suggests that, though it is not essential for the assembly of cytochrome oxidase, it may have a different supporting role to play in mitochondrial function.

There is still a degree of uncertainty in our present model for mitochondrial copper uptake. One of the main questions is whether there may actually be an intermediate acting between Cox17p and Sco1p. We are currently undertaking experiments to answer this question. In addition, preliminary experiments have demonstrated that Sco1p, like Cox17p, binds copper (I). Deciphering the precise function for the Sco proteins, as well as determining the mechanism by which they carry out their roles, is clearly of great importance in our understanding of how procurement of copper fits into the overall assembly pathway for cytochrome oxidase. In addition, it may also provide interesting new paradigms for heavy metal transport in both prokaryotes and eukaryotes.

2.3. Applications to Human COX17 and SCO1 Genes

As mentioned above, we have undertaken our studies of yeast cytochrome oxidase assembly to help provide insight into the lesions involved in inherited human cytochrome oxidase deficiencies. The fact that the copper uptake and delivery system appears to be conserved between yeast and humans (Valentine and Gralla, 1997) provides further support for the notion that our findings will ultimately have a bearing on human COX deficiencies. Since mutations in either *COX17* or *SCO1* elicit an assembly defect in cytochrome oxidase, and most of the human COX deficiencies described to date are assembly defects, both human *COX17* and *SCO1* are potential targets for deleterious mutations resulting in clinical cytochrome oxidase deficiency.

The human *COX17* cDNA was cloned by functional complementation of the yeast *cox17* null mutant (Amaravadi *et al.,* 1997). Attempts to clone the human *SCO1* cDNA by the same method were unsuccessful. However, a cDNA clone for *SCO1* has now been obtained with the help of the Expressed Sequence Tag database (Massimo Zeviani, personal communication). Clearly, functional complementation of the yeast *sco1* null allele will be needed to verify its identity as the *bona fide* human homolog of the yeast *SCO1* gene. Whether there is also a human *SCO2* homolog remains an open question. The availability of cDNAs for both *COX17* and *SCO1* will allow us to map their chromosomal loci and to obtain their full-length genomic sequences. Chromosomal localization studies for the *COX17* gene are currently underway in our laboratory and preliminary results indicate that the *COX17* gene is intron-less (F. Punter, D. Adams and D.M. Glerum, unpublished observations). The cDNAs and genomic information can then be used as tools in screening human COX deficient patients for mutations in these particular components of the cytochrome oxidase assembly pathway.

In addition to characterizing the involvement of Cox17p and/or Sco1p in human COX deficiencies, these proteins may also play important roles in other pathways not directly involved in cytochrome oxidase assembly. An area of increasing interest in the study of mitochondrial diseases is the contribution of reactive oxygen species (ROS) to the clinical pathology of these disorders. Experiments in yeast have demonstrated that the electron transport chain of mitochondria is a major source of ROS (Guido *et al.,* 1993; Longo *et al.,* 1996), and that the principal site of superoxide generation is proximal to the cytochrome oxidase complex (Guido *et al.,* 1993). Recent studies of human mitochondrial defects have shown that in some cases of Complex I deficiency, there is a tremendous increase in the levels of the mangano-superoxide dismutase (MnSOD; Pitkanen and Robinson, 1996), which is the chief mitochondrial scavenger of superoxide radicals. Mutations in either *COX17* or *SCO1* could thus also elicit an excess production of ROS, in addition to causing a deficiency in cytochrome oxidase. Other than by defects or blockages in the respiratory chain, increased generation of ROS can also be caused by an abnormal accumulation of heavy metals such as copper, which is a strong prooxidant (Tkeshelashvili *et al.,* 1991). Though nothing is known regarding the regulation of *COX17* and *SCO1*, one can envisage a situation in which faulty up-regulation of either of these two genes might increase copper loading into mitochondria, with the subsequent generation of deleterious superoxide radicals.

COX17 and *SCO1* may also play potentially deleterious roles in diseases of copper storage and copper toxicity. There is an accumulating body of evidence regarding mitochondrial dysfunction in patients with Wilson disease. A large number of structural abnormalities have been detected in mitochondria from patients with Wilson disease (Sternlieb, 1992). These abnormalities parallel those found in the mitochondria from patients with proven mitochondrial myopathies (Stadhouders and Sengers, 1987). Long-Evans Cinnamon rats are a model for copper overload and have been shown to have reduced respiration and cytochrome oxidase activity in their hepatic mitochondria (Sokol *et al.,* 1993). In addition, the hepatic mitochondria from patients with Wilson disease were shown to accumulate abnormal levels of copper (Sokol *et al.,* 1994). More recent studies have now demonstrated that the hepatic mtDNAs in patients with Wilson disease (Mansouri *et al.,* 1997) accumulate frequent and early deletions. The pathology of Wilson disease results from a failure to transfer intracellular copper to apoceruloplasmin, which ultimately causes hepatic copper toxicity (Danks, 1995). Though we currently know little of how Cox17p and Sco1p function in delivering copper to mitochondria, it is not difficult to imagine that both of these proteins may be involved in the development of the mitochondrial pathology in patients with Wilson disease. Our understanding of how this pathway functions may also allow us to develop suitable therapies to alleviate the mitochondrial symptoms that present an extra burden to Wilson disease patients.

3. SUMMARY

Figure 6 presents an overview of our current knowledge regarding the mitochondrial uptake of copper. Cytosolic copper is transported into the mitochondrial intermembrane space by Cox17p, which then releases it to Sco1p. The Cox17p, relieved of its copper, can shuttle back into the cytosol and is ready to accept another copper molecule. In this manner, Cox17p simply cycles back and forth between the mitochondrial intermembrane space and the cytosol. At the mitochondrial inner membrane, Sco1p accepts copper from Cox17p and can transfer it directly onto the cytochrome oxidase apoenzyme (a). The other possibility, though not favoured by us, is that Sco1p transports copper into the mitochondrial matrix, where its is added to the nascent subunits 1 and 2 polypeptide chains (b).

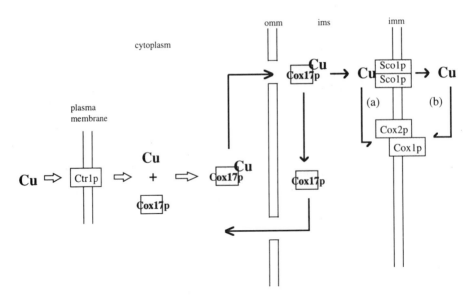

Figure 6. A model for the roles of Cox17p and Sco1p in mitochondrial copper uptake. The (a) and (b) in the model refer to the two possible ways in which Sco1p can supply cytochrome oxidase with copper.

The discovery of a distinct intracellular transport pathway for procuring copper for mitochondria has provided us with new insights into the assembly pathway for cytochrome oxidase and given us potential new targets for causative mutations underlying human COX deficiencies. Clearly, delineating all the components involved in transporting copper to mitochondria and into the cytochrome oxidase active site is of paramount importance, not only in understanding human COX deficiencies, but also in the possible roles that this pathway may play in other pathological processes.

REFERENCES

Adams, P.L., Lightowlers, R.N. and Turnbull, D.M. (1997) Molecular analysis of cytochrome *c* oxidase deficiency in Leigh's syndrome. Ann. Neurol. 41, 268–270.

Aggeler, R. and Capaldi, R.A. (1990) Yeast cytochrome *c* oxidase subunit VII is essential for assembly of an active enzyme. J. Biol. Chem. 265, 16389–16393.

Amaravadi, R., Glerum, D.M. and Tzagoloff, A. (1997) Isolation of a cDNA encoding the human homolog of *COX17*, a yeast gene essential for mitochondrial copper recruitment. Hum. Genet. 99, 329–333.

Arlt, H., Tauer, R., Feldmann, H., Neupert, W. and Langer, T. (1996) The YTA10–12 complex, an AAA protease with chaperone-like activity in the inner membrane of mitochondria. Cell 85, 875–885.

Askwith, C., Eide, D., Van Ho, A., Bernard, P.S., Li, L., Davis-Kaplan, S., Sipe, D.M. and Kaplan, J. (1994) The *FET3* gene of S. cerevisiae encodes a multicopper oxidase required for ferrous iron uptake. Cell 76, 403–410.

Ballabio, A. and Zoghbi, H.Y. (1995) Charcot-Marie-Tooth disease and hereditary neuropathy with liability to pressure palsies. In: Scriver, C.R., Beaudet, A.L., Sly, W.S., Valle, D. (Eds.) "The Metabolic and Molecular Bases of Inherited Disease." McGraw-Hill, New York, pp 4569–4574.

Beers, J., Glerum, D.M. and Tzagoloff, A. (1997) Purification, characterization, and localization of yeast Cox17p, a mitochondrial copper shuttle. J. Biol. Chem. 272, 33191–33196.

Bonne, G., Seibel, P., Possekel, S., Marsac, C. and Kadenbach, B. (1993) Expression of human cytochrome *c* oxidase subunits during fetal development. Eur. J. Biochem. 217, 1099–1107.

Brown, N.G., Costanzo, M.C. and Fox, T.D. (1994) Interactions among three proteins that specifically activate translation of the mitochondrial *COX3* mRNA in *Saccharomyces cerevisiae*. Molec. Cell. Biol. 14, 1045–1053.

Buchwald, P., Krummeck, G. and Roedel, G. (1991) Immunological identification of yeast SCO1 protein as a component of the inner mitochondrial membrane. Mol. Gen. Genet. 229, 413–420.

Bull, P.C., Thomas, G.R., Rommens, J.M., Forbes, J.R. and Cox, D.W. (1993) The Wilson disease gene is a putative copper transporting P-type ATPase similar to the Menkes gene. Nature Genet. 5, 327–337.

Calhoun, M.W., Thomas, J.W. and Gennis, R.B. (1994) The cytochrome oxidase superfamily of redox-driven proton pumps. Trends Biochem. Sci. 19, 325–330.

Capaldi, R.A. (1990) Structure and function of cytochrome c oxidase. Ann. Rev. Biochem. 59, 569–596.

Chelly, J., Tumer, Z., Tonnesen, T., Petterson, A., Ishikawa-Brush, Y., Tommerup, N., Horn, N. and Monaco, A.P. (1993) Isolation of a candidate gene for Menkes disease that encodes a potential heavy metal binding protein. Nature Genet. 3, 14–19.

Chien, C.-T., Bartel, P.L., Sternglanz, R. and Fields, S. (1991) The two-hybrid system: A method to identify and clone genes for proteins that interact with a protein of interest. Proc. Natl. Acad. Sci. USA 88, 9578–9582.

Church, C., Chapon, C. and Poyton, R.O. (1996) Cloning and characterization of PET100, a gene required for the assembly of yeast cytochrome c oxidase. J. Biol. Chem. 271, 18499–18507.

Cooper, C.E. (1990) The steady-state kinetics of cytochrome c oxidation by cytochrome oxidase. Biochim. Biophys. Acta 1017, 187–203.

Costanzo, M.C. and Fox, T.D. (1988) Specific translational activation by nuclear gene products occurs in the 5' untranslated leader of a yeast mitochondrial mRNA. Proc. Natl. Acad. Sci. USA 85, 2677–2681.

Cox, D.W. (1995) Genes of the copper pathway. Am. J. Hum. Genet. 56, 828–834.

Culotta, V.C., Klomp, L.W.J., Strain, J., Casareno, R.L.B., Krems, B. and Gitlin, J. (1997) The copper chaperone of superoxide dismutase. J. Biol. Chem. 272, 23469–23472.

Dancis, A., Haile, D., Yuan, D.S. and Klausner, R.D. (1994a) The Saccharomyces cerevisiae copper transport protein (Ctr1p). J. Biol. Chem. 269, 25660–25667.

Dancis, A., Yuan, D.S., Haile, D., Askwith, C., Eide, D., Moehle, C., Kaplan, J. and Klausner, R.D. (1994b) Molecular characterization of a copper transport protein in S. cerevisiae: An unexpected role for copper in iron transport. Cell 76, 393–402.

Danks, D.M. (1995) Disorders of copper transport. In: Scriver, C.R., Beaudet, A.L., Sly, W.S., Valle, D. (Eds.) "The Metabolic and Molecular Bases of Inherited Disease." McGraw-Hill, New York, pp 2211–2235.

Davis, R.E., Miller, S., Herrnstadt, C., Ghosh, S.S., Fahy, E., Shinobu, L.A., Galasko, D., Thal, L.J., Beal, M.F., Howell, N. and Parker, W.D. Jr. (1997) Mutations in mitochondrial cytochrome c oxidase genes segregate with late-onset Alzheimer disease. Proc. Natl. Acad. Sci. USA 94, 4526–4531.

Decoster, E., Simon, F., Hatat, D. and Faye, G. (1990) The MSS51 gene product is required for the translation of the COX1 mRNA in yeast mitochondria. Mol. Gen. Genet. 224, 111–118.

DiMauro, S. and DeVivo, D.C. (1996) Genetic heterogeneity in Leigh syndrome. Ann. Neurol. 40, 5–7.

DiMauro, S., Nicholson, J.F., Hays, A.P., Eastwood, A.B., Papadimitriou, A., Koenigsberger, R. and DeVivo, D.C. (1983) Benign infantile mitochondrial myopathy due to reversible cytochrome c oxidase deficiency. Ann. Neurol. 14, 226–234.

DiMauro, S., Zeviani, M., Servidei, S., Prelle, A., Miranda, A.F., Bonilla, E. and Schon, E.A. (1988) Biochemical and molecular aspects of cytochrome c oxidase deficiency. Advances Neurol. 48, 93–105.

Dowhan, W., Bibus, C.R. and Schatz, G. (1985) The cytoplasmically-made subunit IV is necessary for assembly of cytochrome c oxidase in yeast. EMBO J. 4, 179–184.

Fields, S. and Song, O.-k. (1989) A novel genetic system to detect protein-protein interactions. Nature 340, 245–246.

Glerum, D.M., Koerner, T.J. and Tzagoloff, A. (1995) Cloning and characterization of COX14, whose product is required for assembly of yeast cytochrome oxidase. J. Biol. Chem. 270, 15585–15590.

Glerum, D.M., Muroff, I., Jin, C. and Tzagoloff, A. (1997) COX15 codes for a mitochondrial protein essential for the assembly of yeast cytochrome oxidase. J. Biol. Chem. 272, 19088–19094.

Glerum, D.M., Robinson, B.H. and Capaldi, R.A. (1989) Fibroblasts and cytochrome c oxidase deficiency. In: Azzi, A., Drahota, Z., Papa, S. (Eds.) "Molecular Basis of Membrane-Associated Diseases." Springer Verlag, Heidelberg, pp 228–238.

Glerum, D.M., Shtanko, A. and Tzagoloff, A. (1996a) Characterization of COX17, a yeast gene involved in copper metabolism and assembly of cytochrome oxidase. J. Biol. Chem. 271, 14504–14509.

Glerum, D.M., Shtanko, A. and Tzagoloff, A. (1996b) SCO1 and SCO2 act as high copy suppressors of a mitochondrial copper recruitment defect in Saccharomyces cerevisiae. J. Biol. Chem. 271, 20531–20535.

Glerum, D.M. and Tzagoloff, A. (1994) Isolation of a human cDNA for heme A: farnesyltranferase by functional complementation of a yeast cox10 mutant. Proc. Natl. Acad. Sci. USA 91, 8452–8456.

Glerum, D.M. and Tzagoloff, A. (1998) Affinity purification of yeast cytochrome oxidase with biotinylated subunits 4, 5, or 6. Anal. Biochem., in press.

Glerum, D.M., Yanamura, W., Capaldi, R.A. and Robinson, B.H. (1988) Characterization of cytochrome-*c* oxidase mutants in human fibroblasts. FEBS Lett. 236, 100–104.

Goto, Y.-i., Nonaka, I. and Horai, S. (1990) A mutation in the tRNA$^{Leu(UUR)}$ gene associated with the MELAS subgroup of mitochondrial encephalomyopathies. Nature 348, 651–653.

Guarente, L. (1993) Synthetic enhancement in gene interaction: a genetic tool come of age. Trends Genet. 9, 362–366.

Guido, D.M., McCord, J.M., Wright, R.M. and Repine, J.E. (1993) Absence of electron transport (rho^0 state) restores growth of a manganese-superoxide dismutase-deficient *Saccharomyces cerevisiae* in hyperoxia. J. Biol. Chem. 268, 26699–26703.

Holt, I.J., Harding, A.E. and Morgan-Hughes, J.A. (1988) Deletions of muscle mitochondrial DNA in patients with mitochondrial myopathies. Nature 331, 717–719.

Iwata, S., Ostermeier, C., Ludwig, B. and Michel, H. (1995) Structure at 2.8 Å resolution of cytochrome *c* oxidase from Paracoccus denitrificans. Nature 376, 660–669.

Kadenbach, B., Stroh, A., Ungibauer, M., Kuhn-Nentwig, L., Buge, U. and Jarausch, J. (1986) Isozymes of cytochrome-*c* oxidase: characterization and isolation from different tissues. Methods Enzymol. 126, 32–45.

Keightley, J.A., Hoffbuhr, K.C., Burton, M.D., Salas, V.M., Johnston, W.S.W., Penn, A.M.W., Buist, N.R.M. and Kennaway, N.G. (1996) A microdeletion in cytochrome c oxidase (COX) subunit III associated with COX deficiency and recurrent myoglobinuria. Nature Genet. 12, 410–416.

Koerner, T.J., Hill, J. and Tzagoloff, A. (1985a) Cloning and characterization of the yeast nuclear gene for subunit 5 of cytochrome oxidase. J. Biol. Chem. 260, 9513–9515.

Koerner, T.J., Homison, G. and Tzagoloff, A. (1985b) Nuclear mutants of *Saccharomyces cerevisiae* with altered subunits 4, 5, and 6 of cytochrome oxidase. J. Biol. Chem. 260, 5871–5874.

LaMarche, A.E.P., Abate, M.I., Chan, S.H.P. and Trumpower, B.L. (1992) Isolation and characterization of *COX12*, the nuclear gene for a previously unrecognized subunit of *Saccharomyces cerevisiae* cytochrome *c* oxidase. J. Biol. Chem. 267, 22473–22480.

Leigh, D. (1951) Subacute necrotizing encephalomyelopathy in an infant. J. Neurol. Neurosurg. Psychiatry 14, 216–22.

Lin, S.-J. and Culotta, V.C. (1995) The *ATX1* gene of *Saccharomyces cerevisiae* encodes a small metal homeostasis factor that protects cells against reactive oxygen toxicity. Proc. Natl. Acad. Sci. USA 92, 3784–3788.

Lin, S.-J., Pufahl, R.A., Dancis, A., O'Halloran, T.V. and Culotta, V.C. (1997) A role for the *Saccharomyces cerevisiae ATX1* gene in copper trafficking and iron transport. J. Biol. Chem. 272, 9215–9220.

Lombes, A., Nakase, H., Tritschler, H.-J., Kadenbach, B., Bonilla, E., DeVivo, D.C., Schon, E.A. and DiMauro, S. (1991) Biochemical and molecular analysis of cytochrome *c* oxidase deficiency in Leigh syndrome. Neurology 41, 491–498.

Longo, V.D., Gralla, E.B. and Valentine, J.S. (1996) Superoxide dismutase activity is essential for stationary phase survival in *Saccharomyces cerevisiae*. J. Biol. Chem. 271, 12275–12280.

Luft, R. (1994) The development of mitochondrial medicine. Proc. Natl.Acad. Sci. USA 91, 8731–8738.

Maarse, A.C., Van Loon, A.P.G.M., Riezman, H., Gregor, I.,Schatz, G. and Grivell, L.A. (1984) Subunit IV of yeast cytochrome *c* oxidase: cloning and nucleotide sequencing of the gene and partial amino acid sequencing of the mature protein. EMBO J. 3, 2831–2837.

Mansouri, A., Gaou, I., Fromenty, B., Berson, A., Letteron, P., Degott, C., Erlinger, S. and Pessayre, D. (1997) Premature oxidative aging of hepatic mitochondrial DNA in Wilson's disease. Gastroenterology 113, 599–605.

Manthey, G.M. and McEwen, J.E. (1995) The product of the nuclear gene *PET309* is required for translation of mature mRNA and stability or production of intron-containing RNAs derived from the mitochondrial *COX1* locus of *Saccharomyces cerevisiae*. EMBO J. 14, 4031–4043.

McEwen, J.E., Hong, K.H., Park, S. and Preciado, G.T. (1993) Sequence and chromosomal localization of two *PET* genes required for cytochrome c oxidase assembly in *Saccharomyces cerevisiae*. Curr. Genet. 23, 9–14.

McEwen, J.E., Ko, C., Kloeckener-Gruissem, B. and Poyton, R.O. (1986) Nuclear functions required for cytochrome *c* oxidase biogenesis in *Saccharomyces cerevisiae*. J. Biol. Chem. 261, 11872–11879.

Merante, F., Petrova-Benedict, R., MacKay, N., Mitchell, G., Lambert, M., Morin, C., DeBraekeleer, M., Laframboise, R., Gagne, R. and Robinson, B.H. (1993) A biochemically distinct form of cytochrome oxidase (COX) deficiency in the Saguenay-Lac-Saint-Jean region of Quebec. Am. J. Hum. Genet. 53, 481–487.

Mita, S., Schmidt, B., Schon, E.A., DiMauro, S. and Bonilla, E. (1989) Detection of "deleted" mitochondrial genomes in cytochrome-*c* oxidase-deficient muscle fibers of a patient with Kearns-Sayre syndrome. Proc. Natl. Acad. Sci. USA 86, 9509–9513.

Miyabayashi, S., Narisawa, K., Tada, K., Sakai, K., Kobayashi, K. and Kobayashi, Y. (1983) Two siblings with cytochrome *c* oxidase deficiency. J. Inher. Metab. Dis. 6, 121–122.

Mulero, J.J. and Fox, T.D. (1993) *PETIII* acts in the 5'-leader of the *Saccharomyces cerevisiae* mitochondrial *COX2* mRNA to promote its translation. Genetics 133, 509–516.

Murakami, T., Reiter, L.T. and Lupski, J.R. (1997) Genomic structure and expression of the human heme A:far-nesyltransferase (*COX10*) gene. Genomics 42, 161–164.

Murtif, V.L., Bahler, C.R. and Samols, D. (1985) Cloning and expression of the 1.3S biotin-containing subunit of transcarboxylase. Proc. Natl. Acad. Sci.USA 82, 5617–5621.

Nobrega, M.P., Nobrega, F.G. and Tzagoloff, A. (1990) *COX10* codes for a protein homologous to the ORF1 product of *Paracoccus denitrificans* and is required for the synthesis of yeast cytochrome oxidase. J. Biol. Chem. 265, 14220–14226.

Odermatt, A., Suter, H., Krapf, R. and Solioz, M. (1993) Primary structure of two P-type ATPases involved in copper homeostasis in *Enterococcus hirae*. J. Biol. Chem. 268, 12775–12779.

Pitkanen, S. and Robinson, B.H. (1996) Mitochondrial complex I deficiency leads to increased production of superoxide radicals and induction of superoxide dismutase. J. Clin. Invest. 98. 345–351.

Poulton, J., Deadman, M.E. and Gardiner, R.M. (1989) Duplications of mitochondrial DNA in mitochondrial myopathy. Lancet 1, 236–240.

Poutre, C.G. and Fox, T.D. (1987) *PETIII*, a *Saccharomyces cerevisiae* nuclear gene required for translation of the mitochondrial mRNA encoding cytochrome *c* oxidase subunit II. Genetics 115, 637–647.

Pufahl, R.A., Singer, C.P., Peariso, K.L., Lin, S.-J., Schmidt, P.J., Fahrni, C.J., Cizewski Culotta, V., Penner-Hahn, J.E. and O'Halloran, T.V. (1997) Metal ion chaperone function of the soluble Cu(I) receptor Atx1. Science 278, 853–856.

Rahman, S., Blok, R.B., Dahl, H.-H.M., Danks, D.M., Kirby, D.M., Chow, C.W., Christodoulou, J. and Thorburn, D.R. (1996) Leigh syndrome: clinical features and biochemical and DNA abnormalities. Ann. Neurol. 39, 343–351.

Reiter, L.T., Murakami, T., Koeuth, T., Gibbs, R.A. and Lupski, J.R. (1997) The human *COX10* gene is disrupted during homologous recombination between the 24 kb proximal and distal CMT1A-REPs. Hum. Molec. Genet. 6, 1595–1603.

Rep, M. and Grivell, L.A. (1996) The role of protein degradation in mitochondrial function and biogenesis. Curr. Genet. 30, 367–380.

Rep, M., van Dijl, J.M., Suda, K., Schatz, G., Grivell, L.A. and Suzuki, C.K. (1996) Promotion of mitochondrial membrane complex assembly by a proteolytically inactive yeast Lon. Science 274, 103–106.

Robinson, B.H., DeMeirleir, L., Glerum, M., Sherwood, G. and Becker, L. (1987) Clinical presentation of mitochondrial respiratory chain defects in NADH-coenzyme Q reductase and cytochrome oxidase: clues to the pathogenesis of Leigh disease. J. Pediatr. 110, 216–222.

Robinson, B.H., Glerum, D.M., Chow, W., Petrova-Benedict, R., Lightowlers, R. and Capaldi, R. (1990) The use of skin fibroblast cultures in the detection of respiratory chain defects in patients with lacticacidemia. Pediatr. Res. 28, 549–555.

Saraste, M. (1990) Structural features of cytochrome oxidase. Quart. Rev. Biophys. 23, 331–366.

Schon, E.A., Rizzuto, R., Moraes, C.T., Nakase, H., Zeviani, M. and DiMauro, S. (1989) A direct repeat is a hotspot for large-scale deletion of human mitochondrial DNA. Science 244, 346–349.

Schulze, M. and Roedel, G. Accumulation of the cytochrome *c* oxidase subunits I and II in yeast requires a mitochondrial membrane-associated protein, encoded by the nuclear *SCO1* gene. Mol. Gen. Genet. 216, 37–43.

Seraphin, B., Simon, M., Boulet, A. and Faye, G. (1984) Mitochondrial splicing requires a protein from a novel helicase family. Nature 337, 84–87.

Seraphin, B., Simon, M. and Faye, G. (1988) *MSS18*, a yeast nuclear gene invovled in the splicing of intron al5 beta of the mitochondrial *cox1* transcript. EMBO J. 7, 1455–1464.

Shoffner, J.M., Lott, M.T., Lezza, A.M.S., Seibel, P., Ballinger, S.W. and Wallace, D.C. (1990) Myoclonic epilepsy and ragged-red fiber disease (MERRF) is associated with a mitochondrial DNA tRNA^{Lys} mutation. Cell 61, 931–937.

Shoffner, J.M. and Wallace, D.C. (1995) Oxidative phosphorylation diseases. In: Scriver, C.R., Beaudet, A.L., Sly, W.S., Valle, D. (Eds.) "The Metabolic and Molecular Bases of Inherited Disease." McGraw-Hill, New York, pp 1535–1609.

Smits, P.H.M., DeHaan, M., Maat, C. and Grivell, L.A. (1994) The complete sequence of a 33 kb fragment on the right arm of chromosome II from *Saccharomyces cerevisiae* reveals 16 open reading frames, including ten new open reading frames, five previously identified genes and a homologue of the *SCO1* gene. Yeast 10, S75-S80.

Sokol, R.J., Devereux, M.W., O'Brien, K., Khandwala, R.A. and Loehr, J.P. (1993) Abnormal hepatic mitochondrial respiration and cytochrome c oxidase activity in rats with long-term copper overload. Gastroenterology 105, 178–187.

Sokol, R.J., Twedt, D., McKim, J.M., Jr., Devereux, M.W., Karrer, F.M., Kam, I., Von Steigman, G., Narkewicz, M.R., Bacon, B.R., Britton, R.S. and Neuschwander-Tetri, B.A. (1994) Oxidant injury to hepatic mitochondria in patients with Wilson's disease and Bedlington terriers with copper toxicosis. Gastroenterology 107, 1788–1798.

Stadhouders, A.M. and Sengers, R.C.A. (1987) Morphological observations in skeletal muscle from patients with a mitochondrial myopathy. J. Inher. Metab. Dis. 10 (Suppl. 1), 62–80.

Sternlieb, I. (1992) Fraternal concordance of types of abnormal hepatocellular mitochondria in Wilson's disease. Hepatology 16, 728–732.

Taanman, J.-W. and Capaldi, R.A. (1992) Purification of yeast cytochrome c oxidase with a subunit composition resembling the mammalian enzyme. J. Biol. Chem. 267, 22481–22485.

Taanman, J.-W. and Capaldi, R.A. (1993) Subunit VIa of yeast cytochrome c oxidase is not necessary for assembly of the enzyme complex but modulates the enzyme activity. J. Biol. Chem. 268, 18754–18761.

Tsukihara, T., Aoyama, H., Yamashita, E., Tomizaki, T., Yamaguchi, H., Shinzawa-Itoh, K., Nakashima, R., Yaono, R. and Yoshikawa, S. (1995) Structures of metal sites of oxidized bovine heart cytochrome c oxidase at 2.8 Å. Science 269, 1069–1074.

Tsukihara, T., Aoyama, H., Yamashita, E., Tomizaki, T., Yamaguchi, H., Shinzawa-Itoh, K., Nakashima, R., Yaono, R. and Yoshikawa, S. (1996) The whole structure of the 13-subunit oxidized cytochrome c oxidase at 2.8 Å. Science 272, 1136–1144.

Tkeshelashvili, L.K., McBride, T., Spence, K. and Loeb, L.A. (1991) Mutation spectrum of copper-induced DNA damage. J. Biol. Chem. 266, 6401–6406.

Tzagoloff, A., Capitanio, N., Nobrega, M.P. and Gatti, D. (1990) Cytochrome oxidase assembly in yeast requires the product of *COXII*, a homolog of the P. denitrificans protein encoded by ORF3. EMBO J. 9, 2759–2764.

Tzagoloff, A. and Dieckmann, C.L. (1990) *PET* genes of *Saccharomyces cerevisiae*. Microbiol. Rev. 54, 211–225.

Tzagoloff, A., Nobrega, M., Gorman, N. and Sinclair, P. (1993) On the functions of the yeast COX10 and COX11 gene products. Biochem. Molec. Biol. Int. 31, 593–598.

Valentine, J.S. and Gralla, E.B. (1997) Delivering copper inside yeast and human cells. Science 278, 817–818.

Wallace, D.C. (1992) Diseases of the mitochondrial DNA. Ann. Rev. Biochem. 61, 1175–1212.

Wallace, D.C., Singh, G., Lott, M.T., Hodge, J.A., Schurr, T.G., Lezza, A.M., Elsas, L.J. II, Nikoskelainen, E.K. (1988) Mitochondrial DNA mutation associated with Leber's hereditary optic neuropathy. Science 242, 1427–1430.

Wright, R.M., Dircks, L.K. and Poyton, R.O. (1986) Characterization of *COX9*, the nuclear gene encoding the yeast mitochondrial protein cytochrome c oxidase subunit VIIa. J. Biol. Chem. 261, 17183–17191.

Wright, R.M., Ko, C., Cumsky, M.G. and Poyton, R.O. (1984) Isolation and sequence of the structural gene for cytochrome c oxidase subunit VI *from Saccharomyces cerevisiae*. J. Biol. Chem. 259, 15401–15407.

Yuan, D.S., Stearman, R., Dancis, A., Dunn, T., Beeler, T. and Klausner, R.D. (1995) The Menkes/Wilson disease gene homologue in yeast provides copper to a ceruloplasmin-like oxidase required for iron uptake. Proc. Natl. Acad. Sci. USA 92, 2632–2636.

18

THE COPPER-TRANSPORTING ATPase DEFECTIVE IN WILSON DISEASE

Diane W. Cox, John R. Forbes, and Manoj S. Nanji

Department of Medical Genetics
University of Alberta
Edmonton, Alberta T6G 2H7

1. INTRODUCTION

Two critical genes have been recognized as essential for the export of copper from cells. The proteins encoded by these genes are very similar, yet perform different functions. Between them, they play a role in controlling copper levels in cells, providing for copper essential for number of enzymes, while preventing the accumulation of toxic levels of copper in cells. The genes for Menkes disease (designated *ATP7A*) and Wilson disease (designated *ATP7B*) were cloned in 1993. Their discovery has increased our knowledge of the basic mechanisms of copper transport, and has also made possible practical applications to diagnosis.

Studies to date indicate that copper is never present as a free ion within the body, but is always bound to a transporter or a storage protein. Copper is transported in the blood by albumin, and by trace amounts of copper histidine, ceruloplasmin, and possibly transcupreine, and other proteins. Details of transport of copper throughout the body are not well established. Protein interactions and cell receptors are important for copper transport. *CTR1* , similar to a yeast gene, is proposed to transport copper into cells (Zhou and Gitschier, 1997). Most ingested copper is transported to the liver, where a portion is incorporated into apoceruloplasmin to form ceruloplasmin, a cerulean blue plasma α2-glycoprotein of a 132 kDa which binds six copper atoms. Ceruloplasmin may be involved in the transport of copper into tissues (Linder, 1991) and, interestingly, plays a role in controlling entry of iron into cells (Mukhopadhyay et al., 1998). A deficiency of ceruloplasmin, caused by a defect in production of apoceruloplasmin (aceruloplasminemia) results in iron accumulation in tissues, demonstrating the critical importance of copper in normal iron transport (Miyajima et al., 1987; Harris et al., 1995). Within the cell, copper is apparently shuttled by a series of specific small proteins or chaperones, first identified in yeast and also present in humans, which direct copper to proteins such as copper-zinc superoxide dismutase, cytochrome oxidase and the copper-trans-

Metals and Genetics, edited by Sarkar.
Kluwer Academic / Plenum Publishers, New York, 1999.

porting ATPases (Valentine and Gralla, 1997). Most of the copper entering the liver is excreted via the bile. The export mechanism can utilize glutathione for a rapid low capacity export component for handling a large bolus of copper. The major, slower export is not dependent on glutathione and involves the copper transporting ATPases (Houwen et al., 1990). The induction of metallothioneins, low molecular weight proteins rich in cysteine residues provides a storage form for copper. Metallothioneins, encoded by a multigene family, are inducible by a variety of heavy metals, particularly zinc, cadmium and copper. Only a small fraction of copper enters the urine, except when an individual is on treatment with chelating agents. Most of the copper transported to the kidney is resorbed, yet kidney tissue has a higher concentration of copper than even the brain and liver (Linder, 1991).

The Menkes disease gene is expressed in many tissues, however expression in the intestine appears to be particularly critical. Copper ingested from the diet, at a level of 1 to 3.5 mg/day, is absorbed into the intestinal cells and exported for use by other parts of the body by the copper transporting ATPase encoded by the *ATP7A* gene. When deleted or mutated, transport out of the intestine does not take place and Menkes disease is the result. In this condition, the lack of copper export after absorption results in copper deficiency as manifest by defective connective tissue (inadequate lysyl oxidase), inadequate ability to scavenge free radicals (inadequate super oxide dismutase), inability to maintain cell energy (impaired transport of copper to cytochrome oxidase) and delayed mental development (a possible consequence of inadequate ß-mono-oxygenase).

Although the gene for Wilson disease (*ATP7B*) is similar to that for Menkes disease (Bull et al., 1993; Tanzi et al., 1993), because of the different expression pattern, different disease symptoms result . *ATP7B* is expressed mainly in liver and kidney. In Wilson disease, copper accumulates in liver, kidney and the brain, the latter accumulation being a result of either secondary accumulation from high levels of copper in plasma, or primary accumulation due to expression of the gene in the brain. The result of an impaired *ATP7B* gene is liver damage, acute or chronic, due to toxic levels of copper in the liver, and impaired neurological function. The disease can be expressed at any time from 3 years of age to 40 or more years of age. Treatment involves removal of copper by chelation or by high oral zinc (Cox and Roberts, 1998).

2. THE MENKES DISEASE (*ATP7A*) AND WILSON DISEASE (*ATP7B*) GENES

The gene for Menkes disease (*ATP7A*) , was isolated by positional cloning (Vulpe et al., 1993; Chelly et al., 1993; Mercer et al., 1993) aided by disruption of the gene by a chromosome translocation present in an affected female. The predicted gene product was identified as a copper transporting ATPase (Vulpe et al., 1993) because of the similarity of the amino acid sequence to bacterial ATPases associated with copper transport (Odermatt et al., 1992). The predicted protein is one of the P-type family of copper transporting ATPases found in many organisms, including yeast and bacteria.

The Wilson disease gene, on chromosome 13 at 13q24.3, was identified in our laboratory by positional cloning, aided by its homology with the Menkes gene (Bull et al., 1993). The resulting gene, *ATP7B*, is 57% identical to the Menkes gene. A shorter transcript was obtained from a brain cDNA library during a search for amyloid like genes (Tanzi et al., 1993). The predicted functionally important regions are similar for the genes in both disorders: 1) a transduction domain (amino acid residues 837–864), containing a Thr-Gly-Glu (TGE) motif involved in transduction of the energy of ATP hydrolysis to cation transport. 2) cation channel and phosphorylation domains (residues 971–1035), containing a highly conserved Asp-Lys-Thr-Gly-Thr (DKTGT) motif. The aspartate residue is predicted to form a phosphorylated intermediate during the cation transport cycle. A cysteine-proline-cysteine cluster, located forty three residues N-terminal to this aspartate, and within the predicted cation channel, may be involved in binding copper during transport across the membrane. 3) An ATP binding domain (residues 1240–1291), is highly conserved. Eight hydrophobic regions are predicted to span the cell membrane. Exon 1 is at least 20 kb 5' of exon 2. The genomic structure of *ATP7B* has been established, and consists of 21 or 22 exons (Petrukhin et al., 1994; Thomas et al., 1995a)

3. DNA MARKERS FLANKING THE WILSON DISEASE GENE

Di- or tri- nucleotide repeats, found throughout the genome are highly variable between individuals. These polymorphisms, easily detectable by PCR, are commonly used as genetic markers for many disorders. They can be useful in linkage studies, either through detecting genetic recombination events, or for searching for a common haplotype (combination of marker alleles on the same chromosome) which can identify the region containing the gene to be cloned. We used highly polymorphic markers to identify the region of the Wilson disease gene as a step in its cloning (Bull et al., 1993). Markers within one or two kb of the gene have been identified.

There are now at least 80 mutations identified in *ATP7B*. Because of the complexity in identifying the variety of different mutations, haplotypes can be used to guide direct mutation testing. A specific combination or haplotype tends to be associated with a specific mutation (Thomas et al., 1995a). A number of highly polymorphic markers have been described which closely flank the gene and are highly variable: D13S315, D13S314-WND-D13S301–214S133-D13S316/D13S129 (the latter two are the same marker) (Petrukhin et al., 1993; Thomas et al., 1993; Thomas et al., 1995a). Testing of at least one parent or child of the patient is necessary in order to obtain the haplotype when markers are different at each locus in a patient. After determining the haplotype, the mutation usually associated with that specific haplotype can be identified, and tested directly. This approach will become less important as additional mutations are identified and rapid screening tests are developed.

3.1. Diagnosis for Presymptomatic Sibs

Currently the main value of polymorphic markers is in the diagnosis of presymptomatic sibs. In some families, an overlap in all parameters can make diagnosis difficult. These microsatellite markers are so variable, usually with 10 to 15 different types of alleles, that parents within any one family will usually carry different lengths or alleles of each marker. The markers are therefore useful for following the segregation of the Wilson disease gene in most families. About 10% of heterozygotes have a decreased concentration of ceruloplasmin. Other diagnostic parameters may also be inconclusive, including urinary copper, hepatic copper, and even liver copper concentrations. While sometimes appearing to be presymptomatic, some have been proven by DNA marker studies to be heterozygotes (Cox and Billingsley, 1989). These atypical heterozygotes are not known to require treatment. DNA markers provide the most reliable method for specific discrimination between heterozygotes and presymptomatic sibs of patients. These markers can be used without specific knowledge of the mutation in a patient.

4. MUTATIONS IN THE GENE

4.1. The Spectrum of Mutations

Patients seldom have the full range of classical symptoms, including chronic liver disease, tremor or dystonia, and Kayser-Fleischer, rings as described by S. A. K. Wilson (Wilson, 1912). No biochemical or clinical sign appears to be specific to Wilson disease only (reviewed in (Danks, 1995)). Therefore the possibility of identifying a specific mutation is an important aid to diagnosis. Ideally, a molecular analysis would quickly identify the majority of mutations. While much progress has been made, this is not the case at present. Direct molecular diagnosis is challenging because of the occurrence of a large number of mutations, more than 80 at present.

In Menkes disease, large gene deletions are found in about 20% of patients (Chelly et al., 1993; Mercer et al., 1993). In the LEC rat, a rodent model for Wilson disease, the terminal approximately 25% of the gene is deleted (Wu et al., 1994). No large gene deletions have been described in humans. The various mutations described to date in *ATP7B* are predominantly missense mutations (Table 1). However the distribution of mutation types is dependent upon the methods of analysis used, which have differed in the various studies reported. Mutation detection has generally been carried out by single strand conformation polymorphism analysis (SSCP), an approach we have found useful in our laboratory (Thomas et al., 1995a). Those samples showing a shift of one or both bands can then be sequenced to identify the exact mutation. In some studies, complete gene sequencing has been carried out in all patients.

The mutations differ in both sequence and frequency between ethnic groups. The most common mutation, histidine1069glutamine (H1069Q, originally reported as H1070Q), occurs in a conserved five amino acid sequence between transmembrane domain 6 and the ATP binding region (Tanzi et al., 1993). This mutation makes up at least 30% of all mutations in populations of European origin, but all other mutations in this population are infrequent. A common missense mutation (arginine778leucine; Arg778Leu, R778L) is found in oriental populations (Thomas et al., 1995a). The first 41 mutations reported are shown, with their position in the gene, in Figure 1. A list of some of the reported variants has been published (Cox, 1996). A HUGO Database for mutations in the

Table 1. Type of mutations in genes for Wilson disease
(*ATP7B)* and Menkes disease (*ATP7A)*

Type of mutation	Wilson disease ATP7B[1] %	Menkes disease ATP7A[2] %
Deletion—large	–	20
Deletion—several bases	29	25
Insertions	7	12
Splice site	8	16
Nonsense	5	20
Missense	51	2

[1]Figures are based on those currently recorded for the HUGO Database, mainly from (Thomas et al., 1995a; Figus et al., 1996; Tanzi et al., 1993; Shah et al., 1997). 'Missense' mutations could in some cases be normal variants.

[2]Figures mainly from (Tumer et al., 1997; Chelly et al., 1993; Mercer et al., 1993) Menkes figures are updated by Tumer in this Symposium.

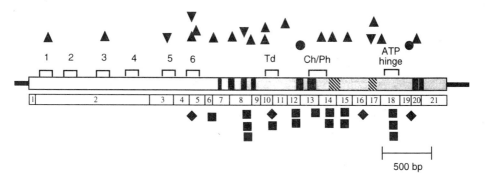

Figure 1. Mutations identified in *WND* (*ATP7B*) gene in patients with Wilson disease. The type of mutation is indicated as follows: ● nonsense; ▲ small deletion; ▼ small insertion; ◆ splice site; ■ missense. (From: Cox, D.W. (1996). Molecular advances in Wilson disease. In Progress in Liver Diseases. J.L. Boyer and R.K. Ockner, eds. (W.B. Saunders), pp. 245–264: with permission.)

Wilson disease gene is in preparation and will appear on the internet. This database now lists 80 mutations.

Sometimes it is difficult to determine if an amino acid change is disease related or a normal variant. If the change is not found in 100 normal chromosomes, it is considered to be disease specific. Not all variants have been tested in this way, so some could actually be normal variants. More than 20 conservative changes believed to be normal variants have been reported (Figus et al., 1996; Thomas et al., 1995a). The missense mutations are clustered in the ATP binding domain and in the transmembrane domains. Missense mutations have not been reported in the copper binding domains. Missense mutations in this region may have little effect because of redundancy of the domain. Current studies are shedding light on the specific domains required for copper binding and the tertiary structure involved. In rat and mouse, the sequence in the fourth copper binding domain has no C-X-X-C motif for binding copper (Wu et al., 1994; Theophilus et al., 1996), suggesting the domain is not critical for gene function.

4.2. Potential Mutations Involving Transcriptional Regulation of *ATP7B*

mRNA analysis of *ATP7B* indicates that the gene is differentially expressed in different tissues, with high levels of transcript in the liver and kidney (Bull et al., 1993). As with many other eukaryotic genes, the regulatory sequence for *ATP7B* transcription may reside at the 5'-upstream region of the gene. Primer extension analysis of total RNA from human liver sample has revealed multiple *ATP7B* transcription start sites (Nanji et al., 1997). The sequence around the start site is highly GC-rich and lacks a TATA or a CAAT box. We also note two metal response elements (MREs) and several potential binding sites for transcription factors including AP-1 and SP-1. Through analysis of DNA samples from 21 Japanese patients with Wilson disease, we identified three sequence changes, 75A->C, 78A->C, 128C->A and a 5 bp deletion (-138delAGCCG) 5' of the first methionine of *ATP7B*. These sequence changes were not observed in the normal chromosomes of the heterozygous parents of the patients, but an adequate number of normal Japanese chromosomes were not available for study. The sequence change 75A->C has recently been reported as a common polymorphism (normal variant) in the Mediterranean population

(Loudianos et al., 1998). Any potential association between these sequence changes in the 5' UTR of *ATP7B* and disease status is as yet unclear and will require direct testing.

4.3. Using a Yeast Assay to Discriminate between Normal and Disease-Causing Mutations

The copper transporter encoded by *ATP7B* is similar to transporters found in a wide range of bacteria (reviewed in (Silver et al., 1989; Bull and Cox, 1994), with high identity in functionally important domains. In bacteria, the copper transporter is frequently encoded on a plasmid. The bacteria can develop copper resistance rapidly by increasing replication of the plasmids. Most of these ATPases transport copper out of the cell.

Yeast, *Saccharomyces cerevisaie*, also has a copper transporting P-type ATPase gene (*CCC2*) similar to the Wilson disease gene, encoding a protein that serves to transport copper and is critical in iron transport. The plasma membrane protein Ctr1p transports copper with high-affinity into the cytoplasm (Dancis et al., 1994). Copper is carried in the cytoplasm by the soluble copper chaperone Atx1p, (Lin et al., 1997) which delivers copper to Ccc2p. Ccc2p is the yeast homologue of ATP7B sharing approximately 29% amino acid identity. Copper is supplied by Ccc2p to the plasma membrane oxidase Fet3p, similar in function to ceruloplasmin in humans, which functions with the high-affinity iron permease Ftr1p to import iron into the yeast cell (Stearman et al., 1996; Yuan et al., 1997). When yeast cells lack Ccc2p, copper is not incorporated into Fet3p, and subsequently the cells lack high-affinity iron uptake (Yuan et al., 1995). In the absence of high-affinity iron uptake, *ccc2* mutant yeast are unable to grow on iron limited medium (Stearman et al., 1996). We have used the high affinity iron uptake phenotype of *ccc2* mutant yeast to create an assay to study the function of ATP7B. ATP7B is able to replace Ccc2p in yeast delivering copper to Fet3p restoring high-affinity iron uptake (J.Forbes, unpublished data). We have used this assay to study ten mutations found in the *ATP7B* gene of patients with Wilson disease. Variant genes were created using site-directed mutagenesis and expressed in *ccc2* mutant yeast from a single-copy vector. Growth curves generated from yeast expressing mutant *ATP7B* cDNA constructs were used to quantitate the relative ability of mutant ATP7B proteins to replace Ccc2p (Figure 2). In the example shown, the common oriental mutation Arg778 Leu construct shows partial capacity to complement ccc2, while the Thr977Met construct is completely inactive. One of the putative normal variants tested was unable to complement *ccc2*.

ATP7B is normally localized in the trans-Golgi network (TGN) (Hung et al., 1997). When cells are exposed to elevated copper levels, ATP7B, similarly to ATP7A (Petris et al., 1996) traffics from the TGN to an undefined vesicular compartment, and appears to be continuously recycled between compartments. The observed movement seems to represent a change in the steady state protein localization in response to copper (Hung et al., 1997). ATP7B has at least two functions in the cell. One role is to deliver copper to ceruloplasmin within the Golgi apparatus (Murata et al., 1995). The other role is to transport excess copper out of the hepatocyte (Danks, 1995). The observed copper-induced trafficking may represent a post-translationally inducible switch from a primarily biosynthetic role in the trans Golgi network to a primarily excretory role under conditions of copper overload. Wilson disease missense mutations may affect the copper dependent trafficking event. Over 50% of patients with Wilson disease have normal levels of ceruloplasmin often in the presence of severe liver damage caused by copper accumulation (Cauza et al., 1997). Therefore a Wilson disease mutant ATP7B protein (even when present as one allele in a compound heterozygote) with partial or normal function may be unable to exit the

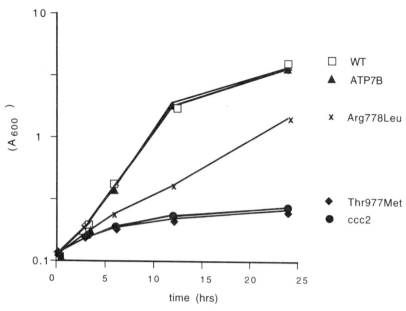

Figure 2. Growth curves of yeast in iron-limited medium at 30°C. Wilson disease ATP7B mutant proteins, designated by mutation, were expressed in *ccc2* mutant yeast from a single-copy integrating expression vector. Growth was monitored spectroscopically for 24 hours.

trans Golgi network in response to copper. This active mutant ATP7B protein could provide copper to ceruloplasmin as normal, but would be unable to perform its excretory function resulting in copper accumulation and hepatic disease. The intracellular localization of Wilson disease mutant ATP7B proteins in mammalian cells must be analyzed to obtain a clear picture of ATP7B function in normal and disease states.

5. COMPARISON OF SPECIFIC MUTATION (GENOTYPE) WITH PHENOTYPE

We have identified mutations in more than 100 patients. The deletions, duplications, and nonsense mutations would be predicted to truncate the mRNA or the protein and severely effect function. The average age of onset is 7.2 years for patients with mutations predicted to destroy the function of the gene (Cox, unpublished). Children as young as 3 years of age and perhaps even earlier, can show signs and symptoms of overt liver disease due to Wilson disease. For mutations which we predict would be less detrimental, the average age at onset is 16.8 years. Missense mutations can impair function to variable degrees. In the mutation, Arg778Leu, transmembrane domain 4 is altered, apparently having a pronounced effect on copper transport. The most common mutation, H1069Q, has been associated with a relatively late onset in Dutch, German and Austrian populations (Houwen et al., 1995; Maier-Dobersberger et al., 1997). In our own studies, the age of onset is relatively high in patients of Eastern European origin, but is variable from 9 or 10 years of age to more than 30 years of age (Thomas et al., 1995a).

The effect of insertions and deletions is not always predictable. Skipping of an exon containing a mutation is possible. We have shown that alternatively spliced forms can be produced from the mutant DNA, suggesting that such forms may have a function and may even modify disease severity (Thomas et al., 1995b). Splice site mutations have not been studied in detail, but results from Menkes disease mutations suggest that some normal product may be formed. Some of the mutations in the *ATP7A* gene lead to exon skipping (Das et al., 1994) and, because of the production of some normal product, these and splice site mutations lead to a milder form of Menkes disease including the occipital horn syndrome, affecting connective tissue (Kaler et al., 1994; Das et al., 1995).

Other genetic factors in addition to the specific mutation also influence the severity and type of expression (phenotype) of the disease. Much of the damage of copper probably occurs because of free radical production, as demonstrated in copper-loaded animals and in patients with Wilson disease (Sokol et al., 1994). This leads to increased lipid peroxidation and impaired mitochondrial respiration (Sokol et al., 1993). Nutritional status, particularly anti-oxidant status, could be an important factor influencing clinical variability. Nutritional factors and genetic status could influence anti-oxidant status of the patient. Other genes in the copper pathway can also play a role in disease variability. This is a fruitful area for further research.

ACKNOWLEDGMENTS

Unpublished studies from our laboratory were supported by grants from the National Science and Engineering Council, Canada, and the Canadian Genetic Diseases Network. J.R.F. holds a graduate scholarship from the Medical Research Council, Canada.

REFERENCES

Bull, P.C., Barwell, J.A., Hannah, H.T-L., Pautler, S.E., Higgins, M.J., Lalande, M., and Cox, D.W. (1993). Isolation of new probes in the region of the Wilson disease locus, 13q14.2–14.3. Cytogenet. Cell Genet. *64*, 12–17.

Bull, P.C. and Cox, D.W. (1994). Wilson disease and Menkes disease: new handles on heavy-metal transport. Trends Genet. *10*, 246–252.

Bull, P.C., Thomas, G.R., Rommens, J.M., Forbes, J.R., and Cox, D.W. (1993). The Wilson disease gene is a putative copper transporting P-type ATPase similar to the Menkes gene [erratum appears in Nat Genet 1994 6:214]. Nature Genet. *5*, 327–337.

Cauza, E., Maier-Dobersberger, T., Polli, C., Kaserer, K., Kramer, L., and Ferenci, P. (1997). Screening for wilson's disease in patients with liver diseases by serum ceruloplasmin. J. Hepatol. *27*, 358–362.

Chelly, J., Tumer, Z., Tonnesen, T., Petterson, A., Ishikawa-Brush, Y., Tommerup, N., Horn, N., and Monaco, A.P. (1993). Isolation of a candidate gene for Menkes disease that encodes a potential heavy metal binding protein. Nature Genet. *3*, 14–19.

Cox, D.W. (1996). Molecular advances in Wilson disease. In Progress in Liver Diseases. J.L. Boyer and R.K. Ockner, eds. (W.B.Saunders), pp. 245–264.

Cox, D.W. and Billingsley, G.D. (1989). The application of DNA markers to the diagnosis of presymptomatic Wilson disease. In Genetics of Neuropsychiatric Diseases. Wenner-Gren International Symposium. L. Wetterberg, ed. (London: Macmillan), pp. 167–177.

Cox, D.W. and Roberts, E.A. (1998). Wilson disease. In Sleisenger and Fordtran's Gastrointestinal and Liver Disease. M. Feldman, B.F. Scharschmidt, and M.H. Sleisenger, eds. (W.B.Saunders), pp. 1104–1111.

Dancis, A., Yuan, D.S., Haile, D., Askwith, C., Eide, D., Moehle, C., Kaplan, J., and Klausner, R.D. (1994). Molecular characterization of a copper transport protein in S. cerevisiae: An unexpected role for copper in iron transport. Cell *76*, 393–402.

Danks, D.M. (1995). Disorders of copper transport. In The Molecular and Metabolic Basis of Inherited Disease. C.R. Scriver, A.L. Beaudet, W.S. Sly, and D. Valle, eds. (New York: McGraw-Hill), pp. 4125–4158.

Das, S., Levinson, B., Vulpe, C., Whitney, S., Gitschier, J., and Packman, S. (1995). Similar splicing mutations of the Menkes/mottled copper transporting ATPase gene in X-linked cutis laxa and the blotchy mouse. Am. J. Hum. Genet. 56, 570–576.

Das, S., Levinson, B., Whitney, S., Vulpe, C., Packman, S., and Gitschier, J. (1994). Diverse mutations in patients with Menkes disease often lead to exon skipping. Am. J. Hum. Genet. 55, 883–889.

Figus, A., Angius, A., Loudianos, G., Bertini, C., Dessi, V., Loi, A., and Deiana, A. (1996). Molecular pathology and haplotype analysis of Wilson disease in Mediterranean populations. Am. J. Hum. Genet. 57, 1318–1324.

Harris, Z.L., Takahashi, Y., Miyajima, H., Serizawa, M., MacGillivray, R.T., and Gitlin, J.D. (1995). Aceruloplasminemia: molecular characterization of this disorder of iron metabolism. Proc. Natl. Acad. Sci. USA 92, 2539–2543.

Houwen, R., Dijkstra, M., Kuipers, I., Smit, E.P., Havinga, R., and Vonk, R.J. (1990). Two pathways for biliary copper excretion in the rat. The role if glutathione. Biomed Pharm 39, 1039–1044.

Houwen, R.H., Juyn, J., Hoogenraad, T.U., Ploos van Amstel, J.K., and Berger, R. (1995). H714Q mutation in Wilson disease is associated with late, neurological presentation. J. Med. Genet. 32, 480–482.

Hung, I.H., Suzuki, M., Yamaguchi, Y., Yuan, D.S., Klausner, R.D., and Gitlin, J.D. (1997). Biochemical characterization of the wilson disease protein and functional expression in the yeast saccharomyces cerevisiae. J. Biol. Chem. 272, 21461–21466.

Kaler, S.G., Gallo, L.K., Proud, V.K., Percy, A.K., Mark, Y., Segal, N.A., Goldstein, D.S., Holmes, C.S., and Gahl, W.A. (1994). Occipital horn syndrome and a mild Menkes phenotype associated with splice site mutations at the MNK locus. Nature Genet. 8, 195–202.

Lin, S.J., Pufahl, R.A., Dancis, A., O'Halloran, T.V., and Culotta, V.C. (1997). A role for the Saccharomyces cerevisiae ATX1 gene in copper trafficking and iron transport. J. Biol. Chem. 272, 9215–9220.

Linder, M.C. (1991). The Biochemistry of Copper (New York: Plenum).

Loudianos, G., Dessi, V., Lovicu, M., Angius, A., Cao, A., and Pirastu, M. (1998). The -75a-->c substitution in the 5' utr of the wilson disease gene is a sequence polymorphism in the mediterranean population [letter]. Am. J. Hum. Genet. 62, 484–485.

Maier-Dobersberger, T., Ferenci, P., Polli, C., Balac, P., Dienes, H.P., Kaserer, K., Datz, C., Vogel, W., and Gangl, A. (1997). Detection of the his1069gln mutation in wilson disease by rapid polymerase chain reaction [see comments]. Ann. Intern. Med. 127, 21–26.

Mercer, J.F.B., Livingstone, J., Hall, B., Paynter, J.A., Begy, C., Chandrasekharappa, S., Lockhart, P., Grimes, A., Bhave, M., Siemieniak, D., and Glover, T.W. (1993). Isolation of a partial candidate gene for Menkes disease by positional cloning. Nature Genet. 3, 20–25.

Miyajima, H., Nishimura, Y., Mizoguchi, K., Sakamoto, M., Shimizu, T., and Honda, N. (1987). Familial apoceruloplasmin deficiency associated with blepharospasm and retinal degeneration. Neurology 37, 761–767.

Mukhopadhyay, C.K., Attieh, Z.K., and Fox, P.L. (1998). Role of ceruloplasmin in cellular iron uptake. Science 279, 714–717.

Murata, Y., Yamakawa, E., Iizuka, T., Kodama, H., Abe, T., Seki, Y., and Kodama, M. (1995). Failure of copper incorporation into ceruloplasmin in the Golgi apparatus of LEC rat hepatocytes [erratum appears in Biochem Biophys Res Commun (1995), 211:348]. Biochem. Biophys. Res. Commun. 209, 349–355.

Nanji, M.S., Nguyen, V.T., Kawasoe, J.H., Inui, K., Endo, F., Nakajima, T., Anezaki, T., and Cox, D.W. (1997). Haplotype and mutation analysis in Japanese patients with Wilson disease. Am. J. Hum. Genet. 60, 1423–1429.

Odermatt, A., Suter, H., Krapf, R., and Solioz, M. (1992). An ATPase operon involved in copper resistance by Enterococcus hirae. Ann. N. Y. Acad. Sci. 484–486.

Petris, M.J., Mercer, J.F., Culvenor, J.G., Lockhart, P., Gleeson, P.A., and Camakaris, J. (1996). Ligand-regulated transport of the Menkes copper P-type ATPase efflux pump from the Golgi apparatus to the plasma membrane: a novel mechanism of regulated trafficking. EMBO J. 15, 6084–6095.

Petrukhin, K., Fischer, S.G., Pirastu, M., Tanzi, R.E., Chernov, I., Devoto, M., Brzustowicz, L.M., Cayanis, E., Vitale, E., and Russo, J.J. (1993). Mapping, cloning and genetic characterization of the region containing the Wilson disease gene. Nature Genet. 5, 338–343.

Petrukhin, K., Lutsenko, S., Chernov, I., Ross, B.M., Kaplan, J.H., and Gilliam, T.C. (1994). Characterization of the Wilson disease gene encoding a P-type copper transporting ATPase: genomic organization, alternative splicing, and structure/function predictions. Hum. Mol. Genet. 3, 1647–1656.

Shah, A.B., Chernov, I., Zhang, H.T., Ross, B.M., Das, K., Lutsenko, S., Parano, E., Pavone, L., Evgrafov, O., Ivanova-Smolenskaya, I.A., Anneren, G., Westermark, K., Urrutia, F.H., Penchaszadeh, G.K., Sternlieb, I., Scheinberg, I.H., Gilliam, T.C., and Petrukhin, K. (1997). Identification and analysis of mutations in the wilson disease gene (atp7b): population frequencies, genotype-phenotype correlation, and functional analyses. Am. J. Hum. Genet. 61, 317–328.

Silver, S., Nucifora, G., Chu, L., and Misra, T.K. (1989). Bacterial resistance ATPases: primary pumps for exporting toxic cations and anions. Trends. Biochem. Sci. *14*, 76–80.

Sokol, R., Twedt, D., McKim, J.M., Devereux, M.W., Karrer, F.M., Kam, I., von Steigman, G., Narkewicz, M.R., Bacon, B.R., Britton, R.S., and Neuschwander-Tetri, B.A. (1994a). Oxidant injury to hepatic mitochondria in patients with Wilson's disease and Bedlington terriers with copper toxicosis. Gastroenterology *107*, 1788–1798.

Sokol, R.J., Devereaux, M.W., O'Brien, K., Khandwala, R.A., Loehr, J.P., Narkewicz, M.R., Bacon, B.R., Britton, R.S., and Neuschwander-Tetri, B.A. (1993b). Abnormal hepatic mitochondrial respiration and cytochrome C oxidase activity in rats with long-term copper overload. Gastroenterology *105*, 178–187.

Stearman, R., Yuan, D.S., Yamaguchi-Iwai, Y., Klausner, R.D., and Dancis, A. (1996). A permease-oxidase complex involved in high-affinity iron uptake in yeast [see comments]. Science *271*, 1552–1557.

Tanzi, R.E., Petrukhin, K.E., Chernov, I., Pellequer, J.L., Wasco, W., Ross, B., Romano, D.M., Parano, E., Pavone, L., Brzustowicz, L.M., Devoto, M., Peppercorn, J., Bush, A.I., Sternlieb, I., Pirastu, M., Gusella, J.F., Evgrafov, O., Penchaszadeh, G.K., Honig, B., Edelman, I.S., Soares, M.B., Scheinberg, I.H., and Gilliam, T.C. (1993). The Wilson disease gene is a copper transporting ATPase with homology to the Menkes disease gene. Nature Genet. *5*, 344–350.

Theophilus, M.B., Cox, D.W., and Mercer, J.F.B. (1996). The toxic milk mouse is a murine model of Wilson disease . Hum Mol Genet (in press)

Thomas, G.R., Forbes, J.R., Roberts, E.A., Walshe, J.M., and Cox, D.W. (1995a). The Wilson disease gene: spectrum of mutations and their consequences [published erratum appears in Nat Genet 1995 Apr;9(4):451]. Nature Genet. *9*, 210–217.

Thomas, G.R., Jensson, O., Gudmundsson, G., Thorsteinsson, L., and Cox, D.W. (1995b). Wilson disease in Iceland: a clinical and genetic study. Hum. Genet. *56*, 1140–1146.

Thomas, G.R., Roberts, E.A., Rosales, T.O., Moroz, S.P., Lambert, M.A., Wong, L.T.K., and Cox, D.W. (1993). Allelic association and linkage studies in Wilson disease. Hum Mol Genet *2*, 1401–1405.

Thomas, G.R., Roberts, E.A., Walshe, J.M., and Cox, D.W. (1995). Haplotypes and mutations in Wilson disease. Am. J. Hum. Genet. *56*, 1315–1319.

Tumer, Z., Lund, C., Tolshave, J., Vural, B., Tonnesen, T., and Horn, N. (1997). Identification of point mutations in 41 unrelated patients affected with Menkes disease. Am. J. Hum. Genet. *60*, 63–71.

Valentine, J.S. and Gralla, E.B. (1997). Delivering copper inside yeast and human cells [comment]. Science *278*, 817–818.

Vulpe, C., Levinson, B., Whitney, S., Packman, S., and Gitschier, J. (1993). Isolation of a candidate gene for Menkes disease and evidence that it encodes a copper transporting ATPase (Correction). Nature Genet. *3*, 273.

Wilson, S.A.K. (1912). Progressive lenticular degeneration: a familial nervous disease associated with cirrhosis of the liver. Brain *34*, 295–509.

Wu, J., Forbes, J.R., Chen, H.S., and Cox, D.W. (1994). The LEC rat has a deletion in the copper transporting ATPase gene homologous to the Wilson disease gene. Nature Genet. *7*, 541–545.

Yuan, D.S., Dancis, A., and Klausner, R.D. (1997). Restriction of copper export in *Saccharomyces cerevisiae* to a late Golgi or post-Golgi compartment in the secretory pathway. J. Biol. Chem. *272*, 25787–25793.

Yuan, D.S., Stearman, R., Dancis, A., Dunn, T., Beeler, T., and Klausner, R.D. (1995). The Menkes/Wilson disease gene homologue in yeast provides copper to a ceruloplasmin-like oxidase required for iron uptake. Proc. Natl. Acad. Sci. USA *92*, 2632–2636.

Zhou, B. and Gitschier, J. (1997). hCTR1: a human gene for copper uptake identified by complementation in yeast. Proc. Natl. Acad. Sci. USA *94*, 7481–7486.

19

TOWARDS THE CHARACTERIZATION OF THE WILSON DISEASE COPPER ATPase METAL BINDING DOMAIN

Michael DiDonato[1] and Bibudhendra Sarkar[1,2*]

[1]Department of Biochemistry
University of Toronto
Toronto, Ontario, Canada, M5S 1A8
[2]Department of Structural Biology and Biochemistry
The Hospital for Sick Children
555 University Ave.
Toronto, Ontario, Canada, M5G 1X8

1. INTRODUCTION

1.1. Identification of the Gene: An Important Step Forward

Wilson disease is one of two major genetic disorders of copper metabolism in humans, the other being Menkes disease (Danks, 1995; DiDonato and Sarkar, 1997; Vulpe and Packman, 1995). It is inherited in an autosomal recessive manner and the gene responsible for the disorder has been identified on chromosome 13 (Bull et al., 1993; Tanzi et al., 1993) and has been predicted to encode a copper transporting P-type ATPase (ATP7B). The gene responsible for Menkes disease was identified and cloned before that for Wilson disease. Both genes codes for a copper transporting ATPase with a high degree of homology to each other. (Chelly et al., 1993; Mercer et al., 1993; Vulpe et al., 1993). Further analysis of the predicted gene products revealed similarities with other metal transporting ATPases both in yeast and bacteria. Besides all the domains expected to be found in a cation transporting P-type ATPase (phosphorylation, transduction, ATP binding, etc), other domains which are conserved among all heavy metal ATPases, are also present (Figure 1). These include a conserved Cys-Pro-Cys motif in transmembrane segment 6, a conserved Ser-Glu-His-Pro-Leu motif in a cytoplasmic loop, and a large N-terminal domain which

*Corresponding author. Fax: +1 416–813–5379.

Metals and Genetics, edited by Sarkar.
Kluwer Academic / Plenum Publishers, New York, 1999.

A

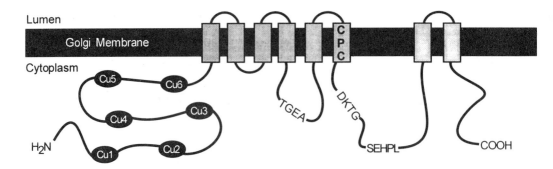

B

WDCu1	ILGMT**C**QS**C**VKSIEDRISNLKGIISMKVSL
WDCu2	VEGMT**C**QS**C**VSSIEGKVRKLQGVVRVKVSL
WDCu3	IDGMH**C**KS**C**VLNIEENIGQLLGVQSIQVSL
WDCu4	IAGMT**C**AS**C**VHSIEGMISQLEGVQQISVSL
WDCu5	IKGMT**C**AS**C**VSNIERNLQKEAGVLSELVAL
WDCu6	ITGMT**C**AS**C**VHNIESKLTRTNGITYASVAL
ATX1	VV-MT**C**SG**C**SGAVNKVLTKLEPDVSKIDIS
HAH1	VD-MT**C**GG**C**AEAVSRVLNKL--GGVKYDID
CopA	ITGMT**C**AN**C**SARIEKELNEQPGVMSATVNL
MerP	ITGMT**C**DS**C**AVHVKDALEKVPGVQSADVCT
CadA	VQGFT**C**AN**C**AGKFEKNVKKIPGVQDAKVNF

Figure 1. Predicted structure of the Wilson/Menkes copper ATPases. A. The Wilson/Menkes copper transporting ATPases contain many motifs common to other cation transporting ATPases (phosphorylation, ATP binding, transduction, phosphatase). Several features are common only to heavy metal transporting ATPases. These being a large N-terminal metal binding domain, a conserved Cys-Pro-Cys motif in the sixth transmembrane segment, and a Ser-Glu-His-Pro-Leu motif of unknown function. B. Alignment of the metal binding regions of various heavy metal carrier/transporter proteins. WDCu1–6, Wilson disease ATPase metal binding domains 1 to 6, ATX1, yeast copper chaperone protein, HAH1, human copper chaperone protein, CopA, *Enterococcus hirae* copper transporting ATPase, MerP, bacterial mercury binding protein, CadA, bacterial cadmium transporting ATPase. The conserved metal binding motif Gly-Xaa-Xaa-Cys-Xaa-Xaa-Cys contains two cysteine residues which are critical for the correct ligation of the metal atom.

contain between one and six copies of the metal binding motif Gly-Xaa-Xaa-Cys-Xaa-Xaa-Cys. The N-terminal domains of both the Menkes and Wilson disease proteins contain six copies of this motif while only one or two are found in yeast or bacterial orthologues.

The Wilson disease gene is expressed at high levels in the liver and to a lesser extent in the kidney, brain, and placenta (Bull et al., 1993; Tanzi et al., 1993). In contrast, the

Menkes disease gene is expressed in all tissues except the liver (Chelly et al., 1993; Mercer et al., 1993; Vulpe et al., 1993). Although it is generally accepted that the large N-terminal metal binding domain in both the Menkes and Wilson disease ATPases are responsible for ligating metal prior to transport it has also been hypothesized that the domain may play a role in the regulation or localization of the protein in response to the copper level within the cell.

1.2. Intracellular Copper Transport: The Picture Comes into Focus

Although various aspects of copper transport have been studied for many years, until recently little was known about the mechanisms by which cells absorbed copper and how intracellular copper levels were regulated. The discovery of both the Wilson and Menkes disease copper ATPases has shed some light on these processes. Further advances have occurred through the elucidation of the copper transport pathways in yeast which share many similarities with the human copper transport pathways (Figure 2). In yeast cellular copper uptake occurs through the membrane bound, high affinity copper transporter CTR1 (Dancis et al., 1994). Through complementation studies in yeast in which the CTR1 gene has been knocked out, a human orthologue of the yeast CTR1 gene has been found (hCTR1) (Zhou and Gitschier, 1997). Once copper is taken up by the cell, it enters one of several recently identified trafficking pathways. Early studies identified glutathione as being one of the major copper chelators in the cytoplasm sequestering it in the Cu(I) oxidation state. Studies on copper loaded hepatoma cells indicated that greater than 60% of the cellular copper was in the Cu(I)-GSH form (Freedman et al., 1989). Furthermore, it has been shown that the Cu(I)-GSH complex is able to fully reconstitute apo-superoxide dismutase (SOD), apo-ceruloplasmin, and apo-metallothionein (Ciriolo et al., 1990; Ferreira et al., 1993; Musci et al., 1996).

However, recent studies have uncovered a family of proteins which are believed to act as copper chaperones delivering copper to specific targets (Figure 2). Three copper chaperone proteins have been identified in yeast and all have been found to have human orthologues which are functional when expressed in yeast defective in the corresponding protein. Cox17/human Cox17 are involved in the transport of copper to the mitochondria (Amaravadi et al., 1997; Glerum et al., 1996), Lys7/human Ccs1 targets copper to CuZn-SOD (Culotta et al., 1997), and Atx1/human Hah1 transports copper to a post-golgi compartment by way of Ccc2, the yeast orthologue of ATP7B (Klomp et al., 1997; Lin et al., 1997; Yuan et al., 1995). Both Atx1 and Hah1 contain one repeat of the metal binding motif Gly-Xaa-Xaa-Cys-Xaa-Xaa-Cys which has been shown in the case of Atx1, through yeast two-hybrid experiments, to directly contact the N-terminus of Ccc2 which contains two copies of the motif (Pufahl et al., 1997). The presence of the metal binding motif in both Atx1 and Ccc2 as well as evidence for a direct interaction strongly suggests that Atx1 transfers copper to the N-terminal domain of Ccc2 by the exchange of thiol ligands between the two motifs. Such a mechanism is most likely operational in humans involving Hah1 and ATP7B, although such an interaction between these two proteins is yet to be delineated.

Immunohistochemical studies have shown that under normal cytosolic copper concentrations the Wilson disease protein (ATP7B) is localized to a post-golgi compartment (Hung et al., 1997; Nagano et al., 1998). These studies have also revealed that under elevated copper conditions the protein is relocalized from the golgi to both the plasma membrane as well as a cytosolic vesicular compartment. When copper levels return to normal the protein is recycled back to its original post-golgi location. The mechanism by which this copper induced relocalization takes place is at present not known. However, it has been hypothesized that the large N-terminal copper binding domain may play a role in 'sensing' the intracellular copper concentration and somehow trigger the relocation event when concentrations are too high. In order to gain a detailed understanding of the function of this ATPase and the precise role of the metal binding domain, we must first characterize the properties of the domain itself. To this end we have expressed the entire N-terminal domain (~70 kDa) as a fusion to glutathione-S-transferase (GST) and have begun to characterize the metal binding properties of this domain.

2. CHARACTERIZATION OF THE N-TERMINAL COPPER BINDING DOMAIN FROM ATP7B

2.1. Expression and Purification

The GST-WCBD (Wilson Copper Binding Domain) expression vector was constructed by inserting the WCBD cDNA into the polylinker region of the GST expression plasmid pGEX-4-T-2 to create the plasmid pGEX-WCBD. The cDNA was obtained by reverse transcription polymerase chain reaction (RT-PCR) in collaboration with John Forbes in the laboratory of Dr. D.W. Cox. Transformation and induction of *E. Coli* harboring this plasmid resulted in the high level expression of an approximately 96 kDa protein corresponding to the GST-WCBD fusion protein. The fusion protein was expressed and purified as previously described and when needed the GST moiety could be cleaved away by treatment with thrombin (DiDonato et al., 1997). The identity of the expressed protein was confirmed by seven cycles of N-terminal amino acid sequence analysis. Approximately 20–30% of the fusion protein was in the soluble fraction and 70–80% was found in inclusion bodies. The insoluble fraction could also be purified after solubilization and refolding from urea. Figure 3 illustrates the results of a typical purification.

2.2. Metal Binding Properties

2.2.1. Immobilized Metal Ion Affinity Chromatography (IMAC). Initial studies using the copper specific chelator bathocuproinedisulfonic acid (BCS) indicated that copper incorporation had taken place during expression of the fusion and that this copper was present in the +1 oxidation state. In addition, neutron activation analysis (NAA) of the copper

→

Figure 2. Human copper transport pathway in the liver. Copper is absorbed from the intestine via the Menkes copper ATPase and enters the exchangeable pool of copper which consists of copper bound to albumin, and small amino acid complexes in the +2 oxidation state. Prior to uptake by hepatocytes copper is reduced to the +1 oxidation state by a membrane bound reductase and taken into the cell by a passive transporter. The human orthologue of the yeast copper uptake protein CTR1 has been implicated to play this role. Once inside the cell copper becomes bound to several recently identified copper chaperones as well as forming a complex with glutathione (GSH). Hah1 is responsible for delivering copper to the Wilson disease copper ATPase which transports it into a post-golgi compartment for incorporation into cupro-enzymes such as ceruloplasmin (Cp). Ccs1 delivers copper specifically to superoxide dismutase (SOD) while Cox17 delivers copper to the mitochondria via unknown intermediates. Cu(I)-GSH has been shown to reconstitute apo-SOD, apo-Cp, and apo-metallothionein both *in vitro* and *in vivo* and its exact role with respect to copper transport in the cell remains unknown. It may function to deliver excess copper to metallothionein for storage.

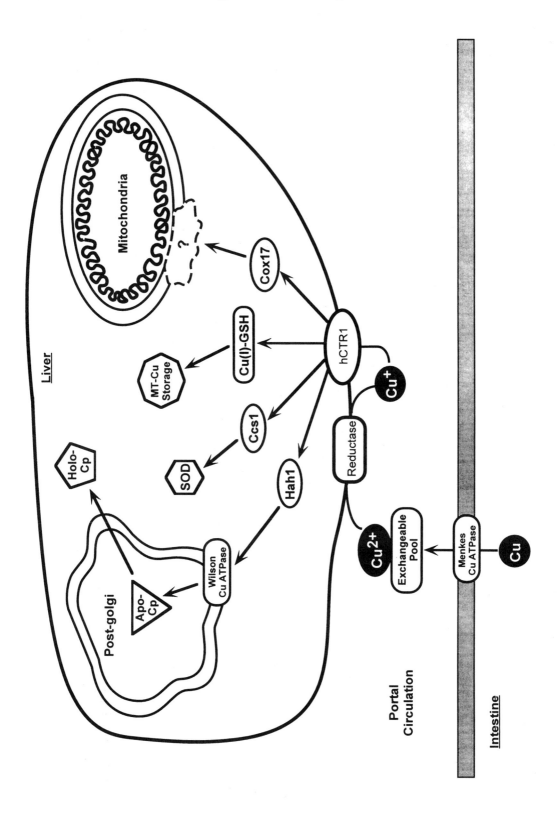

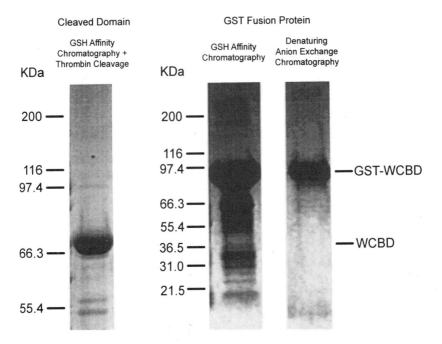

Figure 3. Purification of WCBD and GST-WCBD. Cleaved domain. The fusion protein was purified via glutathione (GSH) affinity chromatography and then treated with thrombin while bound to the matrix. Following cleavage and elution the WCBD is obtained approximately 85–90% pure. GST-Fusion Protein. The fusion protein is first purified by GSH affinity chromatography and is eluted from the affinity resin with urea containing buffer. The protein containing fractions are then applied to an anion exchange resin (Q-Sepharose) under denaturing conditions. The pure fusion protein is eluted with 250 mM NaCl and is obtained at >95% purity. The identity of the purified protein was confirmed by N-terminal amino acid analysis. Copper incorporation occurs during expression and remains intact after GSH affinity chromatography. However, copper is stripped away from the protein during anion exchange chromatography as assessed by treatment with the copper chelator BCS.

reconstituted fusion protein revealed a copper:protein ratio of approximately 6.5:1, indicating that each metal binding motif is responsible for the ligation of one metal atom. To further investigate the metal binding preferences of the domain, the fusion protein was subjected to immobilized metal ion affinity chromatography (IMAC) under both denaturing and native conditions; the results were essentially the same (Figure 4). Examination of GST alone revealed that under native conditions there was some non-specific interactions with the column, while under denaturing conditions no interactions were observed indicating that the major interactions are from the domain itself. No binding to the uncharged resin was observed for either the GST alone or the fusion protein indicating that binding is due to metal protein interactions alone.

The IMAC experiments indicated that the domain has some ability to bind other transition metals with varying degrees of strength. Of the metals tested copper was by far the most strongly bound requiring the use of the copper specific chelator BCS to liberate it from the resin. The weakest binding was observed for cobalt and no binding was observed to columns charged with either iron(II) or iron(III). It is interesting to note that upon elution of the protein from the column charged with copper(II) by BCS a reddish-orange band was formed which eluted with the protein. This color indicates the presence of the $Cu(I)(BCS)_2^-$ complex which has a $\lambda_{max} = 480$ nm (Chen et al., 1996). This would seem to

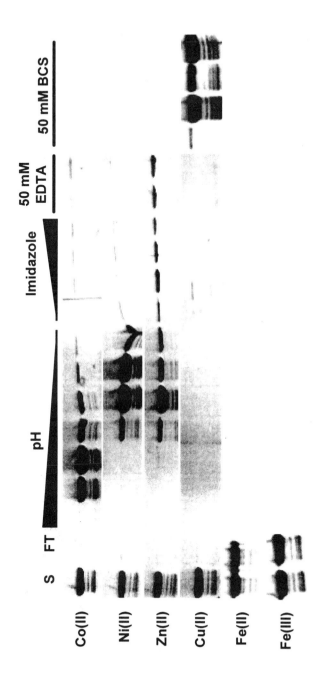

Figure 4. Immobilized Metal Ion Affinity Chromatography (IMAC). Samples (S) of the fusion protein were applied to chelating sepharose fast flow columns charged with the indicated metal under either denaturing or native (data not shown) conditions with nearly identical results. Protein which did not adhere to the column came off in the flow through (FT). The bound protein was then eluted by sequential treatment with low pH (6.5 to 4.0), increasing amounts of imidazole (100 to 500 mM), EDTA (50 mM), and BCS (50 mM). The metals are arranged in order of increased strength of binding from Co(II) to Cu(II). No binding was observed to columns charged with iron (+2 or +3). Elution of the protein from the copper column only occurred after treatment with the copper specific chelator BCS and was accompanied by a reddish-orange band indicative of the presence of Cu(I).

suggest that the domain is able to reduce copper from the +2 to +1 oxidation state upon binding. This could occur through the oxidation of some of the cysteine residues in the domain and subsequently binding to another, reduced pair.

2.2.2. [65]Zinc Blotting and Competition Blotting Analysis.

To confirm further the results obtained from IMAC, [65]Zinc blotting analysis was performed as described previously (DiDonato et al., 1997). Briefly, samples are run on SDS-PAGE and then electroblotted to nitrocellulose. The strips are equilibrated in metal binding buffer and then probed with [65]zinc for one hour, washed and exposed to film. In order to optimize the experimental conditions the effects of pH and dithiothreitol (DTT) were examined. Both the fusion protein and the metal binding domain alone showed very similar binding properties across a pH range from 6.5 to 9.0 (Figure 5). Progressive loss of zinc binding activity was observed at higher pH values for both the fusion protein and the metal domain alone. A small amount of nonspecific binding was observed to GST alone at lower pH values. The estrogen receptor DNA binding domain (2 zinc finger, 4 cys) (ERDBD) was included as a positive control for the assay and shows strong zinc binding affinity near neutral pH.

The effect of DTT on the binding of [65]zinc to the domain was also examined (Figure 6). The zinc blotting analysis was carried out either at pH 6.5 or pH 8.0 in the presence and absence of DTT or the presence of DTT during the equilibration step only. At both pH values no significant amount of binding is observed to either the fusion protein or the free metal domain in the complete absence of DTT. However, if DTT is used in the equilibration step and then removed prior to the addition of [65]zinc, significant binding to both proteins is observed. When DTT is present throughout the assay, the amount of nonspecific binding is significantly decreased. This is a result of the ability of DTT to chelate weakly or non-specifically bound metal atoms. A greater amount of non-specific binding is also observed at higher pH values due to the increased amount of net negative charge on the protein at this pH.

In order to evaluate the relative affinity of different transition metals for the domain, competition [65]zinc blotting was performed. The analysis is identical to the regular zinc blotting protocol except that the amount of radioactive zinc is kept constant while the concentration of a non-radioactive competitor metal is varied. The percent zinc bound is then determined by comparing the signal obtained from the control (no competitor) and that in

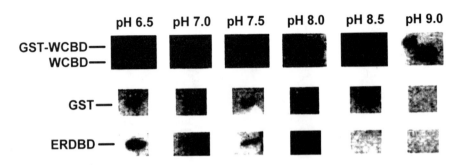

Figure 5. [65]Zinc(II) blotting analysis: effect of pH. Samples of either the fusion protein (GST-WCBD), the metal binding domain alone (WCBD), GST alone (GST), or the estrogen receptor DNA binding domain (ERDBD) were subjected to zinc blotting analysis essentially as described (DiDonato et al., 1997) at various pH values. Strips were autoradiographed overnight and stained with amido black to check for equal protein loading (not shown). The final [65]Zinc(II) concentration in each blot was 30 µM (~10 µCi) and the strips were probed for 1 hr at room temperature.

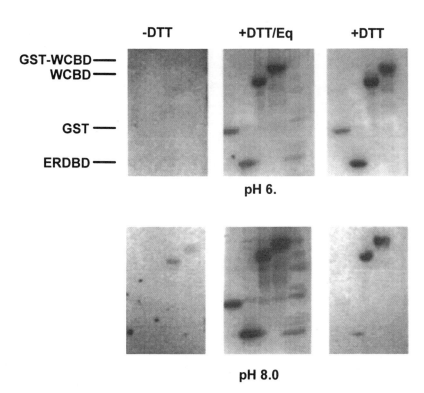

Figure 6. ^{65}Zinc(II) blotting analysis: effect of reducing agents. Samples of either the fusion protein (GST-WCBD), the metal binding domain alone (WCBD), GST alone (GST), or the estrogen receptor DNA binding domain (ERDBD) were subjected to zinc blotting analysis as described in Figure 5, at two pH values (6.5, 8.0). To test the effect of reducing agents, dithiothreitol (DTT) was either omitted from the assay (-DTT), included only during the equilibration step (+DTT/Eq), or included throughout the assay (+DTT). Following the assay the strips were washed and autoradiographed overnight. The strips were stained with amido black to ensure equal amounts of protein were present on each strip (not shown).

which a competitor was present. The strips are then stained with amido black post autoradiography to ensure that equal amounts of protein were present in each lane. The results of such an analysis are summarized in Figure 7. The metal binding domain alone (with GST cleaved away) was used throughout the competition blotting experiments to avoid the effects of non-specific binding to GST. During the assay DTT was only present in the equilibration step and the blotting was carried out at pH 7.0. The results from the competition experiments largely confirm those obtained with IMAC with some notable exceptions (Figure 7A). The fusion protein was able to bind to columns charged with nickel(II) in IMAC; however, nickel(II) was not able to compete with zinc for binding to the domain in competition experiments. This was further confirmed by performing a nickel(II) blotting experiment where radioactive nickel(II) was added alone (data not shown). In addition, iron(III) was able to compete with zinc for binding to the domain in competition experiments, however binding to columns charged with either iron(II) or iron(III) was not observed.

These discrepancies may be the result of the inherent differences between the two methods used. In IMAC steric hindrance from the matrix can be a problem in that the local protein environment may not be able to adopt the correct ligation geometry for the

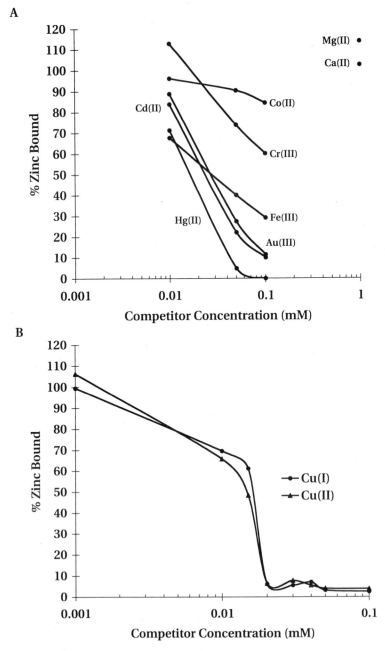

Figure 7. Competition [65]Zinc blotting analysis. **A.** Zinc blotting analysis using the free domain (WCBD) was performed as described except the blots were probed with [65]zinc(II) in the presence of a non-radioactive competitor metal. The [65]zinc(II) concentration was fixed at 30 μM while the competitors were added at the indicated concentration. Successful competition resulted in a decreased signal relative to the control. DTT was only present during the equilibration step and was removed prior to the addition of metal. All metals were added as the chloride salt. **B.** Competition [65]zinc blotting using Cu(I) or Cu(II). Cu(II) was added as $CuCl_2$, while Cu(I) was added as the tetrakis(acetonitrile)Cu(I) hexafluorophosphate complex or generated *in vitro* using $CuCl_2$ and ascorbate (data not shown). All blots were stained with amido black post-autoradiography to ensure equal protein loading.

metal in question. In the competition blotting assay the metal may be competed off if its affinity for the protein is significantly lower than it is for zinc. However, in the case of nickel(II) this does not seem to be the case since nickel(II) blotting did not reveal any binding either. The inability of the protein to ligate iron (+2 or +3) while bound to the column may be a result of steric or geometric constraints imposed by the column matrix as outlined above. Both magnesium and calcium were not able to compete for zinc binding despite the fact that they were present in vast excess of zinc indicating that the binding of zinc is specific and not merely at some charged location on the domain. The 'soft' metals (mercury(II), cadmium(II), and gold(III)) all were able to compete with zinc at equal concentrations which is expected due to their high affinity for sulfur ligands and in the case of cadmium, similarities in their preferred ligation geometries.

The results of the competition experiments performed with copper are very intriguing (Figure 7B). At three times lower concentration, copper is able to compete off about 30 percent of the bound zinc. However, as the concentration of copper is raised the amount of bound zinc detected rapidly decreases. This sharp transition was reproducible and occurred independently of whether copper was presented in the +1 or +2 oxidation state and was not observed for any other metals tested. In these experiments copper in the +1 oxidation state was presented as the tetrakis(acetonitrile)copper(1) hexafluorophosphate complex or generated *in vitro* using $CuCl_2$ and ascorbate. Copper in the +2 oxidation state was presented as $CuCl_2$. Regardless of the form of copper being used, the pattern of competition was the same indicating that the domain may have similar affinities for both copper(I) and copper(II) or that copper is reduced upon binding by the domain from the +2 to +1 oxidation state in agreement with the results obtained in IMAC experiments. The copper competition results also indicate that the domain has a slightly higher affinity for copper relative to zinc. This sharp transition is suggestive of a cooperative binding mode for copper, however a definitive conclusion cannot be made based on these experiments. In addition, because both zinc and the competitor metal are added together, the effects of on/off rates and kinetics for each metal may be influencing the results. Equilibrium binding experiments are now underway to establish whether cooperative binding is taking place.

2.2.3. Structural and Conformational Implications of Metal Binding. To assess whether the binding of copper has any affect on the three dimensional structure of the domain, circular dichrosim (CD) spectroscopy was used to probe any changes in structure upon metal binding. Following denaturing anion exchange chromatography, the fusion protein was found to be essentially free of copper as determined using the copper chelator BCS (data not shown). The fusion protein was then refolded in the presence and absence of copper and the CD spectra were recorded in the secondary structure region (190–300 nm) and the aromatic region (250–400 nm) using a longer pathlength cell. Preliminary results indicate that there are little or no differences in the spectra of the two samples in the secondary structure region suggesting that no gross conformational secondary structure changes are taking place upon metal binding. However, upon examining the aromatic region of the spectra significant changes in the spectra between the two samples is observed. This is suggestive of tertiary structure changes in the domain resulting from the ligation of copper atoms. Alternatively, these changes may be the result of charge transfer absorptions between the copper atoms and the sulfhydryl ligands. Further experiments are underway to confirm this observation using the GST free domain and to determine if these changes occur with the binding of other metals.

X-ray absorption spectroscopy (both Extended X-Ray Absorption Fine Structure (EXAFS) and X-Ray Absorption Near Edge Structure (XANES)) have also been carried out on the copper reconstituted fusion protein in collaboration with Dr. Larry Que to determine the precise ligation environment and geometry of the copper atoms. Preliminary results indicated that the copper atom is in the +1 oxidation state and is ligated most likely by a combination of both sulfur and nitrogen atoms (unpublished observations: L. Que and H-F. Hsu). The sulfur ligands likely originate from the conserved cysteine residues in the domain and nitrogen ligands may be donated from the histidine residues in the domain. This preliminary data did not allow for the unambiguous assignment of the ligation geometry around the copper atoms and will have to be determined by more detailed experiments. Crystallization trials using the copper reconstituted fusion protein are now well underway in an effort to determine the three dimensional structure of the domain. The availability of this structure will greatly enhance our understanding of the metal binding domain in relation to the entire protein.

3. DISCUSSION

We have cloned, expressed, and purified the N-terminal metal binding domain of the Wilson disease ATPase and have characterized several features of its metal binding properties. The domain is found to bind one copper per metal binding motif in the +1 oxidation state (DiDonato et al., 1997; Lutsenko et al., 1997). In addition, the domain seems to have varying affinities for other transition metals with an apparent order of affinity Cu(I)/Cu(II)>>Zn(II)>Ni(II)>Co(II). Although it is clear that the domain is able to bind other transition metals, there is no evidence to suggest that these metals are actually transported. The finding that the domain is able to bind zinc is both expected, and surprising. The *E. Coli* zinc transporting ATPase zntA contains one copy of the metal binding motif Gly-Xaa-Xaa-Cys-Xaa-Xaa-Cys which is repeated six times in the Menkes and Wilson proteins, and in various numbers in the heavy metal transporters of lower organisms (Beard et al., 1997; Rensing et al., 1997). Hence it is not surprising that the WCBD would have some affinity for zinc. However, the binding of zinc to the metal binding region of a copper ATPase does raise some interesting questions about what is conferring the transport specificity of one metal over another in each of these different pumps. This question becomes even more intriguing once it is realized that the conserved metal binding motif is present in many proteins responsible for the transport of a wide variety of metals (copper, cadmium, zinc, and mercury).

It seems unlikely that the two conserved cysteine residues in these domains, while critical for binding, are providing all the specificity needed to select the correct metal to be transported. It is possible that other ligands either external to the domain or from within the domain itself may contribute to the ligation sphere of the metal and provide an additional means of generating the required specificity. Alternatively, the number of metal binding motif repeats, their spacing, and possible interactions may provide a way of modulating the ligation environment of the metal possibly increasing or decreasing the affinity of the domain for one metal over another. It may also be possible that the metal binding domain plays a minor role in determining what metal is actually transported with the selection event occurring elsewhere in the protein, such as the transmembrane channel. Recent evidence has suggested that the six metal binding motifs, at least in the Menkes protein, are not functionally equivalent (Payne and Gitlin, 1998). Sequential mutation of the metal binding motifs does not result in a significant loss of transport activity until the third repeat is mutated, suggesting a critical role for this motif in the binding and/or transport of copper. A copper 'sensing' role has also been hypothesized (DiDonato et al., 1997) for the N-terminal metal binding domain based on the finding that both the Menkes and Wilson disease proteins are relocated to the plasma membrane and a cytosolic vesicular compartment respectively under elevated cytosolic copper levels (Hung et al., 1997; Nagano et al., 1998; Petris et al., 1996).

Competition experiments with copper have suggested the possibility that binding of copper to the domain may be cooperative and that this cooperativity may be restricted to copper alone. A cooperative binding mode would also require structural changes to take place upon successive metal binding which would be transmitted through the protein backbone to the unoccupied sites modulating their ligation environment to increase their affinity for copper. Alternatively, involvement of ligands from other parts of the N-terminal domain in addition to the conserved cysteine residues may allow cooperative binding by allowing distant parts of the domain to 'communicate' with each other. Preliminary results using CD spectroscopy indicate that tertiary but not secondary structural changes take place upon metal binding. Early results from EXAFS and XANES experiments suggest the presence of nitrogen in addition to sulfur residues in the first coordination sphere of the copper atoms likely originating from the side chain of histidine residues. Confirmation of these results as well as equilibrium binding experiments may provide important clues to the origin of metal specificity in these proteins. The three dimensional structure of the copper loaded domain will lead the way to deciphering the function of this domain in the context of the entire copper ATPase protein.

ACKNOWLEDGMENT

Research in this laboratory is supported by a grant from the Medical Research Council of Canada.

REFERENCES

Amaravadi, R., Glerum, D. M., and Tzagoloff, A. (1997). Isolation of a cDNA encoding the human homolog of COX17, a yeast gene essential for mitochondrial copper recruitment. Hum. Genet. 99, 329–333.

Beard, S. J., Hashim, R., Membrillo-Hernandez, J., Hughes, M. N., and Poole, R. K. (1997). Zinc(II) tolerance in Escherichia coli K-12: evidence that the zntA gene (o732) encodes a cation transport ATPase. Mol. Microbiol. 25, 883–891.

Bull, P. C., Thomas, G. R., Rommens, J. M., Forbes, J. R., and Cox, D. W. (1993). The Wilson disease gene is a putative copper transporting P-type ATPase similar to the Menkes gene. Nat. Genet. 5, 327–337.

Chelly, J., Tumer, Z., Tonnesen, T., Petterson, A., Ishikawa-Brush, Y., Tommerup, N., Horn, N., and Monaco, A. P. (1993). Isolation of a candidate gene for Menkes disease that encodes a potential heavy metal binding protein. Nat. Genet. 3, 14–19.

Chen, P., Onana, P., Shaw, C. F. r., and Petering, D. H. (1996). Characterization of calf liver Cu,Zn-metallothionein: naturally variable Cu and Zn stoichiometries. Biochem. J. 317, 389–394.

Ciriolo, M. R., Desideri, A., Paci, M., and Rotilio, G. (1990). Reconstitution of Cu,Zn-superoxide dismutase by the Cu(I).glutathione complex. J. Biol. Chem. 265, 11030–11034.

Culotta, V. C., Klomp, L. W. J., Strain, J., Casareno, R. L. B., Krems, B., and Gitlin, J. D. (1997). The copper chaperone for superoxide dismutase. J. Biol. Chem. 272, 23469–23472.

Dancis, A., Haile, D., Yuan, D. S., and Klausner, R. D. (1994). The Saccharomyces cerevisiae copper transport protein (Ctr1p). Biochemical characterization, regulation by copper, and physiologic role in copper uptake. J. Biol. Chem. 269, 25660–25667.

Danks, D. M. (1995). Disorders of Copper Transport. In Metabolic Basis of Inherited Disease, C. R. Scriver, A. I. Beaudet, W. S. Sly and D. Valle, eds. (New York: McGraw-Hill), pp. 2211–2235.

DiDonato, M., Narindrasorasak, S., Forbes, J. R., Cox, D. W., and Sarkar, B. (1997). Expression, purification, and metal binding properties of the N-terminal domain from the Wilson disease putative copper-transporting ATPase (ATP7B). J. Biol. Chem. 272, 33279–33282.

DiDonato, M., and Sarkar, B. (1997). Copper transport and its alterations in Menkes and Wilson diseases. Biochim. Biophys. Acta (Mol Basis Dis) *1360*, 3–16.

Ferreira, A. M., Ciriolo, M. R., Marcocci, L., and Rotilio, G. (1993). Copper(I) transfer into metallothionein mediated by glutathione. Biochem. J. *292*, 673–676.

Freedman, J. H., Ciriolo, M. R., and Peisach, J. (1989). The role of glutathione in copper metabolism and toxicity. J. Biol. Chem. *264*, 5598–5605.

Glerum, D. M., Shtanko, A., and Tzagoloff, A. (1996). Characterization of COX17, a yeast gene involved in copper metabolism and assembly of cytochrome oxidase. J. Biol. Chem. *271*, 14504–14509.

Hung, I. H., Suzuki, M., Yamaguchi, Y., Yuan, D. S., Klausner, R. D., and Gitlin, J. D. (1997). Biochemical characterization of the Wilson disease protein and functional expression in the yeast Saccharomyces cerevisiae. J. Biol. Chem. *272*, 21461–21466.

Klomp, L. W. J., Lin, S. J., Yuan, D. S., Klausner, R. D., Culotta, V. C., and Gitlin, J. D. (1997). Identification and functional expression of HAH1, a novel human gene involved in copper homeostasis. J. Biol. Chem. *272*, 9221–9226.

Lin, S. J., Pufahl, R. A., Dancis, A., O'halloran, T. V., and Culotta, V. C. (1997). A role for the Saccharomyces cerevisiae ATX1 gene in copper trafficking and iron transport. J. Biol. Chem. *272*, 9215–9220.

Lutsenko, S., Petrukhin, K., Cooper, M. J., Gilliam, C. T., and Kaplan, J. H. (1997). N-terminal domains of human copper-transporting adenosine triphosphatases (the Wilson's and Menkes disease proteins) bind copper selectively in vivo and in vitro with stoichiometry of one copper per metal-binding repeat. J. Biol. Chem. *272*, 18939–18944.

Mercer, J. F., Livingston, J., Hall, B., Paynter, J. A., Begy, C., Chandrasekharappa, S., Lockhart, P., Grimes, A., Bhave, M., Siemieniak, D., and et al. (1993). Isolation of a partial candidate gene for Menkes disease by positional cloning. Nat. Genet. *3*, 20–25.

Musci, G., Dimarco, S., Bellenchi, G. C., and Calabrese, L. (1996). Reconstitution of ceruloplasmin by the Cu(I)-glutathione complex - Evidence for a role of Mg2+ and ATP. J. Biol. Chem. *271*, 1972–1978.

Nagano, K., Nakumura, K., Urakami, K. I., Umeyama, K., Uchiyama, H., Koiwai, K., Hattori, S., Yamamoto, T., Matsuda, I., and Endo, F. (1998). Intracellular distribution of the Wilson's disease gene product (ATPase7B) after in vitro and in vivo exogenous expression in hepatocytes from the LEC rat, an animal model of Wilson's disease. Hepatology *27*, 799–807.

Payne, A. S., and Gitlin, J. D. (1998). Functional expression of the Menkes disease protein reveals common biochemical mechanisms among the copper- transporting P-type ATPases. J. Biol. Chem. *273*, 3765–3770.

Petris, M. J., Mercer, J. F., Culvenor, J. G., Lockhart, P., Gleeson, P. A., and Camakaris, J. (1996). Ligand-regulated transport of the Menkes copper P-type ATPase efflux pump from the Golgi apparatus to the plasma membrane: a novel mechanism of regulated trafficking. EMBO J. *15*, 6084–6095.

Pufahl, R. A., Singer, C. P., Peariso, K. L., Lin, S. J., Schmidt, P. J., Fahrni, C. J., Culotta, V. C., Penner-Hahn, J. E., and O'halloran, T. V. (1997). Metal ion chaperone function of the soluble Cu(I) receptor Atx1. Science *278*, 853–856.

Rensing, C., Mitra, B., and Rosen, B. P. (1997). The zntA gene of Escherichia coli encodes a Zn(II)- translocating P-type ATPase. Proc. Natl. Acad. Sci. USA *94*, 14326–14331.

Tanzi, R. E., Petrukhin, K., Chernov, I., Pellequer, J. L., Wasco, W., Ross, B., Romano, D. M., Parano, E., Pavone, L., Brzustowicz, L. M., and et al. (1993). The Wilson disease gene is a copper transporting ATPase with homology to the Menkes disease gene. Nat. Genet. *5*, 344–350.

Vulpe, C., Levinson, B., Whitney, S., Packman, S., and Gitschier, J. (1993). Isolation of a candidate gene for Menkes disease and evidence that it encodes a copper-transporting ATPase. Nat. Genet. *3*, 7–13.

Vulpe, C. D., and Packman, S. (1995). Cellular copper transport. Annu. Rev. Nutr. *15*, 293–322.

Yuan, D. S., Stearman, R., Dancis, A., Dunn, T., Beeler, T., and Klausner, R. D. (1995). The Menkes/Wilson disease gene homologue in yeast provides copper to a ceruloplasmin-like oxidase required for iron uptake. Proc. Natl. Acad. Sci. USA *92*, 2632–2636.

Zhou, B., and Gitschier, J. (1997). hCTR1: a human gene for copper uptake identified by complementation in yeast. Proc Natl Acad Sci U S A *94*, 7481–7486.

20

MOLECULAR GENETICS OF MENKES DISEASE

Zeynep Tümer and Nina Horn

Department of Medical Genetics
Panum Institute
University of Copenhagen
and The John F. Kennedy Institute
Glostrup, Denmark

1. ABSTRACT

Identification of the gene defective in Menkes disease had an important impact on our understanding of intracellular copper metabolism, and also enabled molecular genetic diagnosis of this lethal disorder. Menkes disease is caused by malfunction of several copper dependent enzymes due to a defect in a membrane bound copper transporter (ATP7A). The mutations leading to Menkes disease show great variety, from cytogenetic abnormalities to partial gene deletions, to single basepair changes. In this review we will mainly discuss the underlying genetic defect in Menkes disease and we will briefly compare the mutation spectrum to that found in the animal model, the mottled mouse, and in Wilson disease, the autosomal recessive disorder of copper metabolism.

2. BACKGROUND

Copper is an essential trace element that is required by almost all living organisms, yet it is toxic in excess amounts. Consequently efficient mechanisms have been developed for intracellular handling of copper. In man deficiency of an energy dependent copper transporting membrane protein (ATP7A) leads to the X-linked lethal disorder, Menkes disease (MD). Clinically the disease is dominated by severe brain degeneration and connective tissue disturbances (Danks, 1995). A conspicuous symptom often leading to the diagnosis is the typical kinky hair. The main symptoms can be ascribed to dysfunction of important copper requiring enzymes, such as cytochrome c oxidase (COX), superoxide dismutase (SOD), lysyl oxidase (LOX) and dopamine beta hydroxylase (DBH). The disease is progressive and most patients die in early childhood, but a significant number of patients (5–10%) have a much longer survival and a less severe clinical course (Horn et al., 1995). The mildest form is the occipital horn syndrome (OHS), where the severe brain damage is absent and connective tissue symptoms become more prominent.

Metals and Genetics, edited by Sarkar.
Kluwer Academic / Plenum Publishers, New York, 1999.

The gene defective in Menkes disease was cloned in 1993 by three groups, including ours, and it was predicted to encode a 1500 amino acid copper translocating P-type AT-Pase (ATP7A) (Vulpe et al., 1993; Chelly et al., 1993; Mercer et al., 1993). ATP7A has recently been localized to the trans-golgi apparatus (Yamaguchi et al., 1996; Petris et al., 1996; Dierick et al., 1997) and it has domains typical for the other members of the P-type ATPase family. These are transmembrane domains, an ATP binding domain, a phosphorylation domain with an invariant aspartate residue (D), and a phosphatase domain (reviewed by Solioz 1998). Besides the domains common to other P-type ATPases, ATP7A has motifs which are now shown to be conserved in copper specific members of this family (reviewed by Solioz 1998). Most prominent of these are the GMXCXXC and the CPC motifs. ATP7A has six repetitive domains with the consensus GMXCXXC motif at the amino terminal and recent studies suggest that these domains are directly involved in selective binding of copper (Lutsenko et al., 1997). CPC is a membrane motif which is at the putative transduction domain of ATP7A and it is suggested to be involved in the translocation of copper through a membrane.

In mouse, about 30 X-linked spontaneous or induced mutations lead to a mottled coat pigmentation in the heterozygous females (Y. Boyd, personal communication). The hemizygous males demonstrate a broad spectrum of neurological and connective tissue abnormalities similar to that observed in MD patients. The most severely affected males die in utero and milder phenotypes are mainly characterized by connective tissue symptoms. Since 1974, when a copper disturbance was demonstrated in several mottled alleles, it has been regarded as a mouse model for MD (Hunt et al., 1974). This was finally confirmed when the mouse homologue *(atp7a)* was cloned using human specific sequences (Levinson et al., 1994; Mercer et al., 1994) and a mutation of *atp7a* was demonstrated in one of the mottled mutants (Das et al., 1995) (see below).

In contrast to Menkes disease, the autosomal recessive Wilson disease (WD) is mainly characterized by toxic symptoms of copper (Danks, 1995). WD presents with hepatic cirrhosis and neurologic or psychiatric symptoms, and it is one of the rare genetic disorders which can be treated almost totally. The clinical features of WD can be ascribed to the toxic accumulation of copper in the liver, and in extrahepatic tissues due to overflow of the metal when the hepatic stores are saturated. WD is caused by a defect in an intracellular protein necessary for incorporation of copper into ceruloplasmin and for the biliary excretion of the metal. Due to the similarities between the cellular pathophysiology of WD and MD, the proteins defective in these two disorders were suggested to be similar (Vulpe et al., 1993). The Wilson disease gene *(ATP7B)* was isolated in 1993 (Bull et al., 1993; Tanzi et al., 1993; Yamaguchi et al., 1993) and two of the groups used sequences specific to *ATP7A* (Bull et al., 1993; Yamaguchi et al., 1993). The putative protein product is (ATP7B) highly homologous to ATP7A and this copper binding P-type ATPase is also localized to trans-golgi network (Hung et al., 1997). Wilson disease has two confirmed animal models, namely LEC rat and toxic milk mouse (see below).

In this review we will mainly discuss the *ATP7A* mutations published previously, and compare these mutations briefly to those found in *ATP7B* and *atp7a*.

3. GENOMIC ORGANIZATION OF *ATP7A*

The 8.5 kb *ATP7A* mRNA consists of 23 exons and spans approximately 145 kb of genomic DNA (Tümer et al., 1995; Dierick et al., 1995). The 5'-untranslated region is GC rich, suggestive of a CpG island, which is a feature of housekeeping genes required by all or most tissues. This is in line with the multisystemic character of Menkes disease. An interesting feature of the promoter region of *ATP7A* is the presence of three 98 bp tandem repeats. The function of these sequences are yet unknown, but deletion of one of these repeats has been observed in one OHS patient (see below) (Levinson et al., 1996). The 134 bp first exon of *ATP7A* contains only untranslated sequences and the first basepair of this exon has been suggested as the transcription start site (Tümer et al., 1995), which is 10 bp upstream to the one suggested by Vulpe et al., (1993). Levinson et al., (1995) have also identified other possible transcription start sites. The ATG translation start codon is within the 141 bp second exon, which also contains part of the 5'-untranslated region and the first metal binding domain. The last exon (4120 bp) contains a 274 bp translated sequence, the TAA translation termination site, the large 3'-untranslated region and a polyadenylation site. Apart from the 23 exons mentioned, the large first intron of *ATP7A* includes a 192 bp exon (exon 1b), which is alternatively spliced and expressed in low amounts in several tissues examined (TW Glover, personal communication).

The exon structure of the mouse gene *(atp7a)* is almost identical to its human counterpart (Cecchi et al., 1996). The 8.0 kb transcribed sequence consists of 23 exons spanning about 120 kb and all the introns interrupt the aligned coding regions of *ATP7A* and *atp7a* at the same positions. Sizes of the exons are thus identical, except for exons 1, 2 and 5, which are shorter in *atp7a* corresponding to the 9 amino acid difference in the respective proteins. Most interestingly in both genes the splice donor site of intron 9 (GT→GC) does not conform to the AG/GT rule indicative of the high conservation of these genes through evolution.

The exon structure of *ATP7B* which consists of 22 exons (Petrukhin et al.,1994; Thomas et al., 1995a) is also highly similar to *ATP7A* (Tümer et al., 1995) and *atp7a* when aligned. The promoter region of *ATP7B* has recently been characterized, but it does not include the tandem repeats found in *ATP7A* (Nanji et al., 1997).

For *ATP7A* the exons seem to correspond to the predicted functional/structural domains of the protein product (Figure 1) and this may be of importance in case of in-frame

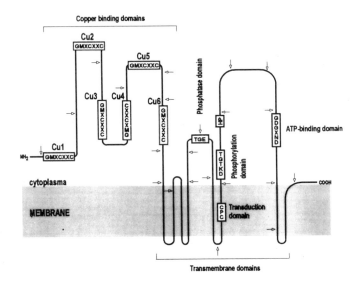

Figure 1. The predicted protein structure of ATP7A with the motifs highly conserved among copper binding P-type ATPases. The approximate positions of the introns are indicated with arrows.

mutations (Tümer et al., 1995). The predicted metal binding domains (Cu1–6) and the transmembrane domains (TMD) are encoded by separate uninterrupted exons, with some exceptions: Cu3 and Cu4 are encoded by a single exon and the exons coding for TMD2 and TMD7 are interrupted by introns. Phosphorylation and ATP-binding domains are also encoded by a single exon, while the exon coding for the phosphatase domain is interrupted by an intron.

Several exons of *ATP7A* are alternatively spliced (Dierick et al., 1995). One of these exons is exon 10 encoding TMD3 and TMD4. Exon 14 coding for TMD5 and/or exon 15 coding for the TMD6 and the putative cation channel with the highly conserved CPC motif are also alternatively spliced. Exon 1b which has been identified in intron 1 is another alternatively spliced exon (TW Glover, personal communication).

4. *ATPA* MUTATIONS IN PATIENTS WITH THE CLASSICAL SEVERE PHENOTYPE

4.1. X-Chromosome Aberrations

In the litterature 4 Menkes disease patients with chromosome abnormalities have been described. Three of them were females and one of them was a male. Two of the female patients had X;autosomal translocations (Kapur et al., 1987; Beck et al., 1995), while the third patient had a 46,XX/45,X karyotype (Barton et al., 1983). The male patient had an intrachromosomal rearrangement (Tümer et al., 1992).

4.2. Partial Gene Deletions

Sixteen patients with gross gene deletions have been partially described in one of the original cloning manuscripts (Chelly et al., 1993), but the extent of these deletions have not been delineated yet (Tümer and Horn, in preparation).

4.3. Point Mutations

Until now 54 point mutations have been described in Menkes disease patients with the classical severe phenotype (Table 1) (Das et al., 1994; Kaler et al., 1995; Tümer et al., 1996; 1997). These were 24 small insertions/deletions, 12 nonsense mutations, 7 missense mutations and 11 splice site mutations. All these mutations were novel except for a nonsense mutation in exon 8 (Arg645Ter) and a missense mutation in exon 10 (Gly727Arg) (Das et al., 1994; Tümer et al., 1997a).

Of the 11 splice site mutations identified, 8 are predicted or shown to affect the splicing efficiency of exon 8, suggesting that these intronic sequences are prone to mutations. The total number of mutations affecting exon 8 is 12, which comprises approximately 25% of the mutations observed in MD patients with the severe phenotype (Das et al., 1994; Tümer et al., 1997). Exon 8 encodes a region between the last metal binding domain and the first transmembrane domain of ATP7A (Figure 1). Though a specific function has not been attributed to this region, it may play an important role in the folding of the protein and serve as a 'stalk' joining the metal binding domains and the ATPase core. Presence of 12 severely affected MD patients with exon 8 mutations is also suggestive for the importance of this domain. Recently a missense mutation within exon 8 has been described in an OHS patient (Table 2) (Ronce et al., 1997). Understanding the function of

Table 1. *ATP7A* mutations in patients with classical Menkes disease

Exon	Mutation	Predicted/observed effect	Ref.
Insertion/deletion mutations			
3	177 ins AAAG	FS	a
4	219-220 del 5bp (ATCTT)	FS	b*
4	220-221 del 5bp (ATCTC)	FS	b*
4	297 del A	FS	c*
4	379 del A	FS	a
4	412 del C	FS	a
7	582-83 del AG	FS	b*
		Skips exon 7	
		Skips exon 7,8,9	
7	590-91 del AG	FS	a
7	594 ins T	FS	a
7	606 del T	FS	a
9	686 del A	FS	a
10	735 ins T	FS	a
10	762 del TT	FS	a
10	779-84 del 14 bp	FS	a
12	836 del G	FS	c*
14	963 del A	FS	a
15	980 del C	FS	a
16	1083 ins A	FS	a
17	1164 del T	FS	a
19	1235 del A	FS	a
20	1294 del A	FS	a
21	1344 ins TGCCA	FS	a
22	1381-84 del GGTTTGGTTT	FS	a
22	1400 del GT	FS	a
Nonsense mutations			
3	Gln167Ter (CAA→TAA)	TP	b*
4	Arg409Ter (CGA→TGA)	TP	a
7	Cys578Ter (TGC→TGA)	TP	a
8	Ser637Ter (TCA→TGA)	TP	a
8	Arg645Ter (CGA→TGA)	TP	a
8	Arg645Ter (CGA→TGA)	TP	b*
		Skips exon 8 (FS)	
9	Glu690Ter (GAA→TAA)	TP	a
10	Arg795Ter (CGA→TGA)	TP	a
13	Gln896Ter (CAG→TAG)	TP	a
13	Gln924Ter (CAA→TAA)	TP	a
14	Trp956Ter (TGG→TAG)	TP	a
16	Glu1081Ter (GAA→TAA)	TP	a
Missense mutations			
8	Ala629Pro (GCT→CCT)	DPS	a
9	Glu724His	No normal splicing	d*
	DS (CAGgca→CATgca)	Skips exon 9 (FS)	
		Skips exon 8,9	
		Skips exon 9,10 (FS)	
		Skips exon 8,9,10	
		Skips exon 7,8,9,10	
10	Gly727Arg (GGA→AGA)	DPS	a
10	Gly727Arg (GGA→AGA)	Missense mutation	b*
15	Leu1006Pro (CTG→CCG)	DPS	a
15	Gly1019Asp (GGT→GAT)	DPS	a
20	Gly1302Arg (GGA→AGA)	Missense mutation	b*
		Skips exon 20	

(Continued)

Table 1. (*Continued*)

Exon	Mutation	Predicted/observed effect	Ref.
Splice site mutations			
IVS7	AS (agAGC→acAGC)	Skips exon 8 (FS)	a
IVS8	DS (ACAgta→ACActa)	Skips exon 8 (FS)	a
IVS8	DS (ACAgtaag→ACAgtaac)	Skips exon 8 (FS)	a
IVS8	DS (ACAgtaagt→ACAgtaagg)	Skips exon 8 (FS)	a
IVS8	DS (ACAg**taa**gt→ACAgagt)	Skips exon 8 (FS)	a
IVS8	DS (ACAgtaag→ACAgtaaa)	Skips exon 8 (FS)	a
IVS8	DS (ACAgt→ACAgc)	Skips exon 8 (FS)	b*
IVS8	AS (ataagAT→ataagATAAGAT)	Skips exon 8 (FS)	b*
IVS12	AS (agGGG→ggGGG)	Skips exon 13 (FS)	a
IVS20	AS (agAAT→ggAAT)	CSS in exon 20 (FS)	b*
		Skips exon 21 (FS)	
IVS22	DS (ACTgt→ACTgc)	Skips exon 22 (FS)	a

a, Tümer et al., 1997a; b, Das et al., 1994; c, Tümer et al., 1996; d, Kaler et al., 1995. *, indicates the cases investigated on the mRNA level. Codon numbers are according to the *ATP7A* cDNA sequence published by Vulpe et al., (1993)., Intron and exon sequences are shown in lowercase and uppercase respectively. FS, frameshift; TP, truncates protein; DPS, disrupts protein structure; AS, acceptor site; DS, donor site.

this domain therefore awaits investigation of the effect of various exon 8 mutations on trancription and translation level.

About 67% of the point mutations observed in *ATP7A* were frameshift or nonsense mutations, which would lead to premature termination of translation resulting in the production of a truncated protein (35/52). Furthermore, all the splice site mutations observed would also lead to a frameshift following the exon skipping. The total of the "truncating" mutations would therefore be about 90% of all the mutations. The protein truncation test (PTT) (Roest et al., 1993), which detects translation terminating mutations by a combination of RT-PCR and *in vitro* transcription/translation reaction, may therefore be a suitable

Table 2. *ATP7A* mutations in patients with atypical phenotypes

Exon	Mutation	Effect	Reference
Promoter	98 bp del	—	Levinson et al., 1997
8	Ser637Leu	Normal splicing	Ronce et al., 1997
		Skips exon 8 (FS)	
		Skips exon 10	
		Skips exon 8-10	
IVS10	DS (AAGgta→AAGgtt)	Skips exon 10	Qi and Byers, 1998
11	Ser833Gly	Normal splicing	Kaler et al., 1994
	DS (CAGgta→CGGgta)	Skips exon 11 (FS)	
		Skips exon 11, 12 (FS)	
IVS14	AS (aaagGGC→gaagGGC)	Normal splicing	Das et al., 1995
		Skips exon 15	
IVS17	DS (CAAgtaag→CAAgtaaa)	Normal splicing	Das et al., 1995
		Skips exon 17 (FS)	
IVS21	DS (CTGgtt→CTGgta)	Normal splicing	Kaler et al., 1994
		Skips exon 21 (FS)	

IVS, intervening sequence; FS, frameshift.

method for screening genetic alterations in *ATP7A*. However, practical difficulties exist with this method compared to easier mutation screening techniques such as SSCP.

Exon skipping due to point mutations within the exons has been observed in *ATP7A* in classical (Das et al., 1994) or in OHS patients (Ronce et al., 1997). These were a 2 bp deletion in exon 7, a nonsense and a missense mutation in exon 8 and a missense mutation in exon 20 (Table 1, 2). In each case besides the abnormal transcript(s) missing one or more exons, a normal size mRNA transcript was also present. However, in other mutations of similar types exon skipping has not been observed suggesting that several different factors are contributing to the exon skipping process.

5. *ATP7A* MUTATIONS IN PATIENTS WITH MILDER PHENOTYPES

Until now 7 point mutations have been reported in patients with OHS (Table 2) (Kaler et al., 1994; Das et al., 1995; Levinson et al., 1996; Ronce et al., 1997; Qi and Byers, 1998). In all these patients, except for the one described by Levinson et al., (1996) the DNA mutation was leading to RNA splicing defects.

In five OHS patients (Kaler et al., 1994; Das et al., 1995; Qi and Byers, 1998) the mutations were at the consensus splice site sequences affecting the normal RNA splicing. In three of these patients (Kaler et al., 1994; Das et al., 1995) besides the abnormal size transcripts, a normal transcript was always present in low amounts. The milder phenotype of the patients could thus be explained by the presence of the normal transcript even in low amounts.

In one patient the mutation was a single basepair substitution within the splice site sequence of exon 11 (Ser833Gly). Besides two abnormal size transcripts a normal size transcript was present but this transcript included the mutation (Kaler et al., 1994). This missense mutation was not within a region coding for a conserved motif and its effect on the function of the protein was probably not severe.

In the patient described by Ronce et al., (1997) the RNA splicing defect was due to a missense mutation in exon 8. In this patient a normal size transcript bearing the mutation and abnormal size transcripts were detected. An explanation for the mild phenotype is the presence of an at least partially functional protein coded by the normal size transcript bearing the missense mutation. Previously a missense mutation within exon 8 was detected in a patient with the classical phenotype, but this mutation has not yet been analysed at the transcription level (Tümer et al., 1997a).

In the OHS patient described by Qi and Byers (1998), a normal size transcript was not present. The abnormal transcript was missing exon 10, which is alternatively spliced out in normal individuals (Dierick et al., 1995). The protein product missing transmembrane domains 3 and 4 encoded by exon 10, was localized to endoplasmic reticulum (Qi and Byers, 1998), instead of trans-golgi where the normal ATP7A resides (Yamaguchi et al., 1996; Petris et al., 1996; Dierick et al., 1997). The milder phenotype of this patient suggests that at least some of the enzymes, other than LOX and DBH, could receive copper eventhough ATP7A is mislocalized.

The type of the mutation described by Levinson et al., (1996) was different from the other OHS mutations. One of the three repeated elements in the regulatory region of *ATP7A* was deleted (Levinson et al., 1996), but the transcription level was normal. In this patient no other DNA mutations could be detected in *ATP7A* and therefore the relation of this deletion to OHS remains obscure.

Table 3. *atp7a* mutations detected in mottled alleles

Localization	Mutation	Allele	Phenotype	Reference
Promoter+exon 1	1.8 deletion	Dappled	Intrauterine death	Levinson et al., 1997a
Exon 11	6bp deletion	Brindled	Classical MNK	Grimes et al., 1997; Reed et al., 1997
Exons 11-14	deletion	Spot	Intrauterine death	Cunliffe et al., in preparation
IVS11	DS a→c +3	Blotchy	Mild phenotype	Das et al., 1995
IVS14	DS g→c +1	1Pub	Intrauterine death	Cechhi et al., 1997
Exon 15	Ala998Thr	Pewter	Mild phenotype	Levinson et al., 1997b
Exon 16	Lys1036Thr	Viable brindled	Mild phenotype	Cechhi et al., 1997; Reed et al., 1997
Exon 21	Ala1364Asp	11H	Intrauterine death	Cechhi et al., 1997
Exon 22	Ser1381Pro	Macular	Classical MNK	Ohta et al., 1997; Mori et al., 1997; Murata et al., 1997

6. *atp7a* MUTATIONS IN MOTTLED MOUSE

In mouse about 30 mutations lead to the mottled phenotype and *atp7a* mutations have been defined in 9 of these alleles (Table 3) (Das et al., 1995; Levinson et al., 1997a,b; Grimes et al., 1997; Cechhi et al., 1997; Reed et al., 1997; Ohta et al., 1997; Mori et al., 1997; Murata et al., 1997; Cunliffe et al., in preparation). Mutation spectum of *atp7a* is rather similar to that of *ATP7A*. Two of the *atp7a* mutations are partial gene deletions identified in mottled dappled and mottled spot (Levinson et al., 1997; Cunliffe et al., in preparation). In mottled dappled a 1.8 kb deletion removed the promoter region and exon 1. In mottled spot exons 11–14 were deleted. Other mutations observed in mottled alleles are, a 6 bp deletion, four missense and two splice site mutations. Small insertions or nonsense mutations have not been identified in mottled mouse, but the present data is not large enough to define the mutation spectrum of *atp7a*.

Mottled blotchy, animal model for OHS, has a splice site mutation affecting the normal mRNA splicing, similar to that observed in several OHS patients (Kaler et al., 1994; Das et al., 1995; Qi and Byers, 1998). Interestingly, Kaler et al., (1994) have described an OHS patient with a mutation occurring at the same splice site (IVS 11) as in mottled blotchy.

7. BRIEF COMPARISON OF *ATP7A* AND *ATP7B* MUTATIONS

The main difference between the mutation spectrums of these two genes is the absence of chromosome mutations or partial gene deletions in *ATP7B*. The lack of chromosome aberrations disrupting *ATP7B* is not unexpected, as WD is inherited as an autosomal recessive trait and chromosome abnormalities are very rare in such disorders. The largest deletions observed in *ATP7B* were a 29 bp deletion in exon 8 (Nanji et al., 1997) and a 24 bp deletion in exon 18 (Figus et al., 1997; Orru et al., 1997). On the other hand in *ATP7A* an approximately 100 kb deletion removes the whole gene except for the first two exons, the leader exon and exon 2 with the translation start site and the first copper binding domain (Tümer and Horn, in preparation).

To our knowledge a total of 101 basepair substitutions or small insertions/deletions of *ATPB* have been described in Wilson disease patients (Bull et al., 1993; Tanzi et al., 1993; Thomas et al., 1995a,b,c; Figus et al., 1995; Houwen et al., 1995; Shimizu et al., 1995; Chuang et al., 1996; Loudianos et al., 1996; Czlonkowska et al., 1997; Nanji et al., 1997; Orru et al., 1997; Shah et al., 1997; Waldenström et al., 1997; Kemppainen et al., 1997; Yamaguchi et al., 1998; Kalinsky et al., 1998; Krawczak and Cooper, 1997). Most of these mutations are missense mutations (49%) or small deletions/insertions (35%) and nonsense or splice mutations comprise only a small percentage. However, some of the missense mutations may represent polymorphisms and whether these basepair substitutions are causative for WD, should be tested by other means, such as complementation analysis in the yeast mutant (DW Cox, personal communication).

LEC rat, one of the two confirmed animal models of WD, has a 900 bp deletion at the 3'-coding region of the rat-*atp7b* gene, which removes the ATP binding domain and downstream of the gene (Wu et al., 1994; Muramatsu et al., 1995; Ono et al., 1995). In the other animal model, the toxic milk mouse, the defect is a missense mutation of the mouse-*atp7b* gene in a region coding for the last transmembrane domain (Theophilos et al., 1996).

8. CONCLUSION

The most important impact of the identification of *ATP7A* mutations in Menkes disease patients is on the diagnosis of the disease. For MD a definitive biochemical diagnosis exists and it is based on the intracellular accumulation of copper in cultured cells due to impaired efflux (Tümer and Horn, 1997). Prenatal diagnosis is carried out by measuring radioactive copper accumulation in cultured amniotic fluid cells in the second trimester and by determining the total copper content in chorionic villi in the first trimester. These analyses demand expertise and are carried out only in a few centers in the world. Demonstration of a defect in *ATP7A* is an alternative diagnosis of Menkes disease, though detection of a mutation may take time due to the large size of the gene and the variety of the genetic defect. A prenatal diagnosis by mutation analysis is therefore possible, only if the mutation of the family has already been identified. In MD carrier identification by measuring ^{64}Cu-uptake in cultured fibroblasts is a possibility. However, this biochemical test is not reliable in case of negative results, due to random X-inactivation. The ultimate proof of heterozygositsy is therefore provided by mutation analyses. In families where the mutation is unknown, intragenic polymorphic markers may also enable carrier diagnosis (reviewed by Tümer and Horn, 1997).

ACKNOWLEDGMENTS

This research was supported by grants from The Danish Medical Research Counsil (12-0058, 12-9292), The Danish Health Insurance Foundation (H 11/210-89, H 11/282-90 and H 11/257-91), The Danish Biotechnological Research and Development Program 1991–1995 (Grant 5.18.03), The Foundation of 1870, and The Novo Nordisk Foundation.

REFERENCES

Barton, N.W., Dambrosia, J.M., and Barranger, J.A. (1983). Menkes' Kinky-Hair Syndrome: Report of a case in a female infant. Neurology 33 [Suppl 2], 154.

Beck, J., Enders, H., Schliephacke, M., Buchwald-Saal, M., and Tümer, Z. (1994). X;1 translocation in a female Menkes patient: characterization by fluorescence in situ hybridization. Clin Genet 46, 295–298.

Bull, P.C., Thomas, G.R., Rommens, J.M., Forbes, J.R., and Cox, D.W. (1993). The Wilson disease gene is a putative copper transporting P-type ATPase similar to the Menkes gene. Nat Genet 5, 327–337.

Cecchi, C. and Avner, P. (1996). Genomic organization of the mottled gene, the mouse homologue of the human menkes disease gene. Genomics 37, 96–104.

Cecchi, C., Biasotto, M., Tosi, M., and Avner, P. (1997). The mottled mouse as a model for human Menkes disease: identification of mutations in the Atp7a gene. Hum. Mol. Genet. 6, 425–433.

Chelly, J., Tümer, Z., Tønnesen, T., Petterson, A., Ishikawa-Brush, Y., Tommerup, N., Horn, N., and Monaco, A.P. (1993). Isolation of a candidate gene for Menkes disease that encodes a potential heavy metal binding protein. Nat Genet 3, 14–19.

Chuang, L.-M., Wu, H.-P., Jang, M.-H., Wang, T.-R., Sue, W.-C., Lin, B.J., Cox, D.W., and Tai, T.-W. (1996). High frequency of two mutations in codon 778 in exon 8 of the ATP7B gene in Taiwanese families with Wilson disease. J Med Genet 33, 521–523.

Czlonkowska, A., Rodo, M., Gajda, J., Ploos van Amstel, H.K., Juyn, J., and Houwen, R.H. (1997). Very high frequency of the His1069Gln mutation in Polish Wilson disease patients [letter] [In Process Citation]. J. Neurol. 244, 591–592.

Danks, D.M. (1995). Disorders of Copper Transport. In The Metabolic Basis of Inherited Disease. J.R. Scriver, A.L. Beaudet, W.S. Sly, and D. Valle, eds. (New York: McGraw-Hill), pp. 2211–2235.

Das, S., Levinson, B., Whitney, S., Vulpe, C., Packman, S., and Gitschier, J. (1994). Diverse mutations in patients with Menkes disease often lead to exon skipping. Am J Hum Genet 55, 883–889.

Das, S., Levinson, B., Vulpe, C., Whitney, S., Gitschier, J., and Packman, S. (1995). Similar splicing mutations of the Menkes/Mottled copper transporting ATPase gene in occipital horn syndrome and the blotchy mouse. Am J Hum Genet 56, 570–576.

Dierick, H.A., Ambrosini, L., Spencer, J., Glover, T.W., and Mercer, J.F.B. (1995). Molecular structure of the Menkes disease gene (ATP7A). Genomics 28, 462–469.

Dierick, H.A., Adam, A.N., Escara-Wilke, J.F., and Glover, T.W. (1997). Immunocytochemical localization of the Menkes copper transport protein (ATP7A) to the trans-Golgi network. Hum. Mol. Genet. 6, 409–416.

Figus, A., Angius, A., Loudianos, G., Bertini, C., Dessi, V., Loi, A., Deliana, M., Lovicu, M., Olla, N., Sole, G., De Virgiliis, S., Lilliu, F., Farci, A.M.G., Nurchi, A., Giacchino, R., Barabino, A., Marazzi, M., Zancan, L., Greggio, N.A., Marcellini, M., Solinas, A., Deplano, A., Barbera, C., Devoto, M., Ozsoylu, S., Kocak, N., Akar, N., Karayalcin, S., Mokini, V., Cullufi, P., Balestrieri, A., Cao, A., and Pirastu, M. (1995). Molecular pathology and haplotype analysis of Wilson disease in Mediterranean populations. Am J Hum Genet 57, 1318–1324.

Grimes, A., Hearn, C.J., Lockhart, P., Newgreen, D.F., and Mercer, J.F. (1997). Molecular basis of the brindled mouse mutant (Mo(br)): a murine model of Menkes disease. Hum. Mol. Genet. 6, 1037–1042.

Horn, N., Tønnesen, T., and Tümer, Z. (1995). Variability in clinical expression of an X-linked copper disturbance, Menkes disease. In Genetic response to metals. B. Sarkar, ed. (New York: Marcel and Dekker), pp. 285–303.

Houwen, R.H.J., Juyn, J., Hoogenraad, T.U., Ploos van Amstel, J.K., and Berger, R. (1995). H714Q mutation in Wilson disease is associated with late, neurological presentation. J Med Genet 32, 180–182.

Hung, I.H., Suzuki, M., Yamaguchi, Y., Yuan, D.S., Klausner, R.D., and Gitlin, J.D. (1997). Biochemical characterization of the Wilson disease protein and functional expression in the yeast Saccharomyces cerevisiae. J Biol Chem 272, 21461–21466.

Hunt, D.M. (1974). Primary defect in copper transport underlies mottled mutants in the mouse. Nature 249, 852–854.

Kaler, S.G., Gallo, L.K., Proud, V.K., Percy, A.K., Mark, Y., Segal, N.A., Goldstein, D.S., Holmes, C.S., and Gahl, W.A. (1994). Occipital horn syndrome and a mild Menkes phenotype associated with splice site mutations at the MNK locus. Nat Genet 8, 195–202.

Kaler, S.G., Buist, N.R.M., Holmes, C.S., Goldstein, D.S., Miller, R.C., and Gahl, W.A. (1995). Early copper therapy in classic Menkes disease patients with a novel splicing mutation. Ann Neurol 38, 921–928.

Kalinsky, H., Funes, A., Zeldin, A., Pel-Or, Y., Korostishevsky, M., Gershoni-Baruch, R., Farrer, L.A., Bonné-Tamir, B. (1998). Novel ATP7B mutations causing Wilson disease in several Israeli ethnic groups. Hum Mutat 11, 145–151.

Kapur, S., Higgins, J.V., Delp, K., and Rogers, B. (1987). Menkes syndrome in a girl with X-autosome translocation. Am J Med Genet 26, 503–510.

Kemppainen, R., Palatsi, R., Kallioinen, M., and Oikarinen, A. (1997). A homozygous nonsense mutation and a combination of two mutations of the Wilson disease gene in patients with different lysyl oxidase activities in cultured fibroblasts. J Invest Dermatol 108, 35–39.

Krawczak, M., Cooper, D.N. (1997). The Human Gene Mutation Database. Trends Genet 13, 121–122.

Levinson, B., Vulpe, C., Elder, B., Martin, C., Verley, F., Packman, S., and Gitschier, J. (1994). The mottled gene is the mouse homologue of the Menkes disease gene. Nat Genet 6, 369–373.

Levinson, B., Conant, R., Schnur, R., Das, S., Packman, S., and Gitschier, J. (1996). A repeated element in the regulatory region of the MNK gene and its deletion in a patient with occipital horn syndrome. Human Molecular Genetics 5, 1737–1742.

Levinson, B., Packman, S., and Gitschier, J. (1997a). Deletion of the promoter region of the *Atp7a* gene of the *mottled dappled* mouse. Nature Genetics 16, 223–224.

Levinson B., Packman, S., and Gitschier, J. (1997b). Mutation analysis of mottled pewter. Mouse Genome 95, 163–165.

Loudianos, G., Dessi, V., Angius, A., Lovicu, M., Loi, A., Deiana, M., Akar, N., Vajro, P., Figus, A., Cao, A., and Pirastu, M. (1996). Wilson disease mutations associated with uncommon haplotypes in Mediterranean patients. Human Genetics 98, 640–642.

Lutsenko, S., Petrukhin, K., Cooper, M.J., Gilliam, C.T., and Kaplan, J.H. (1997). N-terminal domains of human copper-transporting adenosine triphosphatases (the Wilson's and Menkes disease proteins) bind copper selectively in vivo and in vitro with stoichiometry of one copper per metal-binding repeat. J. Biol. Chem. 272, 18939–18944.

Mercer, J.F.B., Livingston, J., Hall, B., Paynter, J.A., Begy, C., Chandrasekharappa, S., Lockhart, P., Grimes, A., Bhave, M., Siemieniak, D., and Glover, T.W. (1993). Isolation of a partial candidate gene for Menkes disease by positional cloning. Nat Genet 3, 20–25.

Mercer, J.F.B., Grimes, A., Ambrosini, L., Lockhart, P., Paynter, J.A., Dierick, H.A., and Glover, T.W. (1994). Mutations in the murine homologue of the Menkes gene in dappled and blotchy mice. Nat Genet 6, 374–378.

Mori, M. and Nishimura, M. (1997). A serine-to-proline mutation in the copper-transporting P-type ATPase gene of the macular mouse. Mamm. Genome 8, 407–410.

Muramatsu, Y., Yamada, T., Moralejo, D.H., Cai, Y., Xin, X., Miwa, Y., Izumi, K., and Matsumoto, K. (1995). The rat homologue of the Wilson's disease gene was partially deleted at the 3' end of its protein-coding region in long-evans cinnamon mutant rats. Res Commun Mol Pathol Pharmacol 89, 421–424.

Murata, Y., Kodama, H., Abe, T., Ishida, N., Nishimura, M., Levinson, B., Gitschier, J., and Packman, S. (1997). Mutation analysis and expression of the mottled gene in the macular mouse model of Menkes disease [In Process Citation]. Pediatr. Res. 42, 436–442.

Nanji, M.S., Nguyen, V.T., Kawasoe, J.H., Inui, K., Endo, F., Nakajima, T., Anezaki, T., and Cox, D.W. (1997). Haplotype and mutation analysis in Japanese patients with Wilson disease. Am. J. Hum. Genet. 60, 1423–1429.

Ohta, Y., Shiraishi, N., and Nishikimi, M. (1997). Occurrence of two missense mutations in Cu-ATPase of the macular mouse, a Menkes disease model [In Process Citation]. Biochem. Mol. Biol. Int. 43, 913–918.

Ono, T., Fukumoto, R., Kondoh, Y., and Yoshida, M.C. (1995). Deletion of the Wilson's disease gene in hereditary hepatitis LEC rats. Jpn J Genet 70, 25–33.

Orru, S., Thomas, G., Cox, D.W., and Contu, L. (1997). 24 bp deletion and Ala$_{1278}$ to Val mutation of the ATP7B gene in a Sardinian family with Wilson disease. Am J Hum Genet 10, 84–85.

Petris, M.J., Mercer, J.F.B., Culvenor, J.G., Lockhart, P., Gleeson, P.A., and Camakaris, J. (1996). Ligand-regulated transport of the Menkes copper P-type ATPase efflux pump from the Golgi apparatus to the plasma membrane: a novel mechanism of regulated trafficking. EMBO J 15, 6084–6095.

Petrukhin, K., Lutsenko, S., Chernov, I., Ross, B.M., Kaplan, J.H., and Gilliam, T.C. (1994). Characterization of the Wilson disease gene encoding a P-type copper transporting ATPase: genomic organization, altrenative splicing, and structure/function predictions. Hum Molec Genet 3, 1647–1656.

Qi, M., and Byers, P.H. (1998). Constitutive skipping of alternatively spliced exon 10 in the ATP7A gene abolishes Golgi localization of the Menkes protein and produces the occipital horn syndrome. Hum Mol Genet 7, 465–469

Reed, V., and Boyd, Y. (1997). Mutation analysis provides additional proof that mottled is the mouse homologue of Menkes' disease. Hum Mol Genet 6, 417–423.

Roest, P.A.M., Roberts, R.G., Sugino, S., van Ommen, G.J.B., den Dunnen, J.T. (1993). Protein truncation test (PTT) for rapid detection of translation-terminating mutations. Hum Mol Genet 2, 1719–1721.

Ronce, N., Moizard, M.P., Robb, L., Toutain, A., Villard, L., and Moraine, C. (1997). A C2055T transition in exon 8 of the ATP7A gene is associated with exon skipping in an occipital horn syndrome family [letter]. Am. J. Hum. Genet. 61, 233–238.

Shah, A.B., Chernov, I., Zhang, H.T., Ross, B.M., Das, K., Lutsenko, S., Parano, E., Pavone, L., Evgrafov, O., Ivanova-Smolenskaya, I.A., Anneren, G., Westermark, K., Urrutia, F.H., Penchaszadeh, G.K., Sternlieb, I., Scheinberg, I.H., Gilliam, T.C., and Petrukhin, K. (1997). Identification and analysis of mutations in the Wilson disease gene (ATP7B): population frequencies, genotype-phenotype correlation, and functional analyses. Am. J. Hum. Genet. 61, 317–328.

Shimizu, N., Kawase, C., Nakazono, H., Hemmi, H., Shimatake, H., and Aoki, T. (1995). A novel RNA splicing mutation in Japanese patients with Wilson disease. Biochem Biophys Res Comm 217, 16–20.

Solioz, M. (1998). Copper homeostasis by CPX-type ATPases: The new class of heavy metal P-type ATPases. Adv Mol Cell Biol 23A, 167–203.

Tanzi, R.E., Petrukhin, K., Chernov, I., Pellequer, J.L., Wasco, W., Ross, B., Romano, D.M., Parano, E., Pavone, L., Brzustowicz, L.M., Devoto, M., Peppercorn, J., Bush, A.I., Sternlieb, I., Pirastu, M., Gusella, J.F., Evgrafov, O., Penchaszadeh, G.K., Honig, B., Edelman, I.S., Soares, M.B., Scheinberg, I.H., and Gilliam, T.C. (1993). The Wilson disease gene is a copper transporting ATPase with homology to the Menkes disease gene. Nat Genet 5, 344–357.

Theophilos, M.B., Cox, D.W., and Mercer, J.F.B. (1996). The toxic milk mouse is a murine model of Wilson disease. Human Molecular Genetics 5, 1619–1624.

Thomas, G.R., Forbes, J.R., Roberts, E.A., Walshe, J.M., and Cox, D.W. (1995a). The Wilson disease gene: spectrum of mutations and their consequences. Nat Genet 9, 210–217.

Thomas, G.R., Jensson, O., Gudmundsson, G., Thorsteinsson, L., and Cox, D.W. (1995b). Wilson disease in Iceland: A clinical and genetic study. Am J Hum Genet 56, 1140–1146.

Thomas, G.R., Roberts, E.A., Walshe, J.M., and Cox, D.W. (1995c). Haplotypes and mutations in Wilson disease. Am J Hum Genet 56, 1315–1319.

Tümer, Z., Tommerup, N., Tønnesen, T., Kreuder, J., Craig, I.W., and Horn, N. (1992). Mapping of the Menkes locus to Xq13.3 distal to the X-inactivation center by an intrachromosomal insertion of the segment Xq13.3-q21.2. Hum Genet 88, 668–672.

Tümer, Z., Vural, B., Tønnesen, T., Chelly, J., Monaco, A.P., and Horn, N. (1995). Characterization of the exon structure of the Menkes disease gene using Vectorette PCR. Genomics 26, 437–442.

Tümer, Z., Horn, N., Tønnesen, T., Christodoulou, J., Clarke, J.T.R., and Sarkar, B. (1996). Early copper-histidine treatment for Menkes disease. Nat Genet 12, 11–13.

Tümer, Z., Lund, C., Tolshave, J., Vural, B., Tønnesen, T., and Horn, N. (1997a). Identification of point mutations in 41 unrelated patients affected with Menkes disease. Am J Hum Genet 60, 63–71.

Tümer Z., and Horn, N. (1997b). Menkes disease: recent advances and new aspects. J Med Genet 34, 265–274.

Vulpe, C., Levinson, B., Whitney, S., Packman, S., and Gitschier, J. (1993). Isolation of a candidate gene for Menkes disease and evidence that it encodes a copper-transporting ATPase. Nat Genet 3, 7–13.

Waldenström, E., Lagerkvist, A., Dahlman, T., Westermark, K., and Landegren, U. (1996). Efficient detection of mutations in Wilson disease by manifold sequencing. Genomics 37, 303–309.

Wu, J., Forbes, J.R., Chen, H.S., and Cox, D.W. (1994). The LEC rat has a deletion in the copper transporting ATPase gene homologous to the Wilson disease gene. Nat Genet 7, 541–545.

Yamaguchi, Y., Heiny, M.E., and Gitlin, J.D. (1993). Isolation and characterization of a human liver cDNA as a candidate gene for Wilson disease. Biochem Biophys Res Comm 197, 271–277.

Yamaguchi, Y., Heiny, M.E., Suzuki, M., and Gitlin, J.D. (1996). Biochemical characterization and intracellular localization of the Menkes disease protein. Proc Natl Acad Sci, USA 93, 14030–14035.

Yamaguchi, A., Matsuura, A., Arashima, S., Kikuchi, Y., Kikuchi, K. (1998). Mutations of ATP7B gene in Wilson disease patients in Japan: Identification of nine mutations and lack of clear founder effect in a Japanese population. Human Mutat [Suppl 1], S320-S322.

21

MUTATIONAL ANALYSIS OF THE COPPER ATPase AFFECTED IN MENKES DISEASE

Daniel Strausak, Loreta Ambrosini, and Julian F. B. Mercer

The Murdoch Institute
Royal Children's Hospital
Flemington Road
Parkville Vic 3052, Australia

1. INTRODUCTION

All organisms require copper as an essential constituent of many biological processes. The chemical properties of copper are ideally incorporated by many enzymatic reactions involved in energy conversion, neurotransmission and the biosynthesis of connective tissues. On the other hand, these redox properties can also cause considerable damage to organisms by forming highly reactive species that unspecifically react with DNA, lipids or proteins. The human hereditary Menkes and Wilson disease well illustrate the requirement for a tightly regulated copper concentration. A failure to deliver copper from the intestine to the blood and defects in the delivery of copper to the brain are the critical steps that result in the severe copper deficiency seen in the Menkes disorder (Danks, 1995). In contrast, patients with Wilson disease have a primary defect of copper homeostasis in the liver. Copper is accumulated in hepatocytes due to defective biliary excretion which ultimately results in liver damage and late onset liver failure (Danks, 1995).

The genes encoding the proteins defective in both diseases were cloned and sequenced five years ago by several research groups (Mercer et al., 1993; Chelly et al., 1993; Vulpe et al., 1993; Bull et al., 1993; Tanzi et al., 1993). The predicted protein structures were 65% identical to each other and had motifs characteristic of members of the family of cation-transport P-type ATPases. They are closely related to Cu-ATPases found in several species, including yeast and bacteria. In addition to features common to P-type ATPases, the Wilson and Menkes proteins, which we refer to as WND and MNK (alternative names are ATP7A and ATP7B), contain a number of prominent conserved amino acid sequences which subdivide them into a subfamily known as CPx-type or soft-metal ATPases (Solioz and Odermatt, 1995). These features include a variable number of repeating Cys-x-x-Cys motifs at the N-terminus ranging from one in bacteria, two in yeast, three in

Metals and Genetics, edited by Sarkar.
Kluwer Academic / Plenum Publishers, New York, 1999.

C. elegans to six in the human MNK and WND proteins. The number of amino-terminal putative metal binding sites (mbs) seems to correlate with the "hierarchy" of the organism in the evolutionary tree in the sense that the higher the organism, the more mbs may be required for an effective copper transporting function.

Further features of CPx-type ATPases include a conserved proline in the putative ion transduction channel flanked by one cysteine at the amino terminal side and either a cysteine, histidine or serine at the carboxyl-side (CPx motif), and a conserved SEHPL domain approximately 40 amino acids C-terminal to the CPx motif (Figure 1). A mutation of the histidine in this motif is the most frequent point-mutation in patients with Wilson's disease. In contrast to the domains common to all the P-type ATPases, the functions of the CPx-type specific amino acid sequences are still obscure.

Most of the work of the CPx-type specific sequences has been performed on the GMxCxxC metal binding sites (here referred to as mbs). In two studies exploring the function of the six GMxCxxC mbs of MNK and WND, both N-termini were over-expressed and purified as fusion proteins. By metal-ion chromatography and ^{65}Zinc-blotting analysis, the ion specificity of the hydrophilic MNK and WND parts were investigated and affinities to Cu(I), Zn, Ni and Co but not Fe (II) or Fe(III) were measured. (DiDonato et al., 1997; Lutsenko et al., 1997). Both reports consistently demonstrated an *in vitro* and *in vivo* stochiometry of one Cu-atom per metal binding site suggesting that each of the mbs is likely to be involved in copper binding.

In whole cell experiments, copper is accumulated in cultured fibroblasts from Menkes patients (Goka et al., 1976; Horn, 1976; Camakaris et al., 1980). Furthermore, copper resistant variants of hamster ovary cell lines (CHO-K1 cells) overexpressing the endogenous MNK homologue accumulate less copper than the parental cell line (Camakaris et al., 1995). These two observations identified the MNK protein as part of the primary mechanism controlling intracellular copper concentration in higher eukaryotes. The regulation of this mechanism occurs at the post-translational level by copper induced

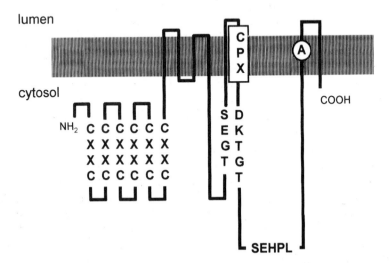

Figure 1. Schematic presentation of the human Menkes protein. Important conserved regions are indicated by their amino acid sequences. *CxxC*, putative metal binding sites; *TGES*, phosphatase domain; *CPX*, putative transduction motif; *DKTGT*, phosphorylation site; *SEHPL*, highly conserved motif of yet unknown function; *alanine* (encircled); alanine converted to valine in a mild Menkes patient discussed in this chapter.

intracellular trafficking of MNK from a *trans*Golgi compartment (TGN) to the plasma membrane (Petris et al., 1996). By immunolabelling of the overexpressed hamster homologue, MNK was localised to the TGN in cells grown in normal media. After the addition of copper, the target protein relocalised via vesicular compartments to the plasma membrane. When cells were returned to normal media, the Cu-transporter was recycled back to the TGN. In bacteria or yeast, the closely related copper ATPases CopA and CCC2 possess only one and two GMxCxxC motifs, respectively (Odermatt et al., 1993; Fu et al., 1995). Unlike *MNK*, the expression of *copA* in *E.hirae* is transcriptionally regulated (Odermatt and Solioz, 1995). A copper-dependent intracellular redistribution of the MNK homologue (CCC2) in yeast cells has never been demonstrated. These observations suggest that the presence of one and two mbs are sufficient for the catalytic activity of the copper transporting protein. The four additional mbs in higher eukaryotes may be involved in intracellular copper sensing resulting in the relocalisation of MNK to the plasma membrane when copper concentrations reach high levels thus permitting the efflux of excess copper. A progressive binding of Cu-atoms to the mbs may lead to a conformational change in the N-terminus that triggers the dislocation of MNK to the plasma membrane. Most recently, the human MNK protein was stably expressed in CHO-K1 cells. The human protein localised correctly to the TGN under copper limiting conditions and a Cu-induced relocation from the TGN to the plasma membrane was observed (La Fontaine et al., 1998). These results formed the basis for most of the experiments presented in this report.

Here, we present data that shows the effect of specifically mutating the mbs on the intracellular redistribution of the protein. In addition, immunochemical localisation of the MNK protein in cultured fibroblasts from a mild Menkes patient and the effect of copper on the intracellular trafficking of the patient protein will be presented. These results suggest that the copper-pumping activity of MNK is directly coupled to the copper-induced intracellular trafficking leading to a novel hypothesis of the regulation of copper homeostasis in the cell.

2. RESULTS

2.1. Analysis of the MNK N-Terminus by Site-Directed Mutagenesis and Construction of a MNK-Glut4 Chimera

The six copper binding domains at the amino terminus of the Menkes protein were mutated from GMxCxxC to GMxSxxS to create MNK mutants with remaining functional copper binding sites as shown in Figure 2. CHO-K1 cells were first transiently transfected with the constructs encoding MNK mutated in mbs1 alone, mbs6 alone, mbs 1–3, mbs 4–6 and mbs 1–6 and the mutant proteins were identified by immunolabelling. In basal media, all the transfected cells showed clear perinuclear staining indicating that the mutated MNK proteins were located to the TGN consistent to the wild type (wt) protein. Cells expressing MNK mutated in either mbs 1 or mbs 6 alone (m1 and m6) grown in elevated copper were predominantly stained in the cytoplasm and at the periphery of the cell similar to cells expressing the wild type protein (Figure 3). This suggests that the disruption of only one mbs does not alter the Cu-induced intracellular redistribution of MNK significantly. MNK mutated in mbs 1–3, mbs 4–6, and mbs 1–6 transiently expressed in CHO-K1 cells and cultured on copper containing media showed some changes in the intracellular localisation compared to wt MNK (data not shown). Therefore, cells were se-

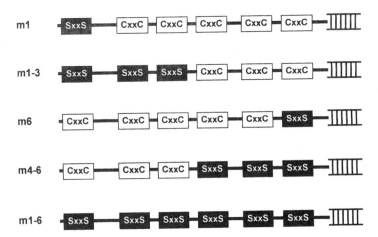

Figure 2. Mutagenesis strategy. By site-directed mutagenesis, putative metal binding sites in the MNK protein were mutated from GMxCxxC to GMxSxxS as indicated.

lected for stable expression of MNK mutated in mbs 1–3, mb 4–6 and mbs 1–6 and further analysed by immunofluorescence.

In basal media, wild type and the mutated MNK proteins clearly localised to the TGN, consistent with previous results observed on copper resistant CHO-K1 cells (Petris et al., 1996). This demonstrates that even multiple mutations in the N-terminus do not disrupt the protein structure in a way that would affect the localisation of the protein. After the addition of copper, MNK mutated in mbs 1–3 (m1–3) redistributed to vesicular structures in the cytoplasm and to the plasma membrane (Figure 4). Under the conditions described, m1–3 seemed to redistribute slightly different compared to the wt protein. It

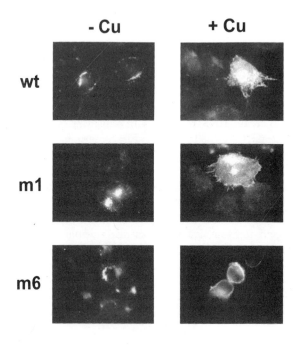

Figure 3. Effect of copper on MNK localisation of transiently transfected CHO-K1 cells expressing wt and mutant MNK proteins. Transiently transfected cells were left overnight to recover and subsequently cultured for 3 h in normal media (- Cu) or in media containing 200µM CuCl$_2$ (+ Cu). Cells were then acetone-fixed, blocked overnight and immunolabelled using an affinity purified antibody directed against the N-terminus of MNK. *wt*, wild type MNK; *m1*, MNK mutated in mbs 1 from GMxCxxC to GMxSxxS; *m6*, same mutation introduced into mbs 6.

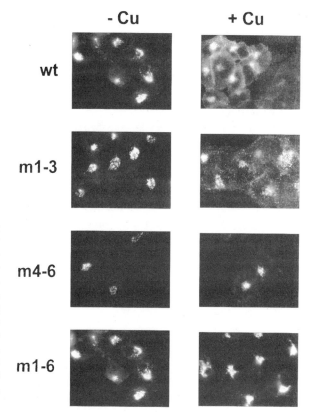

Figure 4. Effect of copper on MNK localisation of stable cell lines expressing wt and mutant MNK proteins. Cells stably expressing wt or mutant MNK were generated starting from a single colony after selection on 400μg/ml of G418 for 12 days. Stable cell lines were incubated with normal media (- Cu) or media containing 200μM CuCl$_2$ (+ Cu). Cells were then fixed and labelled as described in the legend to Fig.3. *wt*, wild type MNK; *m1–3*, MNK mutated in mbs 1 to 3 from GMxCxxC to GMxSxxS; m4–6, same mutations introduced into mbs 4–6; m1–6, same mutations introduced into mbs 1–6.

appeared that the plasma membranes of cells expressing m1–3 was less intensely stained compared to cells expressing wt protein pointing to a reduced exocytic rate of m1–3. However, this difference could as well be an artefact of the overexpression of wt protein, as the expression level of m1–3 is lower than the one of wt MNK. Future work will aim to quantitatively analyse any possible alterations in the copper induced exocytic rate of mutant compared to wt proteins.

In contrast to m1–3, the copper induced redistribution of MNK mutated in mbs 4–6 (m4–6) and MNK mutated in mbs 1–6 (m1–6) was dramatically reduced as shown in Fig 4. From these data it was not clear however, that the recycling of m4–6 and m1–6 was totally abolished. In both cases there could be some residual trafficking activity that is too sensitive to be detected by the indirect immunofluorescence method. These results demonstrate however, that the three mbs closest to the transport channel are most critical for copper induced relocation of MNK from the TGN to the plasma membrane.

In a second attempt to gain evidence that the N-terminus of MNK contains signalling and sorting sequences which direct the mature MNK protein to the secretory pathway, a MNK-Glut4 chimera was constructed and transiently expressed in CHO-K1 cells. Glut4 is an insulin dependent glucose exporting protein of muscle and fat cells. The mechanism by which insulin stimulates glucose transport has been well characterised and found to be similar to the copper dependent recycling of MNK from an intracellular membrane pool (the TGN) to the cell surface (James et al., 1989). Glut4 contains targeting information in the N-terminus as well as in a region spanning *trans*membrane domains 7 and 8 (Asano et al., 1992; Piper et al., 1992). A subsequent report further narrowed the N-terminal targeting sequence down to a phenylalanine at the position 5 (Piper et al., 1993). In our experiments, Glut4 expressed in CHO-K1 cells localised correctly to the TGN. The MNK-Glut4 chimera, however, mislocalised to widespread cytoplasmic compartments, presumably the endoplasmic reticulum (data not shown). Therefore, it was impossible to determine any differences in the intracellular location of MNK-Glut4 grown in basal or copper containing media. Chimeras may not preserve essential structural features leading to aberrant protein conformations that cannot be properly processed. Such misfolded proteins then often remain in the ER where they are subsequently degraded.

2.2. Functional Analysis of a Point Mutation in a Mild Menkes Patient

Immunocytochemical analysis was carried out to investigate the effect of the mutation in a mild Menkes patient on the intracellular location of MNK. The mutation was previously identified as an alanine to valine change at the amino acid position 1362 which is located in *trans*membrane 7 of the Menkes protein (Figure 1) (L. Ambrosini, paper submitted for publication). This region of the protein is not known to contain any specific signalling motifs that could be part of the trafficking mechanism. The mutation causing the Menkes phenotype is located in a *trans*membrane helix and therefore most probably interacts with sites involved in guiding the copper ions through the channel.

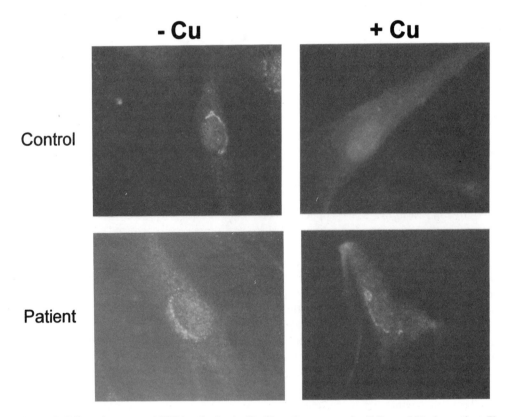

Figure 5. Effect of copper on MNK localisation in fibroblasts from a control cell line and Menkes patient. Fibroblasts from a control and a Menkes patient cell line were essentially prepared as described in Fig. 3. *-Cu*, no copper added; *+ Cu*, cells were supplemented with 189μM CuCl$_2$ for 3 hours.

A copper-induced redistribution to the plasma membrane could be demonstrated in the control fibroblast cell line (Figure 5), consistent with the results obtained on MNK expressed in CHO-K1 cells (this report, labelled as wt in Figures 3 and 4) and the results on copper resistant CHO-K1 cells (Petris et al., 1996). Under normal culturing conditions, fibroblasts from the Menkes patient showed a clear perinuclear signal consistent with TGN localisation of MNK. This observation was confirmed by culturing the cells in the presence of brefeldin A which caused the signal to contract, an effect typical of brefeldin A on MNK and other proteins (Petris et al., 1996; Ladinsky and Howell, 1992; Molloy et al., 1994) (data not shown). However, the addition of copper did not cause the relocalisation of the protein (Figure 5), suggesting that the alanine to valine change in *trans*membrane 7 reduced the ability of MNK to traffic.

3. DISCUSSION

The data presented here give evidence that not all the metal binding sites in MNK fulfill an equally important role in the function of copper dependent intracellular redistribution. Furthermore, we show that a mutation of a mild Menkes patient causes MNK to reside in the TGN of cultured patient fibroblasts. These two observations give rise to the hypothesis that copper induced trafficking of MNK is dependent on the copper-pumping activity of the protein.

3.1. Metal Binding Sites Close to the Membrane Channel Are Essential for Intracellular Trafficking of MNK

Mutations in mbs 1 or mbs 6 alone have no effect on the copper induced intracellular trafficking of MNK (Figure 3) suggesting that possible structural changes that had been introduced by the mutations did not interfere with the trafficking mechanism. Most point mutations observed in Menkes patients occur in exons 7 to 10 comprising metal binding site 6 and the first four *trans*membrane helices. All of the point mutations identified in the N-terminus were either nonsense mutations or insertion/deletion mutations resulting in a truncation of the *MNK* gene product (Tumer et al., 1997). This supports well our findings and suggests that N-terminal point mutations maintain both copper transport activity and the ability of MNK to translocate.

The m1–3 protein expressed in stable cell lines is able to redistribute to the plasma membrane demonstrating that the first three GMxCxxC repeats are not essential for the Cu-induced relocalisation of MNK (Figure 4). However, a slight reduction of intracellular and plasma membrane staining was observed which could be due to a reduced redistribution of the mutated MNK. This difference could not be repeatedly detected and quantitated by laser scanning confocal microscopy which leads to the suggestion that the wt and m1–3 proteins traffic at very similar rates (D. Strausak, unpublished observations). In contrast to m1–3, both m4–6 and m1–6 remained in the perinuclear region even in high copper concentrations (Figure 4). From these observations we can conclude that the metal binding sites closer to the membrane channel are more important for the intracellular trafficking of MNK. However, as discussed above, the function of the protein is fully retained if mbs 6 alone is mutated. This implies that mbs 4 and 5 are most important for the copper dependent translocation of MNK and that these two mbs could be part of a recycling motif. Further mutational analysis of MNK is required to clarify this interesting mechanism.

3.2. Is There a Correlation between Intracellular Trafficking and the Copper Transporting Activity of MNK?

An alanine to valine amino acid change, which results in two additional methyl-groups at the side-chain of the alanine in *trans*membrane 7, inactivates the trafficking function of the protein as shown with a Menkes patient cell line (Figure 5). This minor alteration may have distorted the structure of the Menkes protein, preventing the normal copper-induced translocation from the TGN to the plasma membrane. However, the mutant MNK in the TGN presumably retains some copper transport activity to provide copper to lysyl oxidase, since there are few connective tissue defects in this patient. The mild neurological defects in this patient compared to classical Menkes patients also indicate that some copper efflux activity and copper-trafficking are retained by the mutant molecule (Danks, 1988).

The Menkes patient mutation and the mutations introduced into the mbs discussed earlier in the chapter are the first mutations in MNK affecting the copper induced intracellular redistribution. Identical effects on trafficking caused by mutations in two structurally and functionally distinct parts of the protein strongly indicate that any MNK molecule inactive in copper transport will be unable to relocate to the plasma membrane. The recent publication by Payne et al. (Payne and Gitlin, 1998) adds a very interesting perspective to this point. The function of the mbs in the MNK N-terminus was addressed by expressing wild type and mutant MNK in a yeast strain lacking the WND/MNK homologue CCC2. Menkes protein mutants with one to five non-functional copper binding sites were generated by preogressively mutating the mbs from GMxCxxC to GMxSxxS. Subsequently, the copper transporting activities of the mutated proteins were measured by the incorporation of copper into Fet3, a ceruloplasmin homologue. Wild type MNK fully restored Fet3 oxidase activity, a role normally attributed to the WND protein, suggesting that the two homologues are functionally related. Surprisingly, mutations in mbs 1–3 entirely inhibited Fet3 oxidase activity whereas the protein mutated in mbs 1 and 2 retained Cu-transport activity which suggested a loss of a critical interaction site with a copper chaperone. The results presented by Payne et al. differ somewhat from our results in that we found little effect on copper trafficking of mutations in the first three metal binding sites. Our hypothesis that MNK-trafficking is dependent on Cu-transport activity would suggest that m1–3 will have significant copper transport activity. Further investigations on the N-terminus of MNK and the elaboration of a method to measure direct copper transport activity are required to clarify the outcome of this exciting story.

REFERENCES

Asano, T.K., Takata, K., Katagiri, H., Tsukuda, K., Lin, J.L., Ishihara, H., Inukai, K., Hirano, H., Yazaki, Y., and Oka, Y. (1992). Domains responsible for the intracellular targeting of glucose transport isoforms. J. Biol. Chem. 267, 19636–19641.

Bull, P.C., Thomas, G.R., Rommens, J.M., Forbes, J.R., and Cox, D.W. (1993). The Wilson disease gene is a putative copper transporting P-type ATPase similar to the Menkes gene [published erratum appears in Nat Genet 1994 Feb;6(2):214]. Nat. Genet. 5, 327–337.

Camakaris, J., Danks, D.M., Ackland, M.L., Cartright, E., Borger, P., and Cotton, R.G.H. (1980). Altered copper metabolism in cultured cells from human Menkes' syndrome and mottled mouse mutants. Biochem. Genet. 18, 117–131.

Camakaris, J., Petris, M.J., Bailey, L., Shen, P., Lockhart, P., Glover, T.W., Barcroft, C., Patton, J., and Mercer, J.F. (1995). Gene amplification of the Menkes (MNK; ATP7A) P-type ATPase gene of CHO cells is associated with copper resistance and enhanced copper efflux. Hum. Mol. Genet. 4, 2117–2123.

Chelly, J., Tumer, Z., Tonnesen, T., Petterson, A., Ishikawa-Brush, Y., Tommerup, N., Horn, N., and Monaco, A.P. (1993). Isolation of a candidate gene for Menkes disease that encodes a potential heavy metal binding protein [see comments]. Nat. Genet. 3, 14–19.

Danks, D.M. (1988). The mild form of Menkes disease: progress report on the original case. J. Med. Genet. 30, 859–864.

Danks, D.M. (1995). The metabolic basis of inherited disease. In Disorders of copper transport. J.R. Scriver, A.L. Beaudet, W.S. Sly, and D. Valle, eds. (New York: Mc Graw-Hill), pp. 2211–2235.

DiDonato, M., Narindrasorasak, S., Forbes, J.R., Cox, D.W., and Sarkar, B. (1997). Expression, purification and metal binding properties of the N-terminal domain from the Wilson disease putative copper-transporting ATPase (ATP7B). Journal of Biological Chemistry 272, 33279–33282.

Fu, D., Beeler, T.J., and Dunn, T.M. (1995). Sequence, mapping and disruption of CCC2, a gene that cross-complements the Ca(2+)-sensitive phenotype of csg1 mutants and encodes a P-type ATPase belonging to the Cu(2+)-ATPase subfamily. Yeast. 11, 283–292.

Goka, T.J., Stevenson, R.E., Hefferan, P.M., and Howell, R.R. (1976). Menkes' disease: a biochemical abnormality in cultured human fibroblasts. Proc. Natl. Acad. Sci. U. S. A. 73, 604–606.

Horn, N. (1976). Copper incorporation studies on cultured cells for prenatal diagnosis of Menkes' disease. Lancet 1, 1156–1158.

James, D.E., Strube, M., and Mueckler, M. (1989). Molecular cloning and characterization of an insulin-regulatable glucose transporter. Science 338, 83–87.

La Fontaine, S., Firth, S.D., Lockhart, P.J., Brooks, H., Parton, R.G., Camakaris, J., and Mercer, J.F.B. (1998). Functional analysis and intracellular localization of the human Menkes protein (MNK) stably expressed from a cDNA construct in Chinese hamster ovary cells (CHO-K1). Human Molecular Genetics 7, 1293–1300.

Ladinsky, M.S. and Howell, K.E. (1992). The trans-Golgi network can be dissected structurally and functionally from the cisternae of the Golgi complex by brefeldin A. Eur. J. Cell. Biol. 59, 92–105.

Lutsenko, S., Petrukhin, K., Cooper, M.J., Gilliam, C.T., and Kaplan, J.H. (1997). N-terminal domains of human copper transporting adenosine triphosphatases (the Wilson and Menkes disease proteins) bind copper selectively in vivo and in vitro with stochiometry of one copper per metal-binding repeat. Journal of Biological Chemistry 272, 18939–18944.

Mercer, J.F., Livingston, J., Hall, B., Paynter, J.A., Begy, C., Chandrasekharappa, S., Lockhart, P., Grimes, A., Bhave, M., Siemieniak, D., et, and al. (1993). Isolation of a partial candidate gene for Menkes disease by positional cloning [see comments]. Nat. Genet. 3, 20–25.

Molloy, S.S., Thomas, L., VanSlyke, J.K., Stenberg, P.E., and Thomas, G. (1994). Intracellular trafficking and activation of the furin proprotein convertase: localisation to the TGN and recycling from the cell surface. EMBO J. 13, 18–33.

Odermatt, A. and Solioz, M. (1995). Two trans-acting metalloregulatory proteins controlling expression of the copper ATPases of Enterococcus hirae. Journal of Biological Chemistry 270, 4349–4354.

Odermatt, A., Suter, H., Krapf, R., and Solioz, M. (1993). Primary structure of two P-type ATPases involved in copper homeostasis in Enterococcus hirae. J. Biol. Chem. 268, 12775–12779.

Payne, A.S. and Gitlin, J.D. (1998). Functional expression of the Menkes disease protein reveals common biochemical mechanisms among the copper-transporting P-type ATPases. Journal of Biological Chemistry 273, 3765–3770.

Petris, M.J., Mercer, J.F., Culvenor, J.G., Lockhart, P., Gleeson, P.A., Camakaris, and J. (1996). Ligand-regulated transport of the Menkes copper P-type ATPase efflux pump from the Golgi apparatus to the plasma membrane: a novel mechanism of regulated trafficking. EMBO J. 15, 6084–6095.

Piper, R.C., Tai, C., Kulesza, P., Pang, S., Warnock, D., Baenziger, J., Slot, J.W., Geuze, H.J., Puri, C., and James, D.E. (1993). Glut-4 NH2 terminus contains a phenylalanine-based targeting motif that regulates intracellular sequestration. J. Cell Biol. 121, 1221–1232.

Piper, R.C., Tai, C., Slot, J.W., Hahn, C.S., Rice, C.M., Huang, H., and James, D.E. (1992). The efficient intracellular sequestration of the insulin-regulatable glucose-transporter (Glut4) is conferred by the NH2-terminus. J. Cell Biol. 117, 729–743.

Solioz, M. and Odermatt, A. (1995). Copper and silver transport by CopB-ATPase in membrane vesicles of Enterococcus hirae. J. Biol. Chem. 270, 9217–9221.

Tanzi, R.E., Petrukhin, K., Chernov, I., Pellequer, J.L., Wasco, W., Ross, B., Romano, D.M., Parano, E., Pavone, L., Brzustowicz, L.M., and et al. (1993). The Wilson disease gene is a copper transporting ATPase with homology to the Menkes disease gene. Nat. Genet. 5, 344–350.

Tumer, Z., Lund, C., Tolshave, J., Vural, B., Tonnesen, T., and Horn, N. (1997). Identification of point mutations in 41 unrelated patients affected with Menkes disease. Am. J. Hum. Genet. 60, 63–71.

Vulpe, C., Levinson, B., Whitney, S., Packman, S., and Gitschier, J. (1993). Isolation of a candidate gene for
    Menkes disease and evidence that it encodes a copper-transporting ATPase [published erratum appears in
    Nat Genet 1993 Mar;3(3):273] [see comments]. Nat. Genet. 3, 7–13.

22

CLINICOPATHOLOGICAL AND MOLECULAR GENETIC FEATURES OF HEREDITARY CERULOPLASMIN DEFICIENCY (ACERULOPLASMINEMIA)

Kunihiro Yoshida[*]

Department of Medicine (Neurology)
Shinshu University School of Medicine
3-1-1 Asahi, Matsumoto 390-8621, Japan

1. INTRODUCTION

Ceruloplasmin (EC1.16.3.1) is a multi-copper ferroxidase which plays an important role in copper-iron metabolism in vertebrates (Harris et al., 1997). In vertebrate plasma, ceruloplasmin catalyzes the oxidation of ferrous iron (Fe (II)) to ferric iron (Fe (III)) which binds tightly to apotransferrin to form a major iron-mobilizing protein, transferrin. This enzymatic oxidation of Fe (II) promotes iron egress from some cell types, although the membrane transport system responsible for ferrous iron release into plasma has not yet been clarified (Kaplan and O'Halloran, 1996; Harris et al., 1997). The ferroxidase activity of ceruloplasmin has been proposed by experiments on copper-deficient animals since the 1960's (Osaki et al., 1966; Lee et al., 1968; Ragan et al., 1969; Roeser et al. 1970; Osaki et al., 1971). Recent recognition that mutations in the ceruloplasmin gene cause a neurodegenerative disorder characterized by aceruloplasminemia and neurovisceral storage of iron has further confirmed a biological significance of ceruloplasmin as a ferroxidase (Yoshida et al., 1995; Harris et al., 1995; Daimon et al., 1995b; Takahashi et al. 1996; Okamoto et al., 1996; Harris et al., 1996, Yazaki et al., 1998). This concept is also supported by the findings on FET3 protein in the yeast *Saccharomyces cerevisiae* (Dancis et al., 1994; Askwith et al., 1994; De Silva et al., 1995; Yuan et al., 1995; Kaplan and O'Halloran, 1996). FET3 is a multi-copper oxidase which has extensive homology to a class of copper-binding "blue" oxidases and is necessary for iron transport in *S. cerevisiae*.

---

[*] Tel: +81–263–37–2673; Fax: +81–263–34–0929; e-mail: naokosy@gipac.shinshu-u.ac.jp

Metals and Genetics, edited by Sarkar.
Kluwer Academic / Plenum Publishers, New York, 1999.

Hereditary ceruloplasmin deficiency with hemosiderosis (aceruloplasminemia) is an autosomal recessive disorder of copper-iron metabolism due to mutations in the ceruloplasmin gene (Morita and Yanagisawa, 1997; Kato et al, 1997b). This disease was first described by Miyajima et al. in 1987. Thereafter 15 patients from 8 unrelated families, including the original family (Miyajima et al., 1987), have been reported (Morita et al., 1992; Logan et al., 1994; Morita et al., 1995; Kawanami et al., 1996; Okamoto et al., 1996; Takahashi et al., 1996; Yonekawa et al., 1996; Inoue, 1997). Interestingly, 7 out of 8 families originated in Japan. This may be partly because aceruloplasminemia is more widely recognized in Japan than in other countries. It remains to be answered whether there is an ethnic difference in the predisposition of this disease or not. Although patients with this disease are still very small in number, they have contributed much to the establishment of the framework of a new disease entity.

In this paper, clinicopathological and molecular genetic features of aceruloplasminemia are reviewed. The pathogenesis and pathophysiology of aceruloplasminemia are also briefly discussed, in connection with structural and functional properties of ceruloplasmin.

2. CLINICAL ASPECTS

Clinical signs and symptoms of 15 patients with this disease are summarized in Table 1. The diagnosis of these patients was confirmed by characteristic laboratory, radiological and molecular genetic findings as described below. The age at onset of neurological disorders varied from 38 to 65 years (mean 51 years). About half of the patients showed involuntary movements such as blepharospasm, ataxic gait or slurred speech as initial neurological symptoms. The remaining half of the patients developed mental disturbance or intellectual impairments as initial symptoms. Once the disease was full-blown, various degrees of the neurological triad, cerebellar ataxia, extrapyramidal symptoms and dementia, were seen. Extrapyramidal symptoms included parkinsonism characterized by rigidity and bradykinesia and various types of involuntary movements such as blepharospasm, dystonia, facial grimacing, choreic movements, and oral dyskinesia. Dementia was considered a subcortical type of dementia, which is typically seen in Huntington's disease. Neurological disorders were slowly progressive, but clinical severity varied from patient to patient. Some patients had diabetes mellitus and/or retinal pigment degeneration, but were free of neurological symptoms or signs, although some were wheel-chair bound or bedridden because of severe cerebellar ataxia and parkinsonism. Deep tendon reflexes were often hyperactive, but pathological reflexes were not elicited. Sensory disturbance or autonomic dysfunction was usually absent.

Non-insulin dependent diabetes mellitus was confirmed in all patients. It usually developed in young adulthood and preceded neurological symptoms. Some patients developed diabetes mellitus 10–20 years before neurological manifestations appeared. Retinal pigmentary degeneration was another common feature of this disease. This was generally asymptomatic and was noted at the fundoscopic examination. There were no Kayser-Fleischer rings in the cornea, which is pathognomonic for Wilson's disease. Some neurologically asymptomatic individuals showed abnormal findings on the brain computerized tomography (CT) or magnetic resonance (MR) images, which were suggestive of excessive iron storage in the brain (Miyajima et al., 1987; Morita et al., 1995; Miyajima et al., 1996a, b). In consideration of the time lag between the onset of diabetes and neurological disorders, these individuals may possibly develop neurological symptoms in the future.

Table 1. Clinical summary

Cases	Age at onset (neurological disorders)	Initial symptoms	Diabetes mellitus (age at diagnosis)	Retinal pigment degeneration	Cerebellar ataxia	Extrapyramidal symptoms	Dementia mental disturbance
Miyajima et al. (1987, 1996a, 1996b, 1997, 1998)							
1. 64F	51	blepharospasm	+ (61)	+	+	+	−
2. 59F	56	blepharospasm	+ (51)	+	+	−	−
3. 57M	−	−	IGT	+	−	−	−
Morita et al. (1992, 1995)							
4. 60F*	51	forgetfulness	+ (40)	+	+	+	+
5. 50M	50	oral dyskinesia	+ (30)	+	+	+	−
6. 42M	38	forgetfulness	+ (26)	+	+	+	−
7. 63F	−	−	+ (50)	+	−	−	−
Logan et al. (1994), Harris et al. (1996)							
8. 62M	49	confusion	+ (49)	+	−	−	+
9. 55M	47	mental slowing	+ (49)	?	−	−	+
Daimon et al. (1995), Kawanami et al. (1996), Kato et al. (1997a)							
10. 58M*	56	depressive mode forgetfulness	+ (45)	+	−	+	+
Takahashi et al. (1996)							
11. 47F	45	slurred speech ataxic gait	+ (31)	+	+	−	−
12. 44M	−	−	+ (35-37)	+	−	−	−
Yonekawa et al. (1996)							
13. 54F	49	restlessness	+ (40)	+	+	+	+
Okamoto et al. (1996)							
14. 56M	56	slurred speech ataxic gait	+ (23)	+	+	+	+
Inoue et al. (1997)							
15. 70F	65	gait disturbance	+ (63)	+	?	+	−

Extrapyramidal symptoms include rigidity, parkinsonism and involuntary movements such as dystonia, blepharospasm, facial grimacing, chorea, and oral dyskinesia.

3. BIOCHEMICAL AND RADIOLOGICAL HALLMARKS

Biochemical analysis of the serum is remarkable for complete deficiency of ceruloplasmin, prominent decrease of copper, and marked increase of ferritin in aceruloplasminemia. In contrast to Wilson's disease, urinary copper excretion is usually normal or decreased. Mild microcytic hypochromic anemia associated with decreased serum iron is frequently seen. Total iron-binding capacity and transferrin are normal or slightly decreased. These laboratory findings sometimes provide a first clue to the diagnosis, even in neurologically asymptomatic individuals. There is no laboratory evidence suggestive of hepatic injury. Elevation of total protein in the cerebrospinal fluid was observed in five patients (Miyajima et al., 1987; Morita et al., 1992 and 1995; Inoue, 1997).

Radiological findings are also characteristic. Brain CT revealed high densities symmetrically in the basal ganglia, thalamus, and dentate nucleus of the cerebellum. These regions showed low signal intensities on both the T1- and T2-weighted MR images (Figure 1). The liver showed diffuse high density on the CT scans and hypointensity on both the T1- and T2-weighted MR images. These CT and MR images reflect excessive accumulation of iron.

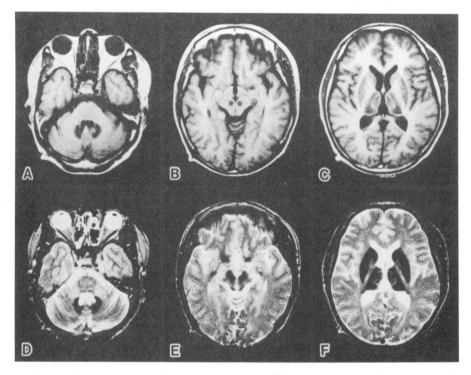

Figure 1. MR images of the brain of a 43-year-old patient at 1.5 tesla. The caudate nucleus, putamen, thalamus, glubus pallidus, red nucleus, and dentate nucleus of the cerebellum show low signal intensities on T1- and T2-weighted images. A-C; TR 400/TE 20, D-F; TR 3000/TE 80.

4. PATHOLOGY

To date, detailed neuropathological reports have been available on two deceased patients with aceruloplasminemia (Morita et al., 1995, Kawanami et al., 1996; Kato et al., 1997a, b). Both patients were Japanese; one was a 60-year-old woman who died of bronchopneumonia (Morita et al., 1995), and the other was a 58-year-old man who was found dead at home (Kawanami et al., 1996; Kato et al., 1997a, b). Pathological findings were essentially the same in both patients, but the severity in each organ or region differed to some extent.

Varying degrees of iron accumulation were observed in visceral organs (Figure 2A-D). The liver and pancreas were most severely affected (Figure 2A and B). In the liver, a heavy iron deposition was found in both hepatocytes and Kupffer's cells, but normal cytoarchitectures were well preserved. There was no evidence of liver cirrhosis. Copper accumulation in the liver was not significant. In the pancreas, iron deposition was more prominent in the exocrine than in the endocrine cells (Morita et al., 1995; Kato et al., 1997a, b). Kato et al. (1997a, b) found a marked depletion in β cells (insulin-containing cells) in the pancreatic islets in spite of no, or only minimal, deposition of iron. Iron deposition was also seen in the heart (Figure 2C), kidney (Figure 2D), thyroid gland, retina, spleen, and bone marrow, but to a lesser degree than in the liver and pancreas.

Macroscopic examination of the brain showed a red-brown discoloration in the basal ganglia, thalamus and dentate nucleus of the cerebellum (Morita et al., 1995, Kawanami et al., 1996; Kato et al., 1997b). In the case of Morita et al. (1995), there was a cavitation in the caudate nucleus and putamen due to severe tissue necrosis. The leptomeningeal and basal arteries were unremarkable. In the central nervous system, iron deposition was seen in both glial and neuronal cells. The extent of neuronal loss well paralleled that of iron deposition. Consisitent with the neuroradiological findings, the most striking lesions were seen in the caudate nucleus and putamen. Both showed spongy softening accompanied by extensive neuronal loss and little evidence of reactive gliosis (Figure 2E). In the other basal ganglia, thalamus, and dentate nucleus, pathological changes were basically similar to those in the striatum, but less severe. The cerebellar cortex showed marked loss of Purkinje cells, (Figure 2F), but the granular cell layer was unaffected. In the cerebral cortex, mild or moderate iron deposition was found in both cases, but neuronal loss was not significant. The spinal cord, peripheral nerves, and skeletal muscles were essentially unaffected.

Electron microscopic examinations revealed many electron-dense inclusions in the cytoplasms of nerve and glial cells in the basal ganglia(Morita et al., 1995). These inclusions were pleomorphic in shape and lacked any consistent relation to the organelle. In the cerebral gray and white matter, large numbers of astrocytes and oligodendrocytes also contained similar intracytoplasmic inclusions. Energy-dispersive X-ray microanalysis showed that iron was a major component of these inclusions in the nerve and glial cells (Morita et al., 1995).

The distribution of neuropathological changes in aceruloplasminemia is similar to that in Wilson's disease, but no copper accumulation was detectable in either case. Kawanami et al. (1996) described spheroids (neuroaxonal dystrophy) in the caudate nucleus, which, in conbination with iron deposition and neuronal loss, is a characteristic feature in Hallervorden-Spatz diaease. However, iron deposition is more widely distributed in the brain in aceruloplasminemia than in Hallervorden-Spatz disease, in which the glubus pallidus, substantia nigra, and red nucleus are preferentially affected. This unique distribution of affected lesions in Hallervorden-Spatz disease is distinct from that in aceruloplasminemia.

Based on a massive iron deposition in visceral organs including the liver, pancreas, and heart, hereditary hemochromatosis should be ruled out in the diagnosis of aceruloplasminemia, especially when the diagnosis is made in the early stage of the disease. The main point is that the central nervous system is usually spared in hereditary hemochromatosis. Interestingly, Nielsen et al. (1995) reported a 50-year-old patient with hereditary hemochromatosis who developed parkisonism in his twenties. He showed an increase of serum ferritin, and revealed excessive iron accumulation in the basal ganglia and dentate nucleus on the MR images. Liver biospy showed a heavy iron overload, but there was no copper

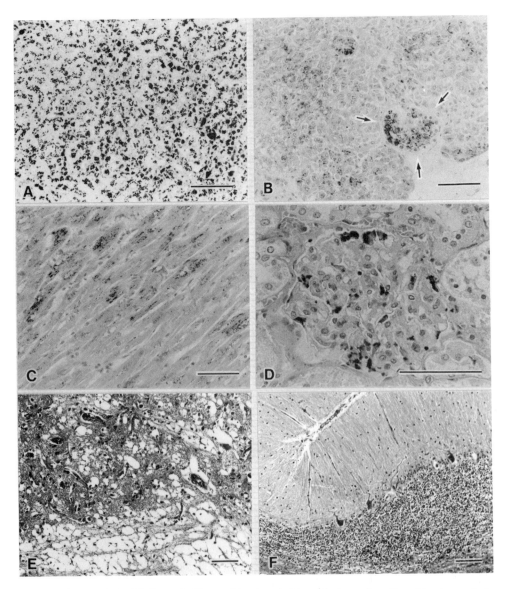

Figure 2. Histopathological findings of a 60-year-old patient. A; Liver: The heavy deposition of iron is seen in the hepatocytes. B; Pancreas: Note that there is iron deposition in the pancreatic islet (arrows). C; Heart (left ventricle): Myocardial fibers show iron deposition. D; Kidney: Iron deposition is seen mainly in the podocytes of the glomerulus. (A-D, Prussian blue stain, bar = 100 μm). E; Putamen: Spondy softening with severe neuronal cell loss is observed. F; Cerebellar cortex: Marked loss of Purkinje cells is noted (E and F, Hematoxylin and eosin stain, bar = 100 μm). (Courtesy of Dr. Hiroshi Morita, Shinshu University School of Medicine.)

excess or evidence of liver cirrhosis. Clinical features of this patient were quite similar to those of aceruloplasminemia, except that he showed a normal concentration of serum ceruloplasmin. At present, no interpretation is available as to what differences exist in iron metabolism between this case and typical hereditary hemochromatosis or aceruloplasminemia. This case indicates that the involvement of iron metabolism in neurodegenerative disorders may be more complicated than expected.

5. MOLECULAR GENETICS

Human ceruloplasmin is composed of a single polypeptide chain of 1,046 amino acids residues, which was first described by Takahashi et al. in 1984. Human ceruloplasmin exhibits a threefold internal homology in amino acid sequence, and each homology unit contains 340–350 amino acids (Takahashi et al., 1983; Takahashi et al., 1984). These data suggest that ceruloplasmin is an evolutionary product of a tandem triplication of ancestral genes. Subsequently, cDNA clones for human ceruloplasmin were isolated and characterized in 1986 (Koschinsky et al., 1986; Mercer et al., 1986; Yang et al., 1986), and the derived amino acid sequence was consistent with the earlier protein works. Rat and murine ceruloplasmin cDNAs were also isolated (Fleming and Gitlin, 1990; Klomp et al., 1996). The amino acid sequences are very homologous between these three species, with complete conservation of amino acid ligands essential for copper binding (Klomp et al., 1996). The genomic structure of the human ceruloplasmin gene was published in 1995 (Daimon et al., 1995a). According to this work, the gene consists of 19 exons spanning approximately 50 kb in size.

Mutations in the ceruloplasmin gene so far identified are summarized in Table 2. To date, 6 mutations have been identified in this disease (Yoshida et al., 1995; Harris et al., 1995; Daimon et al., 1995b; Takahashi et al. 1996; Okamoto et al., 1996; Harris et al., 1996; Yazaki et al., 1998). Gene mutations are heterogeneous in patients with the same ethnic background, though a mutation is shared by two unrelated Japanese families (Daimon et al., 1995b; Takahashi et al. 1996). It is of interest that all mutations are predicted to produce truncated ceruloplasmin by premature termination in translation (Table 2).

Table 2. Mutations identified in the ceruloplasmin gene in aceruloplasminemia

Mutation	Location	Predicted position of PTC	Reference
1-bp (A) insertion	nt 607 (exon 3)	codon 188	Okamoto et al. (1996)
a → g transition	nt 1209-2 (3' splice acceptor site) (intron 6)	codon 388	Yazaki et al. (1998)
5-bp (TACAC) insertion	nt 1287 (exon 7)	codon 446	Harris et al. (1995)
1-bp (G) deletion	nt 2389 (exon 13)	codon 789	Harris et al. (1996)
G → A transition	nt 2630 (exon 15)	codon 858	Daimon et al. (1995b) Takahashi et al. (1996)
g → a transition	nt 3019-1 (3' splice acceptor site) (intron 17)	codon 991	Yoshida et al. (1995)

PTC: premature termination codon.
The numbering of the nucleotide and codon is based upon the previous report (Koschinsky et al. 1986).

Based on high homology in amino acid sequence to the other copper-binding "blue" oxidases such as ascorbate oxidase, C-terminal domain (domain 6) of ceruloplasmin has been considered essential for its oxidase activity as copper-binding sites (Kingston et al., 1979; Ryden, 1982; Takahashi et al., 1983). Recent X-ray structural analysis of ceruloplasmin has further supported this speculation (Zaitseva et al., 1996; Lindley et al., 1997). It has confirmed that ceruloplasmin has six domains arranged in a triangular array and that six copper atoms properly incorporated in the molecule are essential for the oxidase activity of ceruloplasmin as well as the stability of its domain structure (Zaitseva et al., 1996; Lindley et al., 1997). In more detail, the trinuclear copper centre is located at the interface of domains 1 and 6, and each domain contributes four histidine residues to the trinuclear cluster. Furthermore, domain 6 has a ligand (His 975, Cis 1021, His 1026, and Met 1031) for mononuclear copper. Thus, domain 6 is greatly involved in the proper arrangement of copper atoms in the ceruloplasmin molecule. The relative location of the trinuclear copper centre and the nearest mononuclear copper in domain 6 is very similar to that in ascorbate oxidase and probably forms an acitive site as a ferroxidase (Zaitseva et al., 1996). Therefore, it is highly likely that mutations which cause truncated ceruloplasmin result in drastic conformational change and then rapid proteolytic degradation of the molecule, ending in aceruloplasminemia.

6. PATHOGENESIS AND PATHOPHYSIOLOGY

Corresponding to its multi-functional properties, the ceruloplasmin gene is expressd in a wide variety of organs and tissues. The liver is the major site of ceruloplasmin synthesis, however, lower levels of expression are detected in multiple extrahepatic tissues. In the brain, ceruloplasmin mRNA is localized to specific subpopulations of astrocytes and the epithelial cells of the pia-arachnoid and choroid plexus by *in situ* hybridization (Klomp et al., 1996a, b). Transferrin is also synthesized in oligodendrocytes in the brain (Connor and Menzies, 1995; Klomp et al., 1996b). Since neither ceruloplasmin nor transferrin crosses the blood-brain barrier, these proteins synthesized in glial cells should be necessary for iron transport and utilization in the brain.

The precise pathogenesis of neuronal death in this disease has remained unclear, but it is generally accepted that oxygen free-radicals associated with ferrous iron accumulation play an important role in the pathological processes. The brain is highly susceptible to "oxidative stress" because of its high rate of oxygen utilization, high content of lipids, relatively low level of endogenous antioxidants like catalase, and regionally high content of iron and ascorbate (Halliwell, 1992; Gerlach et al, 1994; Connor and Menzies, 1995). Under these circumstances, the antioxidant function of ceruloplasmin as a ferroxidase seems indispensable for neuronal function and survival. If ceruloplasmin is deficient and ferrous iron cannot be oxidized for uptake by transferrin, then free ferrous iron would result in rapid generation of cytotoxic reactive oxigen species. Such "oxidative stress" is thought to increase lipid peroxidation of the cell membrane, resulting in change in membrane fluidity and permeability, and finally cell damage (Halliwell, 1992). In fact, a marked increase in plasma lipid peroxidation was observed in patients with aceruloplasminemia (Miyajima et al., 1996a). Increased very long-chain fatty acids in erythrocyte membranes was also found in this disease, which suggests free-radical-mediated injury in peroxisomal b-oxidation (Miyajima et al., 1998). The generation of oxygen free-radicals is also considered a probable cause of diabetes mellitus (Kato et al., 1997a, b). This is supported by the fact that β cells in the islets are very susceptible to free radicals and that iron chelating therapy can effectively reduce free-radical mediated b cell injury in diabetic mice (Mendola et al., 1989).

7. THERAPEUTIC TRIALS

Effective therapy has not been established in aceruloplasminemia. D-penicillamin therapy was initiated in a patient before the final diagnosis was made, but no clinical or laboratory effects have been reported (Morita et al., 1992, 1995). Miyajima et al. (1997) reported that treatment with an iron chelator, desferrioxamine, effectively reduced lipid peroxidation and also improved neurological symptoms in a patient. Morita et al. (1995) also tried desferrioxamine therapy in two patients with neurological symptoms, but its effectiveness was unclear. Because iron accumulation starts far before neurological disorders develop and increases gradually, a chelating therapy with desferrioxamine may possibly prevent or ameliolate neuronal injury if it is initiated early in the presymptomatic stages. Further trials with desferrioxamine will be needed to confirm its effectiveness in aceruloplasminemia. For this purpose, biochemical and radiological screening for aceruloplasminemia would be warranted in patients who have diabetes mellitus and neurological disorders such as cerebellar ataxia, extrapyramidal symptoms, or dementia.

8. FUTURE PERSPECTIVES

Free-radical mediated tissue injury has recently attracted much attention as a pathogenesis of neuronal death in several neurodegenerative disorders such as Parkinson's disease, Alzheimer's disease and amyotrophic lateral sclerosis (Halliwell, 1992; Gerlach et al., 1994). Iron and copper, two major redox transition metals in biological systems, involve such oxidative tissue injury (De Silva and Aust, 1992; Halliwell, 1992; Gerlach et al., 1994). From this viewpoint, aceruloplasminemia provides a good opportunity to study a direct link between iron-copper metabolism and neurodegeneration. A ceruloplasmin-defficient mouse generated by gene targeting will be required for more detailed analyses of aceruloplasminemia. Further study of the pathogenesis and pathophysiology of aceruloplasminemia will advance our understanding of metal-mediated neuronal injury.

ACKNOWLEDGMENTS

The author thanks Drs. Hiroshi Morita and Shu-ichi Ikeda (Shinshu University School of Medicine) for helpful discussions, and Drs. Takeo Kato (Yamagata University School of Medicine) and Hiroaki Miyajima (Hamamatsu University School of Medicine) for clinical information on the patients.

REFERENCES

Askwith, C., Eide, D., Ho, A.V., Bernard, P.S., Li, L., Davis-Kaplan, S., Sipe, D.M., and Kaplan, J. (1994) The *FET 3* gene of S. cerevisiae encodes a multicopper oxidase required for ferrous iron uptake. Cell 76, 403–410.
Connor, J.R. and Menzies, S.L. (1995) Cellular management of iron in the brain. J. Neurol. Sci. 134 (Suppl.), 33–44.

Daimon, M., Yamatani, K., Igarashi, M., Fukase, N., Kawanami, T., Kato, T., Tominaga, M., and Sasaki, H. (1995a) Fine structure of the human ceruloplasmin gene. Biochem. Biophys. Res. Commun. 208, 1028–1035.

Daimon, M., Kato, T., Kawanami, T., Tominaga, M., Igarashi, M., Yamatani, K., and Sasaki, H. (1995b) A nonsense mutation of the ceruloplasmin gene in hereditary ceruloplasmin deficiency with diabetes mellitus. Biochem. Biophys. Res. Commun. 217, 89–95.

Dancis, A., Yuan, D.S., Haile, D., Askwith, C., Eide, D., Moehle, C., Kaplan, J., and Klausner, R.D. (1994) Molecular characterization of a copper transport protein in S. cerevisiae: an unexpected role for copper in iron transport. Cell 76, 393–402.

De Silva, D.M., and Aust, S.D. (1992) Ferritin and ceruloplasmin in oxidative damage: review and recent findings. Can. J. Physiol. Pharmacol. 71, 715–720.

De Silva, D.M., Askwith, C.C., Eide, D., and Kaplan, J. (1995) The FET 3 gene product required for high affinity iron transport in yeast is a cell surface ferroxidase. J. Biol. Chem. 270, 1098–1101.

Fleming, R.E., and Gitlin, J.D. (1990) Primary structure of rat ceruloplasmin and analysis of tissue-specific gene expression during development. J. Biol. Chem. 265, 7701–7707.

Gerlach, M., Ben-Shachar, D., Riederer, P., and Youdim, M.B.H. (1994) Altered brain metabolism of iron as a cause of neurodegenerative diseases? J. Neurochem. 63, 793–807.

Halliwell, B. (1992) Reactive oxygen species and the central nervous system. J. Neurochem. 59, 1609–1623.

Harris, Z.L., Takahashi, Y., Miyajima, H., Serizawa, M., MacGillivray, R.T.A., and Gitlin, J.D. (1995) Aceruloplasminemia: Molecular characterization of this disorder of iron metabolism. Proc. Natl. Acad. Sci. USA. 92, 2539–2543.

Harris, Z.L., Migas, M.C., Hughes, A.E., Logan, J.I., and Gitlin, J.D. (1996) Familial dementia due to a frameshift mutation in the caeruloplasmin gene. Q. J. Med. 89, 355–359.

Harris, Z.L., Morita, H., and Gitlin, J.D. (1997) The biology of human ceruloplasmin. In: Messerschmidt A (Ed.), Multi-copper oxidases. World Scientific, Singapore, pp. 285–305.

Inoe, Y. (1997) Hereditary ceruloplasmin deficiency with antineuronal antibody presenting with parkinsonism, dystonia, diabetes mellitus and retinal pigmentary degeneration. Neurol. Med. (Tokyo) 47, 74–82 (in Japanese).

Kaplan, J., and O'Halloran, T.V. (1996) Iron metabolism in eukaryotes: mars and venus at it again. Science 271, 1510–1512.

Kato, T., Daimon, M., Kawanami, T., Ikezawa, Y., Sasaki, H., and Maeda, K. (1997a) Islet changes in hereditary ceruloplasmin deficiency. Hum. Pathol. 28, 499–502.

Kato, T., Kawanami, T., Daimon, M., and Sasaki, H. (1997b) Hereditary ceruloplasmin deficiency: a cerebro-retino-diabetic disease. Neuropathology 17, 71–75.

Kawanami, T., Kato, T., Daimon, M., Tominaga, M., Sasaki, H., Maeda, K., Arai, S., Shikama, Y., and Katagiri, T. (1996) Hereditary caeruloplasmin deficiency: clinicopathological study of a patient. J. Neurol. Neurosurg. Psychiatry 61, 506–509.

Kingston, I.B., Kingston, B.L., and Putnam, F.W. (1979) Complete amino acid sequence of a histidine-rich proteolytic fragment of human ceruloplasmin. Proc. Natl. Acad. Sci. USA. 76, 1668–1672.

Klomp, L.W.J., Farhangrazi, Z.S., Dugan, L., and Gitlin, J.D. (1996a) Ceruloplasmin gene expression in the murine central nervous system. J. Clin. Invest. 98, 207–213.

Klomp, L.W.J., and Gitlin, J.D. (1996b) Expression of the ceruloplasmin gene in the human retina and brain: implications for a pathogenic model in aceruloplasminemia. Hum. Molec. Genet. 5, 1989–1996.

Koschinsky, M.L., Funk, W.D., van Oost, B.A., and MacGillivray, R.T.A. (1986) Complete cDNA sequence of human preceruloplasmin. Proc. Natl. Acad. Sci. USA. 83, 5086–5090.

Lee, G.R., Nacht, S., Lukens, J.N., and Cartwright, G.E. (1968) Iron metabolism in copper-deficient swine. J. Clin. Invest. 47, 2058–2069.

Lindley, P.F., Card, G., Zaitseva, I., Zaitsev, V., Reinhammar, B., Selin-Lindgren, E., and Yoshida, K. (1997) An X-ray structural study of human ceruloplasmin in relation to ferroxidase activity. J. Biol. Inorg. Chem. 2, 454–463.

Logan, J.I., Harveyson, K.B., Wisdom, G.B., Hughes, A.E., and Archbold, G.P.R. (1994) Hereditary caeruloplasmin deficiency, dementia and diabetes mellitus. Q. J. Med. 87, 663–670.

Mendola, J., Wright, J.R. Jr., and Lacy, P.E. (1989) Oxygen free-radical scavengers and immune destruction of murine islets in allograft rejection and multiple low-dose streptozocin-induced insulitis. Diabetes 38, 379–385.

Mercer, J.F.B. and Grimes, A. (1986) Isolation of a human ceruloplasmin cDNA clone that includes the N-terminal leader sequence. FEBS Lett. 203, 185–190.

Miyajima, H., Nishimura, Y., Mizoguchi, K., Sakamoto, M., Shimizu, T., and Honda, N. (1987) Familial apoceruloplasmin deficiency associated with blepharospasm and retinal degeneration. Neurology 37, 761–767.

Miyajima, H., Takahashi, Y., Serizawa, M., Kaneko, E., and Gitlin, J.D. (1996a) Increased plasma lipid peroxidation in patients with aceruloplasminemia. Free Rad. Biol. Med. 20, 757–760.

Miyajima, H., Takahashi, Y., Shimizu, H., Sakai, N., Kamata, T., and Kaneko, E. (1996b) Late onset diabetes mellitus in patients with hereditary aceruloplasminemia. Intern. Med. 35, 641–645.

Miyajima, H., Takahashi, Y., Kamata, T., Shimizu, H., Sakai, N., and Gitlin, J.D. (1997) Use of desferrioxamine in the treatment of aceruloplasminemia. Ann. Neurol. 41, 404–407.

Miyajima, H., Adachi, J., Tatsuno, Y., Takahashi, Y., Fujimoto, M., Kaneko, E., and Gitlin, J.D. (1998) Increased very long-chain fatty acids in erythrocyte membranes of patients with aceruloplasminemia. Neurology 50, 130–136.

Morita, H., Inoue, A., and Yanagisawa, N. (1992) A case with ceruloplasmin deficiency which showed dementia, ataxia and iron deposition in the brain. Clin. Neurol. (Tokyo) 32, 483–487 (in Japanese).

Morita, H., Ikeda, S., Yamamoto, K., Morita, S., Yoshida, K., Nomoto, S., Kato, M., and Yanagisawa, N. (1995) Hereditary ceruloplasmin deficiency with hemosiderosis: a clinicopathological study of a Japanese family. Ann. Neurol. 37, 646–656.

Morita, H., and Yanagisawa, N. (1997) The role of ceruloplasmin for iron and copper metabolism in central nervous system disorders. In: Yasui, M., Strong, M.J., Ota, K., Verity, M.A. (Eds.) "Mineral and Metal Neurotoxicology." CRC Press, Boca Raton, pp 405–411.

Nielsen, J.E., Jensen, L.N., and Krabbe, K. (1995) Hereditary haemochromatosis: a case of iron accumulation in the basal ganglia associated with a parkinsonian syndrome. J. Neurol. Neurosurg. Psychiatry 59, 318–321.

Okamoto, N., Wada, S., Oga, T., Kawabata, Y., Baba, Y., Habu, D., Takeda, Z., and Wada, Y. (1996) Hereditary ceruloplasmin deficiency with hemosiderosis. Hum. Genet. 97, 755–758.

Osaki, S., Johnson, D.A., and Frieden, E. (1966) The possible significance of the ferrous oxidase activity of ceruloplasmin in normal human serum. J. Biol. Chem. 241, 2746–2751.

Osaki, S., Johnson, D.A., and Frieden, E. (1971) The mobilization of iron from the perfused mammalian liver by a serum copper enzyme, ferroxidase I. J. Biol. Chem. 246, 3018–3023.

Ragan, H.A., Nacht, S., Lee, G.R., Bishop, C.R., and Cartwright, G.E. (1969) Effect of ceruloplasmin on plasma iron in copper-deficient swine. Am. J. Physiol. 217, 1320–1323.

Roeser, H.P., Lee, G.R., Nacht, S., and Cartwright, G.E. (1970) The role of ceruloplasmin in iron metabolism. J. Clin. Invest. 49, 2408–2417.

Ryden, L. (1982) Model of the active site in the blue oxidases based on the ceruloplamin-plastocyanin homology. Proc. Natl. Acad. Sci. USA. 79, 6767–6771.

Takahashi, N., Bauman, R.A., Ortel, T.L., Dwulet, F.E., Wang, C.-C., and Putnam, F.W. (1983) Internal triplication in the structure of human ceruloplasmin. Proc. Natl. Acad. Sci. USA. 80, 115–119.

Takahashi. N., Ortel, T.L., and Putnam, F.W. (1984) Single-chain structure of human ceruloplasmin: the complete amino acid sequence of the whole molecule. Proc. Natl. Acad. Sci. USA. 81, 390–394.

Takahashi, Y., Miyajima, H., Shirabe, S., Nagataki, S., Suenaga, A., and Gitlin, J.D. (1996) Characterization of a nonsense mutation in the cerulplasmin gene resulting in diabetes and neurodegenerative disease. Hum. Mol. Genet. 5, 81–84.

Yang, F., Naylor, S.L., Lum, J.B., Cutshaw, S., McCombs, J.L., Naberhaus, K.H., McGill. J.R., Adrian, G.S., Moore, C.M., Barnett, D.R., and Bowman, B.H. (1986) Characterization, mapping, and expression of the human ceruloplasmin gene. Proc. Natl. Acad. Sci. USA. 83, 3257–3261.

Yazaki, M., Yoshida, K., Nakamura, A., Furihata, K., Yonekawa M., Okabe, T., Yamashita, N., Ohta, M., and Ikeda, S. (1998) A novel mutation in the ceruloplasmin gene responsible for hereditary ceruloplasmin deficiency with hemosiderosis. J. Neurol. Sci. 156, 30–34.

Yonekawa, M., Okabe. T., Hirano, N., Kikumoto, O., Ohta, M., Doi, H., and Aoki, Y. (1996) A case of hereditary ceruloplasmin deficiency and iron deposition in the brain, associated with chorea, dementia, diabetes mellitus and retinal degeneration. J. Hiroshima Med. Assoc. (Hiroshima) 49, 185–188 (in Japanease).

Yoshida, K., Furihata, K., Takeda, S., Nakamura, A., Yamamoto, K., Morita, H., Hiyamuta, S., Ikeda, S., Shimizu, N., and Yanagisawa, N. (1995) A mutation in the ceruloplasmin gene is associated with systemic hemosiderosis in humans. Nature Genet. 9, 267–272.

Yuan, D.S., Stearman, R., Dancis, A., Dunn, T., Beeler, T., and Klausner, R.D. (1995) The Menkes/Wilson disease gene homologue in yeast provides copper to a ceruloplasmin-like oxidase required for iron uptake. Proc. Natl. Acad. Sci. USA. 92, 2632–2636.

Zaitseva, I., Zaitsev, V., Card, G., Moshkov, K., Bax, B., Ralph, A., and Lindley, P. (1996) The X-ray structure of human serum ceruloplasmin at 3.1 Å: nature of the copper centres. J. Biol. Inorg. Chem. 1, 15–23.

23

COPPER RELATED LIVER DISEASE IN YOUNG CHILDREN

M. S. Tanner

Division of Child Health
Sheffield Children's Hospital
Sheffield S10 2TH, United Kingdom

1. INTRODUCTION

The hepatic copper concentration in normal adults is 15–50 µg/g dry weight. Values higher than this are seen in many situations in childhood:

- Physiologically in the newborn (Reed & Landing, 1972), associated with lower serum caeruloplasmin and copper and higher plasma free copper values than the adult (Barrow & Tanner, 1988).
- In chronic cholestatic disorders such as biliary atresia.
- In chronic inflammatory disorders, for example in autoimmune hepatitis in childhood, liver copper values are raised although usually not greater than 250 µg/g dry weight.
- Indian Childhood Cirrhosis (ICC) and the ICC-like disorders in which liver copper concentrations usually exceed 800 µg/g and values as high as 6,000 µg/g are recorded.
- Wilson's disease, in which liver copper greater than 250 µg/g is one of the diagnostic criteria.
- Cirrhosis associated with raised hepatic copper concentrations but with features neither of Wilson's disease nor ICC.
- Miscellaneous disorders. For example, a child was recently seen in whom tissue from a focal hyper-plastic nodule had a copper concentration of 750 µg/g.

2. INDIAN CHILDHOOD CIRRHOSIS

The features of ICC have been reviewed recently (Pandit & Bhave 1996; Tanner 1998) and therefore are summarised here.

Metals and Genetics, edited by Sarkar.
Kluwer Academic / Plenum Publishers, New York, 1999.

1. Previously it was the commonest cause of liver disease in childhood. Of 375 children with chronic liver disease in the early 1980s in the Pune district of Maharashtra in India, 210 had ICC. ICC was estimated to affect 1 in 4100 live births in the rural community (Tanner, 1986).
2. It has now virtually disappeared (Bavdekar et al. 1996; Pradhan et al. 1995).
3. It had a characteristic histological appearance, quite distinct from other hepatopathies in the same age group (Aikat et al. 1974; Nayak & Ranmalingaswami 1975). The characteristic histological features were:

   a) Necrosis of hepatocytes with ballooning and Mallory's hyaline
   b) Pericellular intralobular fibrosis
   c) Inflammatory infiltrate
   d) Poor hepatocyte regenerative activity
   e) Abscence of fatty change
   f) Granular orcein staining.

These features allowed differentiation from other cirrhoses in Indian children.

4. It was associated with very high liver copper concentrations demonstrated by histochemistry and atomic absorption spectophotometry (Tanner et al. 1979; Popper et al. 1979). It is important to note the difference between ICC and Wilson's disease. In the latter, orcein, rhodanine and rubeanic acid stains are characteristically negative in childhood. Electron probe analysis in ICC demonstrates high concentrations of copper and sulphur in electron dense residual bodies regarded as lysosomes.
5. The copper accumulation was shown to be due to a particular dietary practice, namely the early introduction of animal milk feeds contaminated with copper from brass vessels (Tanner et al. 1983; Bhave et al., 1987).
6. ICC had a homogeneous clinical presentation:

- Occurrence in the Indian sub-continent
- Affected children aged 6 months to 4 years
- Males outnumbered females 4 to 1
- Presented usually with gross abdominal distention
- Found, on examination, to have growth hepatomegaly with an extraordinarily hard liver texture and sharp liver edge
- Without treatment, a rapidly downhill and ultimately fatal clinical course
- ICC was treatable with penicillamine which, whilst without effect in advanced cases, when given to children who had not yet developed jaundice or ascites significantly reduced mortality (Tanner et al. 1987).
- Treated survivors were able to discontinue penicillamine after 3 years and remain clinically normal. They showed a fall in liver copper to near normal values, which contrasts with the situation in Wilson's disease in which penicillamine treatment does not cause liver copper to fall to normal. There was a histological progression in which hepatocyte necrosis and inflammatory infiltrate diminished but nodularity initally increased as hepatocyte regenerative activity improved. This led to a stage of recovery in which an inactive micronodular cirrhosis was apparent. (Pradhan et al. 1995)
- At the time when ICC was common, a number of children with cryptogenic cirrhosis were also seen. It was speculated that some of these were the result of forme fruste ICC. As ICC has declined, so has the number of children with cryptogenic cirrhosis.

7. ICC is, therefore, not an important health problem any longer. It is overshadowed by the large numbers of infants with cholestasis. The reasons why, if it remains of more than historical interest, are:

- The lessons it can teach us about copper toxicity in man
- Continued occurrence of sporadic ICC-like cases in other parts of the world.

3. SPORADIC ICC-LIKE CASES

Walker-Smith & Blomfield (1973) described a child aged one year who developed rapidly progressive and fatal liver disease associated with a liver copper concentration of 3,360 µg/g dry weight. This was attributed to the use of well water with a low pH, which leeched copper from new plumbing within the house so that the copper concentration in the water used to make the infant feeds rose to 6.7 mg/l. At least 10 further isolated cases have been reported (Table 1) common features of which have been:

- A rural household
- Use of water from a private well
- The water was of low pH and acquired increased copper concentrations from plumbing or hot water boilers
- The clinical course was usually fatal without liver transplantation.

4. TYROLEAN CHILDHOOD CIRRHOSIS

In a retrospective review of 138 infants who died in the Austrian Tyrol in 1900–1974 Müller et al. (1996) noted striking clinical and pathological similarities with ICC. The similarities included:

1. *Epidemiology.* The condition was endemic in a defined geographical area, affected rural farming families rather than urban communities and has now virtually disappeared.

Table 1. Cases of infantile cirrhosis in which water used to make up infant feed had high copper concentration

Country	Age	Sex	Outcome	Liver Cu microg/g dry wt	Water Cu mg/l	Reference
Australia	1	M	d in 6w	3360	6.75	Walker-Smith et al. 1973
Germany	7m	F	d in 3m	684[1]	0.430[2]	Muller-Hocker et al. 1987
Germany	9m	F	d in 4m survived	2154	2.2-3.4	Muller-Hocker et al. 1988
	5m	M		2094		
Germany	5-9m	F	died	1240	0.43-3.4	Schramel et al. 1988
		F	died	1870		
		M	survived	1450		
Eire	10m	M	d age 11m	1245	8.0	Baker et al. 1995
	14m	M	d age 15m	1320	6.7	
Germany	13m	M		high	12-28.6	Bent et al. 1995
Australia	20m	M	liver transplantation	10-fold increased	7.8	Price et al. 1996

[1]Liver copper given as micrograms/g wet weight, has been multiplied by 3.
[2]Normal range stated to be <0.01 mg/l.

2. *Clinical features.* TCC affected infants aged 6 months to 4 years, presented with failure to thrive, followed by abdominal distention, hepatomegaly and later jaundice and ascites. The condition was rapidly progressive.

3. *Pathological features.* These were described in a monograph by Gogl (1947). Using haemotoxylin and eosin stains he described appearances identical to ICC. Unfortunately, copper stains were not done and copper measurements not made and so the role of copper in TCC remained inferential.

4. *Feeding history.* Breast feeding was brief or not given at all. The proprietary milk formula was not used but milk from the family's own animal was given to the infant. The animal milk was copper contaminated. There was no evidence implicating copper in drinking water in either condition.

5. *Genetics.* In both ICC and TCC siblings were affected and there was a high rate of parental consanguinity. Some children with the same feeding pattern remained healthy.

6. The two important differences between ICC and TCC were:

   1. The male predominance in ICC whereas sexes were affected equally in TCC
   2. The source of the dietary copper was, in the case of Indian Childhood Cirrhosis, from brass vessels which were poorly tinned and, in the Tryolean cases, from copper vessels in which a formula made of cow's milk, flour, sugar and water were heated.

Müller et al. (1996) concluded that TCC was an ecogenetic disorder, requiring both genetic and environmental factors. In this way it resembled glucose-6 phosphate deficiency in which symptoms are elicited by oxidant challenge or genetic haemochromatosis, the clinical course of which may be modulated by iron loss.

5. INFANTILE COPPER RELATED CIRRHOSIS IN THE EMSLAND

In a further retrospective study Müller et al. examined 8 cases of possible infantile copper toxicosis in a circumscribed and largely rural area of Northern Germany called The Emsland (Data presented at the 5 meeting of the European Copper Research Society, Kinsale, May 1998). The pedigrees of affected families revealed complex inter-relationships with some consanguinity, suggesting autosomal recessive inheritance. The households were served by private wells with water of low pH flowing through copper pipes. In two of these cases copper accumulation in the liver was demonstrated histologically and by analysis.

6. ANIMAL STUDIES

Many attempts have been made to establish an animal model of copper-related cirrhosis.

6.1. Sheep

In an attempt to mimic the feeding scenario in ICC, 4 groups of 5 milk-fed lambs aged 4–17 days were commenced on copper supplementation in a dose of 0.1, 2.5 or 5 mg/kg body weight per day elemental copper as copper sulphate for 13 weeks (Kantarjian et al., unpublished results). All lambs remained healthy for the first 8 weeks with normal liver function tests. Liver biopsies in copper treated lambs at 4 and 8 weeks showed a focal hepatocyte necrosis and inflammation with fine granular orcein staining in periportal hepatocytes. Liver copper concentration was 247 ± 52 µg/g dry weight prior to copper supplementation and rapidly rose 15–20 fold over the first 8 weeks.

A rapid and sudden clinical deterioration occurred in 8 of 15 copper treated lambs between 8 and 13 weeks after starting copper administration. The appearance of jaundice was followed 1–2 days later by massive intravascular haemolysis with gross haemoglobinuria and, in 5 cases, death. Post mortem liver histology showed widespread parenchymal necrosis with grossly copper-laden Kupffer's cells and portal tract macrophages.

Despite the severity of the hepatocyte necrosis, fibrosis was mild. In those lambs surviving the acute haemolytic episode cirrhosis did not occur.

These studies are mirrored in other veterinary and experimental studies of copper toxicosis in sheep. They raise three important questions:

1. What are the mechanisms enabling very large amounts of copper to be stored in the liver in the pre-haemolytic phase? The suggested explanation is that copper is sequestered in lysosomes in an insoluble copper and sulphur rich form presumed to be polymerised metallothionein.
2. What precipitates acute haemolysis and hepatocyte necrosis? The suggested explanation is that the binding mechanisms for copper are overwhelmed, allowing free copper to achieve access to a cell and stimulate the production of reactive oxygen intermediates which damage macromolecules including DNA.
3. Why, unlike the child with ICC, does cirrhosis not develop?

6.1.1. Strain Differences Among Sheep. Sheep who lived for generations on the Orkney island of North Ronaldsey, surviving on a diet very low in copper quickly developed copper toxicosis when fed on a normal diet on the mainland (MacLachlan & Johnston, 1982).

6.2. Rat

In a series of studies Susan Hayward's group established the features of copper toxicosis in the rat (Fuentealba et al., 1989, 1993). Important differences from the sheep are:

1. Physiologically lower liver concentrations
2. Very high doses were needed to elevate liver copper concentration, of the order of 20 mg/kg per day
3. Liver copper levels rose but, with continued administration of copper, plateau'd and then fell again suggesting tolerance
4. No haemolytic crisis occurred and acute liver failure was not seen. Histological abnormalities were limited to focal necrosis and inflammatory infiltration.

Again, in no reported study, has copper alone achieved cirrhosis in the rat liver. Toyokuni et al. (1989) reported that cupric nitriloacetate was able to induce cirrhosis but this could not be replicated in other studies.

6.3. White Perch (Morone Americana) (Bunton TE et al., 1987)

The white perch, like the mute swan and the toad, appears physiologically to have an extremely high liver copper concentration yet suffer no hepatotoxic damage from this.

From these animal data, particularly the sheep data, there are 3 mechanisms by which susceptibility to copper hepatotoxicity might vary.

1. Genetic differences, as shown by the North Ronaldsey sheep
2. Simultaneous ingestion of a second toxin which might be synergistic with copper
3. Simultaneous ingestion of another dietary constituent which might be protective.

7. GENETIC VARIATION

Several lines of evidence point to a genetic factor in infantile copper toxicosis.

- The family data from Tyrolean Childhood Cirrhosis
- The family data from The Emsland study
- In ICC 26% of cases were from consanguinous parents compared with a consanguinity rate of 30% in children with other liver disorders
- Cases of an ICC-like disease in whom there was no history of excessive copper ingestion. (Lefkowitch et al. 1982; Adamson et al. 1992).

8. CANDIDATE GENES

8.1. The Wilson's Gene

Wilson's disease and ICC are quite dissimilar. Nevertheless, it is necessary to exclude the possibility that a mutation at or near the Wilson's locus is responsible for ICC. Since there are now more than 100 described Wilson's disease mutations this cannot be done by direct mutational analysis. Haplotype analysis, however, has been used in ICC families where the parents were consanguinous. In the child whose parents were first cousins the probability of homozygosity at any gene locus is 1/16, the probability depending upon the degree of consanguinity. If ICC is due to a mutation at the Wilson's locus one would expect homozygosity by haplotype at that locus in the child of consanguinous parents.

Polymorphic microsatellite markers adjacent to the WD gene were studied. D13S314 and D13S301 are respectively approximately 900kb and 300kb proximal to ATP7B. D13S296 is approximately 400kb distal to ATP7B. Of 5 families with varying degrees of consanguinity in only one was there homozygosity by haplotype. This is convincing evidence that the Wilson's disease locus cannot be implicated.

In the paediatric liver clinic at KEM Hospital, Pune, 70 families with Wilson's disease and >400 cases of ICC had been seen during the period 1980–1997. No family has been encountered in which both diseases occurred. This makes it vanishingly unlikely that ICC results from the heterozygous state for a Wilson's disease mutation.

8.2. Metallothionein

In one case of an ICC-like disorder occurring in an American child, the defect in metallothionein synthesis was demonstrated in cultured skin fibroblasts (Hahn et al. 1994). However, this defect was not found in three cases of Indian Childhood Cirrhosis from Pune, nor in an Irish child with well-water associated infantile cirrhosis (case 1 from

Baker et al. 1995), nor in an older child with a non-Wilsonian copper associated cirrhosis (Hahn et al. 1995).

Metallothionein is readily identifiable in ICC.

In a study of German children with infantile cirrhosis with copper storage attributed to excessive copper ingestion from contaminated water, Summer et al. (1994) showed a relative impairment of metallothionein production when compared to data from children of a similar age and foetuses at mid-term. This has not been studied in ICC or TCC.

Metallothionein knockout mice do not develop an ICC-like disorder.

8.3. Caeruloplasmin

Since plasma levels of caeruloplasmin are normal or elevated in ICC it seems unlikely that this protein is aetiologically implicable. In the condition acaeruloplasminaemia, excessive iron accumulation occurs but this is not seen in ICC. Indeed, some of the "well-water" cases had an apparently reduced hepatic iron concentration.

8.4. Copper Chaperones

Defects in the recently identified intracellular pathways of delivery of copper to specific proteins are possible causes of disease. These pathways of copper trafficking include a soluble copper carrier, yeast ATX1 and its human homologue HAH1, that specifically function in the delivery of copper to transport ATPases in the secretory pathway (Klomp et al. 1997; Lin et al. 1997). Glerum et al. (1996a,b) have identified a pathway involving a soluble yeast factor, COX17, that specifically delivers copper to cytochrome oxidase in the mitochondria. Culotta et al. (1997) have characterised a "copper chaperone" for superoxide dismutase.

8.5. "Cuprochromatosis?"

Genetic haemochromatosis results from a failure of regulation of iron absorption. Might the primary defect in ICC/TCC lie within the enterocyte? i.e. might unregulated iron uptake cause copper overload? This seems intrinsically unlikely since the primary regulator of copper balance is biliary excretion.

9. SECOND TOXIN

Sheep and cattle grazing on plants containing pyrolizidine alkaloids may develop severe liver damage with copper storage (Howell JMcC et al. 1991; Bull, 1961; Donald & Shanks, 1956; Gooneratne et al., 1980; Habermehl et al., 1988; Seawright et al. 1991; Odriozola et al. 1994). This was studied experimentally by administering copper and/or the pyrolizidine alkaloid retrorsine to male wistar rats (Morris et al., 1994). Rats given both copper and retrorsine developed more severe liver damage than the retrorsine-alone group and more severe copper retention than the copper alone group.

Neonatal rats were particularly susceptible to retrorsine, developing liver damage and copper retention when suckled on dams receiving retrorsine whilst the dams themselves escaped liver injury (Aston et al., 1996, 1998).

A similar synergy was observed between copper and aflatoxin in the guinea pig (Seffner et al., 1997).

By contrast, copper administration partially *prevented* the hepatotoxic effects of galactosamine and of carbon tetrachloride in the rat (Barrow &Tanner, 1987 , 1989).

There is no direct evidence to implicate a second toxin in ICC or in TCC, except for a reported association between aflatoxin and later development of cirrhosis (Amla et al. 1971). Outbreaks of pyrolizidine alkaloid poisoning in India and Afghanistan did not produce an ICC-like picture (Tandon RK, Tandon et al. 1976a, b; 1978; 1984; Mayer et al. 1993). Nevertheless, the possibility must be admitted that milk from a grazing animal in rural India or from a silage-fed cow in the Tyrol might be contaminated with the plant alkaloid or micotoxin.

10. PROTECTIVE DIETARY CONSTITUENTS

Sheep fed on a high molybdenum pasture are resistant to copper poisoning because ammonium tetrathiomolybdate formed in the rumen kelates the copper. Although this effect cannot operate in man, other possible dietary protective agents might include zinc, which impairs copper absorption by enterocyte metallothionein induction, or antioxidants by scavenging reactive oxygen intermediates.

11. MECHANISM OF LIVER INJURY IN INFANTILE COPPER CIRRHOSIS

Fundamental questions remain to be answered concerning the mechanism of liver injury in the copper overload syndromes of infancy. It is possible that the histological appearances hold some clues.

11.1. Hepatocyte Death Is by Necrosis Rather Than by Apoptosis

The hepatocytes in ICC are ballooned with empty vacuolated cytoplasm and eosinophilic hyaline inclusions. They do not have the shrunken appearance of apoptotic cells, with pyknotic nuclei.

This is relevant in view of the recent report from Strand et al. (1988) that liver cell damage in acute Wilson's disease involves CD95 mediated apoptosis. In that study of 4 patients, areas of relatively well preserved integrity did demonstrate apoptosis of individual cells. In these areas the cell surface receptor CD95 was demonstrated immunohistochemically and the CD65 ligand MRNA by in situ hybridisation. Copper treatment of Hep G2 cells produced apoptosis by activation of the CD95 system.

Although similar studies have not been carried out in ICC, the histological absence of apoptosis in ICC suggests that different mechanisms of liver cell injury are operative.

11.2. Fatty Infiltration Is Conspicuously Absent in ICC

This is in contrast to situations of primary mitochondrial injury such as the fatty acid oxidation defects and urea cycle disorders. Experimentally copper is known to produce mitochondrial injury (Sokol et al., 1989, 1993, 1994, 1996). In copper treated Hep G2 cells mitochondrial abnormalities included morphological changes, production of fatty acid oxidation and citrate synthase and reduced uptake of the vital dye rodamine 123 (N Watt, unpublished observations). However, these changes occurred at higher concentrations than those required to inhibit cell growth, see below, and thus may be secondary.

11.3. Poor Regenerative Activity

Nodular regeneration is absent in classical ICC, though it appears during the recovery phase of the penicillamine treated case. This contrasts, for example, with tyrosinaemia which characteristically is associated with hyperplastic nodules at an early stage and risk of malignancy subsequently. In cultured HepG2 cells, Aston et al. (unpublished observations) found that exposure to 16micromolar copper impaired cell growth, prolonged the S phase of the cell cycle, increased the potential doubling time, and impaired colony formation.

11.4. Pericellular Intralobular Collagen Deposition Is a Striking Feature of ICC

Individual hepatocytes or small groups of hepatocytes are surrounded by type 3 collagen, a feature which is seen in some other neonatal hepatitides but to a much less marked extent with the possible exception of congenital syphilis.

12. CLINICAL SIGNIFICANCE

Although copper related cirrhosis in infancy is rare, it does continue to occur and should be borne in mind in an infant with undiagnosed severe liver disease. Penicillamine should be commenced promptly but, in the child who has developed jaundice or ascites, it is unlikely to prevent a fatal outcome. Prompt referral for liver transplantation is, therefore, indicated.

REFERENCES

Adamson, M., Reiner, B., Olson, J.L., Goodman, Z., Plotnick, L., Bernardini, I., Gahl, W.A. (1992). Indian childhood cirrhosis in an American child. Gastroenterology 102,1771–1777.

Aikat, B.K., Bhattacharya, T., Walia, B.N.S. (1974). Morphological features of Indian childhood cirrhosis; the spectrum of changes and their significance. Ind J Med Res 62, 953–963.

Amla, I., Kamala, C.S., Gopalakrishna, G.S., Jayaraj, P., Sreenivasamurthy, V., Parpia, H.A.B. (1971). Cirrhosis in children from peanut meal contaminated by aflatoxin. American Journal of Clinical Nutrition 24, 609–614.

Aston, N., Morris, P., Tanner, S. (1996). Retrorsine in breast milk influences copper handling in suckling rat pups. J Hepatol 25,748–755.

Aston, N., Morris, P., Tanner, S. (1998). An animal model for copper associated cirrhosis. J Path in press.

Baker, A., Gormally, S., Saxena, R., Baldwin, D., Drumm, B., Bonham, J., Portman, B., Mowat, A.P. (1995). Copper-associated liver disease in childhood. J Hepatol 23, 538–43.

Barrow, L., Tanner, M.S. (1987). Copper protects against galactosamine-induced hepatitis. J Hepatol 5,19–26.

Barrow, L., Tanner, M.S. (1988). Copper distribution among serum proteins in paediatric liver disorders and malignancies. Eur J Clin Invest 18,555–560.

Barrow, L., Tanner, M.S. (1989). The effect of carbon tetrachloride on the copper-laden rat liver. Brit J Exp Pathol 70, 9–19.

Bavdekar, A.R., Bhave, S.A., Pradhan, A.M., Pandit, A.N., Tanner, M.S. (1996). Long term survival in Indian childhood cirrhosis treated with d-penicillamine. Arch Dis Child 74, 32–35.

Bent, S., Bohm, K. (1995). Copper-induced liver cirrhosis in a 13-month old boy. Gesundheitswesen 57,667–669.

Bhave, S.A., Pandit, A.N., Tanner, M.S. (1987). Comparison of feeding history of children with Indian Childhood Cirrhosis and paired controls. J Ped Gastroenterol Nutr 6;562–567.

Bull, L.B. (1961). Liver diseases in livestock from intake of hepatoxic substances. Aust Vet J 37, 126–30.

Bunton, T.E., Baksi, S.M., George, S.G., Frazier, J.M. (1987). Abnormal hepatic copper storage in a teleost fish (*Morone americana*) Vet Pathol 24, 515–524.

Culotta, V.C., Klomp, L.W.J., Strain, J., Ruby Leah B. Casareno., Krems, B., Gitlin, J.D. (1997). The Copper Chaperone for Superoxide Dismutase. J Biol Chem 272, 23469–23472.

Donald, L.G., Shanks, P.L. (1956). Ragwort poisoning from silage. Br Vet J 112, 307–11.

Fuentealba, I.C., Haywood, S., Foster, J. (1989). Cellular mechanisms of toxicity and tolerance in the copper-loaded rat. III. Ultrastructural changes and copper localization in the kidney. Br J Exp Pathol 70, 543–556.

Fuentealba, I.C., Davis, R.W., Elmes, M.E., Jasani, B., Haywood, S. (1993). Mechanisms of tolerance in the copper-loaded rat liver. Exp Mol Pathol 59, 71–84.

Gooneratne, S.R., Howell J.McC., Cook, R.D. (1980). An ultrastructural and morphometric study of the liver of normal and copper-poisoned sheep. Amer J Path 99, 429–450.

Habermehl, G.G., Martz, W., Tokarnia, C.H., Dobereiner, J., Mendez, M.C. (1988). Livestock poisoning in South America by species of the Senecio plant. Toxicon 26, 275–286.

Hahn, S.H., Brantly, M.L., Oliver, C., Adamson, M., Kaler, S.G., Gahl, W.A. (1994.) Metallothionein synthesis and degradation in Indian childhood cirrhosis fibroblasts. Pediatric Research 35, 197–204.

Hahn, S.H., Tanner, M.S., Danks, D.M., Gahl, W.A. (1995). Normal metallothionein synthesis in fibroblasts obtained from children with Indian childhood cirrhosis or copper-associated childhood cirrhosis. Biochemical and Molecular Medicine 54, 142–145.

Howell, J.McC., Deol, H.S., Thomas, J.B., Dorling, P.R. (1991). Experimental copper and heliotrope intoxication in sheep: Morphological changes. Journal of Comparative Pathology 105, 49–74.

Klomp, L.W.J., Lin, S.J., Yuan, D., Klausner, R.D., Culotta, V.C., Gitlin, J.D. (1997). Identification and functional expression of HAH1, a novel human gene involved in copper homeostasis J. Biol. Chem 272, 9221–9226.

Lefkowitch, J., Honig, C.L., King, M., Hagstrom, J.W.C. (1982). Hepatic copper overload and features of Indian Childhood Cirrhosis in an American sibship. New England Journal of Medicine 307, 271–277.

MacLachlan, G.K., Johnston, W.S. (1993). Copper poisoning in sheep from North Ronaldsay maintained on a diet of terrestrial herbage. Vet Rec 1982 Sep 25;111(13):299–301 Mayer F, Luthy J. Heliotrope poisoning in Tadjikistan. Lancet 342, 246–247.

Morris, P., O'Neill, D., Tanner, S. (1994). Synergistic liver toxicity of copper and retrorsine in the rat. J Hepatol 21, 735–742.

Muller, T., Feichtinger, H.., Berger, H., Muller, W. (1996). Endemic Tyrolean infantile cirrhosis: An ecogenetic disorder. Lancet 347, 877–880.

Muller-Hocker, J., Weiss, M., Meyer, J. et al. (1987). Fatal copper storage disease of the liver in a German infant resembling Indian childhood cirrhosis. Virchows Archiv A 411, 379–385.

Muller-Hocker, J., Meyer, U., Wiebecke, B., et al. (1988). Copper storage disease of the liver and chronic dietary copper intoxication in two further German infants mimicking Indian Childhood Cirrhosis. Pathology Research & Practice 183, 39–45.

Nayak, N.C., Ramalingaswami, V. (1975). Indian childhood cirrhosis. Clin Gastroenterol 4, 333–349.

Odriozola, E., Campero, C., Casaro, A., Lopez, T., Olivieri, G., Melucci, O. (1994). Pyrrolizidine alkaloidosis in Argentinian cattle caused by Senecio selloi. Vet Human Toxicology 36, 205–208.

Pandit, A., Bhave, S. (1996). Present interpretation of the role of copper in Indian childhood cirrhosis. Am J Clin Nutr 63 in press.

Pradhan, A.M., Bhave, S.A., Joshi, V.V., Bavdekar, A.R., Pandit, A.N., Tanner, M.S. (1995). Reversal of cirrhosis after D-penicillamine therapy in preicteric Indian Childhood Cirrhosis. J Ped Gastroenterol Nutr 20, 28–35.

Popper, H., Goldfisher, S., Sternlieb, I., Nayak, N.C., Madhavan, T.V. (1979). Cytoplasmic copper and its toxic effects. Studies in Indian Childhood Cirrhosis. Lancet 1, 1205–1208.

Price, L.A., Walker, N.I., Clague, A.E., Pullen, I.D., Smits, S.J., Ong, T.H., Patrick, M. (1996). Chronic copper toxicosis presenting as liver failure in an Australian child Pathology 28, 316–320.

Reed, E.M., Landing, B.H. (1972). Copper in childhood liver disease—a histologic, histochemical and chemical survey. Arch Pathol 93, 249.

Schramel, P., Muller-Hocker, J., Meyer, U., Weiss, M., Eife, R. (1988). Nutritional copper intoxication in three German infants with severe liver cell damage (features of Indian Childhood Cirrhosis). J Trace Elem Electrolytes Health Dis 2, 85–89.

Seawrigh, A.A., Kelly, W.R., Hrdlicka, J., McMahon, P., Mattocks, A.R., Jukes, R. (1991). Pyrrolizidine alkaloidosis in cattle due to Senecio species in Australia. Vet Record 129, 198–199.

Seffner, W., Schiller, F., Lippold, U., Dieter, H.H., Hoffmann, A. (1997). Experimental induction of liver fibrosis in young guinea pigs by combined application of copper sulphate and aflatoxin B1. Toxicol Lett 92(3), 161–172.

Sokol, R.J., Devereaux, M.W., Traber, M.G., Shikes, R.H. (1989). Copper toxicity and lipid peroxidation in isolated rat hepatocytes: effect of vitamin E. Pediatr Res 25, 55–62.

Sokol, R.J., Devereaux, M.W., O'Brien, K., Khandwala, R.A., Loehr, J.P. (1993). Abnormal hepatic mitochondrial respiration and cytochrome c oxidase activity in rats with long-term copper overload. Gastroenterology 105, 178–187.

Sokol, R.J., Twedt, D., McKim, J.M Jr., Devereaux, M.W., Karrer, F.M., Kam, I., Von Steigman, G., Narkewicz, M R., Bacon, B.R., Britton, R.S., Neuschwander-Tetri, B.A. (1994). Oxidant injury to hepatic mitochondria in patients with Wilson's disease and Bedlington terriers with copper toxicosis. Gastroenterology 107, 1788–1798.

Sokol, R.J., McKim, J.M., Devereaux, M.W. (1996). Alpha-tocopherol ameliorates oxidant injury in isolated copper-overloaded rat hepatocytes. Pediatr Res 39, 259–263.

Strand, S., Hofmann, W.J., Grambihler, A., Hug, H., Volkmann, M., Otto, G., Wesch, H., Mariani, S.M., Hack, V., Stremmel, W., Krammer, P.H., Galle, P.R. (1998). Hepatic failure and liver cell damage in acute Wilson's disease involve CD95 (APO-1/Fas) mediated apoptosis. Nat Med 4, 588–593.

Tandon, R.K., Tandon, B.N., Tandon, H.D., et al. (1976). An epidemic of veno-occlusive disease in India presumably due to food toxin. Gut 17, 849–55.

Tandon, B.N., Tandon, H.D., Tandon, R.K., et al. (1976). An epidemic of veno-occlusive disease of liver in Central India. Lancet 2, 271–2.

Tandon, H.D., Tandon, B.N., Mattocks, A.R. (1978). An epidemic of veno-occlusive disease of the liver. Am J Gastroenterol 70,607–13.

Tandon, B.N., Joshi, Y.K., Sud, R., Koshy, A., Jain, S.K., Tandon, H.D. (1984). Follow-up of survivors of epidemic veno-occlusive disease in India. Lancet 1, 730.

Tanner, M.S. (1986). Indian Childhood Cirrhosis. In: Meadow, S.R., (Eds). "Recent Advances in Paediatrics" Vol. 8. Churchill Livingstone, Edinburgh, pp 103–120.

Tanner, M.S. (1998). The role of copper in Indian childhood cirrhosis. Amer J Clin Nutr 67(5 Suppl), 1074S-1081S.

Tanner, M.S., Portmann, B., Mowat, A.P., Williams, R., Pandit, A.N., Mills, C.F., Bremner, I. (1979). Increased hepatic copper concentration in Indian Childhood Cirrhosis. Lancet 1, 1203- 1205.

Tanner, M.S., Kantarjian, A.H., Bhave, S.A., Pandit, A.N. (1983). Early introduction of copper-contaminated animal milk feeds as a possible cause of Indian Childhood Cirrhosis. Lancet 2, 992–995.

Tanner, M.S., Bhave, S.A., Pradhan, A.M., Pandit, A.N. (1987). Clinical trials of penicillamine in Indian childhood cirrhosis. Arch Dis Child 62, 1118–1124.

Toyokuni, S., Okada, S., Hamazaki, S., Fujioka, M., Li, J.L., Midorikawa, O. (1989). Cirrhosis of the liver induced by cupric nitrilotriacetate in Wistar rats. An experimental model of copper toxicosis. Amer J Path 134, 1263–1274.

Walker-Smith, J., Blomfield, J. (1973). Wilson's disease or chronic copper poisoning? Arch Dis Child 48, 476–479.

24

REGULATION OF ZINC HOMEOSTASIS IN YEAST BY THE ZAP1 TRANSCRIPTIONAL ACTIVATOR

David Eide, Hui Zhao, Edward Butler, and Jacquelyn Rodgers

Nutritional Sciences Program
University of Missouri-Columbia
Columbia, Missouri 65211

1. INTRODUCTION

The array of roles that zinc plays in cellular biochemistry is an amazing example of the utility of metal ions in biology. Zinc is required for the activity of over 300 enzymes including alcohol dehydrogenase, Cu/Zn superoxide dismutase, carbonic anhydrase, and carboxyproteases (Vallee and Auld, 1990). Moreover, zinc plays an important structural role in many proteins by aiding correct folding of specific subdomains. Certainly the best-known class of proteins that require zinc in this capacity are transcription factors that contain zinc-dependent DNA-binding motifs such as zinc fingers and zinc clusters. These types of proteins are ubiquitous in eukaryotic organisms. For example, almost 2% of the genes in the genome of *Saccharomyces cerevisiae* contain zinc-dependent DNA binding domains (Bohm et al., 1997; Schjerling and Holmberg, 1996). When the zinc-dependent enzymes are also considered, perhaps as many as 5% of all yeast proteins require zinc for some aspect of their function.

Although essential for many different cellular functions, zinc also has its dark side. Excess zinc can be toxic to cells. The precise cause of zinc toxicity is unknown but may be the result of the metal interfering with vital processes through binding to inappropriate intracellular ligands or by competing with other metal ions for enzyme active sites, transporter proteins, etc. Therefore, in the face of changing exogenous zinc availability, cells must maintain an adequate intracellular zinc level to meet cellular requirements and prevent metal ion overaccumulation and toxicity. Mechanisms of regulating the amount or availability of intracellular zinc includes binding of the metal by intracellular macromolecules such as metallothioneins (Hamer, 1986) and phytochelatins (Rauser, 1995), zinc storage in intracellular compartments (Palmiter et al., 1996), and transport of zinc out of the cell (Palmiter and Findley, 1995). In the yeast *Saccharomyces cerevisiae*, the regula-

Metals and Genetics, edited by Sarkar.
Kluwer Academic / Plenum Publishers, New York, 1999.

tion of zinc homeostasis is mediated primarily through the control of zinc uptake across the plasma membrane.

2. IDENTIFICATION OF ZINC TRANSPORTERS IN YEAST

Zinc uptake in *S. cerevisiae* is time-, temperature-, and concentration-dependent and saturable (Fuhrmann and Rothstein, 1968; Mowll and Gadd, 1983; White and Gadd, 1987). Kinetic studies of zinc uptake by cells grown with different amounts of zinc in the medium suggested the presence of at least two uptake systems. One system has a high affinity for zinc with an apparent K_m of 1 μM Zn(II) and is active in zinc-limited cells (Zhao and Eide, 1996a). The second system has a lower affinity for zinc (apparent K_m of 10 μM Zn(II)) and is detectable in zinc replete cells (Zhao and Eide, 1996b). When chelation of the metal in the uptake solution is estimated, these K_m values for total zinc in the assay correspond to K_m's of 10 and 100 nM free Zn(II), respectively.

The *ZRT1* gene encodes the transporter protein of the high affinity system (Zhao and Eide, 1996a). This gene was originally identified in the *S. cerevisiae* genome sequence because of its similarity to the IRT1 Fe(II) transporter from *Arabidopsis thaliana* (see below). Genetic evidence supports the role of ZRT1 in zinc uptake in yeast. First, *zrt1* mutants are defective for growth under zinc-limiting conditions but grow normally when zinc levels are high. Second, *zrt1* mutations eliminate high affinity uptake activity but have no effect on low affinity uptake. Third, overexpressing *ZRT1* increased high affinity uptake. In similar studies, we determined that the *ZRT2* gene encodes the transporter of the low affinity uptake system (Zhao and Eide, 1996b). These results demonstrated that the high and low affinity zinc uptake systems are separate entities and suggested that *ZRT1* and *ZRT2* encode the high and low affinity transporters, respectively. Additional and as yet uncharacterized zinc uptake systems are also present in *S. cerevisiae* as demonstrated by the observation that the *zrt1 zrt2* mutant is viable (Zhao and Eide, 1996b). These additional systems are unlikely to be important for zinc accumulation under any but the most zinc-rich conditions because the *zrt1 zrt2* mutant requires 10^5-fold more zinc to grow than wild type cells.

The ZRT1 and ZRT2 proteins are 376 and 422 amino acids in length, respectively. They are closely related to each other sharing 44% sequence identity and 67% similarity. They each contain eight potential transmembrane domains and have a similar predicted membrane topology in which the amino- and carboxy-terminal ends of the protein are located on the outside surface of the plasma membrane. As stated previously, these proteins are also similar in sequence and predicted topology to the IRT1 Fe(II) transporter from *Arabidopsis* (Eide et al., 1996). In addition to sharing sequence similarity and numbers of potential transmembrane domains, ZRT1, ZRT2, and IRT1 each have a potential metal-binding domain between transmembrane domains three and four that is predicted to be cytoplasmic. In ZRT1, this sequence is HDHTHDE and in IRT1, this motif is HGHGHGH. While the function of this motif is currently unknown, its conserved location in these three proteins and its potential for metal binding suggests that it plays an important role in metal ion uptake or its regulation.

Through DNA sequence database comparisons and additional expression cloning studies, it is now clear that these three metal ion transporters are members of a new family of proteins found in a diverse array of eukaryotic organisms. We refer to this family collectively as the ZIP family for "ZRT, IRT-like Protein." At this time, twenty ZIP members have been identified including eight in plants (seven from *Arabidopsis* and one from rice), two in *S. cerevisiae* (i.e. ZRT1 and ZRT2), one from Trypanosomes, five in nematodes, one in mice, and two in humans (Eng et al., 1998). No prokaryotic members have been identified despite the fact that the genomes of 14 bacteria and archaea have been completely sequenced. This suggests that the ZIP family arose after divergence of eukaryotes from the archaea and eubacterial phyla. It should be emphasized that the ZIP family is completely novel; these transporters bear no significant similarity to other known metal ion transporters including P-type ATPases (Solioz and Vulpe, 1996), CDF metal efflux proteins (Paulsen and Saier, 1997), and SMF/Nramp/DCT1 divalent cation transporters (Gunshin et al., 1997). Moreover, ZIP proteins do not contain ATPase or ATP binding domains indicating that transport is indirectly coupled to energy metabolism. All but two of the known ZIP proteins contain the putative metal-binding domain described above.

We have recently obtained evidence that four of the *Arabidopsis* proteins, called ZIP1- 4, are involved in zinc transport (Grotz et al., 1998). Expressing ZIP1, ZIP2, or ZIP3 in yeast confers increased Zn(II) uptake activity with distinct biochemical properties. Moreover, these transporters are relatively specific for zinc, preferring it as a substrate over several other divalent metal cations. Expression of ZIP4 in yeast did not lead to increased zinc uptake and the reason for this is currently unknown. We have also obtained evidence from plant studies that implicate these genes in zinc transport; i.e., *ZIP1*, *ZIP3*, and *ZIP4* mRNAs are induced in zinc-limited plants. Expression of *ZIP2* has not yet been detected suggesting that it is expressed at an extremely low level or only under specific conditions. To summarize, we have identified a new gene family of potential transporter proteins that are found in a number of different eukaryotes. Experimental evidence has been obtained for seven of these transporters and, in each case, they have been implicated in metal ion transport. Thus, it seems likely that the other proteins in this family are also metal ion transporters. This is an exciting prospect given how little is understood about zinc transport in mammals and other higher eukaryotes.

3. TRANSCRIPTIONAL REGULATION OF YEAST ZINC TRANSPORTERS

Both high and low affinity zinc uptake systems in yeast are regulated by zinc. The activity of the high affinity system is induced more than 30-fold in response to zinc-limiting conditions. Our studies indicated that this rise is due to increased transcription of the *ZRT1* gene in response to declining levels of an intracellular pool of zinc (Zhao and Eide, 1996a). We also determined that transcription of the *ZRT2* gene is also induced in zinc-limited cells, albeit to a lesser extent than *ZRT1* (Zhao and Eide, 1997).

3.1. Isolation of Mutants with Altered *ZRT1* and *ZRT2* Expression

To identify genes whose products regulate *ZRT1* and *ZRT2* transcription, we used a genetic screen to isolate mutants with an elevated level of *ZRT1* transcription during growth on a zinc-replete medium (Figure 1). The design of this screen was similar to a scheme used to isolate yeast mutants with altered iron-responsive gene expression (Dancis et al., 1994). The *ZRT1* promoter is active in zinc-limited cells but inactive in zinc-replete cells. We inserted this promoter upstream of the coding region of the *HIS3* gene, which encodes an enzyme required for histidine biosynthesis. The *ZRT1-HIS3* fusion gene was then integrated into the genome of a *his3Δ* mutant to generate the strain ZHY4. Because of the zinc-responsive expression of the *ZRT1* promoter, ZHY4 cells grow without added histidine on zinc-limiting but not on zinc-replete media.

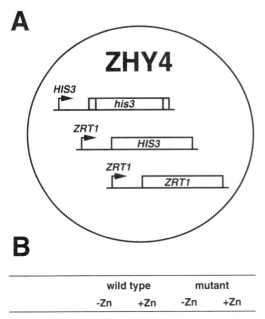

A

ZHY4

HIS3
┌►──┌──his3──┐──┐

ZRT1
┌►──┌───HIS3───┐

ZRT1
┌►──┌────ZRT1────┐

B

	wild type		mutant	
	-Zn	+Zn	-Zn	+Zn
His Phenotype:	+	-	+	⊕
Uptake activity:	High	Low	High	High

Figure 1. The genetic selection used for isolating mutants with altered *ZRT1* expression. A. Relevant genotype of strain ZHY4. ZHY4 bears a deletion removing the ORF of the chromosomal *HIS3* gene. The *ZRT1* promoter (bases -706 to -1; the first base of the translation initiation codon is designated as position +1) was fused to the *HIS3* open reading frame, which retained its own translation initiation codon. This *ZRT1-HIS3* fusion gene was then inserted at another site in the genome (*URA3*). The wild type *ZRT1* gene is also present in ZHY4. B. Histidine auxotrophy and zinc uptake phenotypes of the wild type ZHY4 and mutant derivatives in zinc-limited (-Zn) and zinc-replete (+Zn) media. Mutants were isolated based on their His⁺ phenotype on zinc-replete media (*circled*).

Several spontaneous His⁺ strains were isolated from populations of ZHY4 grown on a zinc-replete medium and one isolate, M18, was characterized further. M18 was His⁺ presumably because of increased expression of the ZRT1-HIS3 fusion gene. To determine if the His⁺ phenotype was due to *trans*-acting effects on the *ZRT1* promoter rather than, for example, mutations within the promoter of the *ZRT1-HIS3* fusion, zinc uptake activity was also assayed in the mutant strain. For M18, zinc uptake activity was increased approximately 10-fold. To determine if this mutation was dominant or recessive, a heterozygous mutant diploid (M18/+) was assessed for histidine prototrophy on zinc-replete plates and found to be semi-dominant. This semi-dominance was also apparent in zinc uptake assays. The uptake rate in the M18/+ diploid, although lower than the rate measured in the M18 haploid, was still significantly higher than the wild type diploid rate (P < 0.01) (6.1 ± 0.1 and 1.1 ± 0.1 pmol/min/10⁶ cells, respectively).

3.2. Cloning of *ZAP1*, the Gene Mutated in M18

We cloned the gene mutated in M18 isolate by virtue of its semi-dominant phenotype. A plasmid library was generated from M18 genomic DNA, transformed into ZHY4 (*his3Δ ZRT1-HIS3*), and transformants were screened for histidine prototrophy. Two plasmids, pZH4 and pZH5, were isolated that conferred both the His⁺ phenotype and increased uptake activity. Sequence analysis of the inserts of these two plasmids indicated that they were overlapping genomic fragments from yeast chromosome X (Galibert et al., 1996) and each contained two open reading frames (ORFs) of unknown function, YJL055W and YJL056C. A frameshift mutation introduced into the *Bss*HII site of YJL056C eliminated

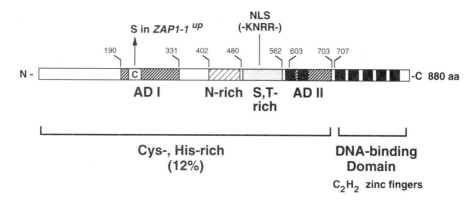

Figure 2. Schematic representation of the *ZAP1* protein summarizing the features described in the text. The individual zinc finger domains are indicated by the black boxes.

both the His$^+$ and uptake phenotypes whereas a similar mutation introduced into the *Mlu*I site of YJL055w had no effect. Based on our results, we have designated YJL056C as the *ZAP1* gene (for "Zinc-responsive Activator Protein"). The M18 mutant allele was designated "*ZAP1–1^{up}*" because of its constitutive effects on *ZRT1* expression.

3.3. The Amino Acid Sequence of ZAP1

The predicted amino acid sequence of wild type ZAP1 is 880 amino acids in length with a molecular mass of 93 kDa. ZAP1 has several features expected in a transcriptional activator protein (Figure 2). First, the carboxy-terminal 174 amino acids (residues 707–880) contain five zinc finger domains of the TFIIIA type, i.e., $C-X_{2-4}-C-X_{12}-H-X_{3-5}-H$. This region has several basic amino acids and a predicted isoelectric point (pI) of 9.9. The predicted pI of the full-length protein is 6.4. We also note the presence of a sixth potential zinc finger in this region with the noncanonical sequence $C-X_2-C-X_{12}-Q-X_3C$ (amino acids 752–772). Data described later in this report indicates that this C-terminal region (i.e., amino acids 687–880) contains the fully intact DNA binding domain. Two additional zinc fingers are located more N-terminal in the protein (i.e., 581–604 and 618–641) but their function is currently unknown. Amino acids 1–706 contain two regions rich in acidic residues (aspartate and glutamate) that could be transcription activation domains. One region (amino acids 190–331) is 142 amino acids in length and has a predicted pI of 4.4. The second region (amino acids 603–703) is 101 amino acids in length and has a predicted pI of 4.6. There is also an asparagine-rich domain (residues 402–480, 27%) and a serine/threonine-rich region (residues 482–564, 29%) and such domains have been noted in other transcription factors (Brugnera et al., 1994; Forsburg and Guarente, 1989; O'Hara et al., 1988). Our studies demonstrate that amino acids 1–687 contains the ZAP1 transcriptional activation domain(s). Amino acids 529–532 (-KNRR-) are similar to a consensus nuclear localization signal (Boulikas, 1994).

Previous results suggested that the *ZRT1* gene is controlled by an intracellular pool of zinc (Zhao and Eide, 1996a). The simplest hypothesis for this regulation is that ZAP1 activity is controlled by zinc binding directly to the protein. Consistent with this proposal, ZAP1 contains a large number of potential zinc ligands. In addition to the zinc finger domains and the acidic residues described above, amino acids 1–706 are highly enriched in cysteine and histidine residues (~12%). This observation suggests that zinc binding in this region may alter ZAP1 function and this hypothesis is supported by our characterization of the *ZAP1–1^{up}* allele. DNA sequencing of the wild type *ZAP1* and *ZAP1–1^{up}* alleles demonstrated that their only difference is that a cysteine in the wild type protein is replaced with a serine in *ZAP1–1^{up}* (Figure 2). The mutated cysteine is at position 203 within one of the potential activation domains.

3.4. Effect of *ZAP1* Alleles on *ZRT1* and *ZRT2* Expression

The His$^+$ prototrophy and increased zinc uptake activity observed in the *ZAP1–1^{up}* mutant suggested that this allele causes increased transcription of *ZRT1* in zinc-replete cells. To characterize the effects of this allele on *ZRT1* transcription more directly, we assessed the zinc-responsiveness of *ZRT1* expression by zinc uptake assays and β-galactosidase activity generated from a *ZRT1-lacZ* reporter gene (Figure 3). In the wild type, as was observed previously, zinc uptake activity and β-galactosidase activity were high in zinc-limited cells and low in zinc-replete cells. In contrast, *ZRT1* expression in an isogenic *ZAP1–1^{up}* strain was fully induced in both zinc-limited and zinc-replete cells. Thus, the *ZAP1–1^{up}* mutation interferes with the zinc-dependent repression of the *ZRT1* promoter but does not prevent full induction of the *ZRT1* gene.

To further examine ZAP1's role in *ZRT1* regulation, we constructed a *zap1Δ* disruption mutation (*zap1Δ::TRP1*) in which the entire *ZAP1* ORF was deleted from its chromo-

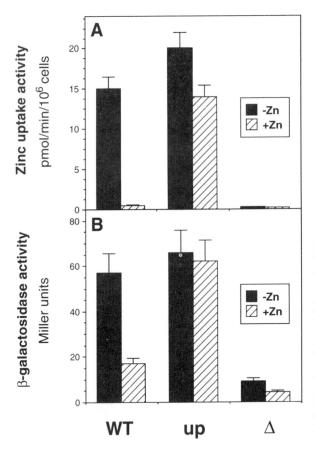

Figure 3. Effect of *ZAP1* alleles on expression of the *ZRT1* gene. Strains used were wild type (WT, DY1457), *ZAP1–1^{up}* (up, ZHY7), and *zap1Δ* (Δ, ZHY6) each of which had been transformed with a *ZRT1-lacZ* fusion gene on plasmid pGI1 (Zhao and Eide, 1996a). The cells were grown to exponential phase in SD medium supplemented with 2% glucose and all necessary auxotrophic supplements either with (zinc-limiting, -Zn) or without (zinc-replete, +Zn) 1 mM EDTA, 50 μM ZnCl$_2$. These cells were then assayed for high affinity uptake activity at 1 μM ^{65}Zn (A) and β-galactosidase activity (B). The error bars represent 1 S. D. (n = 4).

somal site. A haploid strain bearing this mutation was viable indicating that *ZAP1* is not an essential gene. Zinc uptake activity and *ZRT1-lacZ* expression were very low in zinc-limited as well as zinc-replete *zap1Δ* cells (Figure 3). Thus, the *ZAP1* gene is required for transcription of the *ZRT1* gene in response to zinc-limiting growth conditions. Similar experiments performed with a *ZRT2-lacZ* fusion gave nearly identical results to those obtained with *ZRT1*. Thus, the *ZAP1* gene is required for expression of both *ZRT1* and *ZRT2* and a dominant gain-of-function allele causes constitutive expression of both genes. We also noted that the *zap1Δ* strain is defective for growth under zinc-limiting conditions to a level similar to that observed in the *zrt1 zrt2* double mutant. This observation is also consistent with the effects of the *zap1Δ* mutation on *ZRT1* and *ZRT2* expression.

Does the *zap1Δ* strain have defects in generating zinc uptake activity in addition to its inability to express *ZRT1* and *ZRT2*? For example, other proteins may be involved in zinc uptake and their genes could also require ZAP1 for expression. To address this question, we examined uptake activity in cells expressing *ZRT1* or *ZRT2* from the *GAL1* promoter, which does not require activation by ZAP1. A zinc-replete *ZAP1* wild type strain expressing the *ZRT1* gene from the *GAL1* promoter had a zinc uptake rate of 6.0 ± 0.1 pmol/min/10^6 cells when assayed at 40 μM ^{65}Zn. A *zap1Δ* strain grown under similar conditions and also expressing the *ZRT1* gene from the *GAL1* promoter displayed an uptake rate of 7.4 ± 0.1 pmol/min/10^6 cells. The *zap1Δ* vector-only transformant had only 0.04 pmol/min/10^6 cells uptake activity. The *zap1Δ* strain expressing *ZRT2* from the *GAL1* promoter accumulated zinc at a rate of 6.8 ± 0.2 pmol/min/10^6 cells. These results indicated that the zinc uptake defect in the *zap1Δ* strain is solely due to it's inability to express the *ZRT1* and *ZRT2* genes.

3.5. Autoregulation of *ZAP1* Expression

ZAP1 mRNA levels were examined to determine if *ZAP1* expression itself is zinc-responsive (data not shown). Northern blot analysis using the *ZAP1* ORF as the probe detected a single 2.8 kb mRNA species. As expected, no *ZAP1* mRNA was detected in the *zap1Δ* strain. In wild type cells, the abundance of the *ZAP1* transcript appeared slightly higher in zinc-limited cells than in zinc-replete cells. In *ZAP1-1^{up}* cells, *ZAP1* mRNA levels were equally high in both zinc-limited and replete cells. These results suggested that the *ZAP1* gene may itself be regulated by zinc. To assess this regulation more quantitatively, β-galactosidase activity generated by a *ZAP1-lacZ* reporter gene was measured in wild type cells. These results demonstrated that the *ZAP1* gene is also transcriptionally regulated in response to zinc; *ZAP1-lacZ* expression was approximately 3-fold higher in zinc-limited cells than in zinc-replete cells. Moreover, in cells grown under more stringent zinc-limiting conditions (i.e., SD glucose + 1 mM EDTA), *ZAP1-lacZ* β-galactosidase activity was almost 50-fold higher than in zinc-replete cells (270 ± 23 vs. 5.7 ± 0.2 Miller units, respectively). Like *ZRT1* and *ZRT2*, zinc-responsive *ZAP1* regulation also appears to be mediated by ZAP1. *ZAP1-lacZ* expression in *ZAP1-1^{up}* cells was no longer zinc-responsive being elevated in both culture conditions. Furthermore, expression was markedly reduced in both zinc-limited and zinc-replete *zap1Δ* cells. These data suggest that *ZAP1* controls the activity of its own promoter in response to zinc.

3.6. Identification of Zinc-Responsive Elements (ZREs) in ZAP1 Target Promoters

We have identified the cis-acting sequences in the promoters of *ZRT1*, *ZRT2* and *ZAP1* that are responsible for zinc-responsive gene expression. Our previous studies indi-

cated that the promoter region of *ZRT1* extending from -706 to the ATG initiation methionine codon was sufficient to confer wild type regulation of the *ZRT1* gene on a *ZRT1-lacZ* reporter gene (Zhao and Eide, 1996a). [The numbering system used throughout this report to
designate sites within the promoter regions are relative to first base of the ATG sequence.
Initiation of transcription of *ZRT1* was found by primer extension analysis to occur at a single site at position -45.] Likewise, a genomic fragment containing the promoter regions
from -601 to the ATG codon was sufficient to completely complement the mutant defect in
zinc uptake and zinc-limited growth observed in *zrt1* mutant. These results demonstrated
that the full promoter of *ZRT1* is located in this region of the 5' flanking DNA. To map the
specific promoter elements that provide zinc-responsive expression to the *ZRT1* gene, a series of progressive 5' deletion constructs was generated such that increasingly smaller fragments of the *ZRT1* promoter were fused to a *lacZ* reporter gene in an integrating plasmid
vector (Figure 4). These constructs were then transformed into a wild type strain and evaluated for expression in cells grown under zinc-limiting and zinc-replete growth conditions.
Deletion of 5'-flanking sequences from -706 to -521 had no effect on the zinc responsiveness of the reporter gene fusion. Deletion of the region from -521 to -361 reduced β-galactosidase activity in zinc-limited cells by approximately 20% with little effect on expression
in zinc-replete cells. Deletion of sequences from -332 to -305 resulted in an additional loss
of zinc-limited expression, again with no change in the zinc-replete expression, and deletion
from -221 to -201 completely eliminated reporter gene expression in both media conditions.

The results obtained with the 5' deletion series described in Figure 4 are consistent
with the presence of at least three different zinc-responsive promoter elements or "ZREs"

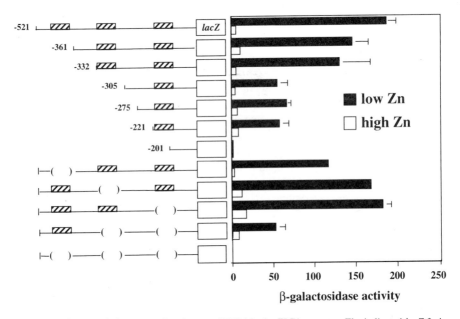

Figure 4. Identification of zinc-responsive elements (ZREs) in the *ZRT1* promoter. The indicated *lacZ* fusion constructs were transformed into a wild type strain (DY1457). Transformants were grown in LZM (Zhao and Eide,
1996a) supplemented with either 10 μM (low Zn) or 1000 μM (high Zn) $ZnCl_2$ prior to β-galactosidase assay. The
relative positions of the three mapped ZREs are shown with the hatched boxes. The location of the internal deletions generated in the *ZRT1* promoter are indicated by parentheses. These deletions span bases -318 to -308
(ZRE1), -203 to -193 (ZRE2), and -454 to -444 (ZRE3), respectively. Each value represents the average of three
independent assays ± one standard deviation.

being present in the *ZRT1* promoter. Analysis of the DNA sequence of this promoter revealed the presence of a conserved eleven base pair sequence in each of the regions implicated by the deletion series. These elements were identified in positions -318 to -308, -203 to -193, and, in the opposite orientation, at -454 to -444 and we refer to these sites as ZRE1, ZRE2, and ZRE3, respectively. To determine if these three conserved elements were in fact required for zinc-responsive gene expression, the eleven base pair sequences were deleted from the -521 to +1 promoter fusion. Deletion of any single ZRE had little or no effect on zinc-responsive gene expression. Deletion of two elements (e.g. ΔZRE1 ΔZRE2), however, drastically reduced expression in zinc-limited cells and deletion of all three elements eliminated both induced and basal expression. These data demonstrate that ZRE1, ZRE2, and ZRE3 are necessary to confer zinc-responsive gene expression on the *ZRT1* promoter. Moreover, they demonstrate that these elements are also required for basal expression of *ZRT1*.

To determine if ZRE1, ZRE2, and/or ZRE3 are sufficient to confer this regulation, oligonucleotides containing these elements were inserted into the promoter of a *lacZ* reporter gene bearing the *CYC1* TATA boxes but lacking a functional upstream activation sequence (UAS) element. Insertion of these elements into this promoter conferred zinc-responsiveness ranging from 15- to 187-fold (data not shown). These results demonstrate that ZREs are sufficient as well as necessary to confer zinc-responsive regulation on a gene. These data also demonstrate that a single ZRE is active and multiple elements are not required for function. Moreover, ZRE3 was inserted in its normal orientation, i.e. opposite to ZRE1 and ZRE2, indicating that ZRE elements are bidirectionally functional. Mutation of bases flanking the eleven base pair core element of ZRE1 had no effect on zinc-responsiveness indicating that the identity of these flanking bases is not important for ZRE function.

A similar analysis of the *ZRT2* and *ZAP1* promoters identified the ZREs in these promoters. Two elements were found in the *ZRT2* promoter (-310 to -300 and -261 to -251) and a single element was found in *ZAP1* (-143 to -133). These data establish the identity of a total of six different zinc-responsive elements. Alignment of these elements has allowed the derivation of a consensus ZRE sequence, 5'-ACCYYNAAGGT-3' (Figure 5). No conserved bases were identified outside of this eleven base pair element and this observation is consistent with our results from a mutational analysis of *ZRT1* ZRE1 indicating that flanking sequences are not important for zinc-responsiveness. Thus, it is our current hypothesis that this consensus sequence represents the complete cis-acting element responsible for ZAP1-responsive gene expression in yeast.

	ZRE1	5'-ACCCTCAAGGT-3'
ZRT1	ZRE2	5'-ACCTTGAAGGT-3'
	ZRE3	5'-ACCCCAAAGGT-3'
ZRT2	ZRE1	5'-ACCCTAAAGGT-3'
	ZRE2	5'-ACCCTAAAGGT-3'
ZAP1	ZRE	5'-ACCCTAAAGGT-3'
Consensus		5'-ACCYYNAAGGT-3'

Figure 5. Derivation of a ZRE consensus sequence. The sequences of the ZREs identified in the *ZRT1*, *ZRT2* and *ZAP1* genes are aligned from 5' to 3'. A consensus ZRE was derived from the identity of bases common to all of the sequences.

3.7. ZAP1 Binds to ZREs in a Sequence-Specific Manner

The genetic evidence described in Section 3.4 demonstrated that ZAP1 plays a central role in zinc ion homeostasis by regulating transcription of the zinc uptake systems. As compelling as those results were, however, they failed to establish if ZAP1 plays a direct or an indirect role in the response. We proposed that ZAP1 was acting as both the primary zinc "sensor" and as the transcriptional activator that induces *ZRT1*, *ZRT2*, and *ZAP1* transcription in response to zinc-limitation. An equally plausible model is that ZAP1 is part of a signal transduction pathway that communicates the cellular zinc status to other proteins that then regulate transcription.

A prediction of the hypothesis that ZAP1 is the direct regulator of zinc-responsive gene expression is that the ZAP1 protein binds directly to ZREs. To test this hypothesis, either full-length ZAP1 protein or a truncated ZAP1 polypeptide containing the five C-terminal zinc fingers of ZAP1 (amino acids 687–880, ZnF_{1-5}) were produced *in vitro* with a coupled transcription/translation system and used for electrophoretic mobility shift assays (EMSAs). The probe used in this experiment was a thirty base pair oligonucleotide containing the ZRE1 sequence from *ZRT1*. No protein/DNA complexes were observed in binding reactions containing either no protein (Figure 6, lane 1) or increasing amounts of vector-programmed transcription/translation product (Figure 6, lanes 2–4). Protein/DNA complexes were observed when either the full-length ZAP1 (Figure 6, lanes 5–7) or ZnF_{1-5} fragment (Figure 6, lanes 8–10) were used in the assay. While ZnF_{1-5} gave only a single EMSA complex, the full length protein gave three bands. These bands were of similar relative intensity with increasing protein added suggesting that there was no cooperativity in binding. We suggest that the three bands correspond to alternative translational start sites used by the *in vitro* synthesis system. Competition studies demonstrated the protein-DNA complexes observed in Figure 6 are sequence-specific; 100-fold molar excess of unlabeled fragments containing functional ZREs competed for complex formation while mutated ZREs or unrelated fragments did not (data not shown).

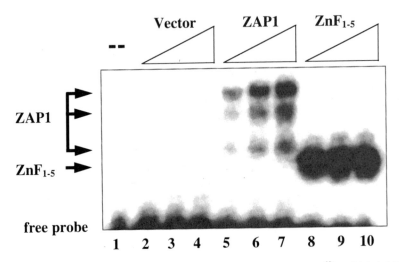

Figure 6. ZAP1 binds to ZRE1. Electrophoretic mobility shift assay (EMSA) using a ^{32}P end-labeled *ZRT1* ZRE1 oligonucleotide. No protein (lane 1) or 1, 2, or 4 μl samples prepared by *in vitro* transcription/translation reactions using either the Bluescript SK[+] vector (lanes 2–4), pSK[+]ZAP1 (i.e. full length ZAP1, lanes 5–7), or pSK[+]ZnF (ZnF_{1-5} truncated polypeptide, lanes 8–10) as the template were used in the binding reactions. The ZAP1- or ZnF_{1-5}-specific DNA-protein complexes are indicated by arrows.

The experiment described in Figure 6 indicated that ZAP1 is a ZRE-binding protein and that this binding is conferred by one or more of the five C-terminal zinc fingers. An independent test of the sequence-specificity of ZAP1 ZRE binding was provided by *in vitro* DNAse I footprint analysis (Figure 7). No difference in the DNAse I digestion products was observed between the no protein and vector-only controls (data not shown). However, full length ZAP1 and the zinc finger truncated fragment gave clear protection to DNAse I cleavage in the regions of the *ZRT1*, *ZRT2*, and *ZAP1* promoters that correspond to the ZREs. These data further demonstrate that the ZAP1 protein binds to ZREs and also indicate that the zinc finger domain retains full DNA binding capacity of the full length protein. Only weak binding of ZAP1 was observed in footprinting on the *ZRT1* ZRE3 site suggesting that this site is recognized by ZAP1 with lower affinity (data not shown).

3.8. Differential Regulation of the *ZRT1*, *ZRT2*, and *ZAP1* Promoters in Response to Zinc

Our previous studies (Zhao and Eide, 1996a; Zhao and Eide, 1996b) indicated that while the ZRT1 high affinity transporter was completely repressed in zinc-replete cells, the ZRT2 low affinity transporter remained active. Surprisingly, the analysis of zinc-responsive transcriptional control described in this and a previous report (Zhao and Eide, 1997) indicated that expression of both *ZRT1* and *ZRT2* was zinc-responsive and regulated by ZAP1.

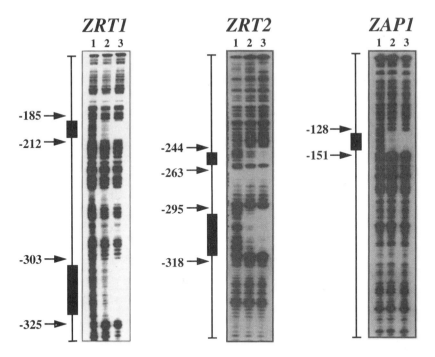

Figure 7. Assessment of ZAP1 DNA binding specificity by *in vitro* DNase I footprint analysis. Four μl volumes of a transcription/translation lysate programmed for synthesis of the vector-only control (lane 1), full length ZAP1 (lane 2), or the ZAP1 ZnF$_{1-5}$ polypeptide (lane 3) were incubated with ^{32}P end-labeled *ZRT1*, *ZRT2* and *ZAP1* promoter fragments and subjected to partial DNase I digestion. The digestion products were resolved on a denaturing polyacrylamide gel and visualized by autoradiography. The boxes indicate the position of ZREs and the numbers refer to the first and the last nucleotide protected from DNase I cleavage. Fragment sizes were determined relative to a DNA sequencing ladder (not shown).

These observations presented a paradox; how can ZRT2 be active in zinc-replete cells yet still be regulated by ZAP1? To address this question, we examined the zinc-responsiveness of the *ZRT1*, *ZRT2*, and *ZAP1* promoter during growth in media containing a range of zinc concentrations. Previous results suggested that repression of *ZRT1* by zinc was in response to an intracellular pool of the metal. Because no methods are currently available to directly measure this pool, we estimated its level by measuring total cell-associated zinc levels. As medium zinc concentrations rose, so did the cell-associated zinc levels (Figure 8). The *ZAP1* and *ZRT1* promoters were extremely zinc responsive. For example, 50% repression of both promoters was observed at 5 µM zinc in the medium which corresponded to a cell-associated zinc level of 75–100 pmol/10^6 cells. In contrast, the *ZRT2* promoter was repressed to a similar level at a medium zinc concentration of 250 µM, i.e. a 50-fold difference, which corresponded to a cell-associated zinc level of approximately 175 pmol/10^6 cells. Therefore, although *ZRT1*, *ZRT2*, and *ZAP1* genes are all zinc-responsive ZAP1 target genes, the *ZRT2* gene is much less responsive than the other two genes. Possible reasons for this differential regulation are considered below (see Discussion).

4. DISCUSSION

The *ZAP1* gene is required for zinc-responsive gene expression in yeast. Target genes regulated by ZAP1 include the *ZRT1* and *ZRT2* zinc transporter genes and the *ZAP1* gene itself. We propose that *ZAP1* encodes a transcriptional activator that binds to the promoters of these genes and activates their transcription when intracellular zinc levels are low. The experiments described in this report directly support this hypothesis, i.e., ZAP1 binds to zinc-responsive elements found in the promoters of these genes in a sequence-specific fashion. Thus, we conclude that ZAP1 does indeed play a direct role in zinc-responsive gene expression. Moreover, the identification of ZAP1 binding sites in its own promoter reinforces our previous hypothesis (Zhao and Eide, 1997) that ZAP1 regulates its own expression in response to zinc via a positive autoregulatory mechanism. An exciting question that still re-

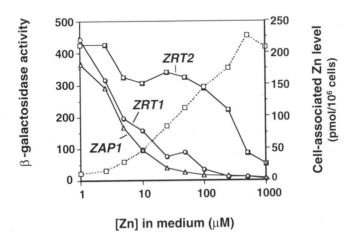

Figure 8. Differential regulation of ZAP1-regulated promoters in response to zinc. Wild type strain DY1457 was transformed with the *lacZ* promoter fusion plasmids pZRT2-lacZ, pZAP1-lacZ, or pGl1 (i.e. ZRT1-lacZ) (Zhao and Eide, 1997). The transformants were inoculated into LZM medium supplemented with the indicated amount of ZnCl₂, grown to exponential phase (approximately 17 hours), harvested, and assayed for β-galactosidase activity and cell-associated zinc levels.

mains to be answered is how zinc regulates ZAP1 activity. Preliminary studies indicate that when ZAP1 is expressed from a zinc-insensitive heterologous promoter, neither the level nor the nuclear localization of protein changes in response to zinc (A. Bird and D. Eide, unpublished result). Therefore, we predict that zinc either alters the ability of ZAP1 to bind to ZREs or the metal influences the activity of the transcription activation domain(s) of ZAP1. This question is the subject of ongoing research in our lab.

Characterization of ZAP1 binding sites in its target promoters has led to the identification of an eleven base pair consensus sequence that we have designated the zinc-responsive element or ZRE. This consensus sequence appears to contain all of the sequence-specific information required for ZAP1 binding. Our studies have illuminated many characteristics of these ZRE sequences. First and foremost, this sequence is both necessary and sufficient for zinc-responsive gene expression. Second we find that the ZRE sequence is bidirectional and can function in either orientation relative to the start site of transcription. We have also found that the ZREs are active in single copy; multiple copies are not required for ZRE function. Moreover, when these elements are present in multiple copies in a promoter, they appear to show simple additivity rather than cooperative effects.

Our studies have also demonstrated that there is a hierarchy of zinc responsiveness among the *ZRT1*, *ZRT2* and *ZAP1* genes; i.e., the *ZRT2* promoter requires significantly more zinc to be down-regulated than does the *ZRT1* or *ZAP1* promoters (Figure 8). This expression pattern fits well with the proposed roles of these proteins. As cells become zinc-limited, their first response is to increase the activity of the ZRT2 low affinity transporter. If zinc limitation becomes more severe, the *ZRT1* gene is induced to provide a higher affinity uptake system for zinc accumulation. The mechanism underlying the differential regulation of these ZAP1 target genes is not yet known but we can suggest some possibilities. First, if zinc controls the affinity of ZAP1 for its ZRE binding sites, the ZRE sequences in *ZRT2* might be of higher affinity than those found in the *ZAP1* and *ZRT1* promoters. By this model, as the level of zinc rises in the cell and the level of active ZAP1 declines, ZAP1 would continue to activate transcription from the high affinity sites in the *ZRT2* promoter after it loses the ability to activate from lower affinity sites. Although attractive, this simple explanation is unlikely given that the two ZRE sequences in *ZRT2* are identical to the *ZAP1* ZRE. A possible variation to the model is that other proteins bind to the *ZRT2* promoter and help stabilize binding of ZAP1 to the ZREs, thus increasing the apparent affinity of ZAP1 for these sites.

The characterization of the ZAP1 binding site in its target promoters is critical to deciphering the regulation of ZAP1 function by zinc. For example, *in vivo* footprinting studies are now possible to determine if zinc alters ZAP1 DNA occupancy. Furthermore, the derivation of a ZRE consensus sequence provides us with a powerful tool to identify other zinc-responsive genes in the yeast genome. This is made possible by the recent completion of the *Saccharomyces* genome sequence. Using the consensus sequence derived in this study, sequence database analysis has identified a total of twenty genes in the yeast genome that contain one or more ZRE-like sequences in their promoters. This list is an exciting resource for the further analysis of how eukaryotic cells respond to zinc limitation and maintain zinc homeostasis.

ACKNOWLEDGMENTS

The authors thank Jim Browning, Thomas Spizzo, and Sara Duesterhoeft for technical assistance during the course of this work. Funding for this research was provided by the U.S. National Institute of Health, the National Science Foundation, and the Department of Energy.

REFERENCES

Bohm, S., Frishman, D., and Mewes, H. W. (1997). Variations of the C_2H_2 zinc finger motif in the yeast genome and classification of yeast zinc finger proteins. Nucl. Acids Res. *25*, 2464–2469.

Boulikas, T. (1994). Putative nuclear localization signals (NLS) in protein transcription factors. J. Cell. Biochem. *55*, 32–58.

Brugnera, E., Georgiev, O., Radtke, F., Heuchel, R., Baker, E., Sutherland, G. R., and Schaffner, W. (1994). Cloning, chromosomal mapping, and characterization of the human metal-regulatory transcription factor MTF-1. Nucl. Acids Res. *22*, 3167–3173.

Dancis, A., Yuan, D. S., Haile, D., Askwith, C., Eide, D., Moehle, C., Kaplan, J., and Klausner, R. D. (1994). Molecular characterization of a copper transport protein in *S. cerevisiae*: an unexpected role for copper in iron transport. Cell *76*, 393–402.

Eide, D., Broderius, M., Fett, J., and Guerinot, M. L. (1996). A novel iron-regulated metal transporter from plants identified by functional expression in yeast. Proc. Natl. Acad. Sci. U.S.A. *93*, 5624–5628.

Eng, B. H., Guerinot, M. L., Eide, D., and Saier, M. H. (1998). Sequence analyses and phylogenetic characterization of the ZIP family of metal ion transport proteins. J. Memb. Biol., in press.

Forsburg, S. L., and Guarente, L. (1989). Identification and characterization of HAP4: a third component of the CCAAT-bound HAP2/HAP3 heteromer. Genes Dev. *3*, 1166–1178.

Fuhrmann, G.-F., and Rothstein, A. (1968). The transport of Zn^{2+}, Co^{2+} and Ni^{2+} into yeast cells. Biochim. Biophys. Acta *163*, 325–330.

Galibert, F., Alexandraki, D., Baur, A., Boles, E., Chalwatzis, N., Chuat, J. C., Coster, F., Cziepluch, C., De Haan, M., Domdey, H., Durand, P., Entian, K. D., Gatius, M., Goffeau, A., Grivell, L. A., Hennemann, A., Herbert, C. J., Heumann, K., Hilger, F., Hollenberg, C. P., Huang, M. E., Jacq, C., Jauniaux, J. C., Katsoulou, C., and Karpfinger-Hartl, L. (1996). Complete nucleotide sequence of *Saccharomyces cerevisiae* chromosome X. EMBO J. *15*, 2031–2049.

Grotz, N., Fox, T., Connolly, E., Park, W., Guerinot, M. L., and Eide, D. (1998). Identification of a family of zinc transporter genes from *Arabidopsis* that respond to zinc deficiency. Proc. Natl. Acad. Sci. USA, in press.

Gunshin, H., Mackenzie, B., Berger, U. V., Gunshin, Y., Romero, M. F., Boron, W. F., Nussberger, S., Gollan, J. L., and Hediger, M. A. (1997). Cloning and characterization of a proton-coupled metal-ion transporter. Nature *388*, 482–487.

Hamer, D. H. (1986). Metallothionein. Ann. Rev. Biochem. *55*, 913–951.

Mowll, J. L., and Gadd, G. M. (1983). Zinc uptake and toxicity in the yeast *Sporobolomyces roseus* and *Saccharomyces cerevisiae*. J. Gen. Microbiol. *129*, 3421–3425.

O'Hara, P. J., Horowitz, H., Eichinger, G., and Young, E. T. (1988). The yeast *ADR6* gene encodes homopolymeric amino acid sequences and a potential metal-binding domain. Nucl. Acids Res. *16*, 10153–10169.

Palmiter, R. D., Cole, T. B., and Findley, S. D. (1996). ZnT-2, a mammalian protein that confers resistance to zinc by facilitating vesicular sequestration. EMBO J. *15*, 1784–1791.

Palmiter, R. D., and Findley, S. D. (1995). Cloning and functional characterization of a mammalian zinc transporter that confers resistance to zinc. EMBO J. *14*, 639–649.

Paulsen, I. T., and Saier, M. H. (1997). A novel family of ubiquitous heavy metal ion transport proteins. J. Memb. Biol. *156*, 99–103.

Rauser, W. E. (1995). Phytochelatins and related peptides. Plant Physiol. *109*, 1141–1149.

Schjerling, P., and Holmberg, S. (1996). Comparative amino acid sequence analysis of the C_6 zinc cluster family of transcriptional regulators. Nucl. Acids Res. *24*, 4599–4607.

Solioz, M., and Vulpe, C. (1996). CPx-type ATPases: a class of P-type ATPases that pump heavy metals. TIBS *21*, 237–241.

Vallee, B. L., and Auld, D. S. (1990). Zinc coordination, function, and structure of zinc enzymes and other proteins. Biochemistry *29*, 5647–5659.

White, C., and Gadd, G. M. (1987). The uptake and cellular distribution of zinc in *Saccharomyces cerevisiae*. J. Gen. Microbiol. *133*, 727–737.

Zhao, H., and Eide, D. (1996a). The yeast *ZRT1* gene encodes the zinc transporter of a high affinity uptake system induced by zinc limitation. Proc. Natl. Acad. Sci. USA *93*, 2454–2458.

Zhao, H., and Eide, D. (1996b). The *ZRT2* gene encodes the low affinity zinc transporter in *Saccharomyces cerevisiae*. J. Biol. Chem. *271*, 23203–23210.

Zhao, H., and Eide, D. J. (1997). Zap1p, a metalloregulatory protein involved in zinc-responsive transcriptional regulation in *Saccharomyces cerevisiae*. Mol. Cell. Biol. *17*, 5044–5052.

25

EMBRYONIC LIVER DEGENERATION AND INCREASED SENSITIVITY TOWARDS HEAVY METAL AND H_2O_2 IN MICE LACKING THE METAL-RESPONSIVE TRANSCRIPTION FACTOR MTF-1

Peter Lichtlen,[1] Çagatay Günes,[1] Rainer Heuchel,[1] Oleg Georgiev,[1] Karl-Heinz Müller,[1] Horst Blüthmann,[2] Silvia Marino,[3] Adriano Aguzzi,[3] and Walter Schaffner[1*]

[1]Institut für Molekularbiologie
Abteilung II der Universität Zürich
Winterthurstrasse 190
CH-8057 Zürich, Switzerland
[2]Hoffmann La-Roche PRTB
CH-4002 Basel
[3]Neuropathologie USZ
Schmelzbergstrasse 12
CH-8091 Zürich, Switzerland

1. ABSTRACT

The expression of metallothioneins, small heavy metal binding proteins, is induced by treatment of cells with heavy metal. This induction depends on the heavy metal-responsive transcriptional activator MTF-1. To investigate the physiological function of MTF-1, we generated null mutant mice by targeted gene disruption. Embryos lacking MTF-1 die in utero at around day 14 of gestation. They show impaired development of hepatocytes and, at later stages, liver decay and generalized oedema. MTF-1$^{-/-}$ embryos fail to transcribe metallothionein I and II genes, and also show diminished transcripts of the gene which encodes the heavy-chain subunit of the gamma glutamylcysteine syn-

[*]Send correspondence to: Prof. Dr. Walter Schaffner, Institut für Molekularbiologie II, Universität Zürich, Winterthurstrasse 190, CH-8057 Zürich, Switzerland.

Metals and Genetics, edited by Sarkar.
Kluwer Academic / Plenum Publishers, New York, 1999.

thetase, a key enzyme for glutathione biosynthesis. Metallothionein and glutathione are involved in heavy metal homeostasis and detoxification processes, such as scavenging reactive oxygen intermediates. Accordingly, primary mouse embryo fibroblasts lacking MTF-1 show increased susceptibility to the cytotoxic effects of cadmium or hydrogen peroxide. We also note that the MTF-1 null mutant phenotype bears some similarity to those of two other regulators of cellular stress response, namely c-Jun and NF-κB (p65/RelA). Furthermore we find MTF-1 binding sites in the p65 and p50 subuntis of NF-κB, the zinc export pump ZnT1 and G6PD. Thus, MTF-1 may help to control metal homeostasis and probably cellular redox state, especially during liver development (Günes et al., 1998).

2. INTRODUCTION

The transcription factor MTF-1 is essential for basal and heavy metal induced expression of metallothioneins (Heuchel et al., 1994). MTF-1 is highly conserved (Radtke et al., 1993; Heuchel et al., 1994; Radtke et al., 1995) and contains six zinc fingers of the C_2H_2-type as a DNA binding domain and at least three distinct domains responsible for transcriptional activation. We previously cloned mouse and human MTF-1, which share a 93% identity in amino acid sequence (Radtke et al., 1993; Brugnera et al., 1994). MTF-1 binds to a number of metal responsive elements (MREs) present in the promoter regions of metallothionein genes I and II (MT-I and MT-II) and regulates their expression. The MREs present in metallothionein gene promoters are short DNA sequence motifs of 9–12 conserved base pairs, whereby a central core of 7 base pairs is strongly conserved (Stuart et al., 1985; Searle et al., 1987; Radtke et al., 1993; Ç. Günes, unpublished results). Binding of MTF-1 to MRE sequences is dependent on zinc cations, as has been shown by in vivo and in vitro experiments (Westin and Schaffner, 1988; Mueller et al., 1988; Searle et al., 1990; Radtke et al., 1993). However, MTF-1 activates the expression of metallothionein genes not only by zinc but also by other heavy metal cations such as cadmium or copper (Heuchel et al., 1994).

There are four types of metallothioneins in mammals, namely MT-I, MT-II, MT-III and MT-IV (Uchida et al. 1991, Palmiter et al., 1992; reviewed in Heuchel et al., 1995). The expression of MT-I and MT-II is ubiquitous and inducible by various stress conditions, while MT-III and MT-IV are tissue specific and only weakly stress-inducible. Consequently MT-I and MT-II have been implicated in heavy metal detoxification, metal homeostasis and radical scavenging (Angel et al., 1986; Kaegi, 1991; Thornally and Vasak, 1985; Thomas et al., 1986; Girotti, 1986). Since constitutive MT-I and MT-II expression is very high in fetal liver (Kern et al., 1981; Ouellette, AJ., 1982; Quaife et al., 1986), it has been suggested that these MTs might play a pivotal role in fetal liver development. However, two groups have shown independently that mouse strains with combined deletion of the MT-I and MT-II genes, as a result of targeted gene disruption, show no altered phenotype under normal conditions. But, such null mutant mice were more sensitive to cadmium exposure, an observation which supports the proposed role of metallothioneins in metal detoxification (Michalska and Choo, 1993; Masters et al., 1994).

We have previously shown that MTF-1 is essential for basal and heavy metal-induced transcriptional activation of MT-I and MT-II genes in cultured cells (Heuchel et al., 1994). Here, using the power of knock-out technology, we show that MTF-1 is essential for embryonic liver function in mice (Günes et al., 1998).

3. RESULTS

3.1. Disruption of the MTF-1 Locus by Gene Targeting

To generate mice lacking functional MTF-1 by targeted disruption of the MTF-1 genomic locus, 129/SV ES cell clones containing one disrupted MTF-1 allele were injected into blastocysts to generate mice heterozygous for the mutated MTF-1 allele (see Materials and methods).

These mice (MTF-1$^{+/-}$) with a mixed 129/SV x C57BL/6 genetic background appeared normal and were intercrossed. Among 41 born pups, none was homozygous for the MTF-1 mutation, implying that MTF-1$^{-/-}$ embryos die *in utero*. Genotype analyses of embryos from MTF-1$^{+/-}$ intercrosses revealed no viable homozygous mutant embryo older than E14.5 days (0 of 25). Before day E13.5, homozygous mutant embryos were obtained at about the expected Mendelian frequency (10 of 41), indicating that implantation and early post-implantation development were not impaired. At E13.5, only one of 14 MTF-1 homozygous mutant embryos was found dead. Conversely, at E14.5, only one homozygous mutant embryo was found alive.

RT-PCR analysis of RNA from 13.5-day old mutant embryos revealed an MTF-1 transcript with an internal deletion giving rise to a stop codon 39 bases after the point of deletion (Figure 1). As expected, total RNA from 13.5-day old embryos did not show an MTF-1 signal with a probe covering the deleted segment . We also analyzed expression of the metallothionein I and II genes in 13.5-day old MTF-1$^{-/-}$ embryos by RNAse protection assay and found no detectable MT-I and MT-II expression whereas expression of the unrelated transcription factor Sp1 was not affected (data not shown). This result agrees with the previous analysis of MTF-1$^{-/-}$ ES cells (Heuchel et al. 1994). Hence, we conclude that homozygous mutant embryos do not produce any functional MTF-1 protein.

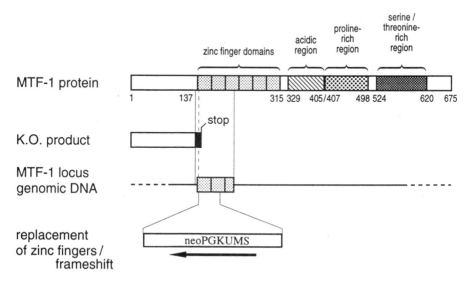

Figure 1. Targeted disruption of the mouse MTF-1 locus (see also Heuchel et al., 1994). A schematic drawing of MTF-1 protein is shown in relation to the genomic locus with the 238 bp exon (which encodes the first two and part of the third zinc finger domains; dotted boxes) and part of the flanking introns. Part of the zinc finger exon was replaced by an inversely oriented neoPGKUMS cassette, as described by Heuchel et al., 1994. Extensive analysis of MTF-1$^{-/-}$ embryos demostrated that the targeted locus expresses only a short transcript encoding an N-terminal protein fragment of MTF-1 which ends in an out-of-frame stop codon (K.O. product).

3.2. Phenotype of Mice Lacking MTF-1

Viable MTF-1$^{-/-}$ embryos at day E13.5 were macroscopically indistinguishable from their heterozygous and wild type littermates. They were uniformly pink in color, well-vascularized and showed no internal bleeding. A histological analysis of whole embryos did not reveal any differences (data not shown) except of a pronounced liver damage (Figure 2).

The liver structure of MTF-1$^{-/-}$ embryos at this stage was, to a variable degree, disrupted with enlarged, congested sinusoids and dissociation of the epithelial compartment (Figure 2D). Furthermore, although individual epithelial cells of MTF-1$^{-/-}$ liver at E13.5

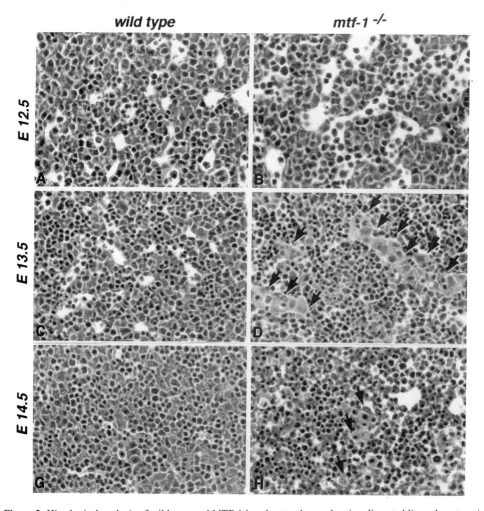

Figure 2. Histological analysis of wild type and MTF-1 knockout embryos showing disrupted liver phenotype in MTF-1$^{-/-}$ embryos. Hematoxylin and eosin-stained sagittal liver sections of wild type embryos at E12.5 (A), at E13.5 (C) and at 14.5 (G) as well as of MTF-1 knockout embryos at E12.5 (B) at E13.5 (D) and at 14.5 ((H) are shown. MTF-1$^{-/-}$ embryonic liver at 12.5 (B) demonstrated no significant differences when compared to the wild type littermate at the same stage (A) whereas MTF-1$^{-/-}$ embryonic liver at 13.5 (D) showed enlarged, congested sinusoids and dissociation of the epithelial compartment and scattered islands of hepotcytes are visible (arrows). Moreover, MTF-1-/- embryonic liver at E14.5 (H) showed an almost complete disruption of the epithelial net. The remaining hepatocytes (arrows) are swollen and show picnotic nuclei.

appeared morphologically normal in conventional histology, immunohistochemical analysis showed that these cells had a reduced cytokeratin expression in comparison to wild type (data not shown). This deficiency was liver-specific since cytokeratin expression in the gut from the same embryo was not affected (data not shown). At E14.5, the only null mutant embryo still alive showed severe liver necrosis (Figure 2H), diffuse bleeding and generalized oedema as well as an almost complete lack of cytokeratin expression in the liver (data not shown). By contrast, α-actin expression in the vessel walls was still indistinguishable from heterozygous or wild type littermates at E14.5 (data not shown). An obvious explanation for the absence of cytokeratin might have been a dependence of liver cytokeratin genes on MTF-1. However, we have not found any binding sites for MTF-1 in the known cytokeratin promoters, and it remains to be seen whether the absence of cytokeratin is a cause or a consequence of liver pathology.

At E13.5, we could not detect any significant increase in the frequency of apoptotic cells in the liver of MTF-1$^{-/-}$ embryos, neither morphologically nor with the TUNEL in situ staining for DNA fragmentation (data not shown). At day E14.5, there was extensive DNA fragmentation, most likely resulting from necrosis rather than apoptosis. We also found that the cell proliferation rate in the MTF-1 homozygous mutant was not affected since bromodeoxyuridine incorporation at E13 revealed no decrease when compared to wild type or heterozygous littermates. Since the liver takes over as a blood-forming organ at this stage, we considered the possibility of a defect in the hematopoetic system in null mutant embryos. However, living E13.5 homozygous mutants showed not only no macroscopic color change indicative of anemia, but also no change in the number or appearance of erythrocytes or hematopoietic precursor cells observable in histological sections (data not shown). At E13.5, the contribution of liver-derived enucleated erythrocytes is small, thus a deficiency would not be easily detectable. However, the only living null mutant embryo at day E14.5 also showed no reduced representation of enucleated erythrocytes, again arguing against a defect in hematopoiesis (data not shown).

Living homozygous MTF-1$^{-/-}$ embryos at stages E13.5 and E14.5 showed no reduced liver size which distinguishes them from the small liver phenotype seen in the knockout of HLX, a developmental control gene for liver formation (Hentsch et al., 1996). Also, the placenta of MTF-1$^{-/-}$ embryos did not reveal any abnormalities (data not shown) in contrast to the combined liver and placental defects seen in hepatocyte growth factor/scatter factor (HGF/SF) knockouts (Uehera et al., 1995; Schmidt et al., 1995). Hence, our findings render unlikely a loss of HLX or HGF/SF function as a downstream effect of MTF-1 deletion.

3.3. MTF-1 Regulates the Expression of Gamma Glutamylcysteine Synthetase Gene

As mentioned before, MTF-1 is essential for basal and heavy metal induced expression of MT-I and MT-II genes. Since mice lacking the metallothionein I and II genes are viable (Michalska and Choo, 1993; Masters et al., 1994) and other MT isoforms cannot account for redundancy in MT-I and MT-II knockout mice because of their temporal and spatial expression (Palmiter et al., 1992; Quaife et al., 1994) we considered it most likely that deregulated expression of at least one additional target gene, perhaps in combination with loss of MT-I and MT-II expression, is responsible for the liver damage phenotype of MTF-1$^{-/-}$ mice.

Recently, metal responsive elements (MREs) have been described in the promoter region of the heavy-chain subunit of gamma glutamylcysteine synthetase (γ-GCS$_{hc}$), the key enzyme in the biosynthesis of glutathione (GSH). We found additional putative MREs in the promoter of γ-GCS$_{hc}$ and tested them for MTF-1 binding. By electrophoretic mobility shift assays (EMSA), we observed a strong, MTF-1-specific complex with one of these MREs (located at position -114) and a specific albeit weak interaction with the MRE located at position +296.

No binding to other putative MREs could be observed in EMSA assays (data not shown).

We then inserted a 1.8 kb fragment of the γ-GCS$_{hc}$ promoter upstream of a β-globin reporter gene (Materials and methods) and tested the expression of the reporter gene in transient transfection assays in the human hepatoma cell line HepG2, which expresses MTF-1 endogenously.

In addition to a basal level of expression (Figure 3C, lane 1), cotransfections with mouse or human MTF-1 expression vectors yielded an increased transcription of the reporter gene (Figure 3C, lanes 2 and 3). Moreover, induction with Zn or Cd further elevated reporter gene expression (Figure 3C, lanes 4–9). These results are in agreement with the recent observation that endogenous γ-GCS$_{hc}$ expression can be induced by zinc or cadmium treatment (Ishikawa et al., 1996).

More importantly, by RT-PCR (data not shown) and Northern blot analysis we found reduced γ-GCS$_{hc}$ gene expression in 13.5-day old MTF-1 knockout embryos compared to wild type embryos (Figure 3D).

The expression of the 36B4 reference gene, which codes for the acidic ribosomal protein P0, was not altered in wild type, heterozygous or knockout embryos whereas the expression of γ-GCS$_{hc}$, encoding the catalytic subunit of the heterodimeric enzyme gamma glutamylcysteine synthetase, was reduced in knockout embryos. In contrast, expression of the gene encoding the light chain subunit (regulatory subunit) of gamma GCS (γ-GCS$_{lc}$) or the gene for glutathione synthetase (GSH-Syn), the second enzyme involved in glutathione biosynthesis, was not altered (data not shown).

3.4. Primary Mouse Embryo Fibroblasts from MTF-1 Null Mutant Embryos Are More Sensitive to Cadmium and Hydrogen Peroxide

To obtain more insights into the physiological role of MTF-1, we cultured primary mouse embryo fibroblasts (MEFs) from 12.5-day old mouse embryos that were either wild type, heterozygous or null mutant for the MTF-1 locus (see Materials and methods). Since we observe a downregulation of genes responsible for metallothionein and glutathione biosynthesis in the MTF-1 null mutant embryos, we decided to test the response of MTF-1 deficient cells against cadmium or H_2O_2-induced cytotoxicity. For this purpose, we treated primary MEFs with cadmium or H_2O_2 and determined the survival of MEFs by neutral red (see Materials and methods for details).

As shown in Figure 4, MTF-1 deficient cells (MTF-1$^{-/-}$) are more susceptible to cadmium or H_2O_2 treatment when compared to wild type cells (MTF-1$^{+/+}$). After 8h of exposure to 40μM cadmium, viability of the MTF-1 deficient cells was about 65% compared to that of wild type cells (Figure 4A). This effect was even more pronounced when the cells were treated with 60μM cadmium: viability of MTF-1 deficient cells was about 40% compared to that of wild type cells (Figure 4B). Similarly, deficient cells were more susceptible to exposure to H_2O_2: after an 8h induction with 20μM and 40μM H_2O_2, about 75% and 60% of the knockout cells, respectively, were viable when compared to wild type cells (Figure 4C, 4D). Primary MEFs heterozygous for the MTF-1 locus (MTF-1$^{+/-}$) were also more susceptible to the cytotoxic effects of cadmium and H_2O_2 when compared to wild type cells. However, these MTF-1$^{+/-}$ cells showed more variation than wild type or knockout cells from experi-

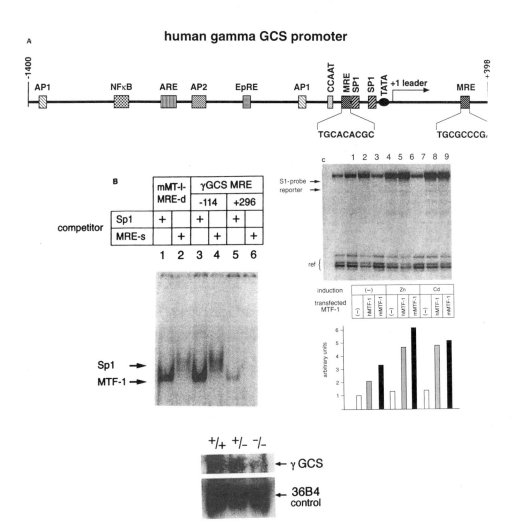

Figure 3. γ-GCS$_{hc}$ as a target gene of MTF-1. (A) The γ-GCS$_{hc}$ gene promoter contains two metal responsive elements (MREs). An 1.8 kb fragment of human γ-GCS$_{hc}$ gene promoter is shown schematically. The putative binding sites for transcription factors are shown as boxes; the TATA-box, as an oval. The transcription start point (tsp) is indicated as +1. Sequences of MTF-1 binding sites (MREs) are shown in the 5' to 3' direction. (B) MTF-1 binds to MREs present in γ-GCS$_{hc}$ gene promoter. Electrophoretic mobilty shift assay (EMSA) with MREs from γ-GCS$_{hc}$ gene promoter using nuclear extracts from HepG2 cells, which were transfected with human MTF-1 expression vectors (see also Radtke et al., 1993). The respective positions within γ-GCS$_{hc}$ promoter of the MREs used are indicated. MREd, the strongest MTF-1 binding site from the MT-I promoter, was used as a positive control. The MREd and γ-GCS$_{hc}$ MRE/-114 sequences share an overlapping Sp1 binding site at their 3' end which results in an additional Sp1-DNA complex in EMSA. Protein-DNA complexes specific for MTF-1 or the unrelated zinc finger factor Sp1 are indicated on the left. About 250- to 500-fold excess of unlabeled competitor oligonucleotides, with binding sites for MTF-1 or for Sp1, were used to show specific binding of the respective proteins. (C) MTF-1 induces expression of a reporter gene from the human γ-GCS$_{hc}$ promoter in transient transfection assays. S1 nuclease mapping of β-globin reporter gene from HepG2 cells. OVEC-ref was used as an internal control for transfection efficiency (see Materials and methods). (D) Northern Blot analysis showing downregulation of the γ-GCS$_{hc}$ gene in MTF-1$^{-/-}$ embryos. A concentration of 2µg mRNA from total embryos (E13.5) was separated on a 1% formaldehyde agarose gel and first probed with human γ-GCS$_{hc}$ cDNA. The film was developed after five days exposure at -70°C. The same filter was stripped, rehybridized with mouse 36B4 control cDNA probe and developed after 36h exposure at -70°C. The coarse and fine grains of γ-GCS$_{hc}$ and 36B4 signals, respectively, are due to the different exposure times. The autoradiograms were recorded by a laser scanner. Positions of γ-GCS$_{hc}$ and 36B4 gene transcripts are indicated by arrows.

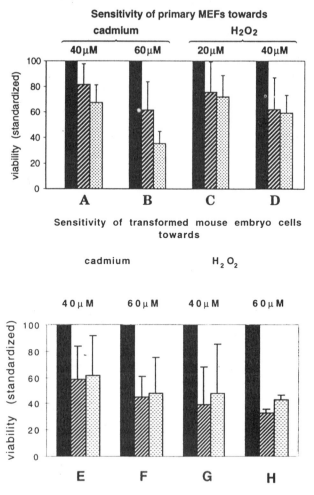

Figure 4. MTF-1 deficient primary mouse embryo fibroblasts are more susceptible to cadmium or H_2O_2-induced cytotoxicity. Primary embryo cells were treated for 8h either by 40μM or 60μM cadmium (A, B) or 20μM or 40μM H_2O_2 (C, D), respectively. Transformed cells were treated by 40μM or 60μM cadmium (E,F) or 40μM or 60μM H_2O_2 (G,H), respectively. Black bars represent wild type (MTF-1$^{+/+}$), striped bars represent heterozygous (MTF-1$^{+/-}$) and dotted bars represent null mutant (MTF-1$^{-/-}$) cells. Each figure shows the average values of at least three independent experiments which were carried out in duplicate. Viability tests with control plates without cadmium or H_2O_2 treatment yielded the same numbers of cells for all three genotypes. Values obtained from experiments with MTF-1$^{+/+}$ cells were set 100% and were compared to those obtained with either MTF-1$^{+/-}$ or MTF-1$^{-/-}$ cells within each experiment.

ment to experiment (see also Discussion). Similar experiments were done with permanent cell lines established from primary embryo fibroblasts. All MTF-1 knockout cell lines were more susceptible to cadmium and H_2O_2 than their wild type counterparts (Figure 4E-H).

4. DISCUSSION

The observed phenotype of liver degeneration in MTF-1 null mutant embryos implies that MTF-1 is particularly important for proper liver development/function already at the embryonic stage. As expected, no expression of the known target genes, metal-lothionein I and II (MT-I and II) could be detected in MTF-1 null mutant embryos. Metal-lothioneins (MTs) are induced via MRE promoter sequences not only by heavy metal load but also by reactive oxygen intermediates (ROI) and thus may help to protect the cell against oxidative stress (Kaegi, 1991; Thornally and Vasak, 1985; Thomas et al., 1986; Girotti, 1986; Dalton et al., 1994; 1997). Null mutant mice for both of the metal-inducible metallothioneins MT-I and MT-II are viable but show greater sensitivity to cadmium toxicity (Michalska and Choo, 1993; Masters et al., 1994). Also, cultured hepatocytes from such mice are highly sensitive to both Cd-induced cytotoxicity (Zheng et al., 1996) or tert-buthylhydroperoxide-induced oxidative stress (Lazo et al., 1995).

In addition to these known MTF-1 target genes, we observed a reduced transcript level of another gene, namely the one coding for the γ-GCS heavy chain. γ-GCS is the first and the rate-limiting enzyme in the biosynthesis of the glutathione (GSH) *de novo* pathway. γ-GCS is an essential enzyme (Meister, 1995; Maellaro et al., 1990). Similarly, mal-function of other enzymes involved in glutathione metabolism in mice and human results in severe phenotypes (Shi et al., 1996; Liebermann et al., 1996). In spite of its ubiquitous presence, glutathione seems of particular importance in the liver, perhaps due to high levels of mitochondrial respiration and lipid peroxidation within this organ. In fact, GSH-homeostasis is mainly regulated by the liver.

Glutathione is essential to maintain the cellular redox-balance and is a scavenger of reactive oxygen intermediates (ROIs), which are mainly generated by respiration (Meister, 1995) or by exogenous cellular stress, such as that caused by heavy metal load (Bagchi et al., 1996). In addition, GSH has been shown to be the first line of defense against cadmium toxicity (Singhal et al., 1987; Shimizu and Morita, 1990; Chan and Cherian, 1992) and its biosynthesis can be induced by heavy metals (Liu et al., 1995; Iszard et al., 1995). A functional link between metallothionein and glutathione in regulating the distribution of cellular zinc was proposed recently (Maret, 1995). This heavy metal-complexing ability of ubiquitous glutathione could explain the viable phenotype of the metallothionein-I and -II null mutant mice (Michalska and Choo, 1993; Masters et al., 1994). Conversely, it has been shown that Zn-MT, at least in part, can substitute for glutathione as a cellular defense against ROI (Srivastava et al., 1993). Thus, glutathione and metallothionein apparently have overlapping functions.

Using primary mouse embryo fibroblasts, we have shown that MTF-1 deficient cells are more susceptible to cadmium or H₂O₂ treatment when compared to their wild type counterparts. Even though mice heterozygous for the MTF-1 locus develop normally and do not show any abnormalities under laboratory conditions, heterozygous (MTF-1$^{+/-}$) cells in culture were also more susceptible to cadmium or H₂O₂ treatment when compared to wild type cells. MTF-1 is ubiquitously expressed, but its known target genes MT-I/MT-II and γ-GCS, though ubiquitously expressed as well, play a particularly important role in the liver. It remains to be seen whether the knockout phenotype can be explained by the existence of a threshold for the accumulation of harmful agents during liver development, or whether MTF-1 has yet another role(s) in liver morphogenesis. Experiments with conditional MTF-1 knockout mice should help to resolve these questions.

Cellular stress response also relies on at least two further ubiquitous transcription factors, AP-1 (Jun/Fos) and NF-κB, which are induced by a number of common stimuli, notably stress signals, and are involved in antioxidant redox regulation (Herrlich et al., 1994; Wilhelm et al., 1995; Angel, 1995; Karin et al., 1997; Pahl and Baeuerle, 1994; Sen and Packer, 1996). It has already been shown that the expression of some genes involved in glutathione metabolism is regulated by AP-1 or NF-κB (Rahman et al., 1996; Friling et al., 1992, Sekhar et al., 1997). Interestingly, the promoter of the γ-GCS$_{hc}$ gene contains two AP-

1 binding sites and one NF-κB binding site (Mulchay and Gipp, 1995; Yao et al., 1995; Tomonari et al., 1997). The knockouts of MTF-1, c-Jun and p65/relA null mutant mice show a similar, but not identical phenotype of embryonic liver decay (Beg et al., 1995; Bladt et al., 1995). This, as well as the similar activation stimuli, suggests that these transcription factors may have overlapping but nonredundant functions during embryogenesis, for example by synergistic regulation of common target genes involved in stress response.

The MTF-1 null mutant phenotype may be a compound effect of the aberrant expression of a number of genes. In fact, besides MT-I, MT-II, and γ-GCS, we also found other candidate target genes involved in glutathione synthesis and/or heavy metal homeostasis. Preliminary experiments suggest that MTF-1 may regulate the expression of the zinc transporter, ZnT-1 which is an ubiquitously expressed zinc export pump. As already described by Palmiter and Findley (1995), the promoter of ZnT-1 contains MTF-1 binding sites (MREs). We find that MTF-1 binds to these MREs and that the expression of the endogenous ZnT-1 transcript can be induced by zinc treatment in primary rat hepatocytes (Ç. Günes, unpublished results). In addition, we have identified MREs in the promoter of the gamma glutamyl transpeptidase gene which is involved in glutathione metabolism and thus is another candidate target gene for MTF-1. Taken together, we consider it likely that MTF-1 plays a general role in cellular stress response.

5. MATERIALS AND METHODS

5.1. Targeted Disruption of the MTF-1 Gene

The construction of the targeting vector used in the targeted disruption of the MTF-1 locus in cultures of embryonic stem cells of mouse strain 129/Sv; the ES-cell line ES-GS was published in Heuchel et al. (1994). ES cells from one clone heterozygous for the mutated MTF-1 allele (26E7) were injected into blastocysts of C57BL/6 mice. Fourteen of the thirty-one male mice born were highly chimeric. Ten of those fourteen were intercrossed with wild type C57BL/6 females and showed germline transmission. Genotype identification was done by PCR with genomic DNA prepared from tail biopsies of three- to six-week old born pups or of genomic DNA prepared from the yolk sac of embryos. The PCR reaction was perfomed as described elsewhere (Heuchel et al., 1994).

5.2. Histological Analysis

Embryos were isolated and fixed in 4% formaldehyde solution overnight. Their yolk sacs were used for genomic DNA preparation and PCR genotyping. After fixation, embryos were embedded in paraffin and processed for conventional histology. Sections were stained with hematoxylin-eosin or used for further treatments.

5.3. RNA Preparations, Cloning of Promoter Constructs, Transfections, Transcript Mappings

Total RNA from embryos was isolated using the guanidium thiocyanate/acidic phenol method-method (Chomczinski and Sacchi, 1987). mRNA from embryos was prepared directly by using the Qiagen® direct mRNA midi/maxi kit. Preparation of cytoplasmic RNA from tissue culture was according to Radtke et al. (1993). RNAase protection assay from MTF-1, Sp1 and metallothionein I and II genes was performed as described by Heuchel et al. (1994). Northern Blot analysis was performed according to standard procedures (Sambrook et al., 1989). To detect the γ-GCS$_{hc}$ mRNA, 50ng of the 2.9-kb human γ-GCS$_{hc}$ cDNA was labelled by random hexanucleotide primers. Expression of 36B4 mRNA was used for RNA integrity. An 800-bp fragment of 36B4 cDNA was labelled by random hexanucleotide primers.

5.4. Transfections, Cloning of Promoter-OVEC Constructs and S1 Nuclease Mapping of Transcripts

Transfections and S1 nuclease mapping of transcripts were performed as described previously (Radtke et al., 1993;). The following primers were used to amplify the 1.8-kb fragment of the γ-GCS heavy chain gene promoter from a human genomic DNA library. The upstream primer (position -1442) is: 5'-GGTCGAGCTCGAGCAC-TATTTAGTGTGGAGC-3' and the downstream primer (position +398) is: 5-TGGCGAC-GTCTGTTCCTCCGGGCTGACGG-CGGTCG-3'. The resulting 1.8-kb fragment was subsequently cloned upstream of the β-globin reporter gene (Westin et al., 1987). The human γ-GCS$_{hc}$ promoter-reporter construct (10μg) were transfected either with or without 2 μg of an expression plasmid carrying mouse MTF-1 or human MTF-1 cDNA. Cells were treated for 4 hours with either 100 μM zinc sulfate or with 40 μM cadmium sulfate before harvest (as described before by Radtke et al., 1993). The intensity of the bands was quantified by phosphorimager.

5.5. Electrophoretic Mobility Shift Assays (EMSA)

EMSA was performed as described by Radtke et al. (1993). Binding reactions were performed by incubating 2–5 fmol end-labeled 31 bp long MRE-containing oligonucleotides with nuclear extracts obtained as described by Schreiber et al., (1989). Identification of the MTF-1 binding was performed by using wild type control extracts in the presence or absence of an MRE-s containing oligonucleotide. As an internal control, Sp1 binding to the oligonucleotides was done.

5.6. Primary Mouse Embryo Fibroblasts and Establishment of Mouse Embryo Fibroblast Cell Lines

Primary embryo fibroblasts were isolated from 12.5-day-old mouse embryos by previously published methods. MTF-1 genotypes of cultured cells were determined by PCR and RT-PCR.

Cells were grown in Dulbecco's modified Eagle's medium (DMEM) supplemented with 10% fetal calf serum (FCS). For the establishment of permanent cell lines wild-type, heterozygous and MTF-1 null mutant primary cells were kept in culture until they reached the border of spontaneous immortalization and were then transfected by lipofection with plasmid CMV-TAG, which directs expression of the simian virus 40 (SV 40) large T-Antigen (T-Ag) and with plasmid p-cHa-ras(A) (Schönthal et al., 1988) which directs expression of the activated Ha-ras proto-oncogene, and immortalized cell lines were derived. Lipofection was performed with Lipofectamine reagent, provided by GIBCO, following the manufacturers protocol.

5.7. Neutral Red Cell Viability Assay

Cytotoxicity of cadmium and H_2O_2 was assessed by determining the uptake of the neutral red dye essentially as described by Sigma, Biosciences^{TM} according to Borenfreund and Puerner (1985). In brief, equal numbers of the indicated MEF cells were grown on 6-well dishes. After 12–16h, media were replaced by fresh media containing the indicated amounts of cadmium or H_2O_2 for 8h before adding the neutral red dye. 1/50 volume of a 0.33% neutral red solution was added to the plates and incubation was continued for 2h in a standard 37°C incubator. Medium containing neutral red was removed and plates were carefully rinsed with 1/100 volume of fixation solution (0.1% $CaCl_2$ in 0.5% formaldehyde). A 1:1 volume of the neutral red solubulisation solution (1% acetic acid in 50% ethanol) was then added and plates were placed on a shaker for about 2h at room temperature. The viabilty of cells was determined spectrophotometrically by measuring the absorbance of the neutral red dye extracted from the cells after solubilisation at a wavelength of 540nm.

ACKNOWLEDGMENTS

We are indebted to Drs. Andrea Kolbus and Peter Angel (DKFZ, Heidelberg) for determining MTF-1 expression in a c-Jun[-/-] cell line. We also thank Drs. Joy Alcedo, Alcide Barberis and Richard Eckner for critical reading of the manuscript and Dr. Ken Zaret (Brown University) for valuable discussions. The human γ-GCS$_{hc}$ cDNA was a gift of R.T. Mulchay (University of Wisconsin-Madison) and the 36B4 reference cDNA was provided by Dr. R. Eckner (Zürich). This work was supported by the Swiss National Science Foundation and by the Kanton Zürich.

REFERENCES AND NOTES

Angel, P., Poting, A., Mallick, U., Rahmsdorf, H.J., Schorpp, M. and Herrlich, P. (1986) Induction of metallothionein and other mRNA species by carcinogens and tumor promoters in primary human skin fibroblasts. Mol. Cell. Biol., 6, 1760–1766.

Angel, P (1995) The role and regulation of the Jun proteins in response to phorbol ester and UV light. Inducible Gene Expression 1, Baeuerle P.A. (Editor), Birkhäuser Boston (Publisher), Vol I, 62–92.

Bagchi, D., Bagchi, M., Hassoun, E.A. and Stohs, S.J. (1996) Cadmium-induced excretion of urinary lipid metabolites, DNA damage, glutathione depletion, and hepatic lipid peroxidation in Sprague-Dawley rats. Biol. Trace Elem. Res., 52, 143–54.

Beg, A.A., Sha, W.C., Bronson, R.T., Ghosh, S. and Baltimore, D. (1995) Embryonic lethality and liver degeneration in mice lacking the relA component of NF-κB. Nature, 376, 167–170.

Bladt, F., Rietmacher, D., Isenmann, S., Aguzzi, A. and Birchmeier, C. (1995) Essential role for the c-met receptor in the migration of myogenic precursor cells into the limb bud. Nature, 376, 768–771.

Borenfreund, E. and Puerner, J. (1985) Toxicity determined in vitro by morphological alterations and neutral red absorption. Toxicol. Lett. 24, 119–124.

Brugnera, E., Georgiev, O., Radtke, F., Heuchel, R., Baker, E., Sutherland, G.R. and Schaffner, W. (1994) Cloning, chromosomal mapping and characterisation of the human metal-regulatory transcription factor MTF-1. Nucl. Acids Res., 22, 3167–3173.

Chan, H.M. and Cherian, M.G. (1992) Protective roles of metallothionein and glutathione in hepatotoxicity of cadmium. Toxicol.., 72, 281–290.

Chomczinski, P. and Sacchi, N. (1987) Single step method of RNA isolation by acid guanidinium thiocyanate-phenol-chloroform extraction. Anal. Biochem., 162, 156–159.

Dalton, T., Palmiter, R.D. and Andrews, G.K. (1994) Transcriptional induction of the mouse metallothionein-I gene in hydrogen peroxide-treated Hepa cells involves a composite major late transcription factor/antioxidant response element and metal response promoter elements. Nucl. Acids Res., 22, 5016–5023.

Dalton, T., Paria, B.C., Fernando, L., Huet-Hudson, Y.M., Dey, S.K. and Andrews, G.K. (1997) Activation of the chicken metallothionein promoter by metals and oxidative stress in cultured cells and transgenic mice. Comp. Biochem. Physiol., 116B, 75–86.

Friling, R.S. Bergelson, S. and Daniel, V. (1992) Two adjacent AP-1-like binding sites form the electrophile-response element of the murine glutathione S-transferase Ya subunit gene. Proc. Natl. Acad. Sci. USA, 89, 668–672.

Günes, C., Heuchel, R., Georgiev, O., Müller, K.H., Lichtlen, P., Blüthmann, H., Marino, S., Aguzzi, A. and Schaffner, W. (1998) Embryonic lethality and liver degeneration in mice lacking the metal-responsive transcriptional activator MTF-1. The EMBO J. (in press).

Hentsch, B., Lyons, I., Hartley, L., Lints, T.J., Adams, J.M. and Harvey, R.P. (1996) Hlx homeo box gene is essential for an inductive tissue interaction that drives expansion of embryonic liver and gut. Genes Dev., 10, 70–79.

Herrlich, P., Sachsenmeier, C., Radier-Pohl, A., Gebel, S., Blattner, C. and Rahmsdorf, H.J. (1994). The mammalian UV response: mechanism of DNA damage induced gene expression. Adv. Enzyme Reg., 34, 381–395.

Heuchel, R., Radtke, F., Georgiev, O., Stark, G., Aguet, M. and Schaffner, W. (1994) The transcription factor MTF-1 is essential for basal and heavy metal-induced metallothionein gene expression. EMBO J., 13, 2870–2875.

Heuchel, R., Radtke, F., and Schaffner, W. (1995) Transcriptional regulation by heavy metals, exemplified at the metallothionein genes. Inducible Gene Expression 1, Baeuerle P.A. (Editor), Birkhäuser Boston (Publisher), Vol I, 206–240.

Hilberg, F., Aguzzi, A., Howells, N. and Wagner, E.F. (1993) c-Jun is essential for normal mouse development and hepatogenesis. Nature, 365, 179–181.

Ishikawa, T., Bao, J-J., Yamane, Y., Akimaru, K., Frindrich, K., Wright, C.D. and Kuo, M.T. (1996) Coordinated expression of MRP/GS-X pump and γ-glutamylcysteine synthetase by heavy metals in human leukemia cells. J. Biol. Chem., 271, 14981–14988.

Iszard, M.B., Liu, J. and Klaassen, C.D. (1995) Effect of several metallothionein inducers on oxidative stress defense mechanisms in rats. Toxicol., 104, 25–33.

Kaegi, J.H.R. (1987) Chemistry and biochemistry of metallothionein. In: Experientia Suppl., 52, 25.

Kaegi, J.H.R. (1991) Overview of metallothionein. In: Methods Enzymol. New York: Academic Press.

Karin, M. (1997) AP-1 function and regulation. Curr. Op. Cell Biol., 9, 240–246.

Kern, S.R., Smith, H.A., Fontaine, D. and Bryan, S.A. (1981) Partitioning of Zinc and Copper in fetal liver subfractions: Appearance of metallothionein-like proteins during development. Toxicol. Appl. Pharmacol., 59, 346–354.

Lazo, J.S., Kundo, Y., Dellapiazza, D., Michalska, A.E., Choo, K.H.A. and Pitt, B.R. (1995) Enhanced sensitivity to oxidative stress in cultured embryonic cells from transgenic mice deficient in metallothionein I and II genes. J. Biol. Chem., 270, 5506–5510.

Lieberman, M.W., Wiseman, A.L., Shi, Z.Z., Carter, B.Z., Barrios, R., Ou, C-N., Chevez-Barrioz, P., Wang, Y., Habib, G.M., Goodman, J.C., Huang, S.L., Lebovitz, R.M. and Matzuk., M.M. (1996) Growth retardation and cysteine deficiency in γ-glutamyltranspeptidase-deficient mice. Proc. Natl. Acad. Sci. USA, 93, 7923–7926.

Liu, J-H., Miyakawa, H., Takano, T., Marumo, F. and Sato, C. (1995) Effects of cadmium on glutathione metabolism in HepG2 cells. Res. Comm. in Mol. Pathol. Pharmacol., 90, 143

Maellaro, E., Casini, A.F., Del Bello, B. and Comporti, M. (1990) Lipid peroxidation and antioxidant systems in the liver injury produced by glutathione depleting agents. Biochem. Pharmacol., 39, 1513–1521.

Masters, B.A., Kelly, E.J., Quaife, C.F., Brinster, R.L. and Palmiter, R.D. (1994) Targeted disruption of metallothionein I and II genes increases sensitivity to cadmium. Proc. Natl. Acad. Sci. USA , 91, 584–588.

Maret, W. (1995) Metallothionein/Disulfide interactions, oxidative stress, and the mobilization of cellular zinc. Neurochem. Int., 27, 111–117.

Meister, A. (1995) Mitochondrial changes associated wtih glutathione deficieny. Biochem. Biophys. Acta, 1271, 35–42.

Michalska, A.E. and Choo, K.H.A. (1993) Targeting and germ-line transmission of a null mutation at the metallothionein I and II loci in mouse. Proc. Natl. Acad. Sci. USA, 90, 8088–8092.

Mueller, P.R., Salser, S.J. and Wold, B. (1988) Constitutive and metal-inducible protein:DNA interactions at the mouse metallothionein I promoter examined by in vivo and in vitro footprinting. Genes Dev., 2, 412–427.

Mulchay, R.T. and Gipp, J.J. (1995) Identification of a putative antioxidant response element in the 5'-flanking region of the human γ-glutamylcysteine synthetase heavy subunit gene. Biochem. Biophys. Res. Com., 209(1), 227–233.

Ouellette, A.J. (1982) Metallothionein mRNA expression in fetal mouse organs. Dev. Biol., 92, 240–246.

Pahl, H.L. and Baeuerle, P.A. (1994) Oxygen and control of gene transcription. Bioassays, 16, 497–502.

Palmiter, R.D., Findley, S.D., Whitmore, T.E. and Durnam, D.M. (1992) MT-III, a brain-specific member of the metallothionein gene family. Proc. Natl. Acad. Sci. USA, 89, 6333–6337.

Palmiter, R.D. and Findley, S.D. (1995) Cloning and characterization of a mammalian zinc transporter that confers resistance to zinc EMBO J., 14, 639–649.

Quaife, C.J. et al., (1994) Induction of a new metallothionein isoform (MT-IV) occurs during differentiation of stratified squamous epithelia. Biochem., 33, 7250–7259.

Quaife, C.J., Hammer, R.E., Mottet, N.K. and Palmiter, R.D. (1986) Glucocorticoid regulation of metallothionein during murine development. Dev. Biol., 118, 549–555.

Radtke, F., Heuchel, R., Georgiev, O., Hergersberg, M., Gariglio, M., Dembic, Z. and Schaffner, W. (1993) Cloned transcription factor MTF-1 activates the mouse metallothionein I promoter. EMBO J., 12, 1355–1362.

Radtke, F., Georgiev, O., Müller, H-P. Brugnera, E. and Schaffner, W. (1995) Functional domains of the heavy metal-responsive transcription regulator MTF-1. Nucl. Acids Res., 23, 2277–2286.

Rahman, I., Smith, C.A.D., Lawson, M.F., Harrison, D.J. and MacNee, W. (1996) Induction of γ -glutamylcysteine synthetase by cigarette smoke is associated with AP-1 in human alveolar epithelial cells. FEBS Lett., 396, 21–25.

Sambrook, J., Fritsch, E.F. and Maniatis, T. (1989) A laboratory manual, second edition. Cold Spring Harbor Laboratory Press 1989.

Schmidt, C. et al., (1995) Scatter factor/hepatocyte growth factor is essential for liver development. Nature , 373, 699–702.

Schönthal, A., Herrlich, P., Rahmsdorf, HJ., Ponta, H., (1988) Requirement for fos gene expression in the transcriptional activation of collagenase by other oncogenes and phorbol esters. Cell 54, 325–334..

Schreiber, E., Matthias, P. Müller, M.M. and Schaffner, W. (1989) Rapid detection of octomer binding proteins with 'mini-extracts', prepared from a small number of cells. Nucl. Acids Res., 17, 6419.

Searle, P.F., Davison, B.L., Stuart, G.W., Wilkie, T.M., Norstedt, G. and Palmiter, R.D. (1984) Regulation, linkage, and sequence of mouse metallothionein I and II genes. Mol. Cell. Biol., 4, 1221–1230.

Searle, P.F., Stuart, G.W. and Palmiter, R.D. (1987) Metal regulatory elements of the mouse metallothionein-I gene. Experientia Suppl., 407–414.

Searle, P.F. (1990) Zinc dependent binding of a liver nuclear factor to a metal response element MRE-a of the mouse metallothionein-I gene and variant sequences. Nucl. Acids Res., 18, 4683–4690.

Sekhar et al., (1997) Expression of glutathione and γ-glutamylcysteine synthetase mRNA is jun dependent. Biochem.Biophys.Res.Com., 234, 588–593.

Sen, C.K. and Packer, L. (1996) Antioxidant and redox regulation of gene transcription. FASEB J ., 10, 709–720.

Shi, Z.Z. Habib, G.M., Rhead W.J., Gahl, W.A., He, X., Sazer, S. and Lieberman, M.W. (1996) Mutations in the glutathione synthetase gene cause 5-oxoprolinuria. Nature genetics 14, 361–365.

Shimizu, M. and Morita, S. (1990) Effects of fasting on cadmium toxicity, glutathione metabolism, and metallothionein synthesis in rats. Toxicol. Appl. Pharmacol., 103, 28–32.

Singhal, R.K., Anderson, M.E. and Meister, A. (1987) Glutathione, a first line of defense against cadmium toxicity. FASEB J., 1, 220–223.

Srivastava, R.C., Hasan, S.K., Gupta, J. and Gupta. S. (1993) Protective role of metallothionein in nickel induced oxidative damage. Biochem. Mol. Biol. Int., 30, 261–270.

Stuart, G.W., Searle, P.F. and Palmiter, R.D. (1985) Identification of multiple metal regulatory elements in mouse metallothionein-I promoter by assaying synthetic sequences. Nature, 317, 828–831.

Thomas, J.P., Bachowski, G.J. and Girotti, A.W. (1986) Inhibition of cell membrane lipid peroxidation by cadmium- and zinc-metallothioneins. Biochem. Biophys. Acta, 884, 448–461.

Thornally, P.J. and Vasak, M. (1985) Possible role for metallothionein in protection against radiation-induced oxidative stress. Kinetics and mechanism of its reaction with superoxide and hydroxyl radicals. Biochem. Biophys. Acta, 827, 36–44.

Tomonari, A. et al. (1997) Identification of cis-acting DNA elements of the human γ-glutamylcysteine synthetase heavy subunit gene. Biochem. Biophys. Res. Com., 232, 522–527.

Uchida, Y., Takio, K., Titani, K., Ihara, Y. and Tomonaga, M. (1991) The growth inhibitory factor that is deficient in the Alzheimer's Disease brain is a 68 amino-acid metallothionein-like protein. Neuron, 7, 337–347.

Uehara, Y., Minowa, O., Mori, C., Shiota, K., Kuno, J., Noda, T. and Kitamura, N. (1995) Placental defect and embryonic lethality in mice lacking hepatocyte growth factor/scatter factor. Nature , 373, 702–705.

Westin, G., Gerster, T., Müller, M.M., Schaffner, G. and Schaffner, W. (1987) OVEC, a versatile system to study transcription in mammalian cells and cell free extracts. Nucl.Acids Res., 15, 6787–6798.

Westin, G. and Schaffner, W. (1988) A zinc responsive factor interacts with a metal-regulated enhancer element (MRE) of the mouse metallothionein-I gene. The EMBO J., 7, 3763–3770.

Wilhelm, D., van Dam, H., Herr, I., Baumann, B., Herrlich, P. and Angel, P. (1995) Both ATF-2 and c-Jun are phosphorylated by stress-activated protein-kinases in response to UV-irradiation. Immunobiol., 193, 143–148.

Yao, K-S., Godwin, A.K., Johnson, S.W., Ozols, R.F., O'Dwyer, P.J. and Hamilton, T.C. (1995) Evidence for altered regulation of γ-glutamylcysteine synthetase gene expression among cisplatin-sensitive and cisplatin-resistant human ovarian cancer cell lines. Cancer Res., 55 , 4367–4374.

Zheng, H., Liu, J., Liu, Y. and Klaasen, C.D. (1996) Hepatocytes from metallothionein-I and II knock-out mice are sensitive to cadmium- and *tert*-butylhydroperoxide-induced cytotoxicity. Toxicol. Lett., 87, 139–145.

26

THE TRANSPORT AND INTRACELLULAR TRAFFICKING OF METAL IONS IN YEAST

Valeria Cizewski Culotta, Xiu Fen Liu, and Paul Schmidt

Department of Environmental Health Sciences
Johns Hopkins University School of Public Health
615 N. Wolfe Street
Baltimore, Maryland 21205

1. SUMMARY

The bakers yeast *Saccharomyces cerevisiae* has proven to be an excellent model organism in which to examine the regulation of metal homeostasis. Numerous genes involved in the transport, intracellular trafficking and detoxification of metals have been conserved between yeast and humans; furthermore, technological advances in yeast genetics and molecular biology have made it possible to rapidly identify such genes and gain an understanding of their function and regulation.

Two fundamental aspects of metal transport will be the focus of this discussion. First, the basis for regulation of the yeast metal transport protein, Smf1p, will be examined. Smf1p is a member of the well conserved Nramp family of eukaryotic metal transporters and it has been shown that yeast Nramp is regulated at the level of protein sorting and protein turn over. When metal ions such as manganese are in abundance, Smf1p is targeted to the vacuole for degradation in a manner dependent on the *S. cerevisiae BSD2* gene product. However when cells are starved for metals, Smf1p fails to arrive at the vacuole, the protein is stabilized and a fraction of Smf1p accumulates at the cell surface to mediate the high affinity uptake of metals. This regulation of Nramp transport activity by metal ions provides an effective method for controlling metal uptake in response to changes in environmental metal conditions. The second focus of study presented herein regards the intracellular trafficking of copper ions in yeast and humans by a family of newly described "copper chaperones". The copper chaperone for superoxide dismutase (SOD) is discussed in detail. This molecule, known as Lys7 in yeast and CCS in humans, has a bifunctional appearance. The amino terminal domain is homologous to the Atx1-family of copper binding proteins and is believed to represent the high affinity copper site needed to capture the metal even under copper limiting conditions. The carboxyl terminal domain

Metals and Genetics, edited by Sarkar.
Kluwer Academic / Plenum Publishers, New York, 1999.

exhibits significant homology to the target of copper transfer, SOD. Surprisingly, this carboxyl terminal region can activate SOD1 in trans on its own, albeit poorly and not at all under copper limiting conditions. We present a model in which the amino terminal domain of the copper chaperone for SOD is needed for efficient binding of copper ions and the carboxyl terminal region is needed to maintain specificity for the target SOD1 molecule.

2. INTRODUCTION

The transport and intracellular trafficking of heavy metals is accomplished at the cellular level by membrane associated and soluble metal binding proteins. Many of these molecules have been well conserved among eukaryotes. We have exploited the bakers yeast *S. cerevisiae* as a model system for the study of genes and proteins that control metal transport and trafficking. This discussion will be divided into two sections, one dealing with the regulated uptake of manganese and copper by the Nramp family of metal transport proteins, and the second dealing with the intracellular partitioning of copper ions.

2.1. The Nramp Family of Metal Transport Proteins

Nramp represents a family of metal transport proteins that have been well conserved in prokaryotes and eukaryotes. These proteins are highly hydrophobic, contain as many as 12 membrane spanning domains and a consensus "transport signature" sequence (Cellier et al., 1995). Nramp1 was originally discovered as the locus affected in the *Bcg/Lsh/Ity* mouse model for increased susceptibility to parasitic infection. Nramp1 is predominantly expressed in the phagolysosome of macrophages, and is believed to aid in the anti-oxidant defense against parasitic infection through the transport of redox active iron or manganese ions (Vidal et al., 1993; Cellier et al., 1996; Atkinson et al., 1997; Gruenheid et al., 1997). A second Nramp gene, Nramp2, has been identified in mammals and is expressed in a wide array of cell types. This transporter is mutated in both the microcytic anemia mouse model and in the anemic Belgrade rat model, demonstrating an important role for Nramp2 in iron transport and utilization (Fleming et al., 1988; Grunheid et al., 1995; Fleming et al., 1997). Recent electrophysiology studies have indicated that the Nramp proteins are largely non-specific for their substrates and can act on an array of essential and toxic heavy metals including iron, zinc, manganese, cobalt, cadmium, copper, nickle and lead (Gunshin et al., 1997).

The bakers yeast *S. cerevisiae* contains at least two members of the Nramp family, denoted Smf1p and Smf2p. These were initially discovered by Horwich and colleagues as multi copy suppressors of the *mif1* mutation (suppressors of **mif1**) affecting the processing of mitochondrial proteins (West et al., 1992). Subsequent studies by Nelson and co-workers identified Smf1p as a high affinity manganese transport protein (Supek et al., 1996). Over-expression of Smf1p suppresses *mif1* mutants by increasing the availability of the intracellular manganese needed for mitochondrial protein processing (Supek et al., 1996). The yeast Smf proteins share roughly 40% identity at the amino acid level with mammalian Nramp proteins; furthermore, Gros and colleagues have noted that mammalian Nramp2 will complement the metal transport defects of a *smf1 smf2* mutant yeast (Pinner et al., 1997). We observed that like their mammalian counterparts, the yeast Nramp proteins are relatively non-specific for their metal ion substrates. In addition to affecting manganese ion uptake, Smf1p effectively transports copper and cadmium and Smf2p is responsible for cobalt accumulation (Liu et al., 1997).

We recently discovered that the yeast Nramp proteins are negatively regulated by the product of the *S. cerevisiae BSD2* gene. *BSD2* was originally discovered as a gene which when mutated will suppress the oxidative damage associated with loss of the copper/zinc superoxide dismutase (SOD1). This suppression of oxidative stress in *bsd2* mutants was associated with hyper-accumulation of redox active metal ions such as copper (Liu and Culotta, 1994). *BSD2* does not encode a metal transporter, but rather a protein restricted to the endoplasmic reticulum that acts at a distance to negatively regulate the uptake of heavy metals from the growth medium. Through a genetic screen, we discovered that the targets of Bsd2p action include the Smf1p and Smf2p transport proteins (Liu et al., 1997). When Bsd2p is present, the contribution of these transporters to metal uptake is negligible, but when Bsd2p is absent, Smf1p and Smf2p are activated and cells accumulate high levels of copper, cadmium, manganese and cobalt from the growth medium (Liu et al., 1997). The precise mechanism by which Bsd2p negatively regulates the Smf proteins was unknown and is the focus of the current study. As described herein, Bsd2p negatively regulates Smf1p at the level of protein stability, by directing Smf1p to the vacuole for degradation by vacuolar proteases. Moreover, the presence of metal ions in the growth medium is the primary signal for the Bsd2-mediated degradation of Smf1p. Overall this provides an effective and rapid regulation for metal transport in response to changes in environmental metal conditions.

2.2. Chaperones for Copper

There are at least four distinct protein targets in the yeast cell that require copper for activity: 1) a cytosolic copper and zinc containing superoxide dismutase (SOD1); 2) a multi-copper oxidase needed for iron uptake (FET3) (Askwith et al., 1994); 3) cytochrome oxidase in the mitochondria and; 4) the Mac1p (Jungmann et al., 1993) and Ace1p (Thiele, 1988; Welch et al., 1989) copper binding transcription factors in the nucleus. All of these copper proteins can readily self-incorporate copper in vitro, however a quite different scenario is predicted for the copper proteins expressed in vivo. In eukaryotic cells, intracellular membranes can act as barriers for copper delivery. Furthermore, available copper is present in limiting concentrations in vivo and is not likely to be in free and exceedingly toxic ionic form. Therefore, accessory factors must exist to properly deliver the metal to the appropriate intracellular target and incorporate the metal into copper requiring proteins.

To overcome physical barriers for copper trafficking, eukaryotic cells have evolved with intracellular transporters that facilitate the delivery of copper across biological membranes. For example, copper delivery to Fet3p in the Golgi is mediated by a copper transporting P-type ATPase, Ccc2p (Yuan et al., 1995), whose human homologues include the WD and MNK proteins affected in Wilson and Menkes diseases of copper metabolism (Bull and Cox, 1994). To deal with limited copper availability, eukaryotes have evolved with a family of small, soluble copper binding proteins that we have referred to as "copper chaperones" (Valentine and Gralla, 1997). Copper chaperones act as escort proteins for the metal to ensure proper delivery of the metal ion to its intracellular target under limiting copper conditions.

The first copper chaperone discovered was Cox17p, a small soluble protein that acts to deliver the metal to the mitochondria where it is essential for the activation of cytochrome oxidase (Glerum et al., 1996). Glerum and Tzagaloff have noted that Cox17p binds two moles of copper and is located in both the cytoplasm and intramembrane space of the mitochondria (Beers et al., 1997). Cox17p presumably works in concert with the mitochondrial proteins Sco1p and Sco2p to charge cytochrome oxidase with copper

(Glerum et al., 1996). A human homologue to Cox17p has been identified and is believed to function in an analogous fashion to deliver copper to the mitochondria of mammalian cells (Amaravadi et al., 1997).

The second copper chaperone discovered was Atx1p (Lin and Culotta, 1995). Atx1p is a 8 kDa soluble protein that delivers copper specifically to the Golgi copper transporter Ccc2p, which in turn deposits copper on the terminal acceptor, Fet3p (Lin et al., 1997). Atx1p is a prototype of a large family of copper binding peptides (the "Atx1p or MerP family") that have in common a MH/TCXXC metal binding motif as well as extended homology throughout a 8 kDa Atx1-like region (Pufahl et al., 1997). Included in this family are the MerP periplasmic mercury carriers of enteric bacteria (Summers, 1986) and the copper binding domains of the P-type copper transporting ATPases (yeast Ccc2p and the human Wilson and Menkes disease proteins).

O'Halloran and colleagues have proposed a novel model for the action of copper delivery by Atx1p (Pufahl et al., 1997). This protein normally binds the metal in a di-thiol linear coordination using the dual cysteines of the MH/TCXXC motif. A third ligand to copper is also evident, and trigonal coordination is believed to represent the transient intermediate of the copper delivery reaction in which an intermolecular bridge is formed between Atx1p and Ccc2p (Pufahl et al., 1997). Consistent with this model, we found that Atx1p physically interacts with the metal binding domains of Ccc2p and that this interaction is dependent on physiological concentrations of copper (Pufahl et al., 1997). The pathway of copper delivery from Atx1p to Ccc2p to Fet3p is highly conserved between yeast and humans. J. Gitlin has isolated the human homologue of Atx1p, HAH1, which is believed to deliver copper to the human Wilson and Menkes copper transporters (the Ccc2p homologues) and ultimately to ceruloplasmin (Fet3p homologue) (Klomp et al., 1997).

Both Atx1p and Cox17p are highly specific for their targets and neither affect the delivery of copper to nuclear copper transcription factors or to the cytosolic copper and zinc containing SOD1. To date, the mechanism of copper transport to the yeast nucleus remains a mystery. However, we have recently uncovered the copper chaperone for SOD1. The identification and characterization of this new member of the copper chaperone family forms the basis for much of the studies presented herein.

3. RESULTS AND DISCUSSION

3.1. Negative Regulation of Nramp Metal Transport by the
S. cerevisiae BSD2 Gene

We have previously shown that the yeast Smf1p and Smf2p Nramp proteins contribute very little to metal uptake in wild type strains of yeast; however, when the *BSD2* gene is inactivated by mutation, the Smf1p and Smf2p transporters are activated and the uptake of copper, cobalt and cadmium from the growth medium increases dramatically (Liu et al., 1997). The focus of the current study was to understand the mechanism underlying the apparent negative control of yeast Nramp transport activity by the *BSD2* gene.

3.1.1. Bsd2p Down Regulates Smf1p at the Level of Protein Stability. First we examined whether *BSD2* controls Nramp activity at the level of gene transcription. By northern blot analysis, we noted that *SMF1* and *SMF2* mRNA levels are unchanged in wild type and *bsd2* mutant strains of yeast. To investigate possible post-transcriptional effects, we employed immuno detection methods to monitor Smf1 and Smf2 proteins levels. A hemagglutinin (HA) tag was introduced at the carboxyl terminus of these proteins and the resultant Smf1-HA and Smf2-HA proteins were found to be fully functional for complementing the corresponding *smf1* and *smf2* mutations. By western blot, we found that Smf1-HA and Smf2-HA levels increased 10–20 fold in a *bsd2* mutant compared to the corresponding wild type strain. To fully understand the mechanism for the Bsd2-repression of the Smf proteins, we focused on Smf1-HA that could be readily detected at physiological levels. We noted that Bsd2p negatively regulates Smf1 at the level of protein stability. Smf1-HA normally has a very short half life (< 10 minutes) in strains containing Bsd2p; however in a *bsd2* mutant, the half life of Smf1-HA increases to 1–2 hours.

There are two major pathways of protein degradation in yeast. One involves ubiquitin and the proteosome and the other involves proteases in the vacuole (yeast lysozome). To discern which of these is responsible for the rapid degradation of Smf1p, we monitored Smf1-HA stability in a *ubc7* mutant defective for the ubiquitin pathway (Jungmann et al., 1993) and in a *pep4* mutant defective for vacuolar proteases (Jones, 1991). We noted that Smf1-HA stability was slightly increased in the *ubc7* mutant, although the effects of combined *bsd2* and *ubc7* mutations were additive, demonstrating that Bsd2p does not work through the ubiquitin pathway to control degradation of Smf1p. In comparison, Smf1-HA was greatly stabilized in a *pep4* mutant and there were no additive effects of *pep4* and *bsd2* mutations. Hence, Bsd2p works through vacuolar proteases to mediate rapid degradation of Smf1p.

3.1.2. Localization of Smf1p. We next monitored the intracellular localization of Smf1-HA in wild type and *bsd2* mutant strains. We noted a very light staining pattern of Smf1-HA in wild type strains consistent with localization to the secretory pathway endoplasmic reticulum and Golgi apparatus. In a *bsd2* mutant, the staining was brilliant. The majority of Smf1-HA was still found in the secretory pathway, however a fraction of Smf1-HA could also be detected at the cell surface.

Since the western blot studies suggested rapid degradation of Smf1-HA in the vacuole, we additionally monitored Smf1-HA localization in a *pep4* mutant defective for vacuolar proteases. Under these conditions, very intense staining of Smf1-HA was observed in the lumen of the vacuole, consistent with the notion that the bulk of this Nramp transporter is targeted to the vacuole for degradation. A very different pattern of staining was observed in a *bsd2 pep4* double mutant. In this case, none of the protein reached the vacuole, and the bulk of Smf1-HA was detected in the secretory pathway and at the cell surface. Overall, these studies demonstrate that Bsd2p negatively regulates Smf1p by targeting this metal transporter to the vacuole for degradation by vacuolar proteases. In cells lacking *BSD2*, Smf1p never reaches the vacuole and a fraction of this transporter accumulates at the cell surface, resulting in elevated uptake of metals from the growth medium.

3.1.3. Post-Translation Regulation of Smf1p by Metal Ions. A question can be raised with regard to the physiological relevance of the rapid turnover of Smf1p. Why would the cell expend so much energy to synthesize this metal transporter just to degrade the protein in the vacuole? Presumably a condition exists in which the cell requires Smf1p, such as metal starvation.

By western blot analysis, we monitored Smf1-HA levels in a wild type strain containing Bsd2p under normal medium conditions and in medium specifically depleted of the heavy metals iron, manganese, zinc, copper and cobalt. We noted a very high level of Smf1-HA accumulation in metal depleted medium. The addition of manganese alone was sufficient to trigger down regulation of Smf1-HA. Iron also diminished Smf1-HA protein levels, while supplementation of zinc, copper and cobalt had no effect. We noted that the regulation of Smf1-HA by metals occurred at the level of protein stability. Smf1-HA was greatly stabilized by the depletion of heavy metals and the supplementation of manganese alone was sufficient to trigger the rapid degradation of Smf1-HA in wild type strains, but not in a *bsd2* mutant. Therefore, cells respond to metal starvation by stabilizing Smf1p, but when metal ions are in ample supply, the Nramp transporter is rapidly degraded in a Bsd2-dependent manner.

3.1.4. A Model for the Regulation of yeast Nramp Metal Transport by Metal Ions and the BSD2 Gene. Based on these recent observations, we propose a model for the regulation of the Smf1p metal transporter by the *BSD2* gene product and metal ions. This model is shown in Figure 1. In wild type strains grown in normal medium, metal ions are in ample supply and the Nramp metal transporter is down regulated to prevent the hyperaccumulation of toxic metals. This down regulation is mediated by the trafficking of Smf1p to the vacuole for degradation by vacuolar proteases. The sorting of Smf1p through the secretory pathway to the vacuole is dependent on the Bsd2 protein located in the endoplasmic reticulum. Bsd2p does not appear to be a general factor for trafficking vacuolar proteins because sorting of the vacuolar carboxyl peptidase Y protein (CPY) is normal in strains lacking Bsd2p (Luo and Chang, 1997). Consequently, we propose that Bsd2p, in an unknown manner specifically recognizes Smf1p and Smf2p and signals their delivery to the vacuole. Although the mechanism of this recognition event is not known, it apparently involves metal ions. Under conditions of metal starvation, Bsd2p fails to down regulate Smf1p and the transporter instead accumulates in the secretory pathway and also the plasma membrane where it mediates the uptake of metals from the growth medium. It is quite possible that Smf1p itself is the sensor for metals. Under metal replete conditions, the metal bound Smf1 protein adopts a conformation that is recognized by Bsd2p, whereas under metal starvation conditions, the apo Smf1 escapes Bsd2p detection and the protein is prevented from entering the vacuole. Regardless of the precise mechanism, this post-translation control of Smf1p allows the cell to respond to rapid changes in environmental metal conditions without the need for new protein synthesis.

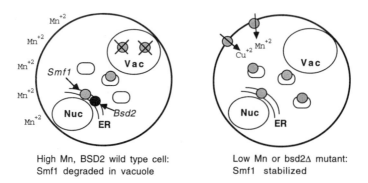

High Mn, BSD2 wild type cell:
Smf1 degraded in vacuole

Low Mn or bsd2Δ mutant:
Smf1 stabilized

Figure 1. Model for regulation of Smf1 metal transport by manganese ions and Bsd2. Under manganese replete conditions in a wild type *BSD2* cell, the bulk of Smf1 is targeted to the vacuole (VAC) for degradation by vacuolar proteases. When cells are starved for metals or when Bsd2 is missing, Smf1 fails to arrive at the vacuole and a fraction of this transport protein is directed to the plasma membrane where it participates in the uptake of heavy metals such as copper and manganese.

3.2. The Copper Chaperone for Superoxide Dismutase

As discussed in Introduction, copper requiring enzymes acquire their metal in vivo through the action of a family of low molecular weight, soluble "copper chaperones". The first of these identified, Atx1p and Cox17p were noted to be highly specific for their targets (Ccc2p and cytochrome oxidase, respectively) and neither affected the delivery of copper to copper/zinc SOD1 in the cytosol.

3.2.1. Cloning of the SOD1 Copper Chaperones from Yeast and Human. To identify the copper chaperone for SOD1, we utilized a genetic approach in yeast. We predicted that a mutant for the corresponding chaperone gene would have the same phenotypes as a *sod1Δ* mutant, since the SOD1 polypeptide is expected to be inactive in these strains. Yeast mutants lacking SOD1 exhibit a number of aerobic growth defects including auxotrophies (growth requirements) for the amino acids lysine and methionine (Bilinski et al., 1985; Chang and Kosman, 1990; Gralla and Kosman, 1992; Liu et al., 1992). We therefore inspected a number of previously characterized *lys* complementation groups for possibly encoding the copper chaperone for SOD1. We noted that one such mutant (*lys7Δ*) was not only auxotrophic for lysine, but also for methionine, and that this dependency on the amino acids was reversed by growth under anaerobic conditions (Culotta et al., 1997). The *LYS7* gene was cloned by Sprague's group (Horecka et al., 1995) and was reported to exhibit no homology to any known lysine biosynthetic enzyme. However, we noted some very intriguing features of the Lys7 polypeptide sequence.

As seen in Figure 2A, the amino terminal portion of Lys7 exhibits strong homology to Atx1p, including the well conserved MT/HCXXC copper binding motif. Therefore,

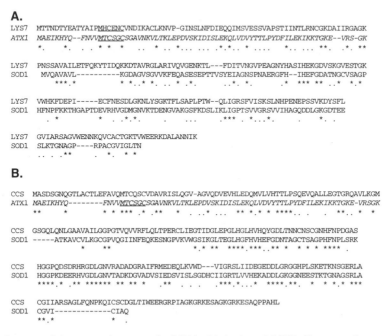

Figure 2. Alignment of the copper chaperones for SOD1 with Atx1p and SOD1. Shown are the yeast Lys7p (A) and human CCS (B) copper chaperone molecules aligned against yeast Atx1p (sequence in italics) and SOD1 from either yeast (A) or humans (B). Asterisks marked identical amino acids and dots indicate amino acid similarities.

Lys7p can be added to the list of the Atx1 family of metal binding proteins. Moreover, the carboxyl terminal portion of Lys7p exhibits homology to its target of copper delivery, SOD1. Therefore, Lys7p appeared to be a bi-functional molecule with a copper chaperone region fused to a SOD-like domain.

To test whether Lys7 is in fact the copper chaperone for SOD1, SOD1 activity was examined in strains lacking Lys7p. We noted that SOD1 polypeptides were present at wild type levels in a *lys7* mutant, however this SOD1 was inactive for superoxide scavenging (Culotta et al., 1997). Furthermore, Gitlin and colleagues noted that SOD1 was apo with regard to copper in strains lacking Lys7p (Culotta et al., 1997). Therefore, the genetic studies were consistent with a model in which Lys7p is the yeast copper chaperone for SOD1 and is absolutely required for SOD1 activation in vivo. These results were recently confirmed by O'Halloran and colleagues in vitro. This group noted that under copper limiting conditions, purified Lys7p will activate SOD1 in vitro without additional cellular factors.

By inspection of available data bases, we noted a partial human cDNA exhibiting homology to Lys7p, a clone previously described as human "SOD4". Gitlin and co-workers isolated the complete cDNA that exhibited roughly 30% identity to Lys7p, including the well conserved MT/HCXXC metal binding motif (Culotta et al., 1997). As seen in Fig. 2b, this human clone denoted CCS for copper chaperone for SOD1, contains both the Atx1 and SOD1 homology domains. The homology to SOD1 is rather striking and includes three out of the four copper binding ligands and all of the zinc binding ligands of SOD1. The human CCS clone expressed in yeast fully complements a *lys7* null mutation and is able to incorporate copper into yeast SOD1. Thus, CCS is indeed a copper chaperone for SOD1.

3.2.2. Separable Proteins Domains for the SOD1 Copper Chaperone. Based on sequence analysis, we predicted that the Lys7 and CCS copper chaperones consist of two distinct protein domains: a copper capturing Atx1-like domain in the amino terminus (referred throughout as "Lys7A") and a SOD1 homology domain in the carboxyl terminus (referred throughout as "Lys7B").

To address the two domain model genetically, we performed a domain swap experiment. We replaced the Lys7A domain of Lys7p with Atx1, creating a copper chaperone fusion molecule. As controls, we expressed the isolated Lys7A and Lys7B domains as independent molecules. All of these Lys7 derivative molecules accumulated in yeast cells to levels seen with full length Lys7p. We first tested whether the Atx1-Lys7B fusion and the isolated Lys7A domain could substitute for Atx1p in delivery of copper to Ccc2p. We found that the Atx1-Lys7B fusion was fully capable of complementing an *atx1* null mutation, demonstrating that Lys7B does not interfere with the ability of Atx1p to recognize its Ccc2p partner. However, the isolated Lys7A domain was unable to complement the *atx1* null mutation. In spite of the apparent homology between Atx1p and Lys7A, the specificity for delivering copper to Ccc2p is encoded by sequences unique to Atx1p.

We next tested whether the Lys7 derivative molecules could complement a *lys7* mutation. We found that the Atx1-Lys7B fusion was capable of complementing the *lys7* mutation, albeit poorly compared to the native Lys7 polypeptide. Surprisingly, the isolated Lys7B molecule also complemented the *lys7* mutation to the same degree. We repeated these experiments under copper limiting conditions using the copper-specific BCS (batho cuprione sulfonate) metal chelator to deplete medium copper. Under these conditions, the full length Lys7p molecule was still capable of complementing a *lys7* mutation, but the isolated Lys7B domain, as well as the Atx1-Lys7B fusion molecule, were completely inactive for supplying copper to SOD1. Hence, the isolated B domain of Lys7 can deliver copper to SOD1, albeit very inefficiently and not at all under copper limiting conditions. The

A domain of Lys7p serves to increase the efficiency of the copper chaperone reaction and insures copper delivery to SOD1 even under copper limiting conditions. In spite of the apparent homology between Atx1p and the A domain of Lys7, Atx1p cannot substitute for Lys7A in cooperating with Lys7B to facilitate copper delivery to SOD1, underscoring the exquisite specificity of the Atx1p and Lys7A protein domains.

3.2.3. A Model for the Copper Chaperone for SOD1. Together, these findings have enabled us to develop a model for the activation of SOD1 by the Lys7 and CCS copper chaperone molecules. The amino terminal A domain of the chaperones contains the high affinity MT/HCXXC copper binding site that efficiently captures copper either directly from the copper transporter Ctr1p (Dancis et al., 1994; Dancis et al., 1994), or from another intermediate protein yet to be determined. This A domain is crucial for binding the copper even under copper starvation conditions. However, the A domain has no specificity for SOD1, as it is unable to independently activate SOD1. Recognition of SOD1 requires the Lys7B domain that exhibits significant homology to SOD1 and is expected to form a transient heterodimer with SOD. The B domain of the chaperones is likely to contain a copper binding site, based on its homology to SOD1 and on the observation that isolated Lys7B can activate SOD1 in the absence of Lys7A. However, this copper binding site of Lys7B may be of poor affinity or may be buried within the molecule to preclude copper chaperoning under conditions of copper limitation. Therefore, we propose that copper first binds the A domain of the chaperone and upon interaction between the B domain and SOD1, the copper is either directly handed to SOD or transferred to SOD1 via the B domain of the chaperone. The validation of this model awaits extensive biochemical and structural analysis of this curious copper chaperone molecule.

ACKNOWLEDGMENTS

We are indebted to members of the Tom O'Halloran lab for sharing unpublished information with us. This work was funded by NIH grants GM50016 and ES08996 awarded to V.C.C. X.F.L. and P.S. are supported by training grant ES07141.

REFERENCES

Amaravadi, R., Glerum, D. M. and Tzagoloff, A. (1997). Isolation of a cDNA encoding the human homolog of COX17, a yeast gene essential for mitochondrial copper recruitment. Hum Genet 99, 329–333.

Askwith, C., Eide, D., V-Ho, A., Bernard, P. S., Li, L., Davis-Kaplan, S., Sipe, D. M. and Kaplan, J. (1994). The *FET3* gene of *S. cerevisiae* encodes a multicopper oxidase required for ferrous iron uptake. Cell 76, 403–410.

Atkinson, P. G., Blackwell, J. M. and Barton, C. H. (1997). Nramp1 locus encodes a 65 kDa interferon-gamma-inducible protein in murine macrophages. Biochem. J. 325, 779–786.

Beers, J., Glerum, D. M. and Tzagoloff, A. (1997). Purification, Characterization, and Localization of Yeast Cox17p, a Mitochondrial Copper Shuttle. J. Biol. Chem. 272, 33191–33196.

Bilinski, T., Krawiec, Z., Liczmanski, L. and Litwinska., J. (1985). Is hydroxyl radical generated by the fenton reaction in vivo? Biochem. Biophys. Res. Comm. 130, 533–539.

Bull, P. C. and Cox, D. W. (1994). Wilson disease and Menkes disease-new handles on heavy-metal transport. Trends Genet. 10, 246–252.

Cellier, M., Belouchi, A. and Gros, P. (1996). Resistance to intracellular infections: comparative genomic analysis of *Nramp*. Trends Genet 12, 201–204.

Cellier, M., Prive, G., Belouchi, A., Kwan, T., Rodrigues, V., Chia, W. and Gros, P. (1995). Nramp defines a family of membrane proteins. Proc. Natl. Acad. Sci. USA 92, 10089–10093.

Chang, E. and Kosman, D. (1990). O_2-dependent methionine auxotrophy in Cu,Zn Superoxide dismutase deficient mutants of *Saccharomyces cerevisiae*. J. Bacteriol. 172, 1840–1845.

Culotta, V. C., Klomp, L., Strain, J., Casareno, R., Krems, B. and Gitlin, J. D. (1997). The copper chaperone for superoxide dismutase. J. Biol. Chem. 272, 23469–23472.

Dancis, A., Haile, D., Yuan, D. S. and Klausner, R. D. (1994). The *Saccharomyces cerevisiae* copper transport protein (Ctr1p). J. Biol. Chem. 269, 25660–25667.

Dancis, A., Yuan, S., Haile, D., Askwith, C., Eide, D., Moehle, C., Kaplan, J. and Klausner, R. (1994). Molecular characterization of a copper transport protein in *S. cerevisiae*; an unexpected role for copper in iron transport. Cell 76, 393–402.

Fleming, M. D., Romano, M. A., Su, M. A., Garrick, L. M., Garrick, M. D. and Andrews, N. C. (1988). Nramp2 is mutated in the anemic Belgrade (b) rat: Evidence of a role for Nramp2 in endosomal iron transport. Proc. Natl. Acad. Sci USA 95, 1148–1153.

Fleming, M. D., Trenor, C. C., Su, M. A., Foernzler, D., Beier, D. R., Dietrich, W. F. and Andrews, N. C. (1997). Microcytic anaemia mice have a mutation in *Nramp2*, a candidate iron transporter gene. Nature Gen. 16, 383–386.

Glerum, D. M., Shtanko, A. and Tzagoloff, A. (1996). Characterization of *COX17*, a yeast gene involved in copper metabolism and assembly of cytochrome oxidase. J. Biol. Chem. 271, 14504–14509.

Glerum, D. M., Shtanko, A. and Tzagoloff, A. (1996). *SCO1* and *SCO2* act as high copy suppressors of a mitochondrial copper recruitment defect in *Saccharomyces cerevisiae*. J. Biol. Chem. 34, 20531–20535.

Gralla, E. B. and Kosman, D. J. (1992). Molecular genetics of superoxide dismutases in yeasts and related fungi. Adv. Genet. 30, 251–319.

Gruenheid, S., Pinner, E., Desjardins, M. and Gros, P. (1997). Natural resistance to infection with intracellular pathogens: The *Nramp1* protein is recruited to the membrane of the phagosome. J. Exp. Med. 185, 717–730.

Grunheid, S., Cellier, M., Vidal, S. and Gros, P. (1995). Identification and characterization of second mouse *Nramp* gene. Genomics 25, 514–525.

Gunshin, H., Mackenzie, B., Berger, U. V., Gushin, Y., Romero, M. F., Boron, W. F., Nussberger, S., Gollan, J. L. and Hediger, M. A. (1997). Cloning and characterization of a mammalian proton-coupled metal-ion transporter. Nature 388, 482–488.

Horecka, J., Kinsey, P. T. and Sprague, G. F. (1995). Cloning and characterization of the *Saccharomyces cerevisiae LYS7* gene: evidence for function outside of lysine biosynthesis. Gene 162, 87–92.

Jones, E. (1991). Three proteolytic systems in the yeast *Saccharomyces cerevisiae*. J. Biol. Chem. 266, 7963–7966.

Jungmann, J., Reins, H., Lee, J., Romeo, A., Hassett, R., Kosman, D. and Jentsch, S. (1993). MAC1, a nuclear regulatory protein related to Cu-dependent transcription factors is involved in Cu/Fe utilization and stress resistance in yeast. EMBO J. 13, 5051–5056.

Jungmann, J., Reins, H. A., Schobert, C. and Jentsch, S. (1993). Resistance to cadmium mediated by ubiquitin-dependent proteolysis. Nature 361, 369–371.

Klomp, L. W. J., Lin, S. J., Yuan, D., Klausner, R. D., Culotta, V. C. and Gitlin, J. D. (1997). Identification and functional expression of HAH1, a novel human gene involved in copper homeostasis. J. Biol. Chem. 272, 9221–9226.

Lin, S. and Culotta, V. C. (1995). The *ATX1* gene of *Saccharomyces cerevisiae* encodes a small metal homeostasis factor that protects cells against reactive oxygen toxicity. Proc. Natl. Acad. Sci. USA 92, 3784–3788.

Lin, S. J., Pufahl, R., Dancis, A., O'Halloran, T. V. and Culotta, V. C. (1997). A role for the *Saccharomyces cerevisiae ATX1* gene in copper trafficking and iron transport. J. Biol. Chem. 272, 9215–9220.

Liu, X. F. and Culotta, V. C. (1994). The requirement for yeast superoxide dismutase is bypassed through mutations in *BSD2*, a novel metal homeostasis gene. Mol. Cell. Biol. 14, 7037–7045.

Liu, X. F., Elashvili, I., Gralla, E. B., Valentine, J. S., Lapinskas, P. and Culotta, V. C. (1992). Yeast lacking superoxide dismutase: isolation of genetic suppressors. J. Biol. Chem. 267, 18298–18302.

Liu, X. F., Supek, F., Nelson, N. and Culotta, V. C. (1997). Negative control of heavy metal uptake by the *Saccharomyces cerevisiae BSD2* gene. J. Biol. Chem. 272, 11763–11769.

Luo, W. and Chang, A. (1997). Novel genes involved in endosomal traffic in yeast revealed by suppression of a targeting-defective plasma membrane ATPase mutant. J. Cell. Biol. 138, 731–746.

Pinner, E., Gruenheid, S., Raymond, M. and Gros, P. (1997). Functional complementation of the yeast divalent cation transporter family *SMF1* by *NRAMP2*, a member of the mammalian natural resistance-associated macrophage protein family. J. Biol. Chem. 272, 28933–28938.

Pufahl, R., Singer, C., Peariso, K. L., Lin, S. J., Schmidt, P., Fahrni, C., Culotta, V. C., Penner-Hahn, J. E. and O'Halloran, T. V. (1997). Metal ion chaperone function of the soluble Cu(I) receptor Atx1. Science 278, 853–856.

Summers, A. O. (1986). Organization, expression, and evolution of genes for mercury resistance. Ann. Rev. Microbiol. 40, 607–634.

Supek, F., Supekova, L., Nelson, H. and Nelson, N. (1996). A yeast manganese transporter related to the macrophage protein involved in conferring resistance to mycobacteria. Proc. Natl. Acad. Sci. USA 93, 5105–5110.

Thiele, D. (1988). Ace1 regulates expression of the *Saccharomyces cerevisiae* metallothionein gene. Mol. Cell. Biol. 8, 2745–2752.

Valentine, J. S. and Gralla, E. B. (1997). Delivering copper inside yeast and human cells. Science 278, 817–818.

Vidal, S. M., Malo, D., Vogan, K., Skamene, E. and Gros, P. (1993). Natural resistance to infection with intracellular parasites: isolation of a candidate for *Bcg*. Cell 73, 469–485.

Welch, J., Fogel, S., Buchman, C. and Karin, M. (1989). The *CUP2* gene product regulates the expression of the *CUP1* gene, encoding yeast metallothionein. EMBO J. 8, 255–260.

West, A. H., Clark, D. J., Martin, J., Neupert, W., Hart, F. U. and Horwich, A. L. (1992). Two related genes encoding extremely hydrophobic proteins suppress a lethal mutation in the yeast mitochondrial processing enhancing protein. J. Biol. Chem. 267, 24625–24633.

Yuan, D. S., Stearman, R., Dancis, A., Dunn, T., Beeler, T. and Klausner, R. D. (1995). The Menkes/Wilson disease gene homologue in yeast provides copper to a ceruloplasmin-like oxidase required for iron uptake. Proc. Natl. Acad. Sci., USA 92, 2632–2636.

27

MECHANISMS OF COPPER CHAPERONE PROTEINS

Robert A. Pufahl[1] and Thomas V. O'Halloran[1,2]

[1]Department of Chemistry
[2]Department of Biochemistry, Molecular, Biology and Cell Biology
Northwestern University
Evanston, Illinois 60208

1. INTRODUCTION

Little is known about the chemistry and biology involved in getting a metal such as copper (Cu) to the right place in a cell. Recently, several members of an emerging class of proteins have been proposed to play a role in guiding and protecting metal ions inside cells, and ultimately delivering these cofactors to specific targets. The copper escort function for one of the yeast proteins, Atx1, has been established (Pufahl et al., 1997). A similar function has been proposed for another yeast protein, Lys7, which shares some sequence similarity to Atx1, but is thought to deliver copper to another target within the cell (Culotta et al., 1997). This review summarizes recent insights into the biological chemistry of these two prototypical metal ion chaperone proteins.

These proteins clearly function in metal ion transport; however, the term *transporter* is the preferred name for membrane-bound proteins that play some role in translocating metal ions across lipid bilayers. We have suggested the term *metal ion chaperones*, or metallochaperones, be used to describe soluble or diffusible proteins with definable functions in intracellular metal ion trafficking. Likewise, many important cofactors in biology must be assembled by a variety of proteins and enzymes. Such proteins will likely constitute a class that is distinct from diffusible metal chaperones that is discussed in detail below. These could be described as *cluster assembly proteins* (Hausinger et al., 1996). These proteins may protect inorganic clusters from their environment and then guide them to their appropriate biological partners. Metal ion chaperones may also serve a protective role for the cell by preventing such clusters from autooxidation or binding to partially folded proteins. While the precise physiological functions are not yet established, examples in the cluster binding or assembly category may include the γ protein, which is involved in the in vitro activation of apodinitrogenase by FeMoCo, an iron-molybdenum-sulfide cluster

Metals and Genetics, edited by Sarkar.
Kluwer Academic / Plenum Publishers, New York, 1999.

365

Metal Ion Transport Proteins
     Membrane-bound receptors, permeases, channels, pumps, cation ATPases

Inorganic Cofactor Assembly Proteins
     FeS cluster assembly
     FeMoco cofactor maturation

Metal Ion Chaperone Proteins (Metallochaperones)
     Diffusable intracellular metal ion receptors: metal ion insertion into target
     proteins (metal ion trafficking)

Figure 1. Classes of proteins involved in the cell biology of metal ions.

(Homer et al., 1995), and HypB, a nickel binding protein required for reconstitution of hydrogenase activity (Fu et al., 1995). Analogous functions may be found for accessory proteins required in assembly of the copper enzyme tyrosinase (Chen et al., 1993) and the nickel enzyme urease (Hausinger, 1997).

We are examining several working models for the function of this emerging class of metallochaperone proteins. Features of any viable model must include the ability of the chaperone protein to (1) bind the metal ion tightly in the presence of competitor ligands and apo-proteins (2) prevent deleterious reactions at the coordinated metal center (3) diffuse through the cytoplasm and bind with moderate specificity to target sites (4) release its cargo preferentially to the target site and (5) finally liberate itself from the target upon unloading and diffuse back to a source of the metal ion. A key function of metallochaperones is not only the accomplishment of its primary goal, i.e. the specific delivery of the metal to the target site, but also the prevention of unintended redox reactions with other cellular components and the inadvertent insertion of metal ions into adventitious binding sites of other metals. The former is controlled by the proper design of a stabilizing metal ligand environment in the protein and the latter by specificity of protein-protein interactions.

2. ATX1: A COPPER CHAPERONE TARGETING INTRACELLULAR COMPARTMENTS

2.1. Copper and Iron Trafficking in Yeast

The past several years have revealed novel systems involving metal ion chaperone proteins to regulate the intracellular levels of copper in yeast. There have been three types of copper metallochaperones identified (Valentine and Gralla, 1997): (a) Atx1, which is involved in intracellular copper trafficking into the secretory pathway (Lin et al., 1997; Pufahl et al., 1997) (b) Lys7/CCS, the proteins required for activation of SOD1 by incorporation of copper into apoSOD (Culotta et al., 1997; Gamonet and Lauquin, 1998)and (c) Cox17, which is involved in copper transport into the mitochondria for cytochrome oxidase (Beers et al., 1997; Glerum et al., 1996; Srinivasan et al., 1998). This review will focus primarily on Atx1 and recent research on the SOD chaperones.

Most information to date has been gathered on copper trafficking into the yeast secretory pathway by a system that is comprised of several proteins that function together to regulate the concentration of not only copper but also iron in the cell (Askwith and

Kaplan, 1998). This is accomplished by copper import through the membrane-bound protein Ctr1 (Dancis et al., 1994). Copper is most likely to be transported into the cell as Cu(I) because of the presence of the cell surface reductases Fre1/Fre2 which can reduce Cu(II) to Cu(I) (Georgatsou et al., 1997; Hassett and Kosman, 1995). Cu(I) is subsequently bound by Atx1. The transfer of copper from Cu-Atx1 to its partner protein, apo-Ccc2, occurs in a step mediated through protein-protein interactions. Ccc2 is a membrane-bound P-type ATPase protein (Yuan et al., 1995) found on the surface of post-Golgi vesicles and contains two MXCXXC motifs in its N-terminal tail that are similar to the one found in Atx1 as well as those in the Menkes (ATP7A) and Wilson (ATP7B) disease proteins. This copper-dependent protein-protein recognition event was observed by yeast two-hybrid experiments using a Ccc2 fragment (Ccc2-a), containing the two N-terminal MXCXXC domains only (Pufahl et al., 1997). Copper is then incorporated into Fet3, a multicopper oxidase (Kosman et al., 1998). Fet3 is ultimately involved in iron uptake after translocation to the plasma membrane and complexation with an iron permease protein Ftr1 in a process not well understood (Stearman et al., 1996).

2.2. Parallels between Mercury and Copper

Studies of bacterial mercury resistance operons over the last 20 years have lead to detailed mechanisms for metal ion recognition, processing, and transport of Hg(II) (Hobman and Brown, 1997; Summers, 1986). These findings are now providing a basis for understanding the trafficking of both essential and toxic metal ions in higher organisms. There are two aspects to these comparisons. First, several domains of copper trafficking proteins in yeast and humans are homologous to the bacterial detoxification proteins (Bull and Cox, 1994; Cox, 1995). Second, Hg(II) coordination chemistry shares many similarities with that of Cu(I) (Utschig et al., 1995) and substitution of Hg into the copper binding sites of proteins can provide fundamental information on structure and coordination.

What can be learned from Hg-binding to these copper sites? There are very few spectroscopic handles for probing the coordination environment of Cu(I) centers in small molecules or proteins. Substitution of apo-forms of metalloproteins with spectroscopically active metals ions has revealed many important mechanistic and structural features for protein containing 'silent' metals. Unfortunately, few good substitutes for Cu(I) centers have emerged. The blue copper proteins provide an exception to this rule. A variety of metals can be substituted into the relatively rigid coordination site of these proteins. For example, spectroscopic and structural studies of Hg-substituted copper proteins have revealed that Hg(II) can directly adopt the geometry typically observed for the native copper protein (Huffman et al., 1997). The chemistry of Cu(I) exhibits many parallels to the chemistry of both Ag(I) and Hg(II). Of these two elements, mercury has the most useful NMR-active isotopes.

2.3. The Metal Binding Site of Atx1

The metal binding site of Atx1 has been studied by x-ray spectroscopy and [199]Hg-NMR spectroscopy (Pufahl et al., 1997). Atx1 was found to bind a single copper in the cuprous state in a thiol-rich ligand environment. K-edge x-ray absorption spectroscopy identified a single transition at 8984 eV, which corresponds to copper in the +1 oxidation state. Additionally, the peak height and shape were indicative of a trigonal environment consisting of sulfur ligands. EXAFS indicated the copper is bound by two cysteines at a short distance of 2.26 Å and a third sulfur ligand at a slightly longer distance of 2.40 Å.

EXAFS are most consistent with this third sulfur atom at a slightly longer distance than the other two. It is unlikely to be the thioether sulfur from the methionine (Met) side chain, although this point is still open. First, mutation of the methionine to leucine does not alter the high copy number suppression of the SOD null phenotype. The second argument against S-Met coordination is based on the NMR structures of MerP (Steele and Opella, 1997) and domain 4 of Menkes protein (Gitschier et al., 1998). Both exhibit significant sequence similarity to Atx1; however, the Met side chains are buried in the hydrophobic interiors of Menkes and MerP. In both of these cases, no ligation between Met and metal has been observed (Gitschier et al., 1998; Steele and Opella, 1997). Our results on Hg-Atx1 are also in agreement with these observations, as shown by EXAFS measurements and ^{199}Hg-NMR experiments (Pufahl et al., 1997). This rational leads to the conclusion that the third ligand in Cu-Atx1 is either a tightly bound small molecule thiol (dithiothreitol) or a cysteine from another Atx1 protein. Although dilute solutions of Atx1 appear to be monomeric in solution, the high concentrations of protein required in EXAFS may alter the oligomeric state of the protein. Of course, the structure of the Cu(I) forms of these proteins could be quite different from the Hg(II)- and Ag(I)-substituted versions. Also, S-Met coordination could be possible if these proteins underwent extensive conformational changes upon binding of Cu(I). There is no evidence to support this hypothesis. Although we are examining these possibilities in continuing studies, the simplest explanation that is consistent with the available data is still the coordination of an exogenous buffer thiolate.

2.4. Mechanism of Copper Transfer

Consideration of both the metal binding properties of the Cu(I)- and Hg(II)-substituted forms of Atx1 and the Cu-dependent interaction of Atx1 with Ccc2a have allowed a mechanistic proposal to be drawn that reflects the inorganic chemistry of the metal site and its biological function. Copper, most likely in the +1 oxidation state, is imported into the cell in a process requiring Ctr1. Cu(I) binds to apo-Atx1 in a subsequent step, although it is not yet clear whether or not Cu(I) is donated directly to Atx1 or whether or not there is the involvement of an intermediary protein or a small thiol-containing molecule such as glutathione. Based on the in vitro reactivity of Cu(I)-Atx1, this complex is envisioned to diffuse through the cytoplasm in a conformation that does not allow copper release or O_2 oxidation of the metal-thiolate center. In this conformation, the Cu(I) is coordinated to two Cys residues and to one additional atom that could come from either an exogenous molecule or from another side chain atom.

Upon encountering its target Ccc2, Cu-Atx1 associates with one or more of the cytosolic Atx1-like domains. Such an association may be stabilized by a copper-induced protein-protein interaction or through a Cu(I) bridge between Atx1 and Ccc2a. As a prelude to ligand exchange, we have proposed that the third S-atom donor discussed above dissociates from the Cu(I) center to give two-coordinate dithiolate metal center in Atx1. The Cu(I) then undergoes a series of rapid associative exchange reactions involving well-precedented two- and three-coordinate intermediates as shown in Figure 2. The coordination chemistry results indicate that the protein fold in Atx1 does not rigidly constrain the coordination environment to one or the other of these metal geometries.

The direction of copper movement along this reaction coordinate are anticipated to involve differences in the binding constants of each domain for Cu(I). Small differences in Cu(I)-affinity could arise in several ways including coupling of protein folding free energy and the metal-binding free energy. We suggest that such a series of increasing Cu-binding

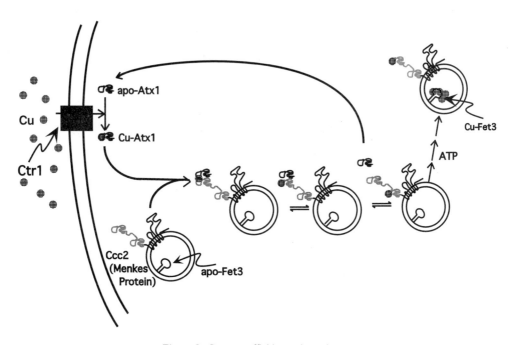

Figure 2. Copper trafficking pathway in yeast.

affinities, when combined with a role for protein-protein recognition between specific Cu-donor and acceptor proteins, constitute a testable mechanism for a diffusion-driven movement of the Cu(I) from one site in the cell to another. Such a flow is basically a vectorial process with the Cu(I) ion being directly and rapidly passed down a local thermodynamic gradient from weaker Cu(I) binding sites to tighter ones, without being released to the cytoplasm. While ATP has not been invoked for these metal transfer steps, it is most likely required in order for the P-type ATPase, Ccc2, or the human homologue the Menkes disease protein, to move the Cu(I) into the secretory vesicle where it is ultimately incorporated into Fet3. If the free energy changes for each of these transfer steps are coupled, ATP hydrolysis then provides the ultimate driving force for movement of Cu(I) into specific sites.

2.5. Implications for the Copper Chemistry of Menkes and Wilson Disease Proteins

The structural and mechanistic information on Atx1-mediated copper transfer to Ccc2 has significant consequences for the potentially more complex systems of the Wilson and Menkes N-terminal domains. The solution structure of a single Atx1-like domain of Menkes disease protein (domain 4) has been solved (Gitschier et al., 1998) and copper binding to a N-terminal peptide containing all six MXCXXC domains has been established (Lutsenko et al., 1997). Similar copper binding experiments have been carried out with Wilson disease protein (DiDonato et al., 1997). Mutations of pairs of Cys in each of the Atx1-like domains has been probed for both humans and yeast forms of the transporter (Iida et al., 1998; Payne and Gitlin, 1998). Not all of the domains are required for ATPase function. At least two roles for the N-terminal domains have been proposed. One proposal is that the Atx1-like domains are copper sensors that function in a regulatory manner that governs trafficking of the protein between the plasma membrane and trans Golgi-like vesicles (Dameron, unpublished results). Another possible role involves the accumulation of copper from chaperone proteins and the subsequent transfer of this copper to the membrane-bound domain of the transporter. This step would be a prelude to the transfer of the copper ion across the membrane. In this scenario, the multiple copper binding domains would accept copper ions donated by the chaperone. In the case of ATP7A, the transporter would receive Cu(I) from the human homologue of Atx1, namely HAH1 (Klomp et al., 1997). This hand-off function would effectively concentrate copper at the actual transport site eliminating the need for copper diffusion to the pumphead. In the simplest model, each of the domains are redundant and would act independently to move copper from the chaperone to the pumphead. More complex scenarios would involve a bucket brigade in which Cu(I) ions move from one domain to another. A higher order structure involving Cu-thiolate bridges between these domains has also been proposed by Dameron and coworkers. Finally, it is possible that subsets of the Atx1-like domains carry out alternate functions, including some of those described above.

Atx1-like domains have also been observed in the copper chaperones for superoxide dismutase (Lys7 and CCS) in yeast and humans (Culotta et al., 1997; Gamonet and Lauquin, 1998). The Cu(I)-thiolate chemistry described for the MXCXXC class of proteins (Atx1, Ccc2, ATP7A, ATP7B) is likely to hold for Lys7 and CCS, although these SOD1 chaperones promise to be additionally intriguing because of the presence of other metal sites described below. In fact, these metal sites provide interesting mechanistic possibilities in the activation chemistry of apoSOD. Mechanistic proposals of these chaperones are consistent with multistep metal transfer mechanisms described below.

3. LYS7: SOD1 COPPER CHAPERONE

3.1. Identification of the *LYS7* Gene

Culotta and coworkers identified Lys7 as a copper chaperone protein for yeast CuZnSOD using a genetic approach (Culotta et al., 1997). They reasoned that a yeast mutant of the chaperone gene would exhibit the same phenotype as a *sodΔ1* mutant Yeast mutants lacking SOD1 have a number of aerobic growth defects including auxotrophies for both lysine and Met (Gralla and Kosman, 1992; Liu et al., 1992). Through complementation studies, they identified one mutant (*lys7Δ*) that satisfied their requirements for a SOD1 chaperone protein. The *LYS7* gene had been previously identified by Sprague and coworkers during their study of genes involved in the lysine biosynthetic pathway (Horecka et al., 1995).

LYS7 was found to be essential for SOD1 activity. SOD1 activity was examined on nondenaturing polyacrylamide gels using nitroblue tetrazolium (NBT) staining. Cell lysates from the *lys7Δ* mutant did not show SOD1 activity, although levels of SOD1 protein in this strain were normal by Western blotting experiments. Next, to determine whether or not the lack of activity was due to a failure of copper incorporation into SOD1, yeast strains were metabolically labeled with radioactive copper. Wild type cells displayed a single band after electrophoresis corresponding to SOD1. However, *lys7Δ* mutants did not contain radioactivity in the SOD1 protein, demonstrating that copper incorporation did not occur in this strain. CCS, the homologous human copper chaperone protein for SOD, was able to restore SOD1 activity (by NBT staining) when *lys7Δ* mutants were provided a plasmid containing the human chaperone gene.

3.2. The Lys7 Protein

LYS7 encodes for a protein of 249 amino acids with a molecular weight of 27.3 kDa. Sequence analysis of the protein reveal homologies to two other copper chaperone proteins found in humans (CCS) and Arabidopsis thaliana. These proteins are 274 and 256 amino acids in length, respectively. Lys7 shows 26% identity and 38% similarity with the human chaperone CCS and 24% identity and 32% similarity with the chaperone from Arabidopsis. Interestingly, Lys7 and the other SOD1 copper chaperones display homologies with Atx1 in their N-termini (domain A) and superoxide dismutases in their C-termini (domain B). The presence of a third region of homology (domain C) is observed at the C-terminus. This short sequence, which is also observed in CCS, is not found in SOD; however, its significance is not known.

The well-conserved MT/HCXXC motif found in the Atx1 family of copper chaperones and in the copper transporting P-type ATPases (Ccc2, Menkes and Wilson proteins) is present in the A-domain of Lys7. The B-domain of Lys7 exhibits a strong similarity with SOD. However, only two out the seven copper and zinc ligands of ySOD are conserved in Lys7: Lys7 H134 and D164 equivalent to SOD H48 and D83. There are, nonetheless, potential metal ligands in Lys7 near the positions of the ligands found in SOD (Figure 3). This contrasts to CCS, in which all of the copper and zinc ligands are conserved except that SOD H120 is D201 in CCS.

Lys7 was overexpressed in *E. coli* and purified to homogeneity. The purified protein runs as a single band (~27 kDa) on SDS-PAGE. The intriguing sequence similarities lead to predicted domain functions in the protein. This possibility was tested using limited proteolytic digests with trypsin to determine if Lys7 displayed a bi-domain structure. Incubation of apo-Lys7 and CuZn-Lys7 with limiting amounts of trypsin gave different results (data not shown). Apo-Lys7 was more resistant to digestion with a significant amount of starting protein undigested even after long incubation periods. Interestingly, the metallated form of Lys7 clipped over a relatively short period. In both experiments, the major digestion product was a protein of approximately 17 kDa (by SDS-PAGE and electrospray spray-mass spectrometry), which roughly corresponded to the SOD domain (domain B).

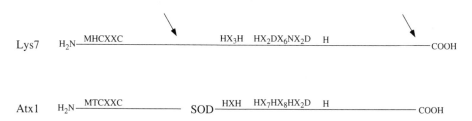

Figure 3. Alignment of Lys7 with Atx1 and ySOD. Lys7 shows a bi-domain structure with a N-terminal Atx1-like domain and a C-terminal SOD-like domain. Potential metal binding residues are shown and proteolytic cleavage sites by trypsin are also indicated by the arrows.

Domain A was also observed on SDS-PAGE, although this product was subsequently degraded over the longer time course, probably due to the basic character of its Atx1-like domain. The digestion products will be characterized further.

4. OVERVIEW OF COPPER TRANSFER MECHANISMS

The mechanism of copper transfer from the chaperone protein to SOD is of great importance and the individual domains of Lys7 will prove invaluable in dissecting this mechanism. In preliminary experiments, Lys7 alone was sufficient to act as a copper chaperone in vivo. We have observed the reconstitution of apoSOD activity in the cytochrome c assay using CuZn-Lys7 full-length protein in the presence of EDTA (data not shown). It is an open question whether or not other accessory proteins are required in this activation. However, the most basic mechanistic questions remain open issues. For instance, does the transfer of copper from a loaded chaperone protein occur via domain A to domain B and then to apoSOD or is only one of the domains of Lys7 sufficient for SOD activation? One working model requires that domain B of the chaperone protein undergoes heterodimer formation with monomeric apo-SOD in order to facilitate Cu(I) transfer. Copper transfer could then either occur directly to apoSOD via domain A or, alternatively, copper could move through domain B and then into SOD. Recent results of Culotta and coworkers indicate that domain B alone can weakly complement a *lys7* mutation and, therefore, it is plausible that copper transfer involves domain B at some point. Domain A provided in trans can augment this process. The chemical steps involved in copper transfer from chaperone protein domain B to SOD are not understood. The role of zinc in this activation process is also unknown. Is zinc present in SOD before copper transfer occurs or is zinc transfer mediated in some manner by Lys7?

This model and others will be examined thoroughly in the future. Undoubtedly, findings from the Atx1-Ccc2 story will continue to drive new thinking in the copper chemistry of SOD activation by Lys7/CCS. But this field deserves merit on its own, not only because of the central role SOD occupies in metallobiochemistry, but due to the new significance of SOD1 as a target of copper chaperones.

ACKNOWLEDGMENTS

This work was funded by NIH grant GM-54111 and post-doctoral grant F32 DK-09305 (RAP). We thank T. Rae, D. Huffman, and C. Singer for helpful discussions.

REFERENCES

Askwith, C., and Kaplan, J. (1998). Iron and copper transport in yeast and its relevance to human disease. TIBS 23, 135–138.

Beers, J., Glerum, D. M., and Tzagoloff, A. (1997). Purification, characterization, and localization of yeast Cox17p, a mitochondrial copper shuttle. J. Biol. Chem. 272, 33191–33196.

Bull, P. C., and Cox, D. W. (1994). Wilson disease and Menkes disease: new handles on heavy-metal transport. TIG 10, 246–252.

Chen, L.-Y., Chen, M.-Y., Leu, W.-M., Tsai, T.-Y., and Lee, Y.-H. W. (1993). Mutational study of *Streptomyces* tyrosinase *trans*-activator MelC1. J. Biol. Chem. 268, 18710–18716.

Cox, D. W. (1995). Genes of the copper pathway. Am. J. Hum. Genet. 56, 828–834.

Culotta, V. C., Klomp, L. W. J., Strain, J., Casereno, R. L. B., Krems, B., and Gitlin, J. D. (1997). The copper chaperone for superoxide dismutase. J. Biol. Chem. 272, 23469–23472.

Dancis, A., Haile, D., Yuan, D. S., and Klausner, R. D. (1994). The *Saccharomyces cerevisiae* copper transport protein (Ctr1p). J. Biol. Chem. 269, 25660–25667.

DiDonato, M., Narindrasorasak, S., Forbes, J. R., Cox, D. W., and Sarkar, B. (1997). Expression, purification, and metal binding properties of the N-terminal domain from the Wilson disease putative copper-transporting ATPase (ATP7B). J. Biol. Chem. 272, 33279–33282.

Fu, C., Olson, J. W., and Maier, R. J. (1995). HypB protein of *Bradyrhizobium japonicum* is a metal-binding GTPase capable of binding 18 divalent nickel ions per dimer. Proc. Natl. Acad. Sci. U.S.A. 92, 2333–2337.

Gamonet, F., and Lauquin, G. J.-M. (1998). The *Saccharomyces cerevisiae LYS7* gene is involved in oxidative stress protection. Eur. J. Biochem. 251, 716–723.

Georgatsou, E., Mavrogiannis, L. A., Fragiadakis, G. S., and Alexandraki, D. (1997). The yeast Fre1p/Fre2p cupric reductases facilitate copper uptake and are regulated by the copper-modulated Mac1p activator. J. Biol. Chem. 272, 13786–13792.

Gitschier, J., Moffat, B., Reilly, D., Wood, W. I., and Fairbrother, W. J. (1998). Solution structure of the fourth metal-binding domain from the Menkes copper-transporting ATPase. Nat. Struct. Biol. 5, 47–54.

Glerum, D. M., Shtanko, A., and Tzagoloff, A. (1996). Characterization of COX17, a yeast gene involved in copper metabolism and assembly of cytochrome oxidase. J. Biol. Chem. 271, 14504–14509.

Gralla, E. B., and Kosman, D. J. (1992). Molecular genetics of superoxide dismutases in yeasts and related fungi. Adv. Genet. 30, 251–319.

Hassett, R., and Kosman, D. J. (1995). Evidence for Cu(II) reduction as a component of copper uptake by Saccharomyces cerevisiae. J. Biol. Chem. 270, 128–134.

Hausinger, R. P. (1997). Metallocenter assembly in nickel-containing enzymes. JBIC 2, 279–286.

Hausinger, R. P., Eichhorn, G. L., and Marzilli, L. G. (1996). Mechanisms of metallocenter assembly. In: Eichhorn, G. L., Marzilli, L. G. (Eds.) "Advances in Inorganic Biochemistry." VCH, New York.

Hobman, J. L., and Brown, N. L. (1997). Bacterial mercury-resistance genes. Metals Ions in Biological Systems. 34, 527–568.

Homer, M. J., Dean, D. R., and Roberts, G. P. (1995). Characterization of the γ protein and its involvement in the metallocluster assembly and maturation of dinitrogenase from *Azotobacter vinelandii*. J. Biol. Chem. 270, 24745–24752.

Horecka, J., Kinsey, P. T., and Sprague, G. F. (1995). Cloning and characterization of the Saccharomyces cerevisiae *LYS7* gene: evidence for function outside of lysine biosynthesis. Gene 162, 87–92.

Huffman, D. L., Utschig, L. M., and O'Halloran, T. V. (1997). Mercury-responsive gene regulation and mercury-199 as a probe of protein structure. In: Sigel, A., Sigel, H. (Eds.) "Mercury and Its Effects on Environment and Biology." Marcel Dekker, Inc., New York, pp. 503–525.

Iida, M., Tereda, K., Sambongi, Y., Wakabayashi, T., Miura, N., Koyama, K., Futai, M., and Sugiyama, T. (1998). Analysis of functional domains of Wilson-disease protein (ATP7B) in *Saccharomyces cerevisiae*. FEBS Lett. 428, 281–285.

Klomp, L. W. J., Lin, S.-J., Yuan, D. S., Klausner, R. D., Culotta, V. C., and Gitlin, J. D. (1997). Identification and functional expression of HAH1, a novel human gene involved in copper homeostasis. J. Biol. Chem. 272, 9221–9226.

Kosman, D. J., Hassett, R., Yuan, D. S., and McCracken, J. (1998). Spectroscopic characterization of the Cu(II) sites in the Fet3 protein, the multinuclear copper oxidase from yeast required for high-affinity iron uptake. J. Am. Chem. Soc. 120, 4037–4038.

Lin, S.-J., Pufahl, R. A., Dancis, A., O'Halloran, T. V., and Culotta, V. C. (1997). A role for the *Saccharomyces cerevisiae ATX1* gene in copper trafficking and iron transport. J. Biol. Chem. 272, 9215–9220.

Liu, X. F., Elashvili, I., Gralla, E. B., Valentine, J. S., Lapinskas, P., and Culotta, V. C. (1992). Yeast lacking superoxide dismutase. Isolation of genetic suppressors. J. Biol. Chem. 267, 18298–18302.

Lutsenko, S., Petrukhin, K., Cooper, M. J., Gilliam, C. T., and Kaplan, J. H. (1997). N-terminal domains of human copper-transporting adenosine triphosphatases (the Wilson's and Menkes disease proteins) bind copper selectively in vivo and in vitro with stoichiometry of one copper per metal-binding repeat. J. Biol. Chem. 272, 18939–18944.

Payne, A. S., and Gitlin, J. D. (1998). Functional expression of the Menkes disease protein reveals common biochemical mechanisms among the copper-transporting P-type ATPases. J. Biol. Chem. 273, 3765–3770.

Pufahl, R. A., Singer, C. P., Peariso, K. L., Lin, S.-J., Schimdt, P., Culotta, V. C., Penner-Hahn, J. E., and O'Halloran, T. V. (1997). Metal ion chaperone function of the soluble Cu(I) receptor Atx1. Science 278, 853–856.

Srinivasan, C., Posewitz, M. C., George, G. N., and Winge, D. R. (1998). Characterization of the copper chaperone Cox17 of Saccharomyces cerevisiae. Biochemistry 37, 7572–7577.

Stearman, R., Yuan, D. S., Yamaguchi-Iwai, Y., Klausner, R. D., and Dancis, A. (1996). A permease-oxidase complex involved in high-affinity iron uptake in yeast. Science 271, 1552–1557.

Steele, R. A., and Opella, S. J. (1997). Structures of the reduced and mercury-bound forms of MerP, the periplasmic protein from the bacterial mercury detoxification system. Biochemistry 36, 6885–6895.

Summers, A. O. (1986). Organization, expression, and evolution of genes for mercury resistance. Ann. Rev. Microbiol. 40, 607–634.

Utschig, L. M., Bryson, J. W., and O'Halloran, T. V. (1995). Mercury-199 NMR of the metal receptor site in MerR and its protein-DNA complex. Science 268, 380–385.

Valentine, J. S., and Gralla, E. B. (1997). Delivering copper inside yeast and human cells. Science 278, 817–818.

Yuan, D. S., Stearman, R., Dancis, A., Dunn, T., Beeler, T., and Klausner, R. D. (1995). The Menkes/Wilson disease gene homologue in yeast provides copper to a ceruloplasmin-like oxidase required for iron uptake. Proc. Natl. Acad. Sci. U.S.A. 92, 2632–2636.

28

TRANSCRIPTIONAL REGULATION OF THE GENE ENCODING MOUSE METALLOTHIONEIN-3

R. Faraonio,[1] P. Moffatt,[1] O. LaRochelle,[1] R. Saint-Arnaud,[2] and C. Séguin[1]

[1]Centre de Recherche en Cancérologie de l'Université Laval, CHUQ
l'Hôtel-Dieu de Québec, Québec, Canada G1R 2J6
[2]Shriners Hospital for Children, Genetics Unit
Montréal, Québec, Canada H3G 1A6

1. INTRODUCTION

Metallothioneins (MTs) are small cysteine-rich proteins that bind transition metal ions such as Cd^{2+}, Zn^{2+}, and Cu^+ (metals) (Andrews, 1990; Suzuki et al., 1993; Moffatt and Denizeau, 1997). All vertebrates examined contain two or more distinct MT isoforms which are grouped into four classes, MT-1 through MT-4. MTs have been identified in a wide range of species and are present in various tissues and cell types from yeast to humans. The genes encoding MTs are inducible at the transcriptional level by a wide variety of agents, including metals, hormones, cytokines and reactive oxygen species. Metals are the most general and potent of these inducers. Metal activation of *MT* gene transcription is dependent on the presence of *cis*-acting metal regulatory elements (MREs) which are present in six non-identical copies (MREa through MREf) in the 5' flanking region of the mouse *MT-1* gene. The highly conserved MRE core sequence, 5'-TGCRCNC-3' (R, purine; N, any nucleotide), is necessary and sufficient for induction by metals (Imbert et al., 1990). At least two MREs are required for efficient metal induction and they can be present either in tandem or opposite orientations. In addition, in the mouse *MT-1* promoter, MREs have different transcriptional efficiencies. MREd is the strongest, MREa and MREc are 50 to 80% weaker, MREb is very weak and MREe and MREf are apparently nonfunctional (Stuart et al., 1985). In addition to MREs, *MT* gene promoters contain binding sites for the transcription factors Sp1, USF/MLTF, AP-1, and AP-2 which contribute to their basal transcriptional efficiency (Imbert et al., 1990).

MTF-1 is a 72.5 kDa Zn-finger protein that binds to the mouse MREd element in a Zn^{2+}-dependent manner (Westin and Schaffner, 1988; Radtke et al., 1993; Radtke et al., 1995). MTF-1 gene knock out by homologous recombination in ES cells showed that

Metals and Genetics, edited by Sarkar.
Kluwer Academic / Plenum Publishers, New York, 1999.

MTF-1 is essential for both basal and metal-induced *MT* gene transcription (Heuchel et al., 1994). Indeed, in MTF-1 null mutant ES cells, the endogenous *MT* genes are silent both before and after treatment of the cells with metals. MTF-1 is required for activation of *MT* gene expression by all inducing metals (Heuchel et al., 1994). Embryos lacking MTF-1 fail to transcribe *MT-1* and *MT-2* and die *in utero* at approximately day 14 of gestation, showing liver decay and generalized edema (Gunes et al., 1998).

MT-3 was first identified in the brain as a growth inhibitory factor, called GIF, which antagonizes the survival and neurite outgrowth of rat cortical neurons cultured in the presence of a brain extract (Uchida et al., 1991). The cloning of human (Palmiter et al., 1992; Tsuji et al., 1992), mouse (Palmiter et al., 1992) and rat (Kobayashi et al., 1993) cDNAs and genes, demonstrated that GIF is a new member of the MT family now called MT-3. Reports of decreased levels of MT-3 protein and RNA in individuals with Alzheimer's disease (AD) (Uchida et al., 1991; Tsuji et al., 1992) suggested that inhibition of *MT-3* gene expression might be involved in pathological processes of this disease. However, other studies showed that neither MT-3 mRNA nor MT-3 protein levels were significantly decreased in the AD patients studied (Erickson et al., 1994; Amoureux et al., 1997) and thus, the potential involvement of MT-3 in AD is still an open question.

Contrary to other *MT* genes, which are expressed in most tissues and which are highly inducible by metals, *MT-3* is predominantly expressed in the central nervous system (Uchida et al., 1991; Palmiter et al., 1992; Tsuji et al., 1992; Kobayashi et al., 1993; Masters et al., 1994; Aschner et al., 1997) and the organs of the reproductive system (Moffatt and Séguin, 1998), and it fails to respond to metals *in vivo* (Palmiter et al., 1992; Dalton et al., 1995; Liang et al., 1996; Kramer et al., 1996; Moffatt and Séguin, 1998). Immunohistochemical studies showed that within the normal brain, MT-3 is found in subset of astrocytes (Uchida et al., 1991; Zheng et al., 1995; Kramer et al., 1996; Yamada et al., 1996; Belloso et al., 1996; Hozumi et al., 1996) and neurons (Masters et al., 1994; Choudhuri et al., 1995; Yamada et al., 1996; Kramer et al., 1996; Belloso et al., 1996; Hozumi et al., 1996). It is abundantly present in a group of neurons of the hippocampus that sequester Zn^{2+} in synaptic vesicles within their terminals (Zn-ergic neurons) (Masters et al., 1994). In human testes, MT-3 mRNA is present in both the Leydig cells and the seminiferous tubules (Moffatt and Séguin, 1998).

The exact function of MT-3 and the mechanism by which it may contribute to the pathogenesis of AD are still unclear. MT-3 null mutant mice appear phenotypically normal (Erickson et al., 1997), which indicates that MT-3 is not essential for development and survival. However, MT-3 null mice are more susceptible to seizure induced by kainic acid and subsequently exhibit greater neuron injury (Erickson et al., 1997), thus suggesting that MT-3 plays important protection role in particular stress conditions. Furthermore, ectopic expression of MT-3 causes pancreatic acinar cell necrosis in transgenic mice (Quaife et al., 1998). In cultured cells, MT-3 displays a number of effects. The growth inhibitory activity of MT-3 on cultured rat cortical neurons has been observed by several groups (Uchida et al., 1991; Tsuji et al., 1992; Erickson et al., 1994; Uchida and Ihara, 1995; Sewell et al., 1995). In addition, MT-3 confers protection against serum-free exposure of CHO-K1 cells (Amoureux et al., 1995), and decreases growth rate of C6-glial cells (Amoureux et al., 1995). *In vivo*, MT-3 levels are reduced following nerve transection (Yuguchi et al., 1995), brain stab wounds (Hozumi et al., 1995; Hozumi et al., 1996), cerebral artery occlusion (Inuzuka et al., 1996), and brain focal ischemia (Yuguchi et al., 1997). This has been interpreted as suggesting that MT-3 plays an important role in the repair following brain injury (Hozumi et al., 1998). Overall, these data show that MT-3, as compared to other MTs, exhibits distinct properties and probably serve distinct functions.

Despite the presence in the promoter of the *MT-3* gene of potential MRE elements with the conserved MRE consensus core sequence, as well as potential sites for a number of transcription factors, transcription of the gene encoding MT-3 is not inducible by any of the stimuli (metals, hormones, etc) that normally increase *MT-1* and *MT-2* gene transcription several fold in liver and brain (Palmiter et al., 1992; Dalton et al., 1995; Liang et al., 1996; Kramer et al., 1996; Moffatt and Séguin, 1998). Metals, dexamethasone, ethanol, interleukin-6, and kainic acid have been reported to reduce MT-3 mRNA levels by 30–60% (Zheng et al., 1995; Kramer et al., 1996; Belloso et al., 1996; Hernandez et al., 1997). The mechanisms that govern the expression of *MT-3* in normal brain, and its repression in non-permissive tissues remain unknown. A CTG triplet repeat present in the promoter of the mouse *MT-3* gene was reported to function as a silencer element contributing to *MT-3* gene negative regulation in non-permissive tissues, as assayed by transient transfection analyses in HepG2 cells (Imagawa et al., 1995). On the other hand, Watabe *et al.*, (1997) reported that DNA sequences homologous to the JC virus (JCV) silencer element, downstream of the CTG repeat, function as a negative element for *MT-3* gene expression. To further characterize the mechanisms controling *MT-3* gene transcription, we performed DNA binding analyses and transient transfection experiments.

2. RESULTS

2.1. The *MT-3* Gene Is Expressed in P19 Embryonal Teratocarcinoma Cells Treated with Retinoic Acid

While the *MT-3* gene is expressed in primary astrocytes and neurons (Masters et al., 1994; Amoureux et al., 1995; Kramer et al., 1996; Kramer et al., 1996; Faraonio et al., 1998), no established cell line expressing MT-3 to a level similar to that present in the brain has been described. Only low MT-3 mRNA levels are present in C6 glial cells (Amoureux et al., 1995; Watabe et al., 1997; Faraonio et al., 1998). We thus searched for a cell line expressing high MT-3 mRNA levels in order to have a cell culture system optimal for transfection studies. No MT-3 transcripts were detected in any of twelve transformed cell lines tested, including CTX TNA$_2$, DI TNC$_1$ astrocytic (Radany et al., 1992), and CEINGE CL3 oligodendrocytic (Russo et al., 1993) cells (Figure 1; Table 1). However, MT-3 mRNA was found in P19 embryonal carcinoma (EC) cells induced to differentiate with retinoic acid (RA). While very low MT-3 mRNA levels were present in the undifferentiated P19 EC cells (Faraonio et al., 1998), treatment with RA first led to a decrease of MT-3 mRNA levels, as no MT-3 transcripts were detected five days after RA treatment (Figure 2A). As P19 cells differentiated into neural cells, MT-3 mRNA levels increased gradually (Figure 2A) to reach, 20 days after treatment, levels similar to those present in the brain (Faraonio et al., 1998). MT-3 transcripts started to increase between five and ten days after RA treatment which correlates with peak levels of differentiated neurons. Thereafter, neurons begin to die and are gradually replaced by proliferating astrocytes (Rudnicki and McBurney, 1987; Philipp et al., 1994). Consequently, the MT-3 mRNA detected in P19 cultures are likely predominately synthesized by P19-derived astrocytic cells, although it can not be excluded that P19-derived neurons also expressed *MT-3*. MT-3 mRNA levels were not increased in response to metal induction in P19 cells at any point after RA treatment (Figure 2A).

P19 EC cells are pluripotent and can differentiate into endomesodermal derivatives upon treatment with dimethyl sulfoxide (DMSO) (Rudnicki and McBurney, 1987). *MT-3*

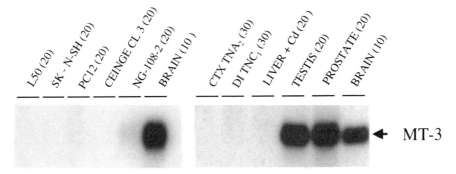

Figure 1. Northern analysis of MT-3 mRNA in different cell lines and adult rat tissues. Total RNA was prepared from untreated male Sprague Dawley rats and was electrophoresed on agarose/formaldehyde gels, transferred to Hybond-N membranes, and hybridized with ^{32}P-labeled cDNA probes corresponding to the rat MT-3 full length cDNA (left panel) or the rat MT-3 3'UTR (Moffatt and Séguin, 1998) (right panel). The membrane on the left was washed and rehybridized with the rat MT-3 3'UTR and similar results were obtained (Faraonio et al., 1998). Numbers in parenthesis indicate the amount in µg of RNA loaded on the gel. Comparable loadings were confirmed by ethidium bromide staining of ribosomal RNA.

gene expression was not induced in P19 cells treated with DMSO (Faraonio et al., 1998), whereas *MT-1* gene expression was induced in both RA- and DMSO-treated cells. While MT-3 transcripts were barely detectable, high MT-1 mRNA levels were present in undifferentiated P19 cells (Figure 2B, lane 0). Five days after RA or DMSO treatment, MT-1 mRNA levels decreased sharply. Thereafter, MT-1 mRNA levels increased gradually to reach a maximum between fifteen and twenty days after both treatments (Figure 2B).

Table 1. Cell lines analyzed for the presence of MT-3 mRNA

Cell lines[1]	Species	Cell type
CTX TNA$_2$	rat	astrocyte type 1
DI TNC$_1$	rat	astrocyte type 1
PC12	rat	pheochromocytoma
CEINGE CL3	rat	oligodendrocyte
NG108-2	murine	neuroblastoma-glioma fusion
SK-N-SH	human	neuroblastoma
C6	rat	glioma
P19	mouse	teratocarcinoma
TM3	mouse	Leydig cell
TM4	mouse	Sertoli cell
DU-145	human	prostate carcinoma
PC3	human	prostate adenocarcinoma
LNCaP	human	prostate carcinoma
T-24	human	bladder carcinoma
HepG2	human	hepatocarcinoma
HeLa	human	fibroblast
L-50	mouse	fibroblast

[1]All these cell lines were obtained from the ATCC, except for the CEINGE CL3 which were obtained from T. Russo (Russo *et al.*, 1993) and the L50 cells from Dean H. Hamer, NIH.

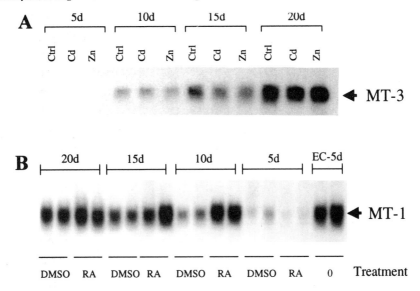

Figure 2. Northern analysis of MT-3 (A) and MT-1 (B) mRNA in P19 EC cells. Neuroectodermal or endo-mesodermal differentiation of P19 cells was induced by treatments with 1 μM all-*trans* RA (A and B) or 1% DMSO (B) (Rudnicki and McBurney, 1987), respectively. Treatment 0 (B) indicates uninduced P19 EC cells. Total RNA was extracted at various time intervals in days (d) after the treatments as indicated over the lanes, and subjected to Northern analysis as described in the legends of Figure 1, using a ^{32}P-labeled cDNA corresponding to the rat MT-3 full length cDNA (A) or the mouse MT-3 3'UTR as probes (B) (Moffatt and Séguin, 1998). The membrane in (A) was washed and rehybridized with the rat MT-3 3'UTR and similar results were obtained (Faraonio et al., 1998). For each time point in (A), different groups of cells were treated or not (Ctrl) with 2 μM CdCl$_2$ (Cd) or 100 μM ZnCl$_2$ (Zn) 8 h before harvesting. Comparable loadings were confirmed by ethidium bromide staining of ribosomal RNA.

2.2. DNaseI Footprinting and Electrophoretic Mobility Shift Analyses

A schematic representation of the mouse *MT-3* promoter is shown in Figure 3 (MT-3, top line). The 5' flanking region of the *MT-3* gene contains three potential MRE elements located at positions -45, -77 (in inverted orientation), and -104, relative to the transcription start point (tsp). In addition, a computer search revealed the presence of putative sites for a number of other transcription factors. Only the sites analyzed in the present study, namely overlapping Sp1 and NF-1 sites and an AP-2 site, are shown in Figure 3. To obtain information on the nuclear proteins interacting with the mouse *MT-3* gene promoter, and obtain clues concerning the sites involved in *MT-3* gene regulation, we performed DNaseI footprinting analyses using an *MT-3* promoter DNA fragment extending from positions -201 to +21. Nuclear extract prepared from mouse liver contained proteins binding to the Sp1/NF-1 site (Figure 4), the putative MREb element and to the tsp region. No major differences were present in the footprints obtained with brain nuclear extracts, apart from a series of DNaseI hypersensitive sites observed over the AP-2 site with the liver extract (Figure 4 and Faraonio et al., 1998). Competition experiments using a 100 fold excess of an oligonucleotide (oligo) corresponding to the Sp1/NF-1 site confirmed the specificity of the binding (Faraonio et al., 1998).

While the putative MRE elements present in the *MT-3* gene promoter display the conserved MRE core sequence TGCRCNC, they do not bind MTF-1, as assayed by electrophoretic mobility shift assay (EMSA) using as the probe a synthetic high affinity MRE oligo

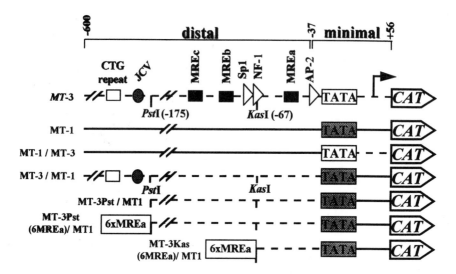

Figure 3. Details of the mouse *MT-3* promoter region and schematic representation of the different plasmid constructions used in the transfection experiments. Approximately 600 bp of the mouse MT-3 (MT-3) or MT-1 (MT-1) 5' flanking regions, including the tsp (bent arrow) were inserted into the CAT reporter plasmid pCATBasic (Stratagene). Open square, the CTG repeat; stippled oval, sequences homologous to the JC virus silencer (JVC); solid square, putative MRE elements; open triangles, binding sites for various transcription factors as indicated; broken line, mouse *MT-3* sequences; solid line, mouse *MT-1* sequences. TATA in an open rectangle indicates the TATA box of the mouse *MT-3* gene, while TATA in a stippled rectangle indicates that of the mouse *MT-1* gene. A synthetic oligo containing six mouse MREa elements, represented by an open rectangle with 6xMREa inside, was inserted at position -175 or -67 of the mouse *MT-3* promoter. The name of each plasmid construction is shown on the left. The experimental procedure used for the construction of these plasmids has been described (Faraonio et al., 1998).

and cold competitor oligos corresponding to the *MT-3* MREa, MREb, and MREc elements (Figure 5). This suggests that the MRE elements in the *MT-3* promoter are very weak or not functional, and that MTF-1 does not interact strongly with these elements *in vivo*.

2.3. Transfection Studies

To further address the question of transcriptional repression of the *MT-3* gene in non-expressing cells, as well as of its transcriptional activation in neural cells, we performed transient transfection experiments in kidney CV-1 and RA-treated P19 EC cells. A 600-bp portion of the mouse *MT-3* gene promoter was inserted into the chloramphenicol acetyltransferase (CAT) reporter plasmid pCATbasic to generate MT3-CAT. The activity of MT3-CAT was compared to that of MT1-CAT in which an equivalent portion of the mouse *MT-1* gene promoter was inserted upstream of the *CAT* coding sequences. While constitutive and metal-inducible CAT activity was readily detected in extracts prepared from CV-1 cells transfected with MT1-CAT, only a low CAT activity was measured in

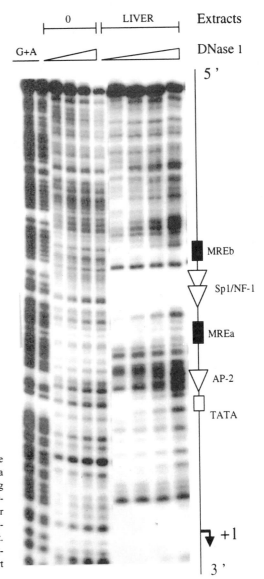

Figure 4. DNaseI footprinting analysis using crude nuclear extract from mouse liver. The probe was a mouse *MT-3* gene promoter DNA fragment extending from positions -201 to +21. The amount of DNaseI varied from 15 to 60 µg as indicated schematically over the lanes. Lanes: G+A, Maxam and Gilbert G+A sequencing reaction; 0, no extract; LIVER, liver extract. The position of the different *cis*-acting elements is indicated on the right as determined by Maxam-Gilbert sequencing. The tsp is indicated by a bent arrow.

cells transfected with the MT3-CAT plasmid (Figure 6). However, in approximately 50% of the experiments, basal CAT activity detected in cells transfected with MT3-CAT was at the background level. Because of these variations it was unclear whether the MT3-CAT plasmid was active in CV-1 cells. However, using more sensitive LUC reporter plasmids, we showed that the *MT-3* promoter was active in both HepG2 and NIH-3T3 cells although at a much lower level than the *MT-1* promoter (Faraonio et al., 1998). In none of these experiments was the *MT-3* promoter activity increased in response to metals.

The presence of a marked DNaseI protection on the tsp in the footprinting analyses suggests that this site may play a role in the regulation of *MT-3* gene expression. Indeed, nuclear proteins such as replication factor A have been shown to bind to the tsp of the human *MT-2A* gene and to repress transcription (Tang et al., 1996). Furthermore, nucleotide se-

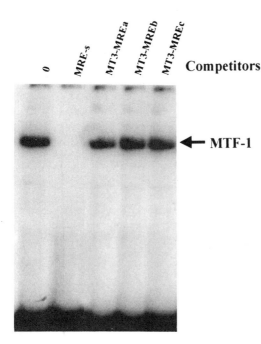

Figure 5. EMSA analysis using nuclear extract from COS cells transfected with an expression vector containing the mouse MTF-1 cDNA (Radtke et al., 1993) and a double-stranded "MRE-s" oligo as the probe. "MRE-s" is a synthetic MRE consensus sequence binding MTF-1 with high affinity (Radtke et al., 1993). Competition was performed with a 100 fold excess of unlabeled double-stranded oligos corresponding to the various mouse *MT-3* gene MRE elements, as indicated above the lanes.

quence comparison of the region around the tsp of a number of *MT* genes of which the transcription is strongly activated in response to metals revealed a consensus nucleotide sequence (Faraonio et al., 1998). Interestingly, this *MT* tsp consensus sequence is not conserved in *MT* genes that are not inducible in response to metals such as the mouse and human *MT-3* genes, or in *MT* genes that are weakly or partially inducible by metals such as the sheep *MT-1B* or the human *MT-1A* genes (Faraonio et al., 1998). Thus, to assess the possible role of the tsp region in the regulation of *MT-3* gene transcription, shuffling experiment were performed. Reporter plasmids containing hybrid promoters were constructed by ligating the upstream promoter sequences of the mouse *MT-1* (-590 to -35) to a mouse *MT-3*

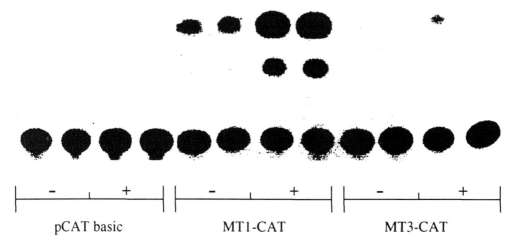

Figure 6. CV-1 cells were transfected with the indicated plasmid constructions. The cells were incubated in the presence (+) or absence (-) of 2.5 μM CdCl₂ for 8 h before harvesting, and CAT activity was determined.

minimal promoter DNA fragment (-35 to +20) and conversely, the upstream *MT-3* promoter was fused to a minimal *MT-1* promoter (Figure 3, constructs MT-1/MT-3 and MT-3/MT-1). These constructs were transfected into CV-1 cells. While CAT activity measured in cells transfected with the plasmid MT3/MT1-CAT was similar to that obtained with the wild type promoter MT3-CAT (Figure 7, construct MT3/MT1-CAT), both constitutive and induced CAT levels were approximately two fold higher than those of MT1-CAT in cells transfected with MT1/MT3-CAT (Figure 7, compare relative CAT activity obtained with MT1/MT3-CAT with that of MT1-CAT). These results showed that the minimal *MT-3* gene promoter is functional, while the region upstream of the TATA box is much less active than the corresponding *MT-1* promoter region (Figures 7A and 7B). Similar results were obtained in HepG2 cells (Faraonio et al., 1998). Thus, the MT-3 tsp region does not appear to be involved in transcriptional repression of the *MT-3* gene in nonpermissive cells.

Next, we investigated the role of the CTG repeat and of the JCV element in the inhibition of *MT-3* gene transcription. Deletion of *MT-3* promoter sequences between -600 and -175 in MT3-CAT (Faraonio et al., 1998) or MT3/MT1-CAT did not appreciably improve

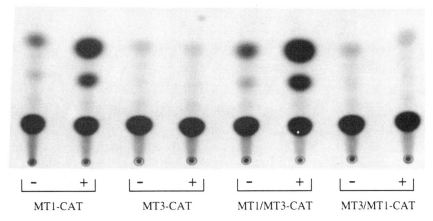

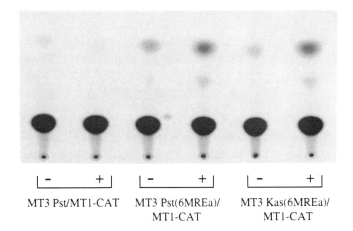

Figure 7. CV-1 cells were transfected with the indicated plasmid constructions. The cells were incubated in the presence (+) or absence (-) of 2.5 μM CdCl₂ for 8 h before harvesting, and CAT activity was determined. (A) The acetylated chloramphenicol derivatives were visualized by autoradiography. (B) Graphic representation of the CAT assay shown in (A). The radioactivity was determined by scintillation counting. Transfections were performed in two separate experiments with similar results.

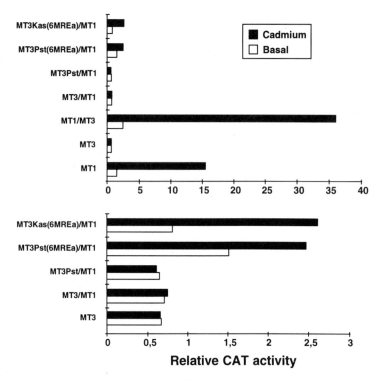

Figure 7. (*Continued*)

either basal or metal-induced expression (Figures 3 and 7, construct MT3Pst/MT1-CAT). This does not support the contention that the CTG repeat and the potential JCV silencer site function as negative regulatory elements for mouse *MT-3* gene expression or represent *bona fide* repressor elements.

The lack of positive regulatory elements in the promoter of the *MT-3* gene may contribute to the absence of transcription of this gene in nonpermissive cells. To test this hypothesis, we inserted into the MT3Pst/MT1-CAT construction six mouse MREa elements. They were placed at -175, a position corresponding to that where MREd and MREc are present in the mouse *MT-1* promoter (Figure 3, construct MT3Pst(6MREa)/MT1). The insertion of functional MREs in the MT-3 promoter led to a two fold increase in basal CAT activity whereas induced levels were increased three fold (Figures 7A and 7B, compare CAT activity of MT3Pst(6MREa)/MT1-CAT with that of MT3Pst/MT1-CAT). This shows that insertion of functional MREs into the mouse *MT-3* promoter confer basal and metal-induced transcriptional activity to the promoter. Deletion of MT-3 promoter sequences to position -67 and insertion of the six MREa elements at this position decreased basal CAT activity close to background levels, but did not affect metal-induced levels (Figures 7A and 7B, construct MT3Kas(6MREa)/MT1-CAT). This suggests that the region between position -67 and -175, which contains the Sp1 and NF-1 sites, is important in maintaining basal expression.

The overall transcriptional activity of the MREa-MT-3 hybrid promoters remained quite low compared to the MT1-CAT reference plasmid. This may indicate that the mouse *MT-3* promoter is lacking some *cis*-acting elements for positively acting transcription factors such as those interspersed in the mouse *MT-1* promoter among the MRE elements. Alternatively, it may indicate that a negative response element is present in the *MT-3* promoter between position -67 and -35. However, the latter possibility appears unlikely because both basal and induced levels of CAT activity measured in cells transfected with a reporter plasmid containing only the TATA box of the mouse *MT-1* gene and six MREa elements were similar to those obtained with the MREa-MT3 hybrid promoters (data not shown). This supports the hypothesis that, in absence of sites for the appropriate general transcription factors, MREs alone cannot confer optimal metal induction to the *MT* genes.

To assess the possible role of the Sp1, NF-1 and AP-2 sites in the regulation of *MT-3* gene expression in neural cells, and in its inhibition in nonexpressing cells, each of these sites were inactivated individually or in groups by site directed mutagenesis, and the resulting plasmids were transfected into CV-1, HepG2, NIH-3T3, and RA-treated P19 cells. Figure 8 shows that the AP-2 site, and to a lesser extent the Sp1 and NF-1 sites, are important for maintaining optimal transcription activity in P19 cells. However, none of these sites appeared to be involved in the repression of *MT-3* gene transcription in CV-1, HepG2, and NIH-3T3 cells (Faraonio et al., 1998). Deletion of the CTG repeat and the JCV repressor-like element did not significantly enhance *MT-3* promoter activity in RA-treated P19 cells (Faraonio et al., 1998). Footprinting experiments using nuclear extracts containing or not containing Sp1 showed that NF-1 preferentially binds to the Sp1/NF-1 region (Faraonio et al., 1998), thus suggesting that *in vivo* this site is preferentially occupied by NF-1. AP-2 levels were evaluated during RA-induced P19 cell differentiation as assayed by EMSA. AP-2 protein rises in parallel with increasing numbers of differentiated neurons (Philipp et al., 1994) and increasing levels of MT-3 mRNA between 3 and 20 days after RA treatment (Figure 9).

3. DISCUSSION

The MT-3 tissue-specific distribution pattern is unique among MT family members in that MT-3 is present mainly in the brain and in the organs of the reproductive system.

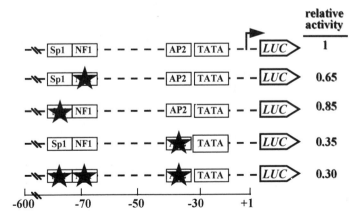

Figure 8. Transfection studies performed with mutant *MT-3* promoter LUC plasmid constructions. Plasmids were transfected into RA-treated P19 cells, and LUC activity was measured. LUC activities relative to the wild type MT-3 promoter taken as 100% are indicated on the right. Stars indicate mutations introduced into the different sites. The numbers at the bottom indicate nucleotide positions relative to the *tsp* (bent arrow). Transfections were performed in three separate experiments with similar results.

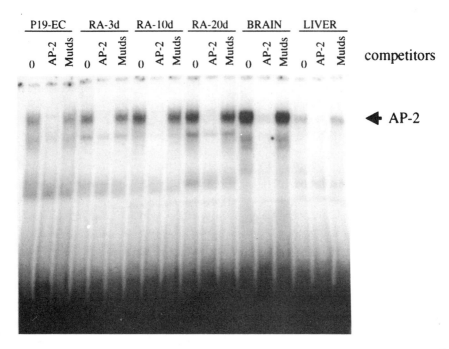

Figure 9. Induction of AP-2 in RA-treated P19 cells. EMSA using nuclear extracts prepared from P19 undifferentiated cells (P19-EC), or from P19 cells 3, 10, and 20 days (d) after RA treatment, and an oligo corresponding to the putative AP-2 site present in the promoter of the mouse *MT-3* gene. For comparison, EMSA reactions were also performed with brain and liver nuclear extracts. The specificity of the retarded band was determined by competition with a 100 fold excess of the same AP-2 oligo used as the probe or with a non specific oligo (Mutds).

MT-3 transcription is activated in mouse P19 EC cells induced to differentiate into neural cells by RA treatments (Figure 2A). This induction of MT-3 in P19 cells was not observed when P19 cells were induced to differentiate into endomesodermal derivatives by DMSO treatments. P19 EC cells are developmentally pluripotent murine cells which can be grown indefinitely in culture. Treatment of P19 cells with RA induces their differentiation into neurons, astrocytes and other cell types derived from neuroectoderm (Bain et al., 1994). The P19-derived neurons express a variety of neuronal markers such as neurofilament proteins, synaptophysin, Wnt-1, and HNK-1 surface binding sites, and are heterogenous, both with respect to morphology and the neurotransmitters they synthesize. Because P19-derived neurons are able to form synapses and acquire the electrophysiological properties of mature neurons, they are an attractive cell culture system for investigating the role of MT-3 in neuronal cell biology.

Contrary to other *MT* genes, *MT-3* gene transcription is not inducible by metal ions. Although it has been suggested that a CTG repeat (Imagawa et al., 1995) and a sequence similar to the JCV silencer element (Watabe et al., 1997) were responsible for the inhibition of *MT-3* transcription in non-permissive tissues, we showed here that deletion of these sites does not restore *MT-3* transcription as assessed by transient transfection analyses in CV-1 cells (Figure 7). This argues against a negative regulatory function for these sequences. Footprinting analyses revealed that few nuclear proteins interact with the mouse *MT-3* promoter between positions +21 and -201. Using crude nuclear extracts prepared from mouse liver (Figure 4) or brain (Faraonio et al., 1998), only the Sp1/NF-1 and the tsp sites were protected from DNaseI digestion. In comparison, the mouse *MT-1* promoter contains binding sites for numerous transcription factors including USF, Sp1, and most importantly MTF-1, all interspersed with each other in a complex array (Faraonio et al., 1998). We also showed that the putative MRE elements present in the *MT-3* promoter do not bind the essential metalloregulatory transcription factor MTF-1 which is required for both basal and metal-induced transcription. The absence in the mouse *MT-3* promoter of functional high affinity MTF-1-binding MRE elements, and of binding sites for other positively acting factors, rather than the existence of specific repressor elements, is likely to contribute to its inactivity in nonpermissive tissues. Indeed, we showed that the insertion of functional *MT-1* MREa (Figure 7) or MREd (Faraonio et al., 1998) elements in the mouse *MT-3* promoter conferred basal and metal-induced transcriptional activity on the promoter.

Mutagenesis and transfection experiments showed that the AP-2 site is important for optimal *MT-3* expression in RA-treated P19 cells. AP-2 and related splicing variants are abundant in the brain, and both mRNA and protein levels are increased during neuroectodermal differentiation of P19 cells (Figure 9) (Philipp et al., 1994; Meier et al., 1995). AP-2 has been shown to regulate the expression of a number of genes in the central nervous system (Greco et al., 1995; Quinn et al., 1995; Baskin et al., 1997; Garcia et al., 1996), and MT-3 appears to be another AP-2-target gene. Although we did not detect any footprint over this site, it is possible that AP-2 is either absent, present in very low levels, or inactive in the nuclear extracts used in this study.

If a repressor is involved in the inhibition of *MT-3* gene transcription in non-permissive tissues, one might expect that nuclear extracts prepared from non-expressing tissues, such as the liver, would contain a specific factor binding to a putative repressor site. The fact that there was no major differences between the footprints obtained with liver and brain extracts does not exclude the existence of the repressor molecule because brain extracts contain proteins from cell types both expressing and not expressing *MT-3*. Thus, brain nuclear extracts are likely to contain the putative repressor, which would make difficult to generate a liver specific footprint. To eliminate the possibility that one of the footprints obtained with liver extracts corresponds to a putative repressor binding site, each of these sites, namely the Sp1, NF-1 and AP-2 sites, were mutated individually or in groups, and each mutant was transfected into RA-treated P19 (Figure 8) and CV-1, HepG2, and NIH-3T3 (Faraonio et al., 1998) cells. The results showed that none of sequences appear to be involved in *MT-3* transcriptional repression. This does not exclude the possibility that other sites in the *MT-3* promoter interact with nuclear proteins, and further *in vitro* and *in vivo* footprint analyses will be required to determine whether other factors bind to the *MT-3* promoter.

In conclusion, our results suggest that differential expression of the *MT-3* gene is partially due to differences in the *cis*-acting control sequences of the *MT-3* gene itself. The absence of functional MRE elements in the promoter of the *MT-3* gene, together with the observation that relatively few nuclear proteins appear to interact with it may partially explain why the *MT-3* gene is transcriptionally inactive in nonpermissive tissues. However, this is unlikely to account for the complete absence of *MT-3* gene transcription in these tissues, and other mechanisms may also be involved. Chromatin structure plays important role in transcription regulation, and small transiently transfected plasmids are not likely to acquire higher order levels of chromatin structure (Smith and Hager, 1997). A more appropriate experimental system using replicative episomal vectors or stable transfection analyses will be required to determine more precisely the mechanisms governing *MT-3* tissue-specific gene expression.

ACKNOWLEDGMENTS

We are grateful to A. Anderson for a critical reading of the manuscript. We thank Richard D. Palmiter for providing the mouse MT-3 plasmid, Walter D. Schaffner for the MTF-1 expression vector, Samuel David and Claude Gravel for the NG108–2 and SK-N-SH cells, and C.F. Deschepper for the CTX TNA$_2$ and DI TNC$_1$ cells. We also thank Gale Stewart for skilful technical assistance. R.F. held a Fellowship from the Ministère de l'enseignement supérieur du Québec, P.M. by the Fédération québecoise des sociétés Alzheimer, and O.L. from the Fonds pour la formation des Chercheurs et l'aide à la recherche du Québec. This study was supported by grants from the NIH and the MRC.

REFERENCES

Amoureux, M.C., Van Gool, D., Herrero, M.T., Dom, R., Colpaert, F.C., and Pauwels, P.J. (1997). Regulation of metallothionein-III (GIF) mRNA in the brain of patients with Alzheimer disease is not impaired. Mol. Chem. Neuropathol. 32, 101–121.

Amoureux, M.C., Wurch, T., and Pauwels, P.J. (1995). Expression of human metallothionein-III confers protection against serum-free exposure of stably transfected Chinese hamster ovary CHO-K1 cells. Neurosci. Lett. 201, 61–64.

Amoureux, M.C., Wurch, T., and Pauwels, P.J. (1995). Modulation of metallothionein-III mRNA content and growth rate of rat C6-glial cells by transfection with human 5- HT1D receptor genes. Biochem. Biophys. Res. Commun. 214, 639–645.

Andrews, G.K. (1990). Regulation of metallothionein gene expression. Prog. Food Nutrition Science 14, 193–258.

Aschner, M., Cherian, M.G., Klaassen, C.D., Palmiter, R.D., Erickson, J.C., and Bush, A.I. (1997). Metallothioneins in brain - The role in physiology and pathology. Toxicol. Appl. Pharmacol. 142, 229–242.

Bain, G., Ray, W.J., Yao, M., and Gottlieb, D.I. (1994). From embryonal carcinoma to neurons: the P19 pathway. Bioessays 16, 343–348.

Baskin, F., Li, Y.P., Hersh, L.B., Davis, R.M., and Rosenberg, R.N. (1997). An AP-2 binding sequence within exon 1 of human and porcine choline acetyltransferase genes enhances transcription in neural cells. Neuroscience 76, 821–827.

Belloso, E., Hernandez,.J., Giralt, M., Kille, P., and Hidalgo, J. (1996). Effect of stress on mouse and rat brain metallothionein I and III mRNA levels. Neuroendocrinology. 64, 430–439.

Choudhuri, S., Kramer, K.K., Berman, N.E.J., Dalton, T.P., Andrews, G.K., and Klaassen, C.D. (1995). Constitutive expression of metallothionein genes in mouse brain. Toxicol. Appl. Pharmacol. 131, 144–154.

Dalton, T., Pazdernik, T.L., Wagner, J., Samson, F., and Andrews, G.K. (1995). Temporalspatial patterns of expression of metallothionein- I and III and other stress related genes in rat brain after kainic acid-induced seizures. Neurochem. Int. 27, 59–71.

Erickson, J.C., Hollopeter, G., Thomas, S.A., Froelick, G.J., and Palmiter, R.D. (1997). Disruption of the metallothionein-III gene in mice: Analysis of brain zinc, behavior, and neuron vulnerability to metals, aging, and seizures. J. Neurosci. 17, 1271–1281.

Erickson, J.C., Sewell, A.K., Jensen, L.T., Winge, D.R., and Palmiter, R.D. (1994). Enhanced neurotrophic activity in Alzheimer's disease cortex is not associated with down-regulation of metallothionein-III (GIF). Brain Res. 649, 297–304.

Faraonio, R., Moffatt, P., Larochelle, O., St-Arnaud, R., Schipper, H.M., and Séguin, C. (1998). Transcriptional activation of the gene encoding metallothionein-3 during neuroectodermal differentiation of P19 mouse embryonal carcinoma cells. Submitted for publication

Garcia, M.A., Vazquez, J., Gimenez, C., Valdivieso, F., and Zafra, F. (1996). Transcription factor AP-2 regulates human apolipoprotein E gene expression in astrocytoma cells. J. Neurosci. 16, 7550–7556.

Greco, D., Zellmer, E., Zhang, Z., and Lewis, E. (1995). Transcription factor AP-2 regulates expression of the dopamine beta-hydroxylase gene. J. Neurochem. 65, 510–516.

Gunes, C., Heuchel, R., Georgiev, O., Muller, K.H., Lichtlen, P., Bluthmann, H., Marino, S., Aguzzi, A., and Schaffner, W. (1998). Embryonic lethality and liver degeneration in mice lacking the metal-responsive transcriptional activator MTF-1. EMBO J. 17, 2846–2854.

Hernandez, J., Molinero, A., Campbell, I.L., and Hidalgo, J. (1997). Transgenic expression of interleukin 6 in the central nervous system regulates brain metallothionein-I and -III expression in mice. Mol. Brain Res. 48, 125–131.

Heuchel, R., Radtke, F., Georgiev, O., Stark, G., Aguet, M., and Schaffner, W. (1994). The transcription factor MTF-1 is essential for basal and heavy metal-induced metallothionein gene expression. EMBO J. 13, 2870–2875.

Hozumi, I., Inuzuka, T., Hiraiwa, M., Uchida, Y., Anezaki, T., Ishiguro, H., Kobayashi, H., Uda, Y., Miyatake, T., and Tsuji, S. (1995). Changes of growth inhibitory factor after stab wounds in rat brain. Brain Res. 688, 143–148.

Hozumi, I., Inuzuka, T., Ishiguro, H., Hiraiwa, M., Uchida, Y., and Tsuji, S. (1996). Immunoreactivity of growth inhibitory factor in normal rat brain and after stab wounds - An immunocytochemical study using confocal laser scan microscope. Brain Res. 741, 197–204.

Hozumi, I., Inuzuka, T., and Tsuji, S. (1998). Brain injury and growth inhibitory factor (GIF) - a minireview. Neurochem. Res. 23, 319–328.

Imagawa, M., Ishikawa, Y., Shimano, H., Osada, S., and Nishihara, T. (1995). CTG triplet repeat in mouse growth inhibitory factor/metallothionein III gene promoter represses the transcriptional activity of the heterologous promoters. J. Biol. Chem. 270, 20898–20900.

Imbert, J., Culotta, V.C., Fürst, P., Gedamu, L., and Hamer, D.H. (1990). Regulation of metallothionein gene transcription by metals. Adv. Inorg. Biochem. 8, 139–164.

Inuzuka, T., Hozumi, I., Tamura, A., Hiraiwa, M., and Tsuji, S. (1996). Patterns of growth inhibitory factor (GIF) and glial fibrillary acidic protein relative level changes differ following left middle cerebral artery occlusion in rats. Brain Res. 709, 151–153.

Kobayashi, H., Uchida, Y., Ihara, Y., Nakajima, K., Kohsaka, S., Miyatake, T., and Tsuji, S. (1993). Molecular cloning of rat growth inhibitory factor cDNA and the expression in the central nervous system. Mol. Brain Res. 19, 188–194.

Kramer, K.K., Liu, J., Choudhuri, S., and Klaassen, C.D. (1996). Induction of metallothionein mRNA and protein in murine astrocyte cultures. Toxicol. Appl. Pharmacol. 136, 94–100.

Kramer, K.K., Zoelle, J.T., and Klaassen, C.D. (1996). Induction of metallothionein mRNA and protein in primary murine neuron cultures. Toxicol. Appl. Pharmacol. 141, 1–7.

Liang, L., Fu, K., Lee, D.K., Sobieski, R.J., Dalton, T., and Andrews, G.K. (1996). Activation of the complete mouse metallothionein gene locus in the maternal deciduum. Mol. Reprod. Dev. 43, 25–37.

Masters, B.A., Quaife, C.J., Erickson, J.C., Kelly, E.J., Froelick, G.J., Zambrowicz, B.P., Brinster, R.L., and Palmiter, R.D. (1994). Metallothionein III is expressed in neurons that sequester zinc in synaptic vesicles. J. Neurosci. 14, 5844–5857.

Meier, P., Koedood, M., Philipp, J., Fontana, A., and Mitchell, P.J. (1995). Alternative mRNAs encode multiple isoforms of transcription factor AP-2 during murine embryogenesis. Dev. Biol. 169, 1–14.

Moffatt, P. and Denizeau, F. (1997). Metallothionein in physiological and physiopathological processes. Drug Metab. Rev. 29, 261–307.

Moffatt, P. and Séguin, C. (1998). Expression of the gene encoding metallothionein-3 in the organs of the reproductive system. DNA Cell Biol. 17, 501–510.

Palmiter, R.D., Findley, S.D., Whitmore, T.E., and Durnam, D.M. (1992). MT-III, a brain-specific member of the metallothionein gene family. Proc. Natl. Acad. Sci. USA 89, 6333–6337.

Philipp, J., Mitchell, P.J., Malipiero, U., and Fontana, A. (1994). Cell type-specific regulation of expression of transcription factor AP-2 in neuroectodermal cells. Dev. Biol. 165, 602–614.

Quaife, C.J., Kelly, E.J., Masters, B.A., Brinster, R.L., and Palmiter, R.D. (1998). Ectopic expression of metallothionein-III causes pancreatic acinar cell necrosis in transgenic mice. Toxicol. Appl. Pharmacol. 148, 148–157.

Quinn, J.P., Mcallister, J., and Mendelson, S. (1995). Multiple protein complexes, including AP2 and Sp1, interact with a specific site within the rat preprotachykinin-A promoter. Bba-Gene. Struct. Express. 1263, 25–34.

Radany, E.H., Brenner, M., Besnard, F., Bigornia, V., Bishop, J.M., and Deschepper, C.F. (1992). Directed establishment of rat brain cell lines with the phenotypic characteristics of type 1 astrocytes. Proc. Natl. Acad. Sci. USA 89, 6467–6471.

Radtke, F., Georgiev, O., Muller, H.P., Brugnera, E., and Schaffner, W. (1995). Functional domains of the heavy metal-responsive transcription regulator MTF-1. Nucleic Acids Res. 23, 2277–2286.

Radtke, F., Heuchel, R., Georgiev, O., Hergersberg, M., Gariglio, M., Dembic, Z., and Schaffner, W. (1993). Cloned transcription factor MTF-1 activates the mouse metallothionein-I promoter. EMBO J. 12, 1355–1362.

Rudnicki, M.A. and McBurney, M.W. (1987). Cell culture methods and induction of differentiation of embryonal carcinoma cell lines. In: Robertson, E.J. (Ed.) "Teratocarcinomas and embryonic stem cells: a practical approach." IRL Press, Oxford, pp 19–49.

Russo, T., Mogavero, A.R., Ammendola, R., Mesuraca, M., Fiore, F., Fatatis, A., Salvatore, G., and Cimino, F. (1993). Immortalization of a cell line showing some characteristics of the oligodendrocyte phenotype. Neurosci. Lett. 159, 159–162.

Sewell, A.K., Jensen, L.T., Erickson, J.C., Palmiter, R.D., and Winge, D.R. (1995). Bioactivity of metallothionein-3 correlates with its novel beta domain sequence rather than metal binding properties. Biochemistry 34, 4740–4747.

Smith, C.L. and Hager, G.L. (1997). Transcriptional regulation of mammalian genes in vivo. A tale of two templates. J. Biol. Chem. 272, 27493–27496.

Stuart, G.W., Searle, P.F., and Palmiter, R.D. (1985). Identification of multiple metal regulatory elements in mouse metallothionein-I promoter by assaying synthetic sequences. Nature 317, 828–831.

Suzuki, K.T., Imura, N., and Kimura, M. (Eds.) (1993). "Metallothionein III: Biological roles and medical implications." Birkauser, Basel.

Tang, C.M., Tomkinson, A.E., Lane, W.S., Wold, M.S., and Seto, E. (1996). Replication protein A is a component of a complex that binds the human metallothionein IIA gene transcription start site. J. Biol. Chem. 271, 21637–21644.

Tsuji, S., Kobayashi, H., Uchida, Y., Ihara, Y., and Miyatake, T. (1992). Molecular cloning of human growth inhibitory factor cDNA and its down-regulation in Alzheimer's disease. EMBO J. 11, 4843–4850.

Uchida, Y. and Ihara, Y. (1995). The N-terminal portion of growth inhibitory factor is sufficient for biological activity. J. Biol. Chem. 270, 3365–3369.

Uchida, Y., Takio, K., Titani, K., Ihara, Y., and Tomonaga, M. (1991). The growth inhibitory factor that is deficient in the Alzheimer's disease brain is a 68 amino acid metallothionein-like protein. Neuron 7, 337–347.

Watabe, M., Gross, S., Lawyer, C., Brewer, G.J., Mashimo, T., and Watabe, K. (1997). Sequence and functional analysis of the 5'-flanking region of the mouse growth inhibitory factor gene. Cell Mol. Neurobiol. 17, 235–243.

Westin, G. and Schaffner, W. (1988). A zinc-responsive factor interacts with a metal-regulated enhancer element (MRE) of the mouse metallothionein-I gene. EMBO J. 7, 3763–3770.

Yamada, M., Hayashi, S., Hozumi, I., Inuzuka, T., Tsuji, S., and Takahashi, H. (1996). Subcellular localization of growth inhibitory factor in rat brain: Light and electron microscopic immunohistochemical studies. Brain Res. 735, 257–264.

Yuguchi, T., Kohmura, E., Sakaki, T., Nonaka, M., Yamada, K., Yamashita, T., Kishiguchi, T., Sakaguchi, T., and Hayakawa, T. (1997). Expression of growth inhibitory factor mRNA after focal ischemia in rat brain. J. Cereb. Blood Flow Metab. 17, 745–752.

Yuguchi, T., Kohmura, E., Yamada, K., Sakaki, T., Yamashita, T., Otsuki, H., Wanaka, A., Tohyama, M., Tsuji, S., and Hayakawa, T. (1995). Changes in growth inhibitory factor mRNA expression compared with those in c-jun mRNA expression following facial nerve transection. Mol. Brain Res. 28, 181–185.

Zheng, H., Berman, N.E.J., and Klaassen, C.D. (1995). Chemical modulation of metallothionein I and III mRNA in mouse brain. Neurochem. Int. 27, 43–58.

CONTRIBUTORS

Dr. Detmar Beyersmann
Department of Biology and Chemistry
University of Bremen NW2
D-28359 Bremen, Germany
Ph# 49-421-218-2377
Fax: 49-421-218-4918
Email: beyers@chemie.uni-bremen.de

Dr. Cynthia J. Burrows
Department of Chemistry
University of Utah
315 S. 1400 E., RM Dock
Salt Lake City
Utah 84112-0850, U.S.A.
Ph# 801-585-7290
Fax: 801-585-7868
Email: burrows@chemistry.chem.utah.edu

Dr. Dipankar Chakraborti
Director, School of Environmental Studies
Jadavpur University
Calcutta - 700 032
West Bengal, India
Ph # 91-33-473-5233
Fax: 91-33-473-4266
Email: dcsoesju@giasc/01.vsnl.net.in

Dr. Max Costa
New York University Medical Center
Nelson Institute of Environmental
   Medicine
57 Old Forge Road
Tuxedo, NY 10987, U.S.A.
Ph# 914-351-2368
Fax: 914-351-2118
Email: costam@charlotte.med.nyu.edu

Dr. Diane Cox
Medical Genetics Department
University of Alberta
8-39 Medical Sciences Building
Edmonton, Alberta
T6G 2H7 Canada
Ph# 403-492-0874
Fax: 403-492-1988
Email: diane.cox@ualberta.ca

Dr. Valeria Culotta
Johns Hopkins University
School of Hygiene and Public Health
615 N Wolfe Street, Room 7032
Baltimore, MD 21205, U.S.A.
Ph# 410-955-3029
Fax: 410-955-0116
Email: vculotta@jhsph.edu

Dr. David Eide
Nutritional Sciences Program
University of Missouri-Columbia
217 Gwynn Hall
Columbia, MO 65211, U.S.A.
Ph# 573-882-9686
Fax: 573-882-0185
Email: deide@showme.missouri.edu

Dr. Seth Frisbie
The Johnson Company, Inc.
100 State St., Suite 600
Montpelier, VT 05602
U.S.A.
Ph # 802-229-4600
Fax: 802-229-5876
Email: erikamitt@yahoo.com

Dr. Moira Glerum
Department of Medical Genetics
8-33A Medical Sciences
University of Alberta
Edmonton, Alberta
T6G 2H7 Canada
Ph# 403-492-4563
Fax: 403-492-1998
E-mail: moira.glerum@ualberta.ca

Dr. Bruce Hammock
Department of Entomology
University of California
1 Shields Avenue
Davis, CA 95616
U.S.A.
Ph# 530-752-7519
Fax: 530-752-1537
Email: bdhammock@ucdavis.edu

Dr. Andrea Hartwig
University of Karlsruhe
Institute of Food Chemistry
76128 Karlsruhe
Germany
Ph# 0049-721-608-2936
Fax: 0049-721-608-7254
Email: Andrea
    Hartwig@chemie.uni-karlsruhe.de

Dr. Jerry Kaplan
Department of Pathology
University of Utah School of Medicine
50 N. Medical Drive Room 5C124
Salt Lake City, Utah 84132
U.S.A.
Ph# 801-581-7427
Fax: 801-581-4517
Email: Kaplan@bioscience.
    biology.utah.edu

Dr. Kazimierz Kasprzak
Laboratory of Comparative Carcinogenesis
National Cancer Institute
NCI-FCRDC, Bldg. 538
Frederick, MD 21702
U.S.A.
Ph# 301-846-5600
Fax: 301 846-5738
E-mail: kasprkaz@ncifcrf.gov

Dr. James Koropatnick
London Regional Cancer Centre
790 Commissioners Road East
London, Ontario
N6A 4L6 Canada
Ph# 519-685-8654
Fax: 519-685-8646
Email: jkoropat@julian.uwo.ca

Dr. Andreas Kortenkamp
Centre for Toxicology
The School of Pharmacy
29/39 Brunswick Square
London WC1N 1AX
United Kingdom
Ph# 44-171-753-5811
Fax: 44-171-753-5908
Email: A.Kortenkamp@cua.ulsop.
    ac.uk

Dr. Maria Linder
Department of Biochemistry
California State University, Fullerton
800 N. State College Blvd.
Fullerton, CA 92834-6866
U.S.A.
Ph# 714-278-2472
Fax: 714-278-5316
Email: mlinder@fullerton.edu

Dr. Julian Mercer
Centre for Cellular and Molecular Biology
Deakin University
662 Blackburn Road
Clayton, Victoria 3168
Australia
Ph# 61-3-9244-7413
Fax: 61-3-9244-7290
Email: jmercer@deakin.edu.au

Professor Robert Nilsson
National Swedish Chemicals Inspectorate
    (KEMI)
Stockholm University
Box 1384
S 17127, Solna
Sweden
Ph# 46-8-730-6721
Fax: 46-8-735-7698
Email: robertn@kemi.se

Dr. Thomas V. O'Halloran
Department of Chemistry
Northwestern University
2145 Sheridan Road
Evanston, IL 60208-3113, U.S.A.
Ph# 847-491-5060
Fax: 847-491-7713
Email: t-ohalloran@nwu.edu

Professor Barry P. Rosen
Department of Biochemistry and
  Molecular Biology
Wayne State University School of
  Medicine
540 E. Canfield
Detroit, MI 48201
U.S.A.
Ph# 313-577-1512
Fax: 313-577-2765
Email: brosen@med.wayne.edu

Dr. Bibudhendra Sarkar
Department of Structural Biology and
  Biochemistry Research
The Hospital for Sick Children
555 University Avenue
Toronto, Ontario M5G 1X8
Ph# 416-813-5921
Fax: 416-813-5379
Email: bsarkar@sickkids.on.ca

Dr. Walter Schaffner
Universität Zürich
Institut Für Molekularbiologie II
Winterthurstrasse 190
CH-8057 Zürich
Switzerland
Ph# 41-1-635-4910
Fax: 41-1-635-6811
Email: wschaffn@molbio2.unizh.ch

Dr. Carl Séguin
Centre de Recherche en Cancérologie de
  l'Universitè Laval
L'Hotel-dieu de Quebec
11 Cote du Palais P.Q.
G1R 2J7 Canada
PH# 418-691-5554
Fax: 418-691-5439
Email: Carl.Seguin@crhdq.ulaval.ca

Dr. David Sigman
University of California Los Angeles
Los Angeles, CA 90095-1570, U.S.A.
Ph# 310-825-8903
Fax: 310-206-7286
Email: SIGMAN@mbi.ucla.edu

Dr. William A.Suk
Chemical Exposure and Molecular
  Biology Branch
N.I.E.H.S., P.O. Box 12233
Research Triangle Park, NC 27709, U.S.A.
Ph# 919-541-0797
Fax: 919-541-2843
Email: suk@niehs.nih.gov

Dr. Stuart M. Tanner
Division of Child's Health
Sheffield Children's Hospital
Sheffield S10 2TH, United Kingdom
Ph# 44-114-271-7228
Fax: 44-114-275-5364
Email: m.s.tanner@sheffield.ac.uk

Dr. Zeynep Tümer
Department of Medical Genetics
Panum Institute
University of Copenhagen
Blegdamsvej 3, 2200 KBH N
Copenhagen, Denmark
Ph# 45-3532-7827
Fax: 45-3532-7845
Email: zeynep@imbg.ku.dk

Professor Joan S. Valentine
Department of Chemistry and Biochemistry
U.C.L.A., 405 Hillgard Avenue
Los Angeles, CA 90095-1569, U.S.A.
Ph# 310-825-9835
Fax: 310-206-7197
Email: jsv@chem.ucla.edu

Dr. Kunihiro Yoshida
Department of Medicine (Neurology)
Shinsu University School of Medicine,
  3-1-1
Asahi, Matsumoto, 390- 8621, Japan
Ph# 81-263-37-2673
Fax: 81-263-34-0929
Email: naokosy@gipac.shinshu-u.ac.jp

INDEX

DATE DUE

RET'D OCT 2 0 2005			
JUL 2 6 2010			